Proof and Proving in Mathematics Education

New ICMI Study Series
VOLUME 15

Published under the auspices of the International Commission on
Mathematical Instruction under the general editorship of

Bill Barton, President   Jaime Carvalho e Silva, Secretary-General

For further volumes:
http://www.springer.com/series/6351

Information on the ICMI Study programme and on the resulting publications can be obtained
at the ICMI website http://www.mathunion.org/ICMI/ or by contacting the ICMI Secretary-
General, whose email address is available on that website.

Gila Hanna • Michael de Villiers

Editors

Proof and Proving
in Mathematics Education

The 19th ICMI Study

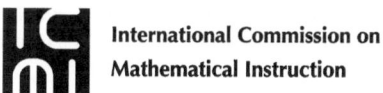

International Commission on
Mathematical Instruction

 Springer

Editors
Gila Hanna
Ontario Institute for Studies
in Education (OISE)
University of Toronto
Toronto, ON, Canada
gila.hanna@utoronto.ca

Michael de Villiers
Faculty of Education
Edgewood Campus
University of KwaZulu-Natal
South Africa
profmd@mweb.co.za

ISSN 1387-6872
ISBN 978-94-007-2128-9   ISBN 978-94-007-2129-6 (eBook)
DOI 10.1007/978-94-007-2129-6
Springer Dordrecht Heidelberg London New York

Library of Congress Control Number: 2011940425

Contents

Contents

Contributors

Ferdinando Arzarello Department of Mathematics, University of Torino, Torino, Italy, ferdinando.arzarello@unito.it

Maria Giuseppina Bartolini Bussi Department of Mathematics, University of Modena and Reggio Emilia (UNIMORE), Modena, Italy, bartolini@unimore.it

Paolo Boero Dipartimento di Matematica, Università di Genova, Genova, Italia, boero@dima.unige.it

Jonathan Michael Borwein Centre for Computer-Assisted Research Mathematics and its Applications, CARMA, University of Newcastle, Callaghan, NSW 2308, Australia, jonathan.borwein@newcastle.edu.au

Richard Cabassut LDAR Laboratoire de Didactique André Revuz, Paris 7 University, Paris, France.

IUFM Institut Universitaire de Formation des Maîtres, Strasbourg University, Strasbourg, France, richard.cabassut@unistra.fr

Karine Chemla CNRS, Université Paris Diderot, Sorbonne Paris Cité, Research Unit SPHERE, team REHSEIS, UMR 7219, CNRS, F-75205 Paris, France, chemla@univ-paris-diderot.fr

Ying-Hao Cheng Department of Mathematics, Taipei Municipal University of Education, Taipei, Taiwan, yinghao.cheng@msa.hinet.net

AnnaMarie Conner Department of Mathematics & Science Education, University of Georgia, Athens, GA, USA, aconner@uga.edu

Michael de Villiers School of Science, Mathematics & Technology Education, University of KwaZulu-Natal, Durban, South Africa, profmd@mweb.co.za

Nadia Douek Institut Universitaire de Formation des Maîtres, Université de Nice, Nice, France, ndouek@wanadoo.fr

Tommy Dreyfus Department of Mathematics, Science and Technology Education, School of Education, Tel Aviv University, Tel-Aviv, Israel, tommyd@post.tau.ac.il

Viviane Durand-Guerrier Département de mathématiques, I3M, UMR 5149, Université Montpellier 2, Montpellier, France, vdurand@math.univ-montp2.fr

Susanna S. Epp Department of Mathematical Sciences, DePaul University, Chicago, IL, USA, sepp@depaul.edu

Fulvia Furinghetti Dipartimento di Matematica, Università di Genova, Genova, Italy, furinghe@dima.unige.it

Judith V. Grabiner Department of Mathematics, Pitzer College, Claremont, CA, USA, jgrabiner@pitzer.edu

Gila Hanna Ontario Institute for Studies in Education, University of Toronto, Toronto, Canada, gila.hanna@utoronto.ca

Patricio Herbst School of Education, University of Michigan, Ann Arbor, MI, USA, pgherbst@umich.edu

Wang-Shian Horng Department of Mathematics, National Taiwan Normal University, Taipei, Taiwan, horng@math.ntnu.edu.tw

Feng-Jui Hsieh Department of Mathematics, National Taiwan Normal University, Taipei, Taiwan, hsiehfj@math.ntnu.edu.tw

Filyet Aslı İşçimen Department of Mathematics and Statistics, Kennesaw State University, Kennesaw, GA, USA, ersozas@yahoo.com

Hans Niels Jahnke Fakultät für Mathematik, Universität Duisburg-Essen, Essen, Germany, njahnke@uni-due.de

Keith Jones Mathematics and Science Education Research Centre, School of Education, University of Southampton, Highfield, Southampton, UK, d.k.jones@soton.ac.uk

Ivy Kidron Applied Mathematics Department, Jerusalem College of Technology (JCT), Jerusalem, Israel, ivy@jct.ac.il

Boris Koichu Department of Education in Technology and Science, Technion – Israel Institute of Technology, Haifa, Israel, bkoichu@technion.ac.il

Margo Kondratieva Faculty of Education and Department of Mathematics and Statistics, Memorial University of Newfoundland, St. John's, Canada, margo.kon@gmail.com

Kyeong-Hwa Lee Department of Mathematics Education, Seoul National University, Seoul, South Korea, khmath@snu.ac.kr

Roza Leikin Department of Mathematics Education, Faculty of Education, University of Haifa, Haifa, Israel, rozal@construct.haifa.ac.il

Allen Yuk Lun Leung Department of Education Studies, Hong Kong Baptist University, Kowloon Tong, Hong Kong, aylleung@hkbu.edu.hk

Fou-Lai Lin Department of Mathematics, National Taiwan Normal University, Taipei, Taiwan, linfl@math.ntnu.edu.tw

Jane-Jane Lo Department of Mathematics, Western Michigan University, Kalamazoo, MI, USA, jane-jane.lo@wmich.edu

Giuseppe Longo CNRS – École Normale Supérieure et CREA, École Polytechnique, Paris, France, longo@di.ens.fr

Maria Alessandra Mariotti Department of Mathematics and Computer Science, University of Siena, Siena, Italy, mariotti21@unisi.it

Francesca Morselli Dipartimento di Matematica, Università di Genova, Genoa, Italy, morselli@dima.unige.it

Elena Nardi School of Education, University of East Anglia, Norwich, UK, e.nardi@uea.ac.uk

Susan D. Nickerson Department of Mathematics and Statistics, San Diego State University, San Diego, CA, USA, snickers@sciences.sdsu.edu

Judy-anne Osborn Centre for Computer Assisted Mathematics and Its Applications, School of Mathematical and Physical Sciences, University of Newcastle, Callaghan, NSW, Australia, Judy-anne.Osborn@anu.edu.au

Frank Quinn Virginia Tech, Blacksburg, VA 24061, USA, quinn@math.vt.edu

Annie Selden Department of Mathematical Sciences, New Mexico State University, Las Cruces, NM, USA, js9484@usit.net

Haw-Yaw Shy Department and Graduate Institute of Mathematics, National Changhua University of Educatio, Changua, Taiwan, shy@cc.ncue.edu.tw

Man Keung Siu Department of Mathematics, University of Hong Kong, Hong Kong SAR, China, mathsiu@hkucc.hku.hk

Ian Stevenson Department of Education and Professional Studies, King's College, London, UK, ian.stevenson@kcl.ac.uk

Andreas J. Stylianides Faculty of Education, University of Cambridge, Cambridge, UK, as899@cam.ac.uk

Gabriel Stylianides Department of Education, University of Oxford, Oxford, UK, gabriel.stylianides@education.ox.ac.uk

Michal Tabach School of Education, Tel-Aviv University, Tel-Aviv, Israel, tabach.family@gmail.com

David Tall Mathematics Education Research Centre, University of Warwick, Coventry, UK, david.tall@warwick.ac.uk

Denis Tanguay Département de mathématiques, Université du Québec à Montréal (UQAM), Montreal, QC, Canada, tanguay.denis@uqam.ca

Dina Tirosh School of Education, Tel-Aviv University, Tel-Aviv, Israel, dina@post.tau.ac.il

Pessia Tsamir School of Education, Tel-Aviv University, Tel-Aviv, Israel, pessia@post.tau.ac.il

Greisy Winicki-Landman Department of Mathematics and Statistics, California State Polytechnic University, Pomona, CA, USA, greisyw@csupomona.edu

Walter Whiteley Department of Mathematics and Statistics, York University, Toronto, Canada, whiteley@mathstat.yorku.ca

Kai-Lin Yang Department of Mathematics, National Taiwan Normal University, Taipei, Taiwan, kailinyang3@yahoo.com.tw

Oleksiy Yevdokimov Department of Mathematics & Computing, University of Southern Queensland, Toowoomba, Australia, oleksiy.yevdokimov@usq.edu.au

Orit Zaslavsky Department of Teaching and Learning, New York University, New York, NY, USA

Department of Education in Technology and Science, Technion – Israel, Institute of Technology, Haifa, Israel, oritrath@gmail.com

Chapter 1
Aspects of Proof in Mathematics Education

Gila Hanna and Michael de Villiers

This volume, *Proof and proving in mathematics education*, is a Study Volume sponsored by the International Commission on Mathematical Instruction (ICMI). ICMI Studies explore specific topics of interest to mathematics educators; they aim at identifying and analysing central issues in the teaching and learning of these topics. To this end, the ICMI convenes a Study Conference on chosen topics: A group of scholars from the conference then prepares a Study Volume that reports on the outcomes of the conference.

The present Study Volume examines several theoretical and practical notions about why and how mathematics educators should approach the teaching and learning of proof and proving. The authors of the chapters here are presenting major themes and subthemes that arose from the presentations and discussions at the 19th ICMI Study Conference.

1   ICMI Study 19

The 19th ICMI Study, intended to examine issues of proof and proving in mathematics education, was officially launched in 2007 with the selection of Gila Hanna and Michael de Villiers as Co-Chairs. In consultation with them, the ICMI Executive invited eight additional experts in the field of proof in mathematics education to

G. Hanna (✉)
Ontario Institute for Studies in Education, University of Toronto, Toronto, Canada
e-mail: gila.hanna@utoronto.ca

M. de Villiers
School of Science, Mathematics & Technology Education, University of KwaZulu-Natal, Durban, South Africa
e-mail: profmd@mweb.co.za

© The Author(s) 2021
G. Hanna and M. de Villiers (eds.), *Proof and Proving in Mathematics Education*,
New ICMI Study Series, https://doi.org/10.1007/978-94-007-2129-6_1

serve on an International Program Committee (IPC). The Co-Chairs prepared a draft Discussion Document, circulated it to the entire IPC, and then revised it in light of the IPC members' input. At its first meeting (Essen, Germany, November 2007), the IPC settled on the themes of the Study and finalised the Discussion Document, which was later published in the Bulletin of the International Commission on Mathematical Instruction as well as in a number of mathematics education journals (see Appendix 1).

Clearly, we could not include in a single ICMI Study all the themes germane to the teaching of proof. Thus the IPC originally selected seven themes that it judged to be most relevant to mathematics education and within the IPC members' realm of expertise. The Discussion Document called for contributions that would address these themes and contained a list of criteria by which the contributions would be assessed.

At its second meeting (Sèvres, France, November 2008), the IPC selected the contributions that had been recommended by reviewers after a strict refereeing process and that were also most closely related to the conference themes. (Unfortunately a few excellent submissions had to be excluded because they treated themes beyond the conference's scope.) The IPC then drew up an invitation list of about 120 contributors.

Taking into account the submissions that had been accepted, the IPC developed a programme that included only six of the original seven themes. Each of these themes was the focus of a Working Group (WG) that met throughout the Study Conference and whose major aim was to prepare one or more chapters for this book.

WG1: Cognitive Development of Proof, co-chaired by David Tall and Oleksiy Yevdokimov, focused on the characteristics of the cognitive development of proof at various school levels, with a view to building an overall picture of the cognitive development of proof.

WG2: Argumentation, chaired by Viviane Durand-Guerrier, focused on the relationship between proof and argumentation from the perspective of opposing qualities such as formal vs. informal, form vs. content, syntax vs. semantics, truth vs. validity, mathematical logic vs. common sense, formal proof vs. heuristics, and continuity vs. discontinuity.

WG3: Dynamic Geometry Software/Experimentation, chaired by Ferdinando Arzarello, focused on the ways in which mathematical investigations using advanced technology and different semiotic resources relate to the formal aspects of mathematical discourse and to the production of proofs.

WG4: Proof in the School Curriculum, Knowledge for Teaching Proof, and the Transition from Elementary to Secondary, chaired by Fou-Lai Lin, focused on the knowledge that teachers need to teach proof effectively and on how proving activities should be designed to best foster successful instruction about proof and proving.

WG5: The Nature of Proof for the Classroom, co-chaired by Tommy Dreyfus, Hans Niels Jahnke, and Wann-Sheng Horng, examined aspects of the teaching of proof from the primary through the tertiary level. It addressed questions about the form, status, and role that proof must assume at each level to ensure success in generating mathematical understanding.

WG6: Proof at the Tertiary Level, chaired by Annie Selden, explored all aspects of the teaching and learning of proof and proving at the tertiary level, including the transition from secondary school to university and the transition from undergraduate to graduate work in mathematics.

Complementary to the Working Groups, the IPC broadened the Study's scope by inviting four distinguished scholars to deliver plenary talks on topics related to proof in mathematics, but not necessarily intimately connected to mathematics education. In their talks, Giuseppe Longo, Jonathan Borwein, Judith Grabiner and Frank Quinn examined proof from the four perspectives of epistemology, experimental mathematics, the history of mathematics, and mathematics itself. The IPC also invited a panel of eminent experts, Karine Chemla, Wann-Sheng Horng and Man Keung Siu to discuss proof as perceived in ancient Chinese mathematics writing.

The ICMI Study Conference itself took place at the National Taiwan Normal University in Taipei, Taiwan, from May 10 to May 15, 2009 (see Appendix 2).

2   Contents of the Volume

A common view of mathematical proof sees it as no more than an unbroken sequence of steps that establish a necessary conclusion, in which every step is an application of truth-preserving rules of logic. In other words, proof is often seen as synonymous with formal derivation. This Study Volume treats proof in a broader sense, recognising that a narrow view of proof neither reflects mathematical practice nor offers the greatest opportunities for promoting mathematical understanding.

In mathematical practice, in fact, a proof is often a series of ideas and insights rather a sequence of formal steps. Mathematicians routinely publish proofs that contain gaps, relying on the expert reader to fill them in. Many published proofs are informal arguments, in effect, but are still considered rigorous enough to be accepted by mathematicians.

This Volume examines aspects of proof that include, but are not limited to, explorations, explanations, justification of conjectures and definitions, empirical reasoning, diagrammatic reasoning, and heuristic devices. The chapter authors, whilst by and large accepting the common view of proof, do diverge on the importance they attach to various aspects of proof and particularly on the degree to which they judge formal derivation as necessary or useful in promoting an understanding of mathematics and mathematical reasoning.

The remainder of the Volume is divided into six parts. These are arranged according to major themes that arose from the conference as a whole, rather than by working groups.

Part I: Proof and Cognition

In Chap. 2, "Cognitive development of proof" David Tall, Oleksiy Yevdokimov, Boris Koichu, Walter Whiteley, Margo Kondratieva and Ying-Hao Cheng examine the development of proof from the child to the adult learner and on to the mature research mathematician. The authors first consider various existing theories and viewpoints relating to proof and proving from education research, brain research, cognitive science, psychology, semiotics, and more, and then go on to offer their own theory of "the broad maturation of proof structures". Their resulting framework for the broad maturation of proof structures consists of six developmental stages which they illustrate with an interesting array of well-chosen examples. They also appropriately elaborate on the novel notion of a "crystalline concept" which they define as "a concept that has an internal structure of constrained relationships that cause it to have necessary properties as part of its context."

In his plenary chapter "Theorems as constructive visions" Giuseppe Longo describes mathematics and proofs as conceptual constructions that, though supported by language and logic, originate in the real activities of humans in space and time. He points out in particular the crucial role of cognitive principles such as symmetry and order in attaining mathematical knowledge and understanding proof, citing several examples to show that in constructing a proof the notions of symmetry and order derived from actual experience are no less essential than logical inference. He concludes that "Mathematics is the result of an open-ended 'game' between humans and the world in space and time; that is, it results from the inter-subjective construction of knowledge made in language and logic, along a passage through the world, which canalises our praxes as well as our endeavour towards knowledge."

Part II: Experimentation: Challenges and Opportunities

Mathematical researcher Jonathan Borwein, in his plenary chapter "Exploratory experimentation: Digitally-assisted discovery and proof" argues that current computing technologies offer revolutionary new scaffolding both to enhance mathematical reasoning and to restrain mathematical error. He shares Pólya's view that intuition, enhanced by experimentation, mostly precedes deductive reasoning. He then gives and discusses some illustrative examples, which clearly show that the boundaries between mathematics and the natural sciences, and between inductive and deductive reasoning, are blurred and getting more blurred.

Borwein points out that the mathematical community faces a great challenge to re-evaluate the role of proof in light of the power of computer systems, the sophistication of mathematical computing packages, and the growing capacity to data-mine on the Internet. As the prospects for inductive mathematics blossom, the need to ensure that the role of proof is properly founded remains undiminished.

The chapter "Experimental approaches to theoretical thinking: Artefacts and proofs" by Ferdinando Arzarello, Mariolina Bartolini Bussi, Andy Leung, Maria Alessandra Mariotti, and Ian Stevenson examines the dynamic tension between the empirical and the theoretical aspects of mathematics, especially in relation to the role of technological artefacts in both. It does so against the background of offering teachers a comprehensive framework for pursuing the learning of proof in the classroom.

The authors discuss and analyse their subject from different linked perspectives: historical, epistemological, didactical and pedagogical. They first present examples of the historical continuity of experimental mathematics from straight-edge and compass construction to the modern use of different dynamic mathematics software. They draw these examples from a few different cultures and epochs in which instruments have played a crucial role in generating mathematical concepts, theorems and proofs.

Second, the authors analyse some didactical episodes from the classroom, where the use of instruments in proving activities makes the aforementioned dynamic tension explicit. Specifically, they examine how this tension regulates students' cognitive processes in solving mathematical problems, first making explorations with technological tools, then formulating suitable conjectures and finally proving them.

The chapter is followed by a commentary "Response to Experimental approaches to theoretical thinking: Artefacts and proofs" by Jonathan Borwein and Judyanne Osborn.

Part III: Historical and Educational Perceptions of Proof

In her plenary address "Why proof? A historian's perspective," historian of mathematics Judith Grabiner traces some of the main aspects of the history of mathematical proof in the Western tradition. She first addresses the birth of logical proof in Greek geometry and why the Greeks moved beyond visualisation to purely logical proof. Then she looks at the use of visual demonstration in Western mathematics since the Greeks, and proceeds to discuss two characteristics of more modern mathematics, abstraction and symbolism, and their power. There follows a discussion of how and why standards of proof change, noting in particular the influence of ideas from philosophy. Finally, the author discusses how proof in mathematics interacts with the 'real world', arguing that proof did not develop in a cultural or intellectual vacuum.

In the chapter "Conceptions of proof – In research and in teaching", Richard Cabassut, AnnaMarie Conner, Filyet Aslı İşçimen, Fulvia Furinghetti, Hans Niels Jahnke and Francesca Morselli describe mathematicians' conceptualisations of proof and contrast them with those of mathematics educators. The authors argue that practising mathematicians do not rely on any specific formal definition of proof but they do seem to know what a proof is. On the other hand, mathematics educators' conceptions of proof derive from the need to teach students to construct proper proofs and to recognise the subtle differences between argumentation and mathematical proof. The authors then discuss the ideas of "genetic", "pragmatic," and "conceptual" proofs. They next examine in detail some epistemological and pedagogical beliefs about the nature and role of proof in mathematics, about the role of proof in school mathematics, about difficulties in proving, about how proof should be taught in school, and about the self as mathematical thinker in the context of proof. The authors conclude by discussing "metaknowledge about proof", its importance and its role in the mathematics curriculum.

Tommy Dreyfus, Elena Nardi and Roza Leikin review diverse forms of proofs in their chapter "Forms of proof and proving in the classroom". Relying on many empirical studies presented at the ICMI 19 Conference and on published empirical research papers, they describe a variety of proofs (e.g., by visual, verbal, and dynamic representations) and an array of mathematical arguments (from example-based, deductive and inductive to generic and general). They discuss different degrees of rigour, where and how these are used, and the contexts in which they appear. The authors also report on students' and teachers' beliefs about various aspects of proof and proving. They discuss the pedagogical importance of multiple-proof tasks and of taking into account the mathematical, pedagogical, and cognitive structures related to the effective teaching of proof and proving. They conclude with a plea for additional empirical research, longitudinal studies, and investigations on the long-term effects of the different approaches to proof.

In the chapter "The need for proof and proving: mathematical and pedagogical perspectives", Orit Zaslavsky, Susan D. Nickerson, Andreas Stylianides, Ivy Kidron and Greisy Winicki-Landman explore three main questions: Why teach proof? What are (or may be) learners' needs for proof? How can teachers facilitate the need for proof?

First, they discuss the connection between different functions of proof in mathematics and the needs those evoke for teaching proof. They briefly explore the epistemology of proof in the history of mathematics in order to illuminate the needs that propelled the discipline's development. Second, the authors take a learner's perspective on the need to prove, and examine categories of intellectual need that may drive learners to prove (i.e., needs for certitude, for understanding, for quantification, for communication, and for structure and connection). Finally, the authors address pedagogical issues involved in teachers' attempts to facilitate learners' need to prove; uncertainty, cognitive conflict or the need for explanation or organised structure may help drive learners to prove.

In his plenary "Contemporary proofs for mathematics education", Frank Quinn argues that the proofs encountered in mathematical practice provide a very high level of reliability, because the proof process creates a record sufficiently detailed to allow easy detection and repair of errors. He therefore recommends the introduction of two key ideas, "potential proof" and "formal potential proof," into school mathematics and undergraduate mathematics education. The first entails asking students to show their work – that is, to provide a detailed record of their solution so that it can be checked for errors. The second means asking students to supply explicit explanations to justify their work. Citing several examples, Quinn argues that students should thereby learn that a train of reasoning leading to a correct conclusion does not count as a proof unless it is a potential proof that has been found to be error-free.

Part IV: Proof in the School Curriculum

Keith Jones and Patricio Herbst, in their chapter "Proof, proving, and teacher-student interaction: Theories and contexts", seek to identify theoretical frameworks that would help understand the teacher's role in proof education. They focus on three theories that might shed light on teacher-student interaction in teaching of proof across diverse contexts. They first discuss the theory of socio-mathematical norms, characterised by inquiry-based mathematics classrooms and the use of classroom interactions to arrive at shared norms of mathematical practice. Second comes the theory of teaching with variation, in which the teacher uses two types of variations: *conceptual variation* (highlighting a new concept by contrasting inadmissible examples), and *procedural variation* (refocusing the learner's attention from a concrete problem to its symbolic representation). Third, the authors examine the theory of instructional exchanges that, borrowing from Brousseau's notion a "didactical contract" presumes that teacher and students are mutually responsible for whatever learning takes place in the classroom.

Feng-Jui Hsieh, Wang-Shian Horng and Haw-Yaw Shy, in "From exploration to proof production", explain how exploration, especially hands-on exploration, is introduced and integrated into the teaching of proof in Taiwan. They describe a conceptual model for the relationship between exploration, problem solving, proving and proof, and illustrate it with two exploratory teaching experiments.

The authors distinguish two different positions in regard to "exploration" as a learning and conceptualising activity. The first position views exploration as a mental process, the second, as an activity that involves manipulating and interacting with external environments (e.g., hands-on or dynamic computer software environments). Exploration generally provides learners with valuable opportunities to construct mathematics objects, transform figures, probe in multiple directions, perceive divergent visual information, and receive immediate feedback on their actions. The authors also give two extracts from a Taiwanese textbook, which demonstrate the integration of exploration in proving. Last, they provide a useful but tentative comparison of dynamic computer and hands-on explorations, and summarise some of the positive and negative issues raised by integrating exploration, as well as suggesting areas for future research.

Fou-Lai Lin, Kyeong-Hwa Lee, Kai-Lin Yang, Michal Tabach and Gabriel Stylianides develop some principles for designing tasks that teach conjecturing and proving in the chapter "Principles of task design for conjecturing and proving". They extract some first principles from design research and the literature for designing tasks for mathematics learning generally. They also briefly reflect on a few historical examples, such as Fermat and Poincaré's conjectures, within the context of Lakatos' model.

They discuss the strategy of promoting 'what-if-not' questions, which encourage students to conjecture the consequence of some change in a statement's premise or conclusion or to explore the transformation and application of algorithms and formulae in other areas. They also explore students' attainment of conviction and ability to refute statements as well as how to scaffold students' progress from an inductive to a symbolic proof schema. They then adapt elements of all the above in order to develop specific principles for formulating a more general framework for conjecturing and proving. They illustrate this model by developing and analysing some practical tasks.

The chapter "Teachers' professional learning of teaching proof and proving" by Fou-Lai Lin, Kai-Lin Yang, Jane-Jane Lo, Pessia Tsamir, Dina Tirosh and Gabriel Stylianides starts with a narrative on the Hanoi Towers activity. From it, the authors draw three essential factors for teachers' competence in teaching proof: knowledge specific to proof content and proof methods; beliefs/values specific to the nature and didactics of proof; and practice specific to motivating, guiding, and evaluating students' argumentation and proof. They then elaborate on these factors with examples related to specific content for the professional learning of primary and secondary mathematics teachers. They distinguish and discuss three further important dimensions from the research literature on mathematics teacher education: establishing conviction, the role of the teacher educator, and the notion of cognitive conflict.

Last the authors discuss the importance of involving teachers in designing task sequences for teaching proof that motivate students' engagement, challenge their mathematical thinking, create cognitive conflict, and encourage argumentation and critical reflection. To sequence such tasks, the teacher as designer needs to take possible learning trajectories into consideration. Testing the instructional tasks with their students engages teachers in productive reflection.

Part V: Argumentation and Transition to Tertiary Level

In their chapter "Argumentation and proof in the mathematics classroom", Viviane Durand-Guerrier, Paolo Boero, Nadia Douek, Susanna Epp and Denis Tanguay adopt the position that the broad concept of argumentation encompasses mathematical proof as a special case. They describe and discuss the complex relationships between argumentation and proof in mathematical practice from various mathematical and educational perspectives. They conclude that students can benefit from the openness of exploration and flexible validation rules typical of argumentation as a prelude to the stricter uses of rules and symbols essential in constructing a mathematical proof. They describe many examples indicating that appropriate learning environments can facilitate both argumentation and proof in mathematics classes.

Next, in "Examining the role of logic in teaching proof", the same authors examine the usefulness of teaching formal logic in school mathematics classes, since many high-school mathematics graduates arrive at the tertiary level deficient in deductive reasoning skills. The authors initially examine the positions taken by cognitive scientists and mathematics educators about the role of formal logic in reasoning, argumentation and proof. This survey reinforces their view that (*pace* some of those commentators) the principles of formal logic operate in all these procedures, even if not explicitly stated. The authors next move to argue the importance of teaching logic along with mathematics, since logic operates in both syntactic and semantic mathematical discourses. Hence, students' constructing proofs requires both logical and mathematical knowledge. Finally, the authors give examples of contexts suitable for fostering students' knowledge of logic and dispelling some of their misconceptions. Familiar contexts provide the best opportunities, they conclude.

Annie Selden's chapter "Transitions and proof and proving at tertiary level" examines some of the changes students experience when moving from secondary school to undergraduate study, or further to graduate studies, in mathematics. For example, they face changes in the didactical contract and a cognitive transition from experiential and intuitive concepts to more abstract ones with formal definitions and properties reconstructed through logical deductions.

More important for Selden, at the tertiary level constructing proofs involves both understanding and using formal definitions and previously established theorems, as well as considerable creativity and insight. Reading and constructing such formal, more rigorous proofs entails a major transition for students, but one that is sometimes supported by relatively little explicit instruction.

In addition, tertiary proofs relate to more complex, abstract structures than those expected of students at primary or secondary level. Comparing typical secondary school geometry proofs with proofs in real analysis, linear algebra, abstract algebra or topology, Selden argues that the objects in geometry are idealisations of real things (points, lines, planes), whereas the objects in the latter subdisciplines (functions, vector spaces, groups, topological spaces) are abstract reifications.

Part VI: Lessons from the Eastern Cultural Traditions

The last two chapters are devoted to a close look at traditional forms of proof in China. In her panel presentation "Using documents from ancient China to teach mathematical proof", Karine Chemla discusses the proofs used in ancient China and shows how they can provide a rich source of ideas worth examining for their relevance to the classrooms of today. Her point of departure is the work of Liu Hui, who in 263 C.E completed a commentary on The Nine Chapters on Mathematical Procedures, the earliest known Chinese mathematical book. Chemla shows that Liu Hui, when investigating the algorithms embedded in this early book, took pains to elaborate how they were developed and how they could best be shown to be correct by way of systematic and detailed analysis. The examples she brings show how important it was for Liu Hui to keep track of the meaning of the operations within an algorithm and of providing evidence for the correctness of these operations. In Chemla's opinion the teaching of algebraic proofs in today's schools could benefit from Liu Hui's insistence, demonstrated in his commentary of The Nine Chapters, on the use of well-defined mathematical procedures and on the need for evidence at each step.

In his panel presentation "Proof in the Western and Eastern traditions: Implications for mathematics education", Man Keung Siu compares and contrasts the Western and Eastern traditions of doing mathematics, whilst maintaining that "there is something about mathematics that is universal, irrespective of race, culture or social context". He states that even if one accepts the over-simplified notion that Western tradition is "dialectic", whereas Eastern tradition is "algorithmic", it can be shown that there are several parallels between these two mathematical traditions. Siu presents several examples of proofs and constructions in which these two approaches can clearly be seen as complementing each other. In Siu's view the Western and Eastern traditions can both play an important role in the teaching of mathematics, because a procedural (algorithmic) approach can be used to help build up a conceptual (dialectic) understanding.

3   Conclusion

The aspects of mathematics proof investigated in this Volume are those that the IPC and ICMI executive representatives judged vital to a better understanding of teaching mathematics in general and teaching proof in particular. We hope they have deepened our understanding of these difficult issues in many ways. The Volume does not address all of the research areas and questions raised in the Discussion Document (Appendix 1) in equal depth and detail. Ideally, the Discussion Document will continue to stimulate new directions for research on proof, for example, the didactical use of the explanatory nature of proof to motivate students to learn proof.

Proof in mathematical research often allows further generalisation and/or specialisation to new results, since proving results usually promotes insight into why they are true. So another under-explored research area encompasses the identification of good problems and the development of effective strategies to help students see and appreciate this 'discovery' function of proof. In addition, the mathematics education community needs to constantly and fundamentally rethink the role of experimentation and proof in the light of rapidly developing computer technologies, dynamic software environments and concurrent advances in cognitive science and in the emerging science of automated proof.

We hope that this book will convince readers that research on proof and proving is indispensable to serious discussions about the place of proof in mathematics education.

Acknowledgements We would like to thank the authors for their spirit of cooperation. They have made our job as editors both a pleasure and a privilege. We are grateful to the referees for their thoughtful and constructive reviewing. Special thanks go to John Holt for his superb work in editing the manuscripts; he has helped many chapters read better than they did before he worked his magic on them. Many thanks go to Sarah-Jane Patterson for her invaluable editorial assistance and her many helpful suggestions. We also thank Gunawardena Egodawatte who provided further editorial assistance. We wish to acknowledge the generous support of the Social Sciences and Humanities Research Council of Canada.

Part I
Proof and Cognition

Chapter 2
Cognitive Development of Proof

**David Tall, Oleksiy Yevdokimov, Boris Koichu, Walter Whiteley,
Margo Kondratieva, and Ying-Hao Cheng**

1 Introduction

In this chapter our aim is to seek how individuals develop ideas of proof appropriate
for their level of maturity at the time, and how these ideas change in sophistication
over the long term, from the young child to the adult user of mathematics and on to
the research mathematician. The exposition focuses on the ways in which the
developing individuals build from real world perceptions and actions to a mental
world of sophisticated mathematical knowledge.

D. Tall (✉)
Mathematics Education Research Centre, University of Warwick, Coventry, UK
e-mail: david.tall@warwick.ac.uk

O. Yevdokimov
Department of Mathematics & Computing, University of Southern Queensland,
Toowoomba, Australia
e-mail: oleksiy.yevdokimov@usq.edu.au

B. Koichu
Department of Education in Technology and Science, Technion – Israel Institute of Technology,
Haifa, Israel
e-mail: bkoichu@technion.ac.il

W. Whiteley
Department of Mathematics and Statistics, York University, Toronto, Canada
e-mail: whiteley@mathstat.yorku.ca

M. Kondratieva
Faculty of Education and Department of Mathematics and Statistics, Memorial University of
Newfoundland, St. John's, Canada
e-mail: mkondra@mun.ca

Y.H. Cheng
Department of Mathematics, Taipei Municipal University of Education, Taipei, Taiwan
e-mail: yinghao.cheng@msa.hinet.net

© The Author(s) 2021
G. Hanna and M. de Villiers (eds.), *Proof and Proving in Mathematics Education*,
New ICMI Study Series, https://doi.org/10.1007/978-94-007-2129-6_2

This chapter is consonant with the four plenaries presented to the ICMI conference on proof. Longo sees the formalism of modern mathematics growing out of the actions and perceptions of the biological human brain. Grabiner reports significant examples of development in history as mathematical experts build on their experience to develop new mathematical constructs. Borwein observes the changing nature of mathematical thought now that we have computer technology to perform highly complex computations almost immediately and to represent information in dynamic visual ways. Quinn underlines the mathematical concern that proof at the highest level needs to be fundamentally based on the precision of the axiomatic method.

Mathematical proof develops in many different forms both in historical time and in the development of any individual. Various degrees of proof are suggested in school mathematics by terms such as 'show', 'justify', 'explain', 'prove from first principles'. Rather than begin by debating the difference between them, we will use the word 'proof' in its widest sense and analyse the changes in its meaning as the individual matures.

The outline of the chapter is as follows. In Section 2 we consider how the nature of proof is envisioned by professional mathematicians and by novices in mathematics. This is followed, in Section 3, by introducing a three-facet conceptual framework based on perceptions, operations and formal structures that enables us to adequately consider the cognitive journey from the child to the adult, and from the novice to professional mathematician. Section 4 deals with the development of proof from human experience. Section 5 considers the development of proof in the context of Euclidean and non-Euclidean geometry. Section 6 details the increasing sophistication of proofs in arithmetic and algebra from proof using specific calculations, generic arguments, algebraic manipulation and on to algebraic proof based on the rules of arithmetic. Section 7 addresses the development of proof in undergraduates and on to research mathematics, followed by a summary of the whole development.

2 Perceptions of Proof

2.1 What Is Proof for Mathematicians?

Mathematics is a diverse and complex activity, spanning a range of contexts from everyday practical activities, through more sophisticated applications and on to the frontiers of mathematical research. At the highest level of mathematical research, discovery and proof of new theorems may be considered to be the summit of mathematical practice. In the words of three mathematicians:

> Proofs are to mathematics what spelling (or even calligraphy) is to poetry. Mathematical works do consist of proofs, just as poems do consist of characters (Arnold 2000, p. 403).
> 'Ordinary mathematical proofs'—to be distinguished from formal derivations—are the locus of mathematical knowledge. Their epistemic content goes way beyond what is summarised in the form of theorems (Rav 1999, p. 5).

> The truth of a mathematical claim rests on the existence of a proof. Stated this way, such a criterion is absolute, abstract, and independent of human awareness. This criterion is conceptually important, but practically useless (Bass 2009, p. 3).

We chose these quotations because they suggest not only the importance of proofs in mathematics, but also reveal the debate on the role and nature of proofs within the mathematical community. We learn from the first quotation that proofs are fundamental to the structure of mathematics. The second tells us that the usual ('ordinary') proofs produced by mathematicians have subtleties of meaning that go beyond the application of logic. The third implies that mathematical proof as an absolute argument is conceptually important but may not be what occurs, or even what is achievable, in practice.

A formal proof, in the sense of Hilbert (1928/1967), is a sequence of assertions, the last of which is the theorem that is proved and each of which is either an axiom or the result of applying a rule of inference to previous formulas in the sequence; the rules of inference are so evident that the verification of the proof can be done by means of a mechanical procedure. Such a formal proof can be expressed in first-order set-theoretical language (Rav 1999). Dawson (2006, p. 271) observed that 'formal proofs appear almost exclusively in works on computer science or mathematical logic, primarily as objects to study to which other, informal, arguments are applied.'

An ordinary mathematical proof consists of an argument to convince an audience of peer experts that a certain mathematical claim is true and, ideally, to explain why it is true (cf. Dawson 2006). Such ordinary proofs can be found in mathematics research journals as well as in school and university-level textbooks. They utilise second or higher-order logic (Shapiro 1991), but often contain conceptual bridges between parts of the argument rather than explicit logical justification (Rav 1999). Sometimes a convincing argument for peer experts does not constitute a formal proof, only a justification that a proof can be constructed, given sufficient time, incentive, and resources (Bass 2009).

From the epistemic point of view, a proof for mathematicians involves thinking about new situations, focusing on significant aspects, using previous knowledge to put new ideas together in new ways, consider relationships, make conjectures, formulate definitions as necessary and to build a valid argument.

In summary, contemporary mathematicians' perspectives on proof are sophisticated yet build on the broad development to 'convince yourself, convince a friend, convince an enemy,' in mathematical thinking at all levels (Mason et al. 1982). To this should be added Mason's further insight that mathematicians are able to develop an 'internal enemy'– a personally constructed view of mathematics that not only sets out to convince doubters, but shifts to a higher level attempting to make sense of the mathematics itself.

2.2 What Is Proof for Growing Individuals?

Children or novices do not initially think deductively. The young child begins by interacting with real-world situations, perceiving with the senses including vision, hearing, touch, acting on objects in the world, pointing at them, picking them up, exploring their properties, developing language to describe them.

In parallel with the exploration of objects, the child explores various operations on those objects: sorting, counting, sharing, combining, ordering, adding, subtracting, multiplying, dividing, developing the operations of arithmetic and on to the generalised arithmetic of algebra. This involves observing regularities of the operations, such as addition being independent of order and various other properties that are collected together and named 'the rules of arithmetic'. Properties that were seen as natural occurrences during physical operations with objects are subtly reworded to become rules that must be obeyed. Children may even establish their own rules which are not necessarily correct yet, nevertheless, help them to make initial steps towards deductive thinking. For instance, a 4-year old child attempting to persuade her parents that, 'I am older (than another child) because I am taller.' (Yevdokimov, in preparation.)

Over the longer term, the convincing power of the emerging rules for the child is often rooted not only in the observation that the rules 'work' in all available situations, but in the external role of authorities such as a parent, a teacher or a textbook (Harel and Sowder 1998).

It is only much later – usually at college level – that axiomatic formal proof arises in terms of formal definitions and deductions. Unlike earlier forms of proof, the axioms formulated express only the properties required to make the necessary deduction, and are no longer restricted to a particular context, but to any situation where the only requirement is that the axioms are satisfied.

This yields a broad categorisation of three distinct forms of proof: using figures, diagrams, transformations and deduction in geometry, using established rules of arithmetic in algebra initially encountered in school, and in axiomatically defined structures met only by mathematics majors in university. Our task is to fit the development of proof in general and these three forms of proof in particular into a framework of cognitive growth.

3 Theoretical Framework

We begin by considering a brief overview of theories of cognitive growth relevant to the development of proof. Then we focus on a single idea that acts as a template for the cognitive development of proof in a range of contexts: the notion of a crystalline concept. This will then be used as a foundation for the development of mathematical thinking over time, in which the cognitive development of proof plays a central role.

3.1 Theories of Cognitive Growth

There are many theories of cognitive growth offering different aspects of development over the longer term. Piaget, the father of cognitive approaches to development, sees the child passing through various stages, from the sensorimotor, through concrete operations, then on to formal operations.

Lakoff (1987) and his co-workers (Lakoff and Johnson 1999; Lakoff and Núñez 2000) claim that all human thought is embodied through the sensorimotor functions of the human individual and builds linguistically through metaphors based on human perception and action.

Harel and Sowder (1998, 2005) describe cognitive growth of proof in terms of the learner's development of proof schemes – relatively stable cognitive/affective configurations responsible for what constitutes ascertaining and persuading an individual of the truth of a statement at a particular stage of mathematical maturation. A broadly based empirical study found a whole range of different proof schemes, some categorised as 'external conviction', some 'empirical' and some 'analytical'.

Van Hiele (1986) focuses specifically on the development of Euclidean geometry, proposing a sequence of stages from the recognition of figures, through their description and categorisation, the more precise use of definition and construction using ruler and compass and on to the development of a coherent framework of Euclidean deductive proof.

There are also theories of development of symbolism through the encapsulation of processes (such as counting) into concepts (such as number) that reveal a different kind of development in arithmetic and the generalised arithmetic of algebra (Dubinsky and McDonald 2001; Gray and Tall 1994; Sfard 1991).

Many theoretical frameworks speak of multiple representations (or registers) that operate in different ways (e.g., Duval 2006; Goldin 1998). The two distinct forms of development through the global visual-spatial modes of operation on the one hand and the sequential symbolic modes of operation on the other can operate in tandem with each supporting the other (Paivio 1991). Bruner's three modes of communication – enactive, iconic, symbolic – also presume different ways of operating: the sensorimotor basis of enactive and iconic linking to the visual and spatial, and the symbolic forms including not only language but the sub-categories of number and logic (Bruner 1966).

The hypothesis about distinct cognitive structures for language/symbolism and for visualisation has received empirical support by means of neuroscience.

For instance, Fig. 2.1 shows the areas of the brain stimulated when responding to the problem '5×7', which coordinates an overall control in the right hemisphere and the language area in the left recalling a verbal number fact, and the response to the problem 'Is 5×7>25?' that uses the visual areas at the back to compare relative sizes (Dehaene 1997). This reveals human thinking as a blend of global perceptual processes that enable us to 'see' concepts as a gestalt (Hoffmann 1998), and sequential operations that we can learn to perform as mathematical procedures.

The brain operates by the passage of information between neurons where connections are excited to a higher level when they are active. Repeated use strengthens the links chemically so that they are more likely to react in future and build up sophisticated knowledge structures. Meltzoff et al. (2009) formulate the child's learning in terms of the brain's implicit recognition of statistical patterns:

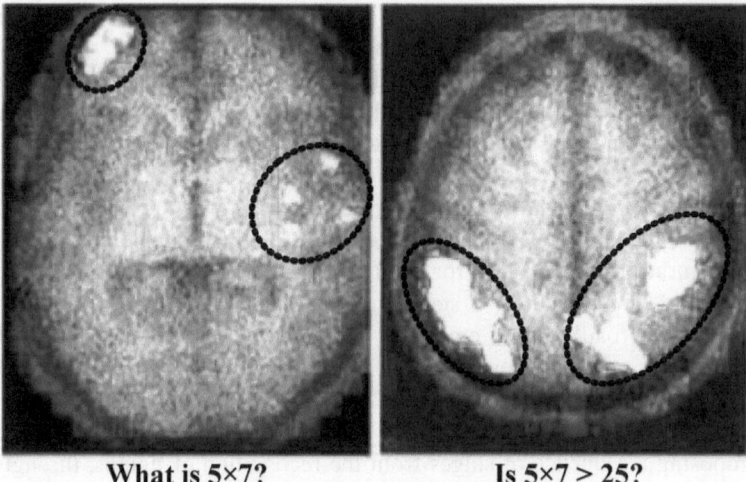

What is 5×7?   **Is 5×7 > 25?**

Fig. 2.1 Areas of the brain recalling a fact and performing a comparison

Recent findings show that infants and young children possess powerful computational skills that allow them automatically to infer structured models of their environment from the statistical patterns they experience. Infants use statistical patterns gleaned from experience to learn about both language and causation. Before they are three, children use frequency distributions to learn which phonetic units distinguish words in their native language, use the transitional probabilities between syllables to segment words, and use covariation to infer cause-effect relationships in the physical world (p. 284).

The facility for building sophisticated knowledge structure is based on a phenomenal array of neuronal facilities for perception and action that are present in the new-born child and develop rapidly through experience in the early years. Tall (2008) refers to these abilities as 'set-befores' (because they are set before birth as part of our genetic inheritance and develop through usage) as opposed to 'met-befores' that arise as a result of previous experience and may be supportive or problematic when that experience is used in new contexts. He hypothesises that three major set-befores give rise to three distinct developments of mathematical thinking and proof through:

• recognition of similarities, differences and patterns through perception,
• repetition: the ability to learn complex sequences of operation through action, and
• the development of language to enable perception and action to be expressed and conceived in increasingly subtle ways.

In the development of mathematical thinking, these three set-befores combine to give the three different ways of constructing mathematical concepts. Language enables the verbal categorisation of the perception of figures in geometry and other aspects of mathematics. It enables the encapsulation of processes as concepts to compress processes that occur in time into manipulable mental objects in arithmetic and algebra. And it enables definition of concepts, both in terms of observed properties of perception and action in school mathematics and also of the proposed set-theoretic properties of axiomatic systems.

As the authors of this chapter reflected together on the total development of mathematical proof and the many cognitive theories available with various links and differences, we came to the view that there is a single broad developmental template that underlies them all, from early perceptions and actions, to the various forms of Euclidean, algebraic and axiomatic proof.

3.2   Crystalline Concepts

The foundational idea that underpins our framework can be introduced using a single specific example that starts simply and becomes more general until it provides a template for the total development of mathematical proof, from the first perceptions and actions of the child to the Platonic concepts of Euclidean geometry, algebraic proofs based on the rules of arithmetic, and formal proofs in axiomatic mathematics.

Our example is the notion of an isosceles triangle as seen by a maturing child. At first the child perceives it as a single gestalt. It may have a shape that is broad at the bottom, narrowing to the top with two equal sides, two equal angles and an axis of symmetry about a line down the middle. If the triangle is cut out of paper, it can be folded over this central line to reveal a complete symmetry. The child can learn to recognise various isosceles triangles and describe some of their properties. However, at this early stage, all these properties occur simultaneously, they are not linked together by cause-effect relationship.

In order to be able to make sense of coherent relationships, the child needs to build a growing knowledge structure (or schema) of experiences involving perceptions and actions that relate to each other. As the child develops, some of the properties described may become privileged and used as a definition of a particular concept. For instance, one may describe an isosceles triangle to be 'a triangle with two equal sides.' Now the child may use this criterion to test whether new objects are isosceles triangles. For the child, the 'proof' that a particular triangle is isosceles is that anyone can see that it has two equal sides.

Subsequently, the child may be introduced to a more sophisticated knowledge structure, such as physically placing one triangle on top of another and verbalising this as the principle of congruence. As the child becomes aware of more and more properties of an isosceles triangle and his or her conceptions of relationships develop, it may become possible to see relationships between properties and to use appropriate principles based on constructions and transformations to deduce some properties from the others. The idea emerges that it is not necessary to include all known properties in a definition. An isosceles triangle defined only in terms of equal sides, with no mention of the angles, can now be proved by the principle of congruence to have equal angles.

Other, more sophisticated, properties may be deduced using similar techniques. For instance, for an isosceles triangle, the perpendicular bisector of the base can be proved to pass through the vertex, or the bisector of the vertex angle can be proved to meet the base in the midpoint at right angles. Further proofs show that any of these deduced properties mentioned may be used as alternative definitions. At this point there may be several different possible definitions that are now seen to be equivalent. It is not that the triangle has a single definition with many consequences, but that it has many equivalent definitions any one of which can be used as a basis for the theory. One of these properties, usually the simplest one to formulate, is then taken as the primary definition and then all other properties are deduced from it.

Then something highly subtle happens: the notion of isosceles triangle – which originally was a single gestalt with many simultaneous properties, and was then defined using a single specific definition – now matures into a fully unified concept, with many properties linked together by a network of relationships based on deductions.

We introduce the term crystalline concept for such a phenomenon. A crystalline concept may be given a working definition as 'a concept that has an internal structure of constrained relationships that cause it to have necessary properties as part of its context.' A typical crystalline concept is the notion of an idealised Platonic figure in Euclidean geometry. However, as we shall see, crystalline concepts are also natural products of development in other forms of proof such as those using symbol manipulation or axiomatic definition and deduction.

The long-term formation of crystalline concepts matures through the construction of increasingly sophisticated knowledge structures, as follows:

- perceptual recognition of phenomena where objects have simultaneous properties,
- verbal description of properties, often related to visual or symbolic representations, to begin to think about specific properties and relationships,
- definition and deduction, to define which concepts satisfy the definition and to develop appropriate principles of proof to deduce that one property implies another,
- realising that some properties are equivalent so that the concept now has a structure of equivalent properties that are related by deductive proof,
- and realising that these properties are different ways of expressing an underlying crystalline concept whose properties are connected together by deductive proof.

Crystalline concepts are not isolated from each other. The deductive network of one crystalline concept may intersect with another. For instance, a child may begin to perceive an isosceles triangle as a representative of a broader class of objects – triangles in general – and compare the definitions and properties involved to find new deductive journeys relating concepts that may not always be equivalent. This leads to the distinction between direct and converse deduction of properties and further developments of deductive relationships (Yevdokimov 2008).

This development represents a broad trend in which successive stages are seen as developing and interrelating one with another, each correlated within the next

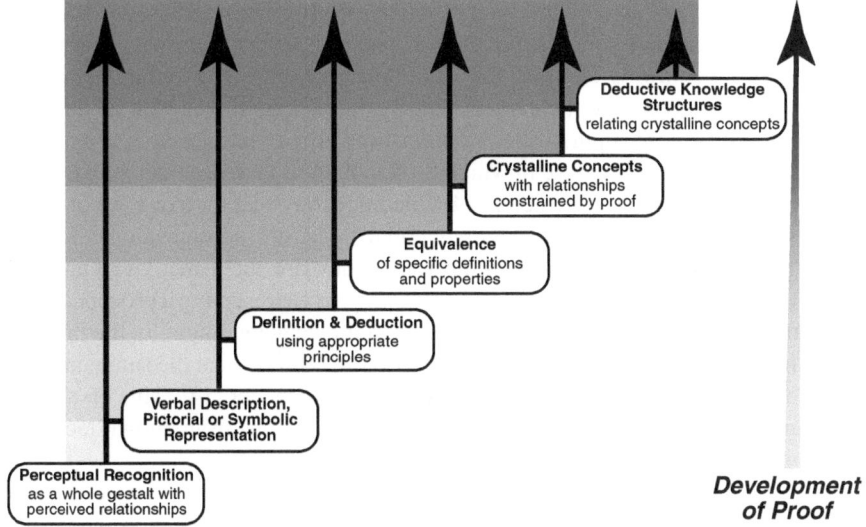

Fig. 2.2 The broad maturation of proof structures

(Fig. 2.2). This is represented by the deepening shades of grey as increasingly sophisticated knowledge structures are connected together as each new stage develops and matures. On the right is a single vertical arrow whose shaft becomes more firmly defined as it is traced upwards. Such an arrow will be used subsequently to denote the long-term development from initial recognition of a phenomenon in a given mathematical context through increasing sophistication to deductive knowledge structures. The development is a natural human growth and should not be seen as a rigid growth of discrete levels, rather as a long-term growth in maturity to construct a full range of mathematical thinking from perceptual recognition to deductive reasoning.

Such a framework is consistent with van Hiele's theory of the teaching and learning of geometry, but it goes further by making explicit the final shift from equivalent figures drawn on the page or in sand to conceiving them as instances of crystalline concepts in the form of perfect platonic objects.

The framework is also consistent with the development of other forms of geometry that arise in new contexts (projective geometry drawing a representation of a three-dimensional scene on a plane, spherical geometry on the surface of a sphere, elliptic and hyperbolic geometries in appropriate contexts, finite geometries, algebraic geometries and so on).

These new frameworks reveal that the 'appropriate principles of deduction' may differ in different forms of geometry. For instance, in spherical geometry, whilst there is a concept of congruence of spherical triangles, there are no parallel lines, and proof is a combination of embodied experience operating on the surface of a sphere coupled with symbolic computations using trigonometry.

In arithmetic and algebra the symbols have a crystalline structure that the child may begin to realise through experience of counting collections that are then put together or taken away. The sum $8+6$ can be computed first by counting, and in this context it may not yet be evident that $8+6$ gives the same result as $6+8$. As the child builds more sophisticated relationships, it may later be seen as part of a more comprehensive structure in which $8+2$ makes 10 and, decomposing 6 into 2 and 4, gives $8+6$ is $10+4$, which is 14. One might say that $8+6$ or $6+8$ or $10+4$ or various other arithmetic expressions equal to 14 are all equivalent, but it is cognitively more efficient to say simply that they are the same. Gray and Tall (1994) referred to such symbols as different ways of representing a procept, where the symbols can stand dually for a process and the concept output by the process. Here the various symbols, $8+6$, $10+4$, and so on, all represent the same underlying procept that operates as the flexible crystalline concept '14'. This crystalline structure is then used to derive more complex calculations from known facts in a deductive knowledge structure.

Algebra arises as generalised arithmetic, where operations having the same effect, such as 'double the number and add six' or 'add three to the number and double the result', are seen as being equivalent; these equivalences give new ways of seeing the same underlying procept written flexibly in different ways as $2 \times x + 6$ and $(x+3) \times 2$. These equivalences can be described using the rules that were observed and described as properties in arithmetic, now formulated as rules to define the properties in algebra. Finally, a crystalline concept, 'an algebraic expression' arises in which equivalent operations are seen as representing the same underlying operation.

Procepts arise throughout the symbolism of mathematics where symbols such as $4+3$, ¾, $2x+6$, dy/dx, $\int \sin x \, dx$, Σu_n dually represent a process of computation and the result of that process. Such procepts, along with the networks of their deduced properties, form crystalline concepts, which allow the human brain to operate flexibly and efficiently in formulating models, solving problems through symbol manipulation and discovering new properties and connections.

Crystalline concepts also operate at the formal-axiomatic level. Mathematicians construct the successive systems $\mathbb{N}$, $\mathbb{Q}$, $\mathbb{Z}$, $\mathbb{R}$, $\mathbb{C}$ formally using equivalence relations, such as defining the integers $\mathbb{Z}$ as equivalence classes of ordered pairs (m,n) where $m,n \in \mathbb{N}$ and $(m,n) \sim (p,q)$ if $m+q = p+n$. The whole number m corresponds to the equivalence class $(m+n,n)$. Successive constructions of $\mathbb{Q}$, $\mathbb{Z}$, $\mathbb{R}$ and $\mathbb{C}$ formulate each as being isomorphic to a substructure of the next. Cognitively, however, it is more natural to see the successive number systems contained one within another, with $\mathbb{N}$, $\mathbb{Q}$, $\mathbb{Z}$, $\mathbb{R}$ seen as points on the x-axis and $\mathbb{C}$ as points in the plane. This is not a simple cognitive process, however, as such extensions involve changes of meaning that need to be addressed, such as subtraction always giving a smaller result in $\mathbb{N}$ but not in $\mathbb{Q}$, and a (non-zero) square always being positive in $\mathbb{R}$ but not in $\mathbb{C}$.

The various number systems may be conceived as a blend of the visual number line (or the plane) and its formal expression in terms of axioms. For example, the real numbers $\mathbb{R}$ have various equivalent definitions of completeness that the expert recognises as equivalent ways of defining the same underlying property. The real numbers $\mathbb{R}$ now constitute a crystalline concept whose properties are constrained by the axioms for a complete ordered field.

More generally, any axiomatic system, formulated as a list of specific axioms uses formal proof to develop a network of relationships that gives the axiomatic system the structure of a crystalline concept. Whilst some may be unique (as in the case of a complete ordered field), others, such as the concept of group, have different examples that may be classified by deduction from the axioms.

3.3   A Global Framework for the Development of Mathematical Thinking

The full cognitive development of formal proof from initial perceptions of objects and actions to axiomatic mathematics can be formulated in terms of three distinct forms of development (Tall 2004, 2008):

- a development of the conceptual embodiment of objects and their properties, with increasing verbal underpinning appropriate for the maturation of Euclidean geometry;
- a translation of operations into proceptual symbolism, where a symbol such as $2x + 3$ represents both a process of evaluation, 'double the number and add three', and a manipulable concept, an algebraic expression;
- the development of axiomatic formalism, in which set-theoretic axioms and definitions are used as a basis of a knowledge structure built up through mathematical proof.

In each form of development, the idea of proof builds through a cycle pictured in Fig. 2.2 represented by a vertical arrow to give the overall framework in Fig. 2.3 as proof develops in the geometric embodiment, algebraic symbolism, and axiomatic formalism. The framework represents the child at the bottom left playing with physical objects, reflecting on their shapes and relationships to build an increasingly sophisticated development of Euclidean geometry through constructions, verbalised definitions and Euclidean proof.

In the centre, as the child reflects on his or her actions on objects, there is a blend of embodiment and symbolism in which properties such as addition are seen visibly to be independent of order (even though the counting procedures and visual representations may be different) and are translated into verbal rules such as the commutative law of addition. Specific pictures may also be seen as generic representations of similar cases, building up generalisations. The culmination of this central overlap includes any proof blending embodiment and symbolism.

On the right-hand side, the operation of counting is symbolised as the concept of number and, in proceptual symbolic terms, specific instances of arithmetic such as $10 + 7 = 7 + 10$ are generalised into an algebraic statement $x + y = y + x$, and such generalities are formulated as rules to be obeyed in algebra.

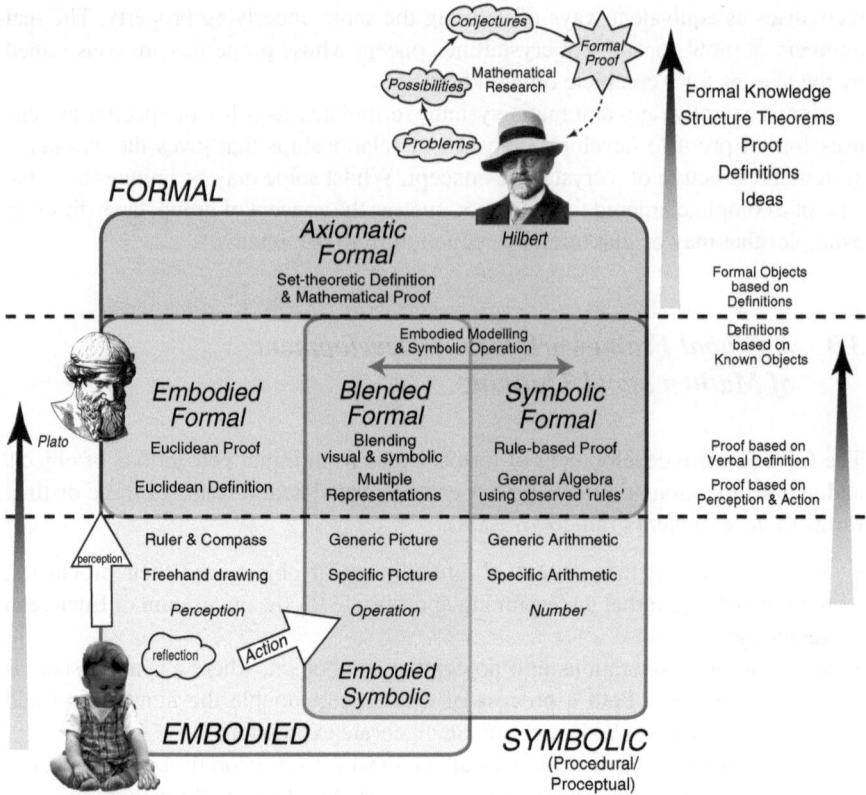

Fig. 2.3 Three strands of conceptual development: embodied, symbolic, formal

As the child matures, he or she is introduced to the idea that conceptions based in perception and action can be transformed into proof by verbal definition – a significant cognitive change that leads to Euclidean proof in geometry and rule-based proof in algebra. The figure of Plato here represents a view of crystalline conceptions that are so perfect that they seem to be independent of the finite human brain.

There is a major cognitive change denoted by the horizontal dotted line from inferences based on perception and action to proof based on a definition (of figures in geometry and rules of arithmetic in algebra). For instance, in algebra, the power rule $a^m \times a^n = a^{m+n}$ for whole numbers m and n can be embodied directly by counting the factors, but when m and n are taken as fractions or negative numbers, then the idea of counting the factors no longer holds. Now the power rule is used as a definition from which the meanings of $a^{\frac{1}{2}}$ and a^{-1} are deduced. This requires a significant shift of meaning from properties based on perception to properties deduced from rules.

The formal world includes all forms of proof based on an appropriate form of definition and agreed processes of deduction. These include Euclidean proof based on geometric principles such as congruence, algebraic proof based on the rules of arithmetic, and axiomatic formal proof. In applications, applied proof builds on embodiment and symbolism, developing refined strategies of contextual reasoning appropriate to the context.

A major development in formal proof is the shift to the axiomatic method of Hilbert. Now mathematical objects are defined as lists of set-theoretic axioms and any other properties must be deduced from the axioms and subsequent definitions by formal mathematical proof.

This gives a second major cognitive change, from definitions based on familiar objects or mental entities – as in the thought experiments of Plato – to formal definitions where proofs apply in any context where the required axioms hold – as in the formal theory of Hilbert.

At the level of axiomatic formalism, the top right arrow in the figure represents the desired development of the student maturing from a range of familiar ideas to their organisation as formal definitions and proof. The student learning the axiomatic method for the first time is faced with a list of axioms from which she or he must make initial deductions, building up a knowledge structure of formally deduced relationships that leads to the proof of successive theorems.

Some theorems, called 'structure theorems' have special qualities that prove that the axiomatic structures have specific embodiments and related embodied operations. For instance the structure theorem that all finite dimensional vector spaces over F are isomorphic to a coordinate space F^n reveals that such a vector space can be represented symbolically using coordinates and operations on vectors that can be carried out symbolically using matrices. When $F = \mathbb{R}$, the formalism is then related to embodiments in two and three-dimensional space.

In this way, embodiment and symbolism arise once more, now based on an axiomatic foundation. Mathematicians with highly sophisticated knowledge structures then reflect on new problems, think about possibilities, formulate conjectures and seek formal proofs of new theorems in a continuing cycle of mathematical research and development. At each stage, this may involve embodiment, symbolism and formalism as appropriate for the given stage of the cycle.

4 The Development of Proof from Embodiment

4.1 From Embodiment to Verbalisation

Young children have highly subtle ways of making sense of their observations. For instance, Yevdokimov (in preparation) observed that young children playing with wooden patterns of different shapes sense symmetry even in quite complicated forms and may build their own conceptions of symmetry without any special emphasis and influence from adults. However, they experience enormous difficulties when attempting to describe a symmetric construction verbally.

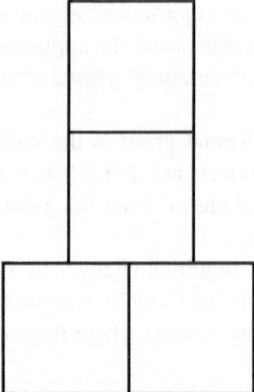

Fig. 2.4 Simon's triangle

The young child may develop ways to recognise and name different shapes in ways that may be quite different from an adult perspective. Two-year old Simon learned to recognise and say the name 'triangle' for a shape he recognised (Tall 2012, under review). He learnt it through watching and listening to a television programme in which Mr. Triangle was one of several characters, including Mr. Square, Mr. Rectangle and Mr. Circle where each character had the named shape with a face on it and hands and feet attached.

When Simon saw a triangle, he named it, but then, when playing with some square table mats, he put together a 3 by 2 rectangle, a 2 by 2 square, and then reorganised four squares into an upside down T-shape that he called a triangle (Fig. 2.4). Although the figure lacks three sides, it is fat at the bottom, thin at the top and symmetric about a vertical axis. It is the nearest word in his vocabulary to describe what he sees, being more like a triangle than any other shape that he can name.

4.2 From Embodiment and Verbalisation to Pictorial and Symbolic Representations

A case of interest is the 5-year long study of Maher and Martino (1996), which followed the developing ideas of a single child, called Stephanie. This revealed the emergence of statements that are precursors of quantifiers (such as 'there is an A such that B occurs' or 'for all A, B occurs'). During her third and fourth grade, Stephanie and her classmates were given variations of the following question, which will be termed the four-cube-tall Tower Problem:

How many different four-cube-tall towers can be built from red and blue cubes?

When Stephanie and her partner approached this problem for the first time, she started from the search for different four-cube-tall towers using a trial-and-error strategy. Namely, she built a tower, named it (e.g. 'red in the middle' or 'patchwork'), and compared it with all the towers that had been constructed so far to see whether it was new or a duplicate. In several minutes, Stephanie began to spontaneously notice

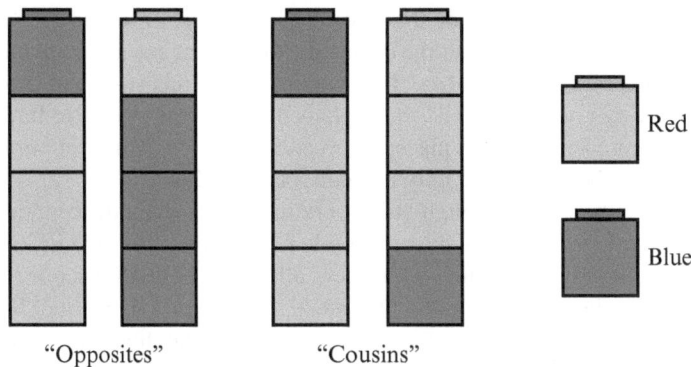

"Opposites"   "Cousins"

Fig. 2.5 'Opposites' and 'cousins'

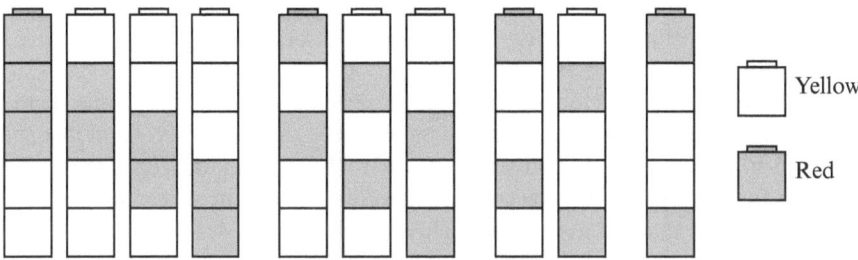

Fig. 2.6 Towers of five with two *red* and three *yellow*

relationships between pairs of towers and put them together. She then called the towers such as the left pair on Fig. 2.5 as 'opposites', and the right pair as 'cousins'. Maher and Martino interpreted this as the beginning of Stephanie's classification of towers into sets by a local criterion. No global organisational criterion emerged at this stage, and Stephanie, who eventually constructed all 16 towers, did not know whether or not she found them all. When asked about it, she explained that she 'continued to build towers until [she] couldn't find any that were different' (p. 204).

Eighteen months later, in the fourth grade, Stephanie was presented with the problem concerning the number of five-cube-tall towers that can be built from red and yellow cubes. This time she constructed 28 original combinations organised in the sets consisting of four elements: some tower, its 'opposite', its 'cousin' and 'the opposite of the cousin' (see Fig. 2.5). Thus, though a global organisational principle still remained murky for Stephanie, she progressed from the pure trial-and-error strategy to trial-and-error strategy combined with the local classification strategy.

A crucial event occurred when the teacher asked the class about the number of five-cube-tall towers with exactly two red cubes. In response, Stephanie argued that all towers in this category can be accounted by the following organisational criterion: there are towers in which two red cubes are separated by no yellow cube, one yellow cube, two yellow cubes or three yellow cubes (see Fig. 2.6). It is notable that at this stage Stephanie began to draw pictures of the towers she produced rather than building them with plastic cubes.

In this episode Stephanie for the first time arrived at an existential algebraic statement that required proving. She declared: 'there exist exactly 10 five-cube-tall towers with exactly two red cubes.' Her justification of this statement involved an indirect proof by contradiction as she explained: 'it is not possible to have towers with four or more yellow cubes placed between the two red cubes without violating the requirement that a tower be only five cubes tall' (p. 205).

One can see here that, though still operating in her embodied mathematical world, Stephanie succeeds in substituting her previous local organisational criterion ('opposites' and 'cousins') with a new, all-inclusive or global, one (the number of yellow cubes separating the red ones). This invention enabled Stephanie to construct – though only for a sub-problem of the five-cube-tall Towers Problem – a mathematically valid argument. An important sign of the development of the idea of proof in Stephanie is that she was able to mentally represent not only what is possible to do with the cubes, but also what is impossible. The spirit of a crystalline concept – a concept that has an internal structure of constrained relationships that cause it to have necessary properties as part of its context – enters here!

Two weeks later, two more cognitive advances emerged when Stephanie shared with the interviewers her thinking about the six-cube-tall Towers Problem that she assigned to herself. The first advance occurred when she introduced a letter-grid notation for representing the towers, instead of drawing them or assembling from the plastic cubes that she used before. This is an important step on her way to the symbolic mathematical world. Interestingly, the authors mentioned that, when in first grade, Stephanie had used a letter-grid notation for solving another combinatorial problem, and then, it seemed, she forgot about it. As Lawler (1980) noted, it may seem that a child regresses in knowledge, whereas he or she may in fact be attempting to insert new knowledge into an existing knowledge structure. (See also Pirie and Kieren 1994.)

Our point here is that the pictorial and later symbolic representations that she used were indeed rooted in Stephanie's embodied world. The second advance was that Stephanie started to fluently consider different organisational principles for producing the towers in a systematic way. In particular, she introduced a method of holding the position of one colour when varying positions for another. Apparently, the second advance is related to the first one: convenient notation makes consideration of various patterns more accessible. These important advances eventually led Stephanie to construct a full proof for the four-cube-tall Tower Problem (by classifying the towers into five categories: towers with no white cube, towers with exactly one cube, etc.). We interpret Stephanie's last insight as a verbalisation of the embodied skill to infer cause-effect relationships from statistical patterns: Stephanie found that the answers for three-, four-, and five-cube-tall towers were 8, 16, and 32, respectively, and just expressed her belief that this numerical pattern will work forever. Such a guess is statistically justified for Stephanie, and is appropriate at her stage of development. In the fifth grade she is only just beginning to transfer embodied proof concepts into a symbolic mathematical form.

4.3 From Embodiment, Verbalisation and Symbolism to Deduction

The experiences just described offer an example of a learner building up increasingly sophisticated knowledge structures by physical experiment with objects, finding ways to formulate similarities and differences, then representing the data observed using drawings and symbols that have an increasingly meaningful personal interpretation. In this section we discuss the general principles underpinning a learner's path towards deductive reasoning from its sensorimotor beginnings, through the visual-spatial development of thought and on to the verbal formulation of proof, particularly in Euclidean geometry.

As the child matures, physical objects, experienced through the senses, become associated with pictorial images, and develop into more sophisticated knowledge structures that Fischbein (1993) named figural concepts. These are

> mental entities [...] which reflect spatial properties (shape, position, magnitude) and, at the same time, possess conceptual qualities like ideality, abstractness, generality, perfection (Fischbein 1993, p. 143).

Figural concepts reflect the human embodiments that underlie our more abstract formal conceptions.

Cognitive science sees the human mental and physical activity underlying all our cognitive acts (Johnson 1987; Lakoff and Johnson 1999; Lakoff and Núñez 2000). For example, the schema of 'containment' where one physical object contains another underlies the logical principle of transitivity $A \subset B$, $B \subset C$ implies $A \subset C$ and operates mentally to infer that if A is contained in B and B is contained in C, then A must be contained in C. Even Hilbert – on the occasion declaring his famous 23 problems at the International Congress of 1900 – noted the underpinning of formalism by visual representation, picturing transitivity as an ordering on a visual line:

> Who does not always use along with the double inequality $a > b > c$ the picture of three points following one another on a straight line as the geometrical picture of the idea 'between'? (Hilbert 1900).

Empirical evidence in support of this can be found throughout the literature. Byrne and Johnson-Laird (1989) studied adults responding to tasks in which they were given verbal evidence of relative positions of objects placed on a table and found that the subjects used visual mental models rather than a pure logical approach to produce their deductions.

The other side of the coin suggests that a mismatch between images (schemata) and formal definitions of mathematical concepts can be a source of difficulty in the study of mathematics and mathematical reasoning. For instance, Tall and Vinner (1981) illustrated how students may interpret real analysis based on their concept imagery of earlier experiences rather than on formal definitions. Hershkowitz and Vinner (1983) similarly revealed how particular attributes of pictures interfere with the general conceptualisation process in geometry.

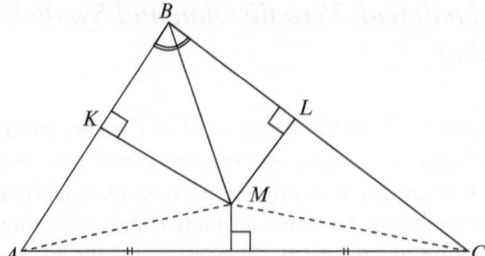

Fig. 2.7 Proof that all triangles are isosceles

In a similar way, Núñez et al. (1999) suggested that the epsilon-delta approach to continuity is problematic for students because it conflicts with natural embodied conceptions of continuity. The problem, however, is more subtle, because formalism only captures specific explicit properties, such as the 'closeness' of natural continuity formulated in the epsilon-delta definition. Natural continuity also involves other relevant aspects such as the completeness of the real numbers and the connectedness of the domain.

Sometimes, the practice of generalising from empirical findings or building argument from intuitively appealing images leads to a possibility that happens to be wrong. As an example of a situation where valid deductive argument is applied to a plausible, but wrong image, we give a 'proof' that every triangle is isosceles. Kondratieva (2009) used this example in order to illustrate the process of learning the art of deduction through the analysis of unexpected or contradictory results.

Consider an arbitrary triangle ABC. Let the point of intersection of the bisector of the angle B and the perpendicular bisector of the side AC meet in the point M. For simplicity, assume that M lies inside of the triangle (Fig. 2.7).

> Let points K and L be the feet of perpendiculars dropped from M to the sides adjacent to vertex A. In the triangles BMK, BML, the side BM is common; the angles KBM and LBM are equal, as are the right angles BKM, BLM. Therefore, triangles BMK, BLM are congruent, and $BK=BL$. The right-angled triangles AKM and CLM are also congruent because the legs $MK=ML$, and their hypotenuses are equal $AM=CM$ by the property of the perpendicular bisector. Thus, $AK=CL$. Finally, $AB=AK+KB=CL+LB=CB$. Hence the triangle ABC is isosceles. QED.

Students often perceive the figure to be legitimate and concentrate on looking for mistakes in the argument. The deductions concerning congruency in this proof are supported by reference to appropriate theorems and appear to be correct. It is often a surprise for a student to realise that the actual fallacy arises through drawing M inside the triangle when it should be outside, more precisely on the triangle's circumcircle.

Such examples have the potential to prepare the learner for the need to doubt representations and arguments that seem intuitively valid. This can lead to several different ways of emphasising different aspects of proof.

The first is to realise that correct reasoning based on a misleading (incorrect) diagram can lead to false conclusions. Thus, in mathematical derivations we must attend to both the deduction and the assumption(s).

The second is to illustrate the idea of a proof by contradiction. For instance, consider the statement: 'The intersection point of an angular bisector and the right bisector of the opposite side in any triangle lies outside the triangle.' Drawing a picture purporting to represent the contrary situation where the intersection lies inside the triangle, as in Fig. 2.7, can then be seen as leading to the impossible statement that 'all triangles are isosceles'.

The third possibility is to introduce the concept of a direct constructive proof by inviting the student to perform the construction using dynamic software and arguing why the point M must lie outside the triangle.

The need for a more reliable proof leads to the idea of deductive reasoning, which may be seen as the ability of an individual to produce new statements in the form of conclusions from given information, particularly in areas where the subject has no prior knowledge other than the information given. The new statements must be produced purely by reasoning with no simultaneous access to hand-on materials and experimentation. However, such methods require the individual to build a knowledge structure that enables the use of logical forms of deduction. As we analyse Euclidean geometry, we find that its deductive methods build on ways of working that are themselves rooted in human embodiment.

5 Euclidean and Non-Euclidean Proof

5.1 The Development of Euclidean Geometry

From a mathematician's viewpoint, the study of Euclidean geometry in school has often been considered as providing a necessary basis for the formal notion of proof and, in particular, the building of a succession of theorems deductively from basic assumptions. From a cognitive viewpoint it is our first example of the long-term development of crystalline concepts. Beginning from personal perceptions and actions, the learner may build personal knowledge structures relating to the properties of space and shape, then develop definitions and deductions to construct the crystalline platonic objects of Euclidean geometry.

Even though the books of Euclid produce a sequence of successively deduced propositions based on specified common notions, definitions and postulates, a closer inspection reveals the use of principles based on human perception and action. For example, the notion of congruence involves the selection of certain minimal properties that enable triangles to be declared to have all their properties in common, in terms of requiring only three corresponding sides (SSS), two sides and included angle (SAS), or two angles and corresponding side (AAS). All are based on an embodied principle of superposition of one triangle upon another (perhaps turning it over). This concept of congruence of triangles is not endowed with a specific name in the Books of Euclid. It is used as an established strategy to formulate minimal conditions under which different triangles are equivalent and consequently have all the same properties.

Parallel lines are another special concept in Euclid, defined to be straight lines which, being in the same plane and being produced indefinitely in both directions, do not meet one another in either direction (Euclid, Book I, Definition 23). Using this definition, Euclid establishes various (equivalent) properties of parallel lines, such as alternate angles being equal to one another, corresponding angles being equal, and the sum of the interior angles being equal to two right angles (Proposition 27 et seq.). Once more the concept of parallel lines is a crystalline concept with a range of interlinked properties each of which can be used to furnish a Euclidean proof of the others.

The development of geometry, starting at Euclid Book I, has been declared inappropriate for young children:

> The deductive geometry of Euclid from which a few things have been omitted cannot produce an elementary geometry. In order to be elementary, one will have to start from a world as perceived and already partially globally known by the children. The objective should be to analyze these phenomena and to establish a logical relationship. Only through an approach modified in this way can a geometry evolve that may be called elementary according to psychological principles (van Hiele-Geldof 1984, p. 16).

Some enlightened approaches to geometry have attempted to integrate it with broader ideas of general reasoning skills including the need for clear definitions and proof. For example, Harold Fawcett, the editor of the 1938 NCTM Yearbook on Proof, developed his own high school course that he taught at the University School in Ohio State entitled The Nature of Proof. His fundamental idea was to consider any statement, to focus on words and phrases, to ask that they be clearly defined, to distinguish between fact and assumption, and to evaluate the argument, accepting or rejecting its arguments and conclusion, whilst constantly re-examining the beliefs that guided the actions. Speaking about the course at an NCTM meeting in 2001, Frederick Flener observed:

> Throughout the year, the pupils discussed geometry, creating their own undefined terms, definitions, assumptions and theorems. In all he lists 23 undefined terms, 91 definitions, and 109 assumptions/theorems. The difference between an assumption and a theorem is whether it was proved. Briefly, let me tell you a few of the undefined terms which the pupils understood, but were unable to define. For example, as a class they couldn't come up with 'the union of two rays with a common endpoint' as the definition of an angle, so they left it as undefined. Nor could they define horizontal or vertical, or area or volume. Yet they went on to define terms like dihedral angle, and the measure of a dihedral angle—which I assume involved having rays perpendicular to the common edge (Flener 2001).

A detailed study of the students who attended this course revealed significant long-term gains in reasoning skills over a lifetime (Flener 2006). This suggests that making personal sense of geometry and geometric inferences can have long-term benefits in terms of clarity of thinking and reasoning skills.

Modern trends in teaching have led to encouraging young children to build a sense of shape and space by refining ideas through experience, exploring the properties of figures and patterns. They experiment with geometrical objects, seek to recognise the properties of card shapes, sketch the faces of a box when pulled out into a flat figure, fold paper, measure the angles and sides of a triangle, and discuss their ideas with friends or teachers.

They may have experiences in constructing and predicting what occurs in geometric software such as Logo in ways that give figural meanings to geometric ideas. For instance, a 'turtle trip' round a (convex) polygon may enable the child to sense that 'If the turtle makes a trip back to its starting state without crossing its own path, the total turn is 360°'. According to Papert (1996), this is a theorem with several important attributes: first, it is powerful; second, it is surprising; third, it has a proof. It is not a theorem in the formal mathematical sense, of course, but it is a meaningful product of human embodiment, sensing a journey round a circuit and ending up facing the same direction, thus turning through a full turn.

These experiences provide young children with preliminary background on which to develop ideas of proof in a variety of ways. For example, it is possible to experience many ways in which children may attempt to provide a proof for the statement that the sum of the interior angles of a triangle is 180°.

In the studies of Lin et al. (2003), Healy and Hoyles (1998), and Reiss (2005), eight distinct proofs were collected from students before they had any formal introduction to Euclidean proof as a deductive sequence of propositions.

The first two are pragmatic actions applied to specific cases:

Proof 1: by physical experiment (Fig. 2.8).

Take a triangle cut out of paper, tear off the corners and place them together to see that they form a straight line. Do this a few times for different triangular shapes to confirm it.

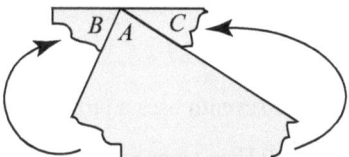

Fig. 2.8 A proof by physical experiment

Proof 2: by practical measurement (Fig. 2.9).

Draw a triangle. Measure the angled to check that the sum is equal to 180°. Repeat the same process on other triangles.

A third proof is a dynamic embodiment, which arises in Logo:

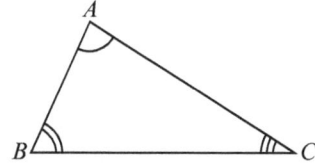

Fig. 2.9 A proof by practical measurement

Proof 3: the turtle-trip theorem: imagine walking round a triangle (Fig. 2.10).

Start at any point P and walk all the way round. This turns through 360°. At each vertices, the sum of exterior and interior angles is 180° so the sum of the three exterior and interior angles is 540°. Subtract the total 360° turn to leave the sum of the interior angles as 540°−360° = 180°.

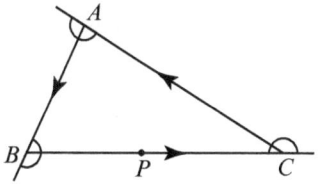

Fig. 2.10 A proof by the turtle-trip

A fourth proof uses a known fact about triangles to infer another fact.

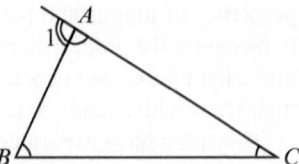

Fig. 2.11 A proof using exterior angles

Proof 4: Use the fact that the exterior angle equals the sum of the two interior opposite angles (Fig. 2.11).

Extend the segment CA; the exterior angle $\angle 1$ is equal to the sum of the two interior opposite angles. Because the exterior angle and $\angle BAC$ make a straight line, the sum of all three angles is $180°$.

The fifth and sixth proofs introduce additional parallel lines.

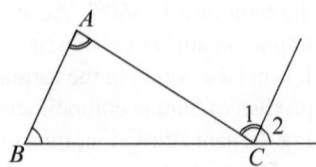

Fig. 2.12 A proof using parallel lines

Proof 5: A proof using parallel lines (Fig. 2.12).
1. Draw a line parallel to AB through point C.
2. $\angle A = \angle 1$ (alternate angles) and $\angle B = \angle 2$ (corresponding angles).
3. $\angle A + \angle B + \angle C = \angle 1 + \angle 2 + \angle C = 180°$.

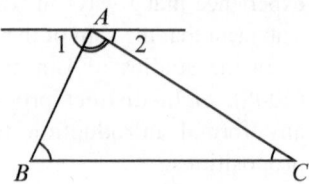

Fig. 2.13 A second proof using parallel lines

Proof 6: A second proof using parallel lines (Fig. 2.13).
1. Draw a line L parallel to BC through point A. Then $\angle B = \angle 1$ and $\angle C = \angle 2$ (alternate angles).
2. Hence

$$\angle B + \angle BAC + \angle C = \angle 1 + \angle BAC + \angle 2 = 180°.$$

The seventh uses a property of the circle.

Proof 7: Using a property of angles subtended by a chord (Fig. 2.14).

The angle at the circle is half the angle at the centre, so

$$\angle A = \tfrac{1}{2}\angle BOC, \quad \angle B = \tfrac{1}{2}\angle AOC,$$
$$\angle C = \tfrac{1}{2}\angle AOB.$$

Adding these together:
$$\angle A + \angle B + \angle C = \tfrac{1}{2} 360° = 180°.$$

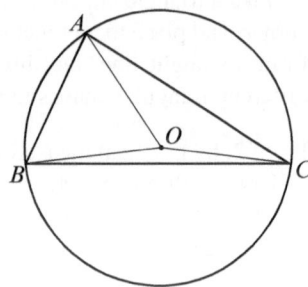

Fig. 2.14 A proof using a property of angles subtended by a chord

Proof eight appeals to the more general form of the angle sum of an n-sided polygon, either as a consequence of the turtle trip theorem, or simply by substitution in the general formula.

Proof 8: An n-sided (convex) polygon has angle sum $n \times 180° - 360°$ (Fig. 2.15).

Put $n = 3$ in the general formula to get the angle sum for a triangle is $180°$.

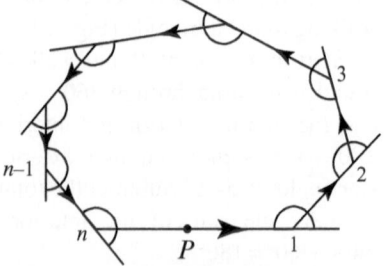

Fig. 2.15 A proof using a formula for an n-sided (convex) polygon

These eight solutions reveal a broad hierarchy. Proofs 1 and 2 are embodied approaches, the first by a physical process of putting the angles together (in a manner that may have been suggested to them earlier), the second by measuring a few examples. Proof 1 contains within it the seeds of more sophisticated proofs 5 and 6 using the Euclidean idea of parallel lines. Proof 3 (and its generalisation to a polygon used in proof 8) are dynamic proofs that do not arise in the static formal geometry of Euclid and yet provide a dynamic embodied sense of why the theorem is true. Proof 4, relating to the exterior angle property, nicely links two properties of a triangle and yet, an expert may know that these are given as equivalent results from a single theorem of Euclid (Book 1, proposition 32, as given in Joyce 1998):

> In any triangle, if one of the sides is produced, the exterior angle is equal to the two interior and opposite angles, and the three interior angles of the triangle are equal to two right angles (Joyce 1998).

Though some experts, looking at this proof from a formal viewpoint, may see it relating two equivalent properties in a circular manner, it is a natural connection for a student to make in the early stages of building a knowledge structure of relationships in geometry (Housman and Porter 2003; Koichu 2009).

Proofs 5 and 6 are in the spirit of Euclid, constructing a parallel line and using established propositions concerning parallel lines to establish the theorem. However, here they are more likely to involve an embodied sense of the properties of parallel lines than the specific formal sequence of deductions in Euclid Book I.

Proofs 7 and 8 both use more sophisticated results to prove simple consequences and have a greater sense of a general proof. And yet one must ask oneself, how does one establish the more general proof in the first place? Whilst networks of theorems may have many different paths and possible different starting points, the deeper issues of sound foundations and appropriate sequences of deductions remain.

It has long been known that students have difficulty reproducing Euclidean proof as a sequence of statements where each is justified in an appropriate manner. Senk (1985) showed that only 30% of students in a full-year geometry course reached a 70% mastery on a set of six problems in Euclidean proof.

Given the perceived difficulties in Euclidean geometry, the NCTM Standards (2000) suggested that there should be decreased attention to the overall idea of geometry as an axiomatic system and increased attention on short sequences of theorems. These can in themselves relate to Papert's notion that a theorem should be powerful, surprising, and have a proof. Two examples include the theorem that two parallelograms on the same base and between the same parallels have the same area, even though they may not look the same, and the theorem that the angles subtended by a chord in a circle are all equal (Fig. 2.16).

Here, short sequences of construction are possible. For instance, in the area of parallelograms, one may prove that the triangles ADD' and BCC' are congruent and that each parallelogram is given by taking one of these triangles from the whole polygon $ABC'D$. The equalities of the angles subtended by a chord is established by constructing the angle at the centre of the circle and proving that it is twice the angle at the circle.

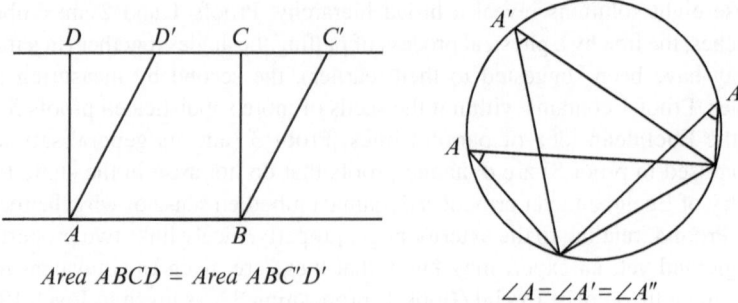

Area ABCD = Area ABC′D′

∠A = ∠A′ = ∠A″

Fig. 2.16 Interesting theorems

5.2 The Beginnings of Spherical and Non-Euclidean Geometries

The teaching of non-Euclidean geometries is not central in the current curriculum, but practical experience in such geometries is becoming part of the development of college mathematics in the USA.

> College-intending students also should gain an appreciation of Euclidean geometry as one of many axiomatic systems. This goal may be achieved by directing students to investigate properties of other geometries to see how the basic axioms and definitions lead to quite different – and often contrary – results (NCTM Standards 1989, p. 160).

Although spherical geometry goes back to the time of Euclid, it does not share the axiomatic tradition developed in plane geometry. Spherical geometry may be approached as a combination of embodiment and trigonometric measurement. One may begin with the physical experience of operating on the surface of a sphere (say an orange, or a tennis ball with elastic bands to represent great circles). This produces surprising results quite different from the axiomatic geometry of Euclid. When exploring spherical triangles whose sides are great circles, the learner may find a new geometry that shares properties predicting congruence (SAS, SSS, AAS), but is fundamentally different from plane Euclidean geometry, in that parallel lines do not occur and the angles of a triangle always add up to more than 180°.

This can be proved by a combination of embodiment and trigonometry. Figure 2.17 shows a spherical triangle ABC produced by cutting the surface with three great circles. It has a corresponding triangle $A'B'C'$ where the great circles meet on the opposite side of the sphere having exactly the same shape and area.

The total area of the shaded parts of the surface between the great circles through AB and AC can be seen by rotation about the diameter AA' to be α / π of the total area, where α denotes the size of the angle A measured in radians. This area is $4\pi r^2 \times \alpha / \pi = 4\alpha r^2$. The same happens with the slices through B and through C with area $4\beta r^2$ and $4\gamma r^2$. These three areas cover the whole surface area of the

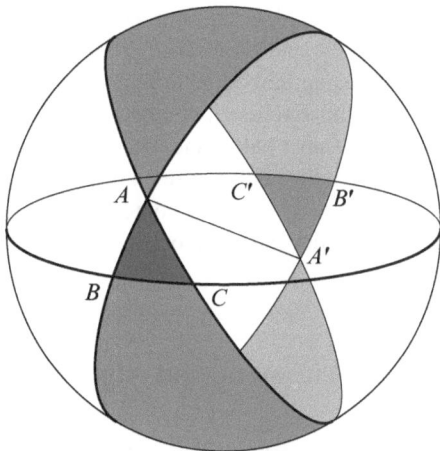

Fig. 2.17 Area of a spherical triangle

sphere and all three overlap over the triangles ABC and $A'B'C'$. Adding all three together, allowing for the double overlap gives the surface area of the sphere as

$$4\pi r^2 = 4\alpha r^2 + 4\beta r^2 + 4\gamma r^2 - 4\Delta$$

where Δ is the area of the spherical triangle ABC. This gives the area Δ as $(\alpha + \beta + \gamma - \pi)r^2$ and the sum of the angles as

$$\alpha + \beta + \gamma = \pi + \frac{\Delta}{r^2}.$$

The experience challenges learners to rethink the 'deductive arguments' given above that the sum of the angles of a triangle is 180° when the sum of the angles of a spherical triangle is always more by a quantity proportional to its area.

In long-term implementations with in-service teachers and university students, learners report that they could not have developed concepts and the arguments without access to materials to handle and dynamic geometry sketches to explore the tasks. In short, this provides evidence for the continuing role for embodied experience in the cognitive development of proof in adults. Examples of such an approach arise in the book Experiencing Geometry (Henderson and Taimina 2005) and the work of Lénárt (2003), encouraging comparison of concepts and reasoning in spherical geometry and plane geometry, through practical activities such as handling spheres or folding paper. These explorations give an emphasis to transformations and symmetry, matching the 'modern' definition of geometry of Klein (1872) and offer a setting for increasingly sophisticated reflections on two distinct embodied geometries. The similarities and contrasts between the two structures provoke a reflective reworking of unexamined concepts such as 'straight lines' and 'angles' and also of principles related to 'congruence of triangles' and the properties of 'parallel lines'.

Other geometries may be studied, perhaps as axiomatic systems, but more often as a combination of embodiment and symbolism. For instance, projective geometry of the plane can be studied using embodied drawing or symbolic manipulation of homogeneous coordinates. Non-Euclidean geometries include the Poincaré model of hyperbolic geometry in the upper half plane where 'points' are of the form (x, y) for $y > 0$ and 'lines' are semi-circles with centre on the x-axis; two 'lines' are said to be 'parallel' if they do not meet. In this new context, students must reflect on new meanings to make deductions dependent upon the definitions in the new context (Neto et al. 2009).

6   Symbolic Proof in Arithmetic and Algebra

As the mathematics becomes more sophisticated, increasingly subtle forms of proof develop in arithmetic and algebra. They build from demonstrations or calculations for single examples, to considering a specific example to represent a generic proof that applies to all similar cases, and then to general proofs expressed algebraically. Induction proofs can operate at two distinct levels, one the potentially infinite process of proving a specific case and repeating a general step as often as is required, the second involving a three stage proof that compresses the potentially infinite repetition of steps to a single use of the induction axiom in a more formal setting.

6.1   The Increasing Sophistication of Proof in Arithmetic and Algebra

An example of such successively sophisticated forms of proof is the Gauss Little Theorem that is reputed to have been produced by the schoolboy Gauss when his teacher requested the class to add up all the whole numbers from 1 to 100 (Fig. 2.18).

These successive proofs are not all as successful in giving meaning to students. Rodd (2000) found that both the generic pictorial and algebraic proofs made more sense to the students because they gave a meaningful explanation as to why the proof is true, whilst the formal proof by induction is more obscure because it seems to use the result of the proof (assuming $P(n)$ for a specific n to prove $P(n+1)$) during the proof itself. In addition, the finite proof by induction (Proof 6) may cause further problems because, although it has only three steps, the set defined by the Peano axioms must itself be infinite.

6.2   Proof by Contradiction and the Development of Aesthetic Criteria

On the one hand, it is well established that proving by contradiction is problematic for many students (e.g. Antonini 2001; Epp 1998; Leron 1985; Reid and Dobbin

Approach 1: specific arithmetical calculation (only works for even numbers): $1 + \ 2 + \ 3 + ... + 50$ $\underline{100 + \ 99 + 98 + ... + \ 51}$ $101 + 101 + 101 + ... + 101 = 101 \cdot 50 = 5050$	Approach 2: generic arithmetical argument ($n = 100$) (works for all numbers even and odd): $1 + \ 2 + \ 3 + ... + 100$ $\underline{100 + 99 + 98 + ... + \ \ 1}$ $101 + 101 + 101 + ... + 101 = 101 \cdot 100 = 2 \cdot 5050$
Approach 3: generic pictorial argument 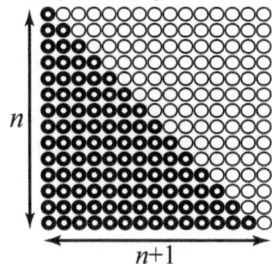 $n+1$ The sum $1 + ... + n$ is half of $n(n+1)$	Approach 4: algebraic proof (for any n) $\begin{array}{ccccccc} 1 & + \ 2 & + & 3 & + ... + & n \\ n & + (n-1) & + (n-2) & + ... & + & 1 \\ \hline (n+1) & + (n+1) & + (n+1) & + ... & + (n+1) \end{array}$ $= (n+1) \cdot n$ $= 2 \cdot \dfrac{(n+1) \cdot n}{2}$

Approach 5: a potentially infinite proof by induction (for any n so far)

$1 + ... + n = \frac{1}{2}n(n+1)$ is true for $n = 1$, because $1 = \frac{1}{2} \times 1 \times (1+1)$.

If it is true for $n = k$, use the formula $1 + ... + k = \frac{1}{2}k(k+1)$ and add $k + 1$ to both sides to deduce the truth for $n = k + 1$, and then repeat this general step as often as is required:

It is true for $n = 1$, hence it is true for $n = 2$, hence for $n = 3$, hence for $n = 4$, ... and so on for any specific whole number ... (ad infinitum).

Approach 6: a finite proof by the Peano axioms (for all n)

The Peano axioms define natural numbers as a set $\mathbb{N}$ satisfying the following conditions:

- There is a natural number 1.

- Every natural number a has a natural number successor, denoted by $s(a)$.

- Distinct natural numbers have distinct successors: if $a \neq b$, then $s(a) \neq s(b)$.

- 1 is not the successor to any natural number.

- If A is a subset of natural numbers containing 1 and if the successor of any number in A is also in A, then A contains all the natural numbers.

Let us now prove the formula.

Let A be a subset of natural numbers such that for any $n \in A$
$1 + ... + n = \frac{1}{2}n(n+1)$.
1. Prove, by substitution in the formula, that $1 \in A$.
2. If $k \in A$ then use the formula for k as above to show that $k + 1 \in A$.
3. Observe that A satisfies the Peano axioms, and thus is the whole of $\mathbb{N}$.

Fig. 2.18 Proofs of the formula for the sum of the first n whole numbers

1998; Tall 1979). On the other hand, Freudenthal (1973, p. 629) notes that indirect proof arises spontaneously in young children in statements such as 'Peter is at home since otherwise the door would not be locked'. Such a phenomenon occurred with Stephanie (see Sect. 4.2) where she observed that it is not possible to have a tower of five cubes to have four or more yellow cubes placed between two red cubes.

The full notion of proof by contradiction (to prove P, assume P is false and deduce that this leads to a contradiction) is an altogether more problematic mode of proving as the prover must simultaneously hold a falsehood to be true and attempt to argue why it is false whilst in a state of stress. Leron (1985) suggests that it is preferable to reorganise a contradiction proof so that it initially involves a positive construction and the contradiction is postponed to the end. For instance, in order to prove there are an infinite number of primes, start by proving positively that given any finite number of primes one can always construct another one, then – and only then – deduce that, if there are only a finite number of primes, then this would lead to a contradiction.

A similar technique is to use a generic proof. For instance, rather than prove that $\sqrt{2}$ is irrational by contradiction, first prove that if one squares a rational number where denominator and numerator are factorised into different primes, then its square has an even number of each prime factor in the numerator or denominator. Then, and only then, deduce that $\sqrt{2}$ cannot be a rational because its square is 2, which only has an odd number of occurrences of the prime 2. (At a formal level, we note that this proof is dependent on the uniqueness of factorisation into primes, but this is not a concern that occurs to students on first acquaintance with the proof.)

Tall (1979) presented a choice of two proofs: the contradiction proof that $\sqrt{2}$ is irrational and the generic proof that $\sqrt{(5/8)}$ is irrational because 5/8 contains an odd number of 5s and also an odd number of 2s in its prime factorisation. The generic proof has explanatory power that generalises easily whilst the contradiction proof is both problematic and not easily generalised. Students significantly preferred the generic proof over the contradiction proof and even amongst students who were already familiar with the contradiction proof, their preference for the generic proof increased over a period of days.

The process of turning all contradiction proofs to more direct proofs with a later contradiction, however, may not be profitable in the long-term as proofs by contradiction are central to mathematical analysis. This suggests that direct proofs will be helpful in the early stages, but there is a need to shift to the use of contradiction proofs to enable the student to build more powerful knowledge structures that are sufficiently robust for more advanced mathematical analysis.

Dreyfus and Eisenberg (1986) found that mathematicians comparing different proofs of the irrationality of $\sqrt{2}$ ranked the proofs by contradiction that lack the need for prerequisite knowledge as elegant and appropriate for their teaching. They also found that college students had not yet developed the sense of aesthetics of a proof and proposed that such a sense should be encouraged.

Koichu and Berman (2005) found that gifted high school students, who were asked to prove the Steiner-Lehmus theorem ('If the bisectors of two angles of a triangle are equal, then it is an isosceles triangle'), could fluently operate in the mode of proving by contradiction. In addition, they manifested the developed aesthetic sense, by their incentive to find the most parsimonious proof. For instance, they realised that it was possible to build a concise proof that used the minimum of prerequisite knowledge. However, in the pressure of a contest, they found that the brute force of using well-rehearsed procedures could prove to be more efficient if less aesthetic.

The development of more sophisticated insight into proof reveals the fact that the use of contradiction requires the propositions involved to be either true or false with no alternative. More general forms of logic are possible, for instance allowing an extra alternative that a theorem might be undecideable (it could be true, but the truth cannot be established in a finite number of steps), or there may be different gradations between absolute truth and absolute falsehood. This might occur in multivalent logic allowing in-between possibilities between 0 (false) and 1 (true). It also occurs when a conjecture is formulated, which may be considered 'almost certain', 'highly probable', 'fairly likely' or some other level of possibility prior to the establishment of a formal proof.

7   Axiomatic Formal Proof

Formal proof, as introduced earlier in Sect. 2.1, refers either to a precise logical form as specified by Hilbert or to forms of proof used by mathematicians to communicate to each other in conversation and in journal articles. We begin by considering the undergraduate development in formal proof as part of the long-term cognitive growth of proof concepts from child to adult.

7.1   Student Development of Formal Proof

Initial student encounters with formal proof can occur in a number of ways. At one end of the spectrum is the Moore method in which students are given basic definitions and theorems and encouraged to seek proofs for themselves. R. L. Moore reckoned that 'That student is taught the best who is told the least' (Parker 2005). He produced a rich legacy of graduates in mathematics who advanced the frontiers of research, producing a phenomenon consistent with the long-term success of Fawcett's open-ended approach to geometry in school.

When students are introduced to formal proof in a university pure mathematics course, the objective is for them to make sense of formal definitions in a way that can be used for deduction of theorems. Pinto (1998) found that there was a spectrum of approaches in an analysis course, which she classified in two main groups: those who gave meaning to the definition from their current concept imagery, and those who extracted meaning from the definition by learning to reproduce it fluently and studying its use in proofs presented in class. She found that both approaches could lead to the building of successful formal knowledge structures but that either could break down as a result of conflict between previous experience and the theory, or through the sheer complexity of the definitions that could involve three or more nested quantifiers.

These observations are consistent with the work of Duffin and Simpson (1993, 1995) who distinguished between 'natural' approaches to describe a new experience that fits a learner's current mental structures, 'alien' approaches when the learner finds no connection with any of his or her internal structures, and 'conflicting' when the learner realises the experience is inconsistent with them.

Pinto and Tall (1999) used the terms 'natural' to describe the process of extracting meaning and 'formal' for the process of giving meaning by working with the formal deductions. To the categories 'natural' and 'formal', Weber (2004) added the notion of 'procedural' learning for those students who simply attempted to cope with formal definitions and proof by learning them by rote.

Alcock and Weber (2004) later classified the responses of students into 'semantic' and 'syntactic', using terms from linguistics that essentially refer to the meaning of language (its semantic content) and the structure of the language (its syntax). They described a syntactic approach as one in which 'the prover works from a literal reading of the involved definitions' and a semantic approach 'in which the prover also makes use of his or her intuitive understanding of the concepts.' These are broadly consistent (though not identical) with the extracting and giving of meaning and the related categories of 'formal' and 'natural'.

This reveals the complexity that students face in attempting to make sense of formal mathematical theory. Overall, it may be seen as a journey making sense of axiomatic systems by natural or formal means, beginning with specific axioms and seeking to construct a more flexible knowledge structure. For instance, one may begin with the definition of completeness in terms, say, of an increasing sequence bounded above having a limit, and then deduce other properties such as the convergence of Cauchy sequences or the existence of a least upper bound for a non-empty subset bounded above. All of these properties may then be seen to be equivalent ways of defining completeness, with the ultimate goal being the conception of a complete ordered field as a crystalline concept having a tight coherent structure with the various forms of completeness being used flexibly as appropriate in a given situation.

In this way, expertise develops as hypothesised in Fig. 2.3, from the students' previous growing experience, shifting from description to formal definition, constructing links through formal proof, establishing the equivalence of various definitions, and then building increasingly flexible links to construct crystalline concepts that can be mentally manipulated in flexible ways.

Evidence for this further development was given by Weber (2001) who investigated how four undergraduates and four research students responded to formal problems, such as the question of whether there is an isomorphism between the group of integers $\mathbb{Z}$ under addition and the group $\mathbb{Q}$ of rational numbers under addition. None of the four undergraduates could provide a formal response but all four graduates were able to do so. Whilst the undergraduates attempted to deal with the proof in terms of definition to find a bijection that preserved the operation, the graduates used their more flexible knowledge structures to consider whether an isomorphism was even possible.

Some undergraduates focused on their memory that $\mathbb{Z}$ and $\mathbb{Q}$ have the same cardinal number and so already had a bijective correspondence between them. They had part of the idea – a bijection – but not a bijection that respected the operation. Meanwhile the graduates used their wider knowledge structures, including a wider repertoire of proving strategies, to suggest possible ways of thinking about the problem. For instance, one immediately declared that $\mathbb{Q}$ and $\mathbb{Z}$ could not be isomorphic, first by speculating that $\mathbb{Q}$ is dense but $\mathbb{Z}$ is not, then that '$\mathbb{Z}$ is cyclic, but $\mathbb{Q}$ s not' (meaning that the element 1 generates the whole of $\mathbb{Z}$ under addition but no element in $\mathbb{Q}$ does so).

7.2   Structure Theorems and New Forms of Embodiment and Symbolism in Research Mathematics

In the process of building formal knowledge structures, certain central theorems play an essential role. These are structure theorems. They state that an axiomatic structure has certain essential properties which can often involve enriched visual and symbolic modes of operation that are now based not on naïve intuition but on formal proof. Typical examples are:

- An equivalence relation on S corresponds to a partition of S.
- A finite group is isomorphic to a subgroup of a group of permutations.
- A finite dimensional vector space over a field F is isomorphic to F^n.
- A field contains a subfield isomorphic to $\mathbb{Q}$ or to $\mathbb{Z}_n$.
- An ordered field contains a subfield isomorphic to $\mathbb{Q}$.
- An ordered field extension K of $\mathbb{R}$ contains finite and infinite elements and any finite element is of the form $c + \varepsilon$ where $c \in \mathbb{R}$ and $\varepsilon \in K$ is infinitesimal.

Such theorems enable research mathematicians to develop personal knowledge structures that operate in highly flexible ways. Some build formally, others naturally, as observed by the algebraist Saunders MacLane (1994) speaking of a conversation with the geometer Michael Atiyah:

> For MacLane it meant getting and understanding the needed definitions, working with them to see what could be calculated and what might be true, to finally come up with new 'structure' theorems. For Atiyah, it meant thinking hard about a somewhat vague and uncertain situation, trying to guess what might be found out, and only then finally reaching definitions and the definitive theorems and proofs (pp. 190–191).

This division between those 'preoccupied with logic' and those 'guided by intuition' was noted long ago by Poincaré (1913), citing Hermite as a logical thinker who 'never evoked a sensuous image' in mathematical conversation and Riemann as an intuitive thinker who 'calls geometry to his aid' (p. 212).

Such 'intuition' used by mathematicians relates to deeply embedded subtle linkages in their own personal knowledge structures that suggest likely relationships before they may be amenable to formal proof. Formal proof is the final stage of research in which the argument is refined and given in a deductive manner based on precise definitions and appropriate mathematical deductions. Making sense of mathematics requires more:

> Only professional mathematicians learn anything from proofs. Other people learn from explanations. A great deal can be accomplished with arguments that fall short of proofs (Boas 1981, p. 729).
>
> If mathematics were formally true but in no way enlightening this mathematics would be a curious game played by weird people (Rota 1997, p. 132).

Burton (2002) interviewed 70 research mathematicians (with equal numbers of males and females) to study a range of aspects under headings such as thinking styles, socio-cultural relatedness, aesthetics, intuition and connectivities. She initially hypothesised that she would find evidence of the two styles of thinking formulated above (which she described as visual and analytic) and that mathematicians would move flexibly between the two. However, her analysis of the data showed a more complex situation that she organised into three categories – visual (thinking in pictures, often dynamic), analytic (thinking symbolically, formalistically) and conceptual (thinking in ideas, classifying). The majority of those interviewed (42/70) embraced two styles, a small number (3/70) used all three and the rest referred only to one (15 visual, 3 analytic and 7 conceptual). Rather than a simple dualism, this suggests a range of thinking styles used in various combinations.

This finer analysis remains consistent with the broad framework of embodiment, symbolism and formalism, whilst revealing subtle distinctions. The research cycle of development builds on the experiences of the researchers and involves exploration in addition to formal proof. The visual category is concerned with embodiment in terms of 'thinking in pictures, often dynamic'. The analytic category specifically describes 'thinking symbolically, formalistically'. The conceptual category refers to 'thinking in ideas', which occurs at an exploratory phase of a cycle of research, whilst 'classifying' relates to classifying structures that satisfy an appropriate definition. These definitions may vary in their origins, as verbally defined embodiments, symbolic concepts related to rules of operation, or formally defined concepts that may then be proved to have a given structure. For instance simple groups are defined formally but their classification may be performed using structural properties such as generators and relations rather than formal proof as a sequence of quantified statements.

Applied mathematicians develop contextually based inference in their area of expertise, often based on embodiment of situations, translated into some kind of symbolism (such as systems of algebraic or differential equations) to manipulate and predict a solution.

Research mathematicians have a variety of ways of thinking creatively to develop new theorems. Byers (2007) observed that true creativity arises out of paradoxes, ambiguities and conflicts that occur when ideas from different contexts come into contact. This encourages mathematicians to attempt to make sense of the problem, thinking about possibilities, suggesting possible hypotheses, making new definitions, and seeking mathematical proofs.

In creating new mathematical theorems, mathematicians will work not only according to their personal preferences, but also with respect to the particular mathematical context. The possible combinations of embodiment, symbolism and formalism lead to a variety of techniques in which a particular activity may lean towards one or more modes of operation appropriate to the situation.

Mathematical proof at the highest level is an essential part of the story of development, with differently oriented mathematicians having different ways of thought but sharing common standards as to the need for proof to establish a desired result. However, a mathematical proof is not the end of the story, it is the full stop at the end of a paragraph, a place to pause and celebrate the proof of a new theorem which, in turn, becomes the launching pad for a new cycle of research and development.

8   Summary

In considering the development of proof from the child to the adult and the mathematical researcher, we have embraced a wide range of viewpoints that relate directly to the ideas of proof and proving. Overall, however, we see the cognitive development beginning with the perceptions, actions and reflections of the child, developing out of the sensorimotor foundation of human thought, as the child observes regularities, builds mental concepts and makes links between them.

Different developments of proof occur in the visuospatial development of geometry and the symbolic development of arithmetic and algebra. In geometry, perceptions are described and knowledge structures are developed that construct relationships, develop definitions of figures, deduce equivalent properties and build up a coherent structure of Euclidean geometry. In arithmetic, general properties of operations are observed, described, then later formulated as rules that should be obeyed in algebra, enabling a form of proof based on algebraic manipulation subject to the rules of arithmetic.

The general population builds mainly on the physical, spatial and symbolic aspects of mathematics. Pure mathematicians in addition develop formally defined set-theoretic entities and formal proof. This leads to structure theorems that give the entities a rich combination of embodiment and symbolism, now supported by formal deduction. Applied mathematicians, develop contextual ways of using structures to model embodied situations as symbolic representations, solved by symbolic operations.

In each case the strand of development begins from human perception and action, through experience of objects and properties, that are described, with meanings that are refined and defined, then built into more sophisticated thinkable concepts that have a rich knowledge structure. These are connected together through relationships constructed by appropriate forms of deduction and proof to lead to sophisticated crystalline concepts whose properties are constrained by the given context.

In Euclidean geometry the crystalline concepts are platonic figures that represent the essentials underlying the physical examples, in arithmetic and algebra they are flexible procepts that enable fluent calculation and manipulation of symbols, and in formal mathematics they are the total entities arising through deduction from axioms and definitions, with their essential structure revealed by structure theorems.

Proof involves a lifetime of cognitive development of the individual that is shared within societies and is further developed in sophistication by successive generations of mathematicians. Mathematical proof is designed to furnish theorems that can be used in a given context as both the culmination of a process of seeking certainty and explanation and also as a foundation for future developments in mathematical research.

References*

Alcock, L., & Weber, K. (2004). Semantic and syntactic proof productions. *Educational Studies in Mathematics, 56*, 209–234.

Antonini, S. (2001). Negation in mathematics: Obstacles emerging from an exploratory study. In M. van den Heuvel-Panhuizen (Ed.), *Proceedings of the 25th Conference of the International Group for the Psychology of Mathematics Education* (Vol. 2, pp. 49–56). Utrecht, The Netherlands: Utrecht University.

Arnold, V. I. (2000). Polymathematics: Is mathematics a single science or a set of arts? In *Mathematics: Frontiers and perspectives* (pp. 403–416). Providence: American Mathematical Society.

Bass, H. (2009). *How do you know that you know? Making believe in mathematics.* Distinguished University Professor Lecture given at the University of Michigan on March 25, 2009. Retrieved from the internet on January 30, 2011, from http://deepblue.lib.umich.edu/bitstream/2027.42/64280/1/Bass-2009.pdf.

Boas, R. P. (1981). Can we make mathematics intelligible? *The American Mathematical Monthly, 88*(10), 727–773.

Bruner, J. S. (1966). *Towards a theory of instruction.* Cambridge: Harvard University Press.

Burton, L. (2002). Recognising commonalities and reconciling differences in mathematics education. *Educational Studies in Mathematics, 50*(2), 157–175.

Byers, W. (2007). *How mathematicians think.* Princeton: Princeton University Press.

Byrne, R. M. J., & Johnson-Laird, P. N. (1989). Spatial reasoning. *Journal of Memory and Language, 28*, 564–575.

Dawson, J. (2006). Why do mathematicians re-prove theorems? *Philosophia Mathematica, 14*(3), 269–286.

Dehaene, S. (1997). *The number sense: How the mind creates mathematics.* New York: Oxford University Press.

Dreyfus, T., & Eisenberg, T. (1986). On the aesthetics of mathematical thoughts. *For the Learning of Mathematics, 6*(1), 2–10.

Dubinsky, E., & McDonald, M. A. (2001). APOS: A constructivist theory of learning in undergraduate mathematics education research. In D. Holton (Ed.), *The teaching and learning of*

---

*NB: References marked with * are in F. L. Lin, F. J. Hsieh, G. Hanna, & M. de Villiers (Eds.) (2009). *ICMI Study 19: Proof and proving in mathematics education.* Taipei, Taiwan: The Department of Mathematics, National Taiwan Normal University.

mathematics at university level: An ICMI study (New ICMI study series, Vol. 7, pp. 273–280). Dordrecht: Kluwer.

Duffin, J. M., & Simpson, A. P. (1993). Natural, conflicting and alien. *The Journal of Mathematical Behavior, 12*(4), 313–328.

Duffin, J. M., & Simpson, A. P. (1995). A theory, a story, its analysis, and some implications. *The Journal of Mathematical Behavior, 14*, 237–250.

Duval, R. (2006). A cognitive analysis of problems of comprehension in a learning of mathematics. *Educational Studies in Mathematics, 61*, 103–131.

Epp, S. (1998). A unified framework for proof and disproof. *Mathematics Teacher, 91*(8), 708–713.

Fischbein, E. (1993). The theory of figural concepts. *Educational Studies in Mathematics, 24*, 139–162.

Flener, F. (2001, April 6). *A geometry course that changed their lives: The Guinea pigs after 60 years.* Paper presented at the Annual Conference of The National Council of Teachers of Mathematics, , Orlando. Retrieved from the internet on January 26, 2010, from http://www.maa.org/editorial/knot/NatureOfProof.html

Flener, F. (2006). *The Guinea pigs after 60 years.* Philadelphia: Xlibris Corporation.

Freudenthal, H. (1973). *Mathematics as an educational task.* Dordrecht: Reidel Publishing Company.

Goldin, G. (1998). Representational systems, learning, and problem solving in mathematics. *The Journal of Mathematical Behavior, 17*(2), 137–165.

Gray, E. M., & Tall, D. O. (1994). Duality, ambiguity and flexibility: A proceptual view of simple arithmetic. *Journal for Research in Mathematics Education, 26*(2), 115–141.

Harel, G., & Sowder, L. (1998). Student's proof schemes: Results from exploratory studies. In A. Schoenfeld, J. Kaput, & E. Dubinsky (Eds.), *Research in collegiate mathematics education III* (pp. 234–283). Providence: American Mathematical Society.

Harel, G., & Sowder, L. (2005). Advanced mathematical thinking at any age: Its nature and its development. *Mathematical Thinking and Learning, 7*, 27–50.

Healy, L., & Hoyles, C. (1998). *Justifying and proving in school mathematics.* Summary of the results from a survey of the proof conceptions of students in the UK (Research Report, pp. 601–613). London: Mathematical Sciences, Institute of Education, University of London.

Henderson, D. W., & Taimina, D. (2005). *Experiencing geometry: Euclidean and Non-Euclidean with history* (3rd ed.). Upper Saddle River: Prentice Hall.

Hershkowitz, R. & Vinner, S. (1983). The role of critical and non-critical attributes in the concept image of geometrical concepts. In R. Hershkowitz (Ed.), *Proceedings of the 7th International Conference of the International Group for the Psychology of Mathematics Education* (pp. 223–228). Weizmann Institute of Science: Rehovot.

Hilbert, D. (1900). *The problems of mathematics. The Second International Congress of Mathematics.* Retrieved from the internet on January 31, 2010, from http://aleph0.clarku.edu/~djoyce/hilbert/problems.html

Hilbert, D. (1928/1967). The foundations of mathematics. In J. Van Heijenoort (Ed.), *From Frege to Gödel* (p. 475). Cambridge: Harvard University Press.

Hoffmann, D. (1998). *Visual Intelligence: How we change what we see.* New York: W.W Norton. 1998.

Housman, D., & Porter, M. (2003). Proof schemes and learning strategies of above-average mathematics students. *Educational Studies in Mathematics, 53*(2), 139–158.

Johnson, M. (1987). *The body in the mind: The bodily basis of meaning, imagination, and reason.* Chicago: Chicago University Press.

Joyce, D. E. (1998). *Euclid's elements.* Retrieved from the internet on January 30, 2011, from http://aleph0.clarku.edu/~djoyce/java/elements/elements.html

Klein, F. (1872). *Vergleichende Betrachtungen über neuere geometrische Forschungen.* Erlangen: Verlag von Andreas Deichert. Available in English translation as a pdf at http://math.ucr.edu/home/baez/erlangen/erlangen_tex.pdf (Retrieved from the internet on January 30, 2011).

Koichu, B. (2009). What can pre-service teachers learn from interviewing high school students on proof and proving? (Vol. 2, 9–15).*

Koichu, B., & Berman, A. (2005). When do gifted high school students use geometry to solve geometry problems? *Journal of Secondary Gifted Education, 16*(4), 168–179.

Kondratieva, M. (2009). Geometrical sophisms and understanding of mathematical proofs (Vol. 2, pp. 3–8).*

Lakoff, G. (1987). *Women, fire, and dangerous things: What categories reveal about the mind.* Chicago: Chicago University Press.

Lakoff, G., & Johnson, M. (1999). *Philosophy in the flesh.* New York: Basic Books.

Lakoff, G., & Núñez, R. (2000). *Where mathematics comes from: How the embodied mind brings mathematics into being.* New York: Basic Books.

Lawler, R. W. (1980). The progressive construction of mind. *Cognitive Science, 5*, 1–34.

Lénárt, I. (2003). *Non-Euclidean adventures on the Lénárt sphere.* Emeryville: Key Curriculum Press.

Leron, U. (1985). A direct approach to indirect proofs. *Educational Studies in Mathematics, 16*(3), 321–325.

Lin, F. L., Cheng, Y. H. et al. (2003). The competence of geometric argument in Taiwan adolescents. In *Proceedings of the International Conference on Science and Mathematics Learning* (pp. 16–18). Taipei: National Taiwan Normal University.

MacLane, S. (1994). Responses to theoretical mathematics. *Bulletin (new series) of the American Mathematical Society, 30*(2), 190–191.

Maher, C. A., & Martino, A. M. (1996). The development of the idea of mathematical proof: A 5-year case study. *Journal for Research in Mathematics Education, 27*(2), 194–214.

Mason, J., Burton, L., & Stacey, K. (1982). *Thinking mathematically.* London: Addison Wesley.

Meltzoff, A. N., Kuhl, P. K., Movellan, J., & Sejnowski, T. J. (2009). Foundations for a new science of learning. *Science, 325*, 284–288.

National Council of Teachers of Mathematics. (1989). *Curriculum and evaluation standards for school mathematics.* Reston: NCTM.

National Council of Teachers of Mathematics. (2000). *Principles and standards for school mathematics.* Reston: NCTM.

Neto, T., Breda, A., Costa, N., & Godino, J. D. (2009). Resorting to Non–Euclidean plane geometries to develop deductive reasoning: An onto–semiotic approach (Vol. 2, pp. 106–111).*

Núñez, R., Edwards, L. D., & Matos, J. F. (1999). Embodied cognition as grounding for situatedness and context in mathematics education. *Educational Studies in Mathematics, 39*, 45–65.

Paivio, A. (1991). Dual coding theory: Retrospect and current status. *Canadian Journal of Psychology, 45*, 255–287.

Papert, S. (1996). An exploration in the space of mathematics educations. *International Journal of Computers for Mathematical Learning, 1*(1), 95–123.

Parker, J. (2005). *R. L. Moore: Mathematician and teacher.* Washington, DC: Mathematical Association of America.

Pinto, M. M. F. (1998). *Students' understanding of real analysis.* PhD thesis, University of Warwick, Coventry.

Pinto, M. M. F., & Tall, D. O. (1999). Student constructions of formal theory: Giving and extracting meaning. In O. Zaslavsky (Ed.), *Proceedings of the 23rd Conference of the International Group for the Psychology of Mathematics Education* (Vol. 4, pp. 65–73). Haifa, Israel: Technion - Israel Institute of Technology.

Pirie, S., & Kieren, T. (1994). Growth in mathematical understanding: How can we characterize it and how we can represent it? *Educational Studies in Mathematics, 26*(2–3), 165–190.

Poincaré, H. (1913/1982). *The foundations of science* (G. B. Halsted, Trans.). The Science Press (Reprinted: Washington, DC: University Press of America).

Rav, Y. (1999). Why do we prove theorems? *Philosophia Mathematica, 7*(1), 5–41.

Reid, D., & Dobbin, J. (1998). Why is proof by contradiction difficult? In A. Olivier & K. Newstead (Eds.), *Proceedings of the 22nd Conference of the International Group for the Psychology of Mathematics Education* (Vol. 4, pp. 41–48). Stellenbosch, South Africa: University of Stellenbbosch.

Reiss, K. (2005) *Reasoning and proof in geometry: Effects of a learning environment based on heuristic worked-out examples*. In 11th Biennial Conference of EARLI, University of Cyprus, Nicosia.

Rodd, M. M. (2000). On mathematical warrants. *Mathematical Thinking and Learning, 2*(3), 221–244.

Rota, G. C. (1997). *Indiscrete thoughts*. Boston: Birkhauser.

Senk, S. L. (1985). How well do students write geometry proofs? *Mathematics Teacher, 78*(6), 448–456.

Sfard, A. (1991). On the dual nature of mathematical conceptions: Reflections on processes and objects as different sides of the same coin. *Educational Studies in Mathematics, 22*, 1–36.

Shapiro, S. (1991). *Foundations without foundationalism: A case for second-order logic*. Oxford: Clarendon.

Tall, D. O. (1979). Cognitive aspects of proof, with special reference to the irrationality of √2. In *Proceedings of the 3rd Conference of the International Group for the Psychology of Mathematics Education* (pp. 203–205).Warwick, UK: University of Warwick.

Tall, D. O. (2004). Thinking through three worlds of mathematics. In M. J. Hoines and A. Fuglestad (Eds.), *Proceedings of the 28th Conference of the International Group for the Psychology of Mathematics Education* (Vol.4, pp. 281–288). Bergen, Norway, Bergen University College.

Tall, D. O. (2008). The transition to formal thinking in mathematics. *Mathematics Education Research Journal, 20*(2), 5–24.

Tall, D. O. (2012, under review). *How humans learn to think mathematically*.

Tall, D. O., & Vinner, S. (1981). Concept image and concept definition in mathematics with particular reference to limits and continuity. *Educational Studies in Mathematics, 12*(2), 151–169.

Van Hiele, P. M. (1986). *Structure and insight*. New York: Academic.

Van Hiele-Geldof, D. (1984). The didactics of geometry in the lowest class of secondary school. In D. Fuys, D. Geddes, & R. Tischler (Eds.), *English translation of selected writings of Dina van Hiele-Geldof and Pierre M. van Hiele*. Brooklyn: Brooklyn College (Original work published 1957).

Weber, K. (2001). Student difficulty in constructing proofs: The need for strategic knowledge. *Educational Studies in Mathematics, 48*(1), 101–119.

Weber, K. (2004). Traditional instruction in advanced mathematics courses: A case study of one professor's lectures and proofs in an introductory real analysis course. *The Journal of Mathematical Behavior, 23*, 115–133.

Yevdokimov, O. (2008). Making generalisations in geometry: Students' views on the process. A case study. In O. Figueras, J. L. Cortina, S. Alatorre, T. Rojano & A. Sepulova (Eds.), *Proceedings of the 32nd Conference of the International Group for the Psychology of Mathematics Education* (Vol. 4, pp. 441–448). Morelia, Mexico: Cinvestav-UMSNH.

Yevdokimov, O. (in preparation). *Mathematical concepts in early childhood*.

Chapter 3
Theorems as Constructive Visions

Giuseppe Longo*

1   The Constructive Content of Euclid's Axioms

From the time of Euclid to the age of super-computers, Western mathematicians have continually tried to develop and refine the foundations of proof and proving. Many of these attempts have been based on analyses logically and historically linked to the prevailing philosophical notions of the day. However, they have all exhibited, more or less explicitly, some basic cognitive principles – for example, the notions of symmetry and order. Here I trace some of the major steps in the evolution of notion of proof, linking them to these cognitive basics.

I take as a starting point Euclid's *Aithemata* (Requests), the *minimal constructions required* to do geometry:

1. To draw a straight line from any point to any point.
2. To extend a finite straight line continuously in a straight line.
3. To draw a circle with any centre and distance.
4. That all right angles are equal to one another.
5. That, if a straight line falling on two straight lines makes the interior angles on the same side less than two right angles, the two straight lines, if produced indefinitely, meet on that side on which the angles are less than the two right angles. (Heath 1908; pp. 190–200).

These "Requests" are *constructions performed by ruler and compass*: an abstract ruler and compass, of course, not the carpenter's tools but tools for a dialogue with the Gods. They provide the minimal "construction principles" the geometer should be able to apply.

---

*(Longo's articles can be downloaded from: http://www.di.ens.fr/users/longo/)

G. Longo (✉)
CNRS – École Normale Supérieure et CREA, École Polytechnique, Paris, France
e-mail: longo@di.ens.fr

© The Author(s) 2021   51
G. Hanna and M. de Villiers (eds.), *Proof and Proving in Mathematics Education*,
New ICMI Study Series, https://doi.org/10.1007/978-94-007-2129-6_3

Note that these Requests follow a "maximal symmetry principle". Drawing a straight line between two points, one obtains the most symmetric possible structure: any other line, different from this one, would introduce asymmetries by breaking at least the axial symmetry of the straight line. The same can be said for the second axiom, where any other extension of a finite line would yield fewer symmetries. Similarly, the third, a complete *rotation symmetry*, generates the most symmetric figure for a line enclosing a point. In the fourth, equality is defined by congruence; that is, by a *translation symmetry*. Finally, the fifth construction again is a matter of drawing, intersecting and then extending. The most symmetric construction occurs when the two given lines do *not* intersect: then the two inner angles are right angles on both sides of the line intersecting the two given lines. The other two cases, as negations of this one (once the theorem in Book I n. 29 in Euclid's Elements has been shown), would reduce the number of symmetries. Their equivalent formulations (more than one parallel in one point to a line, no parallel at all) both yield fewer symmetries, *on a Euclidian plane*, than having exactly one parallel line.

Euclid's Requests found geometry by actions on figures, implicitly *governed by symmetries*. Now, "symmetries" are at the core of Greek culture, art and science. They refer to "balanced" situations or, more precisely, "measurable" entities or forms. But the meaning we give to symmetries today underlies Greek "aesthetics" (in the Greek sense of the word) and their sensitivity, knowledge and art, from sculpture to myth and tragedy. Moreover, loss of symmetry (symmetry-breaking) originated the world as well as human tragedy; as a breaking of equilibrium between the Gods, it underlies the very sense of human life. As tools for mathematical construction, symmetry-breakings participate in the "original formation of sense", as Husserl would say (see below and Weyl 1952).

Concerning the axioms of geometry, the formalist universal-existential version ("For *any* two points on a plane, there *exists one and only* one segment between these points" etc.) misses the constructive sense and misleads the foundational analysis into the anguishing quest for formal, thus finitistic, consistency proofs.[1] We know how this quest ended: by Gödel's theorem, there is no such proof for the paradigm of finitism in mathematics, formal arithmetic.

2 From Axioms to Theorems

"Theorem" derives from "*theoria*" in Greek; it means "vision", as in "theatre": a theorem *shows*, by constructing. So, the first theorem of Euclid's first book shows how to take a segment and trace the (semi-)circles centred on the extremes of the segment, with the segment as radius. These intersect in one point. Draw straight lines from the extremes of the segment to that point: this produces an equilateral triangle.

---

[1] In my interpretation, existence, in the first axiom, is by *construction* and unicity by *symmetry*.

For a century critics have told us that this is not a proof (in Hilbert's sense!): One must formally prove the existence of the point of intersection. These detractors could use more of the Greeks' dialogue with their Gods.[2] Both points and continuous lines are founding notions, but the conceptual path relating them is the inverse of the point-wise constructions that have dominated mathematics since Cantor. Lines are ideal objects, they are *a cohesive continuum with no thickness*. Points, in Euclid, are obtained as a *result* of an intersection of lines: two *thickless* (one-dimensional) lines, suitably intersecting, produce a point, *with no parts* (no dimension) The immense step towards abstraction in Greek geometry is the invention of continuous lines with no thickness, as abstract as a divine construction. As a matter of fact, how else can one propose a general *Measure Theory* of surfaces, the aim of "geo-metry"? If a plane figure has thick borders, which is the surface of the figure?

Thus came this amazing conceptual (and metaphysical) invention, done within the Greek dialogue with the Gods: the continuous line with no thickness. Points – with no dimension, but nameable, as Euclid defines them[3] – are then produced by intersecting lines or sit at the extremes of a line or segment (definition γ). But lines are not composed of signs-points. A line, either continuous or discrete, is a *gestalt*, not a set of points.

Greek geometric figures and their theatrical properties derive by constructions from these fundamental gestalts, signs-points and lines, in a game of rotations and translations, of constructing and breaking symmetries. These gestalts inherently penetrate proofs even now.

3 On Intuition

Mathematical intuition is the result of an historical praxis; it is a constituted frame for active constructions, grounded on action in space, stabilised by language and writing in inter-subjectivity.

A pure intuition refers to *what can be done*, instead of to *what is*. It is the seeing of a mental construction; it is the appreciation of an active experience, of an active (re-) construction of the world. We can intuit, because we *actively construct* (math-ematical) knowledge on the phenomenal screen between us and the world.

As for that early and fundamental gestalt, the continuous line, our evolutionary and historical brain sets contours that are not in the world, beginning with the activity of the primary cortex. There is a big gap – actually, an abyss – between the biological-evolutionary path and the historical-conceptual construction; yet, I'll try to bridge it in a few lines.

---

[2] Schrödinger stresses that a fundamental feature of Greek philosophy is the absence of 'the unbearable division, which affected us for centuries… : the division between science and religion' (quoted in Fraisopi 2009).

[3] Actually "signs" (σημεια, definition α): Boetius first used the word and the meaning of "point". Note that a sign-point (σημειον) in Euclid is *identified* with the letter that names it (see Toth 2002).

The neurons of the primary cortex activate by *contiguity* and *connectivity* along non-existent lines and "project" non-existing continuous contours on objects (at most, contours are singularities). More precisely, recent analyses of the primary cortex (see Petitot 2003) highlight the role of intra-cortical synaptic linkages in the perceptual construction of edges and of trajectories. In the primary cortex, neurons are sensitive to "directions": they activate when oriented along the tangent of a detected direction or contour. More precisely, the neurons which activate for *almost* parallel directions, possibly along a straight line, are more connected than the others. In other words, neurons whose receptive field, *approximately* and *locally*, is upon a straight line (or along parallel lines) have a larger amount of synaptic connections amongst themselves. Thus, the activation of a neuron stimulates or prepares for activation neurons that are *almost* aligned with it or that are *almost* parallel – like tangents along a continuous virtual line in the primary cortex. We detect the continuity of an edge by a global "gluing" of these tangents, in the precise geometrical (differential) sense of gluing. More exactly, our brain "imposes" by continuity the unity of an edge by relating neurons which are structured and linked together in a continuous way and locally *almost* in parallel. Their "integral" gives the line (Petitot 2003).

The humans who first drew the *contours* of a bison on the walls of a cavern (as in Lascaux) instead of painting a brown or black body, communicated to other humans with the same brain cortex and life experience. A real bison is not made just of thick contours as in some drawings on those walls. Yet, those images evoke the animal by a reconstruction of it on that phenomenal screen which is the constructed interface between us and the world. The structures of mathematics originate also from such drawings, through their abstract lines. The Greek "limit" definition and construction of the ideal line with no thickness is the last plank of our long bridge: a constructed but "critical" transition to the pure concept (see Bailly and Longo 2006), far from the world of senses and action, well beyond all we can say by just looking at the brain, but grounded on and made possible by our brain and its action in this world.

Consider now the other main origin of our mathematical activities: the counting and ordering of small quantities, a talent that we share with many animals (see Dehaene 1997). By language we learn to iterate further; we stabilise the resulting sequence with names; we propose numbers for these *actions*. These numbers were first associated, by common names, with parts of the human body, beginning with the fingers. With writing, their notation departed from just iterating fingers or strokes; yet, in all historical notations, we still write with strokes up to 3, which is given by three parallel segments interconnected by continuous drawing, like 2, which is given by two connected segments. However, conceptual iteration has no reason to stop: it may be "*apeiron*" – "without limit", in Greek. Thus, since that early conceptual practice of potential infinity, we started seeing the endless number line, a discrete gestalt, because we iterate an action-schema in space (counting, ordering …) and we *well order* it by this very conceptual gesture. For example, we look at that discrete endless line, which goes from left to right (in our Western culture, but opposite for Arabs; Dehaene 1997), and observe "a generic non-empty subset has a least element." (A sufficiently mathematically minded reader should pause here and envision this.) This is the principle of *well-ordering* as used every day by mathematicians. It is a consequence of the discrete spatial construction, a geometric invariant resulting from different practices of discrete ordering and counting into mental spaces. It originates in the simple, small counting and ordering that we share with many animals (Dehaene 1997; Longo and Viarouge 2010) Further on, in a long path, via language, arithmetical (logico-formal) induction follows from those early active forms of ordering and counting objects rather than founding them – contrary to Frege's and Hilbert's views (see below). The mathematical construction, induction, is the result of these ancient practices, by action and language; then it organises the world and allows proofs. Yet, it is grounded on a "gestalt", the discrete well-ordering where individual points make no sense without their ordered context.

4 Little Gauss' Proof

At the age of 7 or 8, Gauss was asked by his school teacher to produce the result of the sum of the first n integers (or, perhaps, the question was slightly less general…), Gauss (1801).[4] He then proved a theorem, by the following method. He wrote on the first line the increasing sequence 1,… , n, then, below it and inverted, the same sequence; finally, he added the vertical lines:

$$1 \qquad 2 \dots n$$
$$n \quad (n-1) \dots 1$$
$$\text{-------}$$
$$(n+1) \dots (n+1)$$

Then the result is obvious: $\sum_{1}^{n} i = n(n+1)/2$

This proof is *not* by induction. Given n, it proposes a uniform argument which works for *any* integer n. Following Herbrand (Longo 2002), we may call this kind of proof a *prototype*: it provides a (geometric) prototype or schema for any intended parameter of the proof. Of course, once the formula $\sum_{1}^{n} i = n(n+1)/2$ is given, we can very easily prove it by induction as well. But one must know the formula or, more generally, the "induction load". Little Gauss did not know the formula; he had to construct it as a result of the proof. On the contrary, we have the belief induced by the formalist myth: that proving a theorem is proving an *already given formula*! We learn, more or less implicitly, from the formal approach, that mathematics is

---

[4] This section is partly borrowed from the introduction to Longo (2002).

"the use of the axioms to prove a given formula" – an incomplete foundation and a parody of mathematical theorem proving.

Except in a few easy cases, even when the formula to be proved is already given (best known example: Fermat's last theorem), the proof requires the invention of an induction load and of a novel deductive path which may be very far from the formula. In Fermat's example, the detour requires the invention of an extraordinary amount of new mathematics. The same is true also in Automatic Theorem Proving, where human intervention is required even in inductive proofs because, except in a few trivial cases, the assumption required in the inductive step (the induction load) may be much stronger than the thesis or have no trivial relation to it. Clearly, *a posteriori* the induction load may be generally described within the formalism, but its "choice", out of infinitely many possibilities, may require some external heuristics (typically: analogies, symmetries, symmetry-breaking, etc.).

More generally, *proving a theorem is answering a question*, like Gauss' teacher's question, about a property of a mathematical structure or about relating different structures; *it is not proving an already given formula*.

Consider a possible way to Gauss' proof. Little Gauss "saw" the discrete number line, as we all do, well ordered from left to right. But then he had a typical hit of mathematical genius: He dared to invert it, to force it to go backwards in his mind, an amazing step. This is a paradigmatic mathematical invention: constructing a new symmetry, in this case by an audacious space rotation or order inversion. That reverse-reflection (or mirror) symmetry gives the equality of the vertical sums. The rest is obvious.

In this case, order and symmetries both *produce* and *found* Gauss' proof. Even *a posteriori*, the proof cannot be founded on formal induction, as this would assume the knowledge of the formula.

4.1   Arithmetic Induction and the Foundation of Mathematical Proof

Above, I hinted at an understanding of the ordering of numbers with reference to a mental construction in space (or time). Frege would have called this approach "psychologism" – Herbart's style, according to Frege (1884). Poincaré (1908) instead could be a reference for this view on the certainty and meaning of induction as grounded on intuition in space. In Brouwer's (1948) foundational proposal, the mathematician's intuition of the sequence of natural numbers, which founds mathematics, relies on another phenomenal experience: It should be grounded on the "discrete falling apart of time", as "twoness" ("the falling apart of a life moment into two distinct things, one which gives way to the other, but is retained by memory"; Brouwer 1948). Thus, "Brouwer's number line" originates from (a discrete form of) phenomenal time and induction derives meaning and certainty from it.

Intuition of the (discrete and increasing) ordering in space and time contributes to establishing the well-ordered number line as an *invariant* of these different active phenomenal experiences: Formal induction follows from and is *founded* on this intuition in Poincaré's and Brouwer's philosophy. Recent scientific evidence (see Longo and Viarouge 2010) suggests that we use extensively, in reasoning and computations, the "intuitive" number line as an order in space; those remarkable neuropsychological investigations take us beyond the "introspection" that the founding fathers used as the only way to ground mathematics on intuition. We are probably in the process of transforming the analysis of intuition from naive introspection to a scientific investigation of our cognitive performances, in which the "Origin of Geometry" and the intuition of numbers blend in an indissoluble whole.

I return now to the sum of the first n integers and induction. About 80 years later, Peano (1889) and Dedekind (1888), by their work on Arithmetic, (implicitly) suggested that little Gauss' proof was certainly a remarkable achievement (in particular for a child), but that adults had to prove theorems in Number Theory by a "formal and uniform method", defined as a "potentially mechanisable" one by Peano (1889) and Padua (1900). Then Peano definitely specified "formal induction" as *the* proof principle for arithmetic, thus defining Peano Arithmetic, or PA (Kennedy 2006).

Frege set induction at the basis of his logical approach to mathematics; he considered it a founding (and absolute) logical principle, and thus gave PA the foundational status that it still has. Of course, Frege thought that logical induction (or PA) was "categorical" (in modern terms); that is, that induction exactly captured the "theory of numbers" or that everything was said within PA: This logical theory simply coincided, in his view, with the structure and properties of numbers. (Frege didn't even make the distinction "theory vs. model" and never accepted it; the logic was exactly the mathematics, for him.)

In *The Foundation of Geometry* (1899), Hilbert set the formal foundation for geometry, as a solution for the incredible situation where many claimed that rigid bodies could be not so rigid and that light rays could go along curved geodetics. Riemann's (1854) work (under Gauss' supervision) had started this "delirium", as Frege called the intuitive-spatial meaning of the new geometry (1884, p. 20). Later, Helmholtz, Poincaré and Enriques (see Boi 1995; Bottazzini and Tazzioli 1995) developed both the geometry and Riemann's epistemological approach to mathematics as a "genealogy of concepts", partly grounded on action in space.

For these mathematicians, *meaning*, as a reference to phenomenal space and its mathematical structuring, preceded rigour and provided "foundation". Thus, through mathematics, geometry in particular, Poincaré and Enriques wanted to make the physical world intelligible (see Boi 1995). For them, proving theorems by rigorous tools and conceptual constructions did not coincide with a formal/mechanical game of symbols. Hilbert (1899) had a very different foundational attitude: For the purposes of foundations (but only for these purposes), one has to forget the meaning in the physical space of the axioms of non-Euclidean geometries and interpret their purely formal presentation in PA. In his 1899 book, Hilbert fully formalised a unified approach to geometry and "interpreted" it in

PA. Formal rigour and proof principles as effective-finitistic reduction lie at the core of his analysis.[5]

On the one hand, that geometrisation of physics, from Riemann (1854) to Einstein (1915) and Weyl (1949) (via Helmholtz, Clifford and Poincaré; see Boi 1995), brought a revolution in that discipline, originating in breathtaking physico-mathematical theories (and theorems). On the other hand, the attention to formal, potentially mechanisable rigour, independent of meaning and intuition, provided the strength of the modern axiomatic approach and fantastic logico-formal *machines*, from Peano and Hilbert to Turing and digital computers (Longo 2009).

At the 1900 Paris conference, Hilbert contributed to giving PA (and formal induction) its central status in foundation by suggesting one could prove (formally) the consistency of PA. In his analytic interpretation, the consistency of the geometric axiomatisations would have followed from that of formal Number Theory, with no need of reference to meaning.

Moreover, a few years later, Hilbert proposed a further conjecture, the "ultimate solution", to all foundational problems, a jump into perfect rigour: Once shown the formal consistency of PA by finitistic tools, prove the completeness of the formal axioms for arithmetic. Independent of its heuristics, a proof's certainty had to ultimately be given by formal induction.

However, the thought of many mathematicians at the time (and even now) proposed more than that. That is, in addition to its role as a foundation for "*a posteriori* formalisation", they dreamed the "potential mechanisation" of mathematics was not only a locus for certainty but also a "complete" method for proving theorems. The Italian logical school firmly insisted on this with their "pasigraphy": a universal formal language that was a mechanisable algebra for all aspects of human reasoning. Now the "sausage machine" for mathematics (and thought), as Poincaré ironically called it (Bottazzini 2000), could be put to work: Provide pigs (or axioms) as input and produce theorems (or sausages) as output. (Traces of this mechanisation may still be found in applications of AI or in teaching.) The story of complete *a posteriori* formalisation and, *a fortiori*, of potential mechanisation of deduction ended badly. Hilbert's conjectures on the formally provable consistency, decidability and completeness of PA turned out to be all wrong, as Gödel (1931) proved. Gödel's proof gave rise to (incomplete but) fantastic formal machines by the rigorous definition of "computable function". More precisely, Gödel's negative result initiated a major expansion of logic: Recursion Theory (in order to prove undecidability, Gödel had to define precisely what decidable/computable means), Model Theory (the fact that not all models of PA are elementarily equivalent strongly motivates further investigations) and Proof Theory (Gentzen) all got a new start. (Negative results matter immensely in science, see Longo 2006.) The latter

---

[5] For more on the connections between "proof principles" and "construction principles" in mathematics and physics, see Bailly and Longo (2006).

led to the results, amongst others, of Girard and Friedman (see Longo 2002). For Number Theory, the main consequence is that formal induction is incomplete and that one cannot avoid infinitary machinery in proofs (e.g., in the rigorous sense of Friedman 1997).

4.2   Prototype Proofs

In some cases, the incompleteness of formal induction can be described in terms of the structure of "prototype proofs" or of "geometric judgements" with no explicit reference to infinity. As Herbrand proposed, "…when we say that a theorem is true for all x, we mean that for each x individually it is possible to iterate its proof, which may just be considered a *prototype* of each individual proof" (see Goldfarb and Dreben 1987). Little Gauss' theorem is an example of such a prototype proof. But any proof of a universally quantified statement over a structure that does not realise induction constitutes a "prototype".

For example, consider Pythagoras' theorem: one *needs* to draw, possibly on the sand of a Greek beach, a right triangle, with a specific ratio amongst the sides. Yet, at the end of the proof, one makes a fundamental remark, the true beginning of mathematics: Look at the proof; it does not depend on the specific drawing, but only on the existence of a right angle. The right triangle is *generic* (it is an invariant of the proof) and the proof is a *prototype*. There is no need to scan all right triangles. By a similar structure of the proof, one has to prove a property to hold for any element of (a sub-set of) real or complex numbers; that is, for elements of non-well ordered sets. However, in number theory, one has an extra and very strong proof principle: induction.

In a prototype proof, one must provide a reasoning which uniformly holds for all arguments; this uniformity allows (and is guaranteed by) the use of a generic argument. Induction provides an extra tool: The intended property doesn't need to hold for the same reasons for all arguments. Actually, it may hold for different reasons for each argument. One only has to give a proof for 0, and then provide a uniform proof from x to $x+1$. That is, uniformity of reasoning is required only in the inductive step. This is where the prototype proof steps in again: the argument from x to $x+1$. Yet, the situation may be more complicated: In the case of nested induction, the universally quantified formula of this inductive step may be given by induction on x. However, after a finite number of nestings, one has to get to a prototype proof going from x to $x+1$ (i.e., the rule of induction is logically well-founded).

Thus, induction provides a rigorous proof principle, which, over well-orderings, holds in addition to uniform (prototype) proofs, though sooner or later a prototype proof steps in. However, the prototype/uniform argument in an inductive proof allows one to derive, from the assumption of the thesis for x, its validity for $x+1$, in any possible model. On the other hand, by induction one may inherit properties from x to $x+1$ (e.g., totality of a function of x; see Longo 2002).

As we already observed, in an inductive proof, one must know in advance the formula (the statement) to be proved: little Gauss did not know it. Indeed, (straight) induction (i.e., induction with no problem in the choice of the inductive statement or load) is closer to proof-checking than to "mathematical theorem proving"; if one already has the formal proof, a computer can check it.

5 Induction *vs.* Well-Ordering in Concrete Incompleteness Theorems

Since the 1970s several examples of "concrete incompleteness results" have been proved.[6] That is, some interesting properties of number theory can be shown to be true, but their proofs cannot be given within number theory's formal counterpart, PA. A particularly relevant case is Friedman's Finite Form (FFF) of Kruskal's Theorem (KT), a well-known theorem on sequences of "finite trees" in infinite combinatorics (and with many applications).[7] The difficult part is the proof of unprovability of FFF in PA. Here, I am interested only in the proof that FFF holds over the structure of natural numbers (the standard model of PA). FFF is easily derived from KT, so the problem of its formal unprovability lies somewhere in the proof of KT. Without entering into the details even of the statements of FFF or KT (see Gallier 1991; Harrington and Simpson 1985; Longo 2002), I shall skip to where "meaning," or the geometric structure of integer numbers in space or time (the gestalt of well-ordering) steps into the proof.

The set-theoretic proof of KT (Gallier 1991; Harrington and Simpson 1985) goes by a strong non-effective argument. It is non-effective for several reasons. First, one argues "*ad absurdum*"; that is, one shows that a certain set of possibly infinite sequences of trees is empty by deriving an absurd if it were not so ("not empty implies a contradiction; thus it is empty"). More precisely, one first assumes that a certain set of "bad sequences" (sequences without ordered pairs of trees, as required in the statement of KT) is not empty and then defines a minimal bad sequence from this assumption. Then one shows that *that* minimal sequence cannot exist, as a smaller one can be easily defined from it. This minimal sequence is obtained by using a quantification on a set that is going to be proved to be empty, a rather non-effective procedure. Moreover, the to-be-empty set is defined by a $\sum_1^1$ predicate, well outside PA (a proper, impredicative second-order quantification over sets). For a non-intuitionist who accepts a definition *ad absurdum* of a mathematical object (a sequence in this case), as well as an impredicatively defined set, the proof poses

---

[6] Concerning "concrete" incompleteness: An analysis of the nonprovability of normalisation for non-predicative Type Theory, Girard's system F, in terms of prototype proofs is proposed in Longo (2002).

[7] For a close proof-theoretic investigation of KT, see Harrington and Simpson (1985), Gallier (1991). I borrow here a few remarks from Longo (2002), which proposes a further analysis.

no problem. It is abstract, but very convincing (and relatively easy). The key non-arithmetisable steps are in the $\sum_{1}^{1}$ definition of a set and in the definition of a new sequence by taking, iteratively, the least element of this set.

Yet, the mathematically minded readers (and the graduate students to whom I lecture) have no problem in applying their shared mental experience of the "number line" to accept this formally non-constructive proof: From the assumption that the intended set is non-empty, one understands ("sees") that it has a least element, without caring about its formal (infinitary, $\sum_{1}^{1}$) definition. If the set is assumed to contain an element, then the way the rest of the set "goes to infinity" doesn't really matter; the element supposed to exist (by the non-emptiness of the set) must be somewhere in the finite, and the least element will be amongst the finitely many preceding elements, even if there is no way to present it explicitly. This is well-ordering. Finally, the sequence defined *ad absurdum*, in this highly non-constructive way, will never be used; it would be absurd for it to exist. So its actual "construction" is irrelevant. Of course, this is far from PA, but it is convincing to anyone accepting the "geometric judgement" of well-ordering: "A *generic* non-empty subset of the number line has a least element". This vision of a property, a fundamental judgement, is grounded in the gestalt discussed above.

An intuitionistically acceptable proof of KT was later given by Rathjen and Weierman (1993). This proof of KT is still not formalisable in PA, of course, but it is "constructive", at least in the broad sense of "infinitary inductive definitions" as widely used in the contemporary intuitionist community. It is highly infinitary, because it uses induction beyond the first impredicative ordinal Γ_0. Though another remarkable contribution to the ordinal classification of theorems and theories, this proof is in no way "more evident" than the one using well-ordering given above. In no way does it "found" arithmetic more than that geometric judgement, as the issue of consistency is postponed to the next ordinal on which induction would allow one to derive the consistency of induction up to Γ_0.

6   The Origin of Logic

Just as for geometry or arithmetic, mathematicians have to pose the epistemological problem of logic itself. That is, we have to stop viewing formal properties and logical laws as meaningless games of signs or absolute laws preceding human activities. They are not a linguistic description of an independent reality; we have to move towards understanding them as a result of a *praxis* in analogy to our *praxes* in and of space and time, which create their geometric intelligibility by their own *construction*.

The logical rules or proof principles have constituted the invariants of our practice of discourse and reasoning since the days of the Greek Agora; they are organised also, but not only, by language. Besides the geometry of figures with their borders with no thickness, which forced symmetries and order in space (our bodily symmetries, our need for order), the Greeks extracted the regularities of discourse. In the novelty of democracy, political power in the Agora was achieved by arguing and convincing. Some patterns of that common discourse were then stabilised and later theoretised, by Aristotle in particular, as rules of reasoning (Toth 2002). These became established as invariants, transferable from one discourse to another (even in different areas: politics and philosophy, say). The Sophistic tradition dared to argue *per absurdum*, by insisting on contradictions, and, later, this tool for reasoning became, in Euclid, a method of proof. All these codified rules made existing arguments justifiable and provided a standard of acceptability for any new argument, while nevertheless being themselves the *a posteriori* result of a shared activity in history.

Much later, the same type of social evolution of argument produced the practice of actual infinity, a difficult achievement which had required centuries of religious disputes in Europe over metaphysics (see Zellini 2005). Actual infinity became rigorous mathematics (geometry) after developing first as perspective in Italian Renaissance painting. Masaccio first used the convergence point at the horizon in several (lost) Annunciations (see Vasari 1998); Piero della Francesca followed his master and theoretised this practice in a book on painting, the first text on projective geometry.[8] The advance of discourse helped to restore infinity, initially conceived as a metaphysical commitment, in space as a projective limit, a very effective tool to represent three-dimensional finite spaces in (two-dimensional) painting. Mathematicians later dared to manipulate the linguistic-algebraic representations of such inventions, abstracted from the world that originated them but simultaneously making that same world more intelligible. The conception of actual infinity enabled mathematics to *better organise the finite*. For example, infinity became an analytic tool which Newton and Leibniz used for understanding finite speed and acceleration through an asymptotic construction. In the nineteenth century, Cantor made the extremely audacious step (see Cantor 1955) and turned infinity into an algebraic and logically sound notion: He objectivised infinity in a sign and dared to compute on it. A new praxis, the arithmetic of infinity (both on ordinal and cardinal numbers) started a new branch of mathematics. Of course, this enrichment of discourse would have been difficult without the rigorous handling of quantification proposed in Frege's foundation of logic and arithmetic (Frege 1884).

That fruitful resonance between linguistic constructions and the intelligibility of space contributed to the geometrisation of physics. Klein's and Clifford's algebraic treatment of non-Euclidean geometries (see Boi 1995) was crucial for the birth of Relativity Theory.[9] Since his 1899 work, Hilbert's axiomatic approach was also fundamental in this, despite his erroneous belief in the completeness and (auto-)consistency

---

[8] Masaccio and Piero invented the modern perspective, in Annunciations first (1400–1450), by the explicit use of points of converging parallel lines. As a matter of fact, the Annunciation is the locus of the encounter of the Infinity of God with the Madonna, a (finite) woman (see Panovsky 1991). Later, "infinity in painting", by the work of Piero himself, became a general technique to describe finite spaces better.

[9] Klein and Clifford also stressed the role of symmetries in Euclidean Geometry: It is the only geometry which is closed under homotheties. That is, its group of automorphisms, and only its group, contains this form of symmetry.

of the formal approach. In addition, physicists, like Boltzmann, conceived limit constructions, such as the thermodynamic integral, which asymptotically unified Newton's trajectories of gas particles and thermodynamics (Cercignani 1998). Statistical physics, or re-normalisation methods, play an important role in today's physics of criticality, where infinity is crucial (Binney et al. 1992). Logicians continued to propose purely linguistic infinitary proofs of finitary statements.

The development of infinity is but one part of the never-ending dialogue between geometric construction principles and logical proof principles. It started with projective geometry, as a mathematisation of the Italian invention of perspective in painting, first a *praxis*, a technique, in art. It is a subset of the ongoing historical interaction between invariants of action in space and time and their linguistic expressions, extended also by metaphysical discussions (on infinity), originating in human inter-subjectivity, including the invariants of historical, dialogical reasoning (logic). These interactions produce the constitutive history and the evolving, cognitive and historical foundations of mathematics.

7 Conclusion

In my approach, I ground mathematics and its proofs, as conceptual constructions, in humans' "phenomenal lives" (Weyl 1949): Concepts and structures are the result of a cognitive/historical knowledge process. They originate from our actions in space (and time) and are further extended by language and logic. Mathematics, for example, moved from Euclid's implicit use of connectivity to homotopy theory or to the topological analysis of dimensions. Symmetries lead from plane geometry to dualities and adjunctions in categories, some very abstract concepts. Likewise, the ordering of numbers is formally extended into transfinite ordinals and cardinals.

In this short essay, I have tried to spell-out the role of prototype proofs and of well-ordering vs. induction. I insisted on the role of symmetries both in our understanding of Euclid's axioms and in proofs; I stressed the creativity of the proof, which often requires the invention of new concepts and structures. These may be, in most cases, formalised, but *a posteriori* and each in some *ad hoc* way. However, there is no Newtonian absolute Universe; nor a Zermelo-Fraenkel unique, absolute and complete set theory; nor any ultimate foundations: This is a consequence of incompleteness (see Longo 2011). More deeply, evidence and foundation are not completely captured by formalisation, beginning with the axioms: "The primary evidence should not be interchanged with the evidence of the 'axioms'; as the axioms are mostly the result already of an original formation of meaning and they already have this formation itself always behind them," (Husserl 1933). This is the perspective applied in my initial sketchy analysis of the symmetries "lying behind" Euclid's axioms.

Moreover, recent concrete incompleteness results show that the reference to this underlying and constitutive meaning cannot be avoided in proofs or in foundational analyses. The consistency issue is crucial in any formal derivation and cannot be solved within formalisms.

After the early references to geometry, I focused on arithmetic as foundational analyses have mostly done since Frege. Arithmetic has produced fantastic logico-arithmetical machines – and major incompleteness results. I have shown how geometric judgements penetrate proofs even in number theory; I argue, *a fortiori*, their relevance for general mathematical proofs. We need to ground mathematical proofs also on geometric judgements which are no less solid than logical ones: "Symmetry", for example, is at least as fundamental as the logical "*modus ponens*"; it features heavily in mathematical constructions and proofs. Physicists have long argued "by symmetry". More generally, modern physics extended its analysis from the Newtonian "causal laws" – the analogue to the logico-formal and absolute "laws of thought" since Boole (1854) and Frege (1884) – to understanding the phenomenal world through an active geometric structuring. Take as examples the conservation laws as symmetries (Noether's theorem) and the geodetics of Relativity Theory.[10] The normative nature of geometric structures is currently providing a further understanding even of recent advances in microphysics (Connes 1994). Similarly, mathematicians' foundational analyses and their applications should also be enriched by this broadening of the paradigm in scientific explanation: from *laws* to *geometric intelligibility*. (I discussed symmetries, in particular, but also the geometric judgement of "well-ordering".) Mathematics is the result of an open-ended "game" between humans and the world in space and time; that is, it results from the inter-subjective construction of knowledge made in language and logic, along a passage through the world, which canalises our praxes as well as our endeavour towards knowledge. It is effective and objective exactly because it is constituted by human action in the world, while by its own actions transforming that same world.

Acknowledgements Many discussions with Imre Toth, and his comments, helped to set my (apparently new) understanding of Euclid on more sound historical underpinnings. Rossella Fabbrichesi helped me to understand Greek thought and the philosophical sense of my perspective. My daughter Sara taught me about "infinity in the painting" in the Italian Quattrocento. The editors proposed a very close revision for English and style.

References

Bailly, F. & Longo, G. (2006). *Mathématiques et sciences de la nature. La singularité physique du vivant*. Paris: Hermann (Translation in English : Imperial College Press/ World Sci., London, 2010).

Bailly, F., & Longo, G. (2008). Phenomenology of incompleteness: From formal deductions to mathematics and physics. In Lupacchini (Ed.), *Deduction, computation, experiment*. Berlin: Springer.

Binney, J., Dowrick, N., Fisher, A., & Newman, M. (1992). *The theory of critical phenomena: An introduction to the renormalization group*. Oxford: Oxford University Press.

---

[10] See Weyl (1949) for an early mathematical and philosophical insight into this. For recent reflections, see van Fraassen (1993); and Bailly and Longo (2006).

Boi, L. (1995). *Le problème mathématique de l'espace*. Berlin: Springer.

Boole, G. (1854). *An investigation of the laws of thought*. London: Macmillan.

Bottazzini, U. (2000). Poincaré, *Pour la Science*, N. Spécial, 4.

Bottazzini, U., & Tazzioli, R. (1995). Naturphilosophie and its role in Riemann's mathematics. *Revue d'Histoire des Mathématiques, 1*, 3–38.

Brouwer, L. (1948). Consciousness, philosophy and mathematics. In Heyting (Ed.), *Collected works* (Vol. 1). Amsterdam: North-Holland.

Cantor, G. (1955). *Contributions to the founding of the theory of transfinite numbers. Collected works*. New York: Dove.

Cercignani, C. (1998). *Ludwig Boltzmann: The Man Who trusted atoms*. Oxford: Oxford University Press.

Connes, A. (1994). *Non-commutative geometry*. New York: Academic.

Dedekind, R. (1996). What are numbers and what should they be? In E. William (Ed.), *From kant to hilbert: A source book in the foundations of mathematics* (pp. 787–832). Oxford: Oxford University Press.

Dehaene, S. (1997). *The number sense*. Oxford: Oxford University Press. Longo's review downloadable.

Einstein, A. (1915). *Die feldgleichungen der gravitation (The field equations of gravitation)* (pp. 844–847). Berlin: Königlich Preussische Akademie der Wissenshaften.

Fraisopi, F. (2009). *Besinnung*. Roma: Aracne.

Frege, G. (1884, [1980]). *The foundations of arithmetic* (Engl. transl. Evanston, London).

Friedman, H. (1997). Some historical perspectives on certain incompleteness phenomena. Retrieved May 21, 1997, 5 pp. Draft: http://www.math.ohio-state.edu/~friedman/manuscripts.html

Gallier, J. (1991). What is so special about Kruskal's theorem and the ordinal Γ0? *Annals of Pure and Applied Logic, 53*, 132–187.

Gauss, C. (1986). *Disquisitiones Arithmeticae (1801)* (A. A. Clarke, Trans.). New York: Springer.

Girard, J. (2001). Locus Solum. *Mathematical Structures in Computer Science, 11*(3), 323–542.

Girard, J., Lafont, Y., & Taylor, P. (1989). *Proofs and types*. Cambridge: Cambridge University Press.

Gödel, K. (1992). *On formally undecidable propositions of principia mathematica and related systems* (1931) (B. Meltzer, Trans.). New York: Dover.

Goldfarb, H., & Dreben, B. (Eds.). (1987). *Jacques Herbrand: Logical writings*. Boston: Harvard University Press.

Harrington, L., & Simpson, S. (Eds.). (1985). *H. Friedman's research on the foundations of mathematics*. Amsterdam: North-Holland.

Heath, T. (1908). *The thirteen books of Euclid's elements (Heath's translation and comments)*. Cambridge: Cambridge University Press.

Hilbert, D. (1899). *Grundlagen der Geometrie* Leipzig: Teubner (Translated by L. Unger as Foundations of Geometry, Open Court, La Salle, 1971).

Husserl, E. (1933). *The Origin of Geometry* (Appendix III of *Krysis*) (trad. fran. by J. Derida, Paris: PUF, 1962).

Kennedy, H. C. (2006). *Life and works of Giuseppe Peano*. Dordrecht, Holland: D. Reidel Publishing Company.

Longo, G. (2002). Reflections on incompleteness, or on the proofs of some formally unprovable propositions and prototype proofs in type theory. Invited lecture. In: P. Callaghan (Ed.), *Types for Proofs and Programs, Durham, (GB), December 2000, Lecture Notes in Computer Science 2277*. Berlin: Springer.

Longo, G. (2005). The cognitive foundations of mathematics: Human gestures in proofs and mathematical incompleteness of formalisms. In M. Okada & P. Grialou (Eds.), *Images and reasoning* (pp. 105–134). Tokyo: Keio University Press.

Longo, G. (2006). Sur l'importance des résultats négatifs. *Intellectica, 40*(1), 28–43. see http://www.di.ens.fr/users/longo/ for a translation in English.

Longo, G. (2009). Critique of computational reason in the natural sciences. In E. Gelenbe & J.-P. Kahane (Eds.), *Fundamental concepts in computer science* (pp. 43–70). London: Imperial College Press/World Scientific.

Longo, G. (2011). Interfaces of incompleteness. Italian version in *La Matematica* (Vol. 4), Einuadi.

Longo, G. & Viarouge, A. (2010). Mathematical intuition and the cognitive roots of mathematical concepts. In L. Horsten & I. Starikova (Eds.), Invited paper, *Topoi,* Special issue on *Mathematical Knowledge: Intuition, Visualization, and Understanding, 29*(1), 15–27.

Padua, A. (1967). Logical introduction to any deductive theory. In J. van Heijenoort (Ed.), *A source book in mathematical logic* (pp. 118–123). Boston: Harvard University Press.

Panovsky, E. (1991). *Perspective as symbolic form.* New York: Zone Books.

Peano, G. (1967). The principles of arithmetic presented by a new method. In J. van Heijenoort (Ed.), *A source book in mathematical logic* (pp. 83–97). Boston: Harvard University Press.

Petitot, J. (2003). The neurogeometry of pinwheels as a sub-Riemannian contact structure. *The Journal of Physiology, 97*(2–3), 265–309.

Poincaré, H. (1908). *Science et Méthode.* Paris: Flammarion.

Rathjen, M., & Weiermann, A. (1993). Proof-theoretic investigations on Kruskal's theorem. *Annals of Pure and Applied Logic, 60,* 49–88.

Riemann, B. (1854, [1873]). *On the hypothesis which lie at the basis of geometry* (English trans. by W. Clifford). *Nature 8*(1873), 14–17, 36–37.

Tappenden, J. (1995). Geometry and generality in Frege's philosophy of arithmetic. *Synthese, 102*(3), 25–64.

Toth, I. (2002). *Aristotele ed i fondamenti assiomatici della geometria.* Milano: Vita e Pensiero.

Turing A. M. (1950) Computing machines and intelligence, *Mind,* LIX. Reprinted in M. Boden (Ed.), Oxford University Press, 1990.

Turing, A. M. (1952). The chemical basis of morphogenesis. *Philosophical Transactions of the Royal Society, B237,* 37–72.

van Fraassen, B. (1993). *Laws and symmetry.* Oxford: Oxford University Press.

Vasari, A. (1998). *The lives of the artists.* Oxford: Oxford University Press.

Weyl, H. (1949). *Philosophy of mathematics and of natural sciences.* Princeton: Princeton University Press.

Weyl, H. (1952). *Symmetry.* Princeton: Princeton University Press.

Zellini, P. (2005). *A brief history of infinity.* New York: Penguin.

Part II
Experimentation: Challenges and Opportunities

Chapter 4
Exploratory Experimentation: Digitally-Assisted Discovery and Proof

Jonathan Michael Borwein

1 Digitally-Assisted Discovery and Proof

> [I]ntuition comes to us much earlier and with much less outside influence than formal arguments which we cannot really understand unless we have reached a relatively high level of logical experience and sophistication.
>
> Therefore, I think that in teaching high school age youngsters we should emphasise intuitive insight more than, and long before, deductive reasoning.—George Pólya (1887–1985) (1981, (2) p. 128)

1.1 Exploratory Experimentation

I share Pólya's view that intuition precedes deductive reasoning. Nonetheless, Pólya also goes on to say, proof should certainly be taught in school. I begin with some observations many of which have been fleshed out in *The Computer as Crucible* (Borwein and Devlin 2009), *Mathematics by Experiment* (Borwein and Bailey 2008), and *Experimental Mathematics in Action* (Bailey et al. 2007). My musings here focus on the changing nature of mathematical knowledge and in consequence ask the questions such as "How do we come to believe and trust pieces of mathematics?", "Why do we wish to prove things?" and "How do we teach what and why to students?"

J.M. Borwein (✉)
Centre for Computer-Assisted Research Mathematics and its Applications, CARMA,
University of Newcastle, Callaghan, NSW 2308, Australia
e-mail: jonathan.borwein@newcastle.edu.au

© The Author(s) 2021
G. Hanna and M. de Villiers (eds.), *Proof and Proving in Mathematics Education*,
New ICMI Study Series, https://doi.org/10.1007/978-94-007-2129-6_4

While I have described myself in Bailey et al. (2007) and elsewhere as a "computationally assisted fallibilist", I am far from a social-constructivist. Like Richard Brown, I believe that Science "at least attempts to faithfully represent reality" (2008, p. 7). I am, though, persuaded by various notions of embodied cognition. As Smail writes:

[T]he large human brain evolved over the past 1.7 million years to allow individuals to negotiate the growing complexities posed by human social living. (Smail 2008, p. 113)

In consequence, humans find various modes of argument more palatable than others, and are prone to make certain kinds of errors more than others. Likewise, Steve Pinker's observation about language as founded on

...the ethereal notions of space, time, causation, possession, and goals that appear to make up a language of thought (Pinker 2007, p. 83)

remains equally potent within mathematics. The computer offers to provide scaffolding both to enhance mathematical reasoning and to restrain mathematical error.

To begin with let me briefly reprise what I mean by discovery and by proof in mathematics. The following attractive definition of *discovery* has the satisfactory consequence that a student can certainly discover results whether those results are known to the teacher or not.

In short, discovering a truth is coming to believe it in an independent, reliable, and rational way. (Giaquinto 2007, p. 50)

Nor is it necessary to demand that each dissertation be original (only independently discovered).

A standard definition[1] of *proof* follows.

PROOF, n. a sequence of statements, each of which is either validly derived from those preceding it or is an axiom or assumption, and the final member of which, the conclusion, is the statement of which the truth is thereby established.

As a working definition of mathematics itself, I offer the following, in which the word "proof" does not enter. Nor should it; mathematics is much more than proof alone:

MATHEMATICS, n. a group of subjects, including algebra, geometry, trigonometry and calculus, concerned with number, quantity, shape, and space, and their inter-relationships, applications, generalisations and abstractions.
DEDUCTION, n. 1. the process of reasoning typical of mathematics and logic, in which a conclusion follows necessarily from given premises so that it cannot be false when the premises are true.
INDUCTION, n. 3. (Logic) a process of reasoning in which a general conclusion is drawn from a set of particular premises, often drawn from experience or from experimental evidence. The conclusion goes beyond the information contained in the premises and does not follow necessarily from them. Thus an inductive argument may be highly probable yet lead

[1] All definitions below are taken from the *Collin's Dictionary of Mathematics* which I co-authored. It is available as software—with a version of Student *Maple* embedded in it—at http://www.math-resources.com/products/mathresource/index.html.

to a false conclusion; for example, large numbers of sightings at widely varying times and places provide very strong grounds for the falsehood that all swans are white.

It awaited the discovery of Australia to confound the seemingly compelling inductive conclusion that all swans are white. Typically, mathematicians take for granted the distinction between *induction* and *deduction* and rarely discuss their roles with either colleagues or students. Despite the conventional identification of Mathematics with deductive reasoning, in his 1951 Gibbs Lecture Kurt Gödel (1906–1978) said:

> If mathematics describes an objective world just like physics, there is no reason why induc-
> tive methods should not be applied in mathematics just the same as in physics.

He held this view until the end of his life, despite the epochal deductive achievement of his incompleteness results. And this opinion has been echoed or amplified by logicians as different as Willard Quine and Greg Chaitin. More generally, one discovers a substantial number of great mathematicians from Archimedes and Galileo—who apparently said "All truths are easy to understand once they are discovered; the point is to discover them."—to Poincaré and Carleson who have emphasised how much it helps to "know" the answer. Over two millennia ago Archimedes wrote to Eratosthenes in the introduction to his long-lost and recently re-constituted *Method of Mechanical Theorems*:

> I thought it might be appropriate to write down and set forth for you in this same book a
> certain special method, by means of which you will be enabled to recognise certain math-
> ematical questions with the aid of mechanics. I am convinced that this is no less useful for
> finding proofs of these same theorems.
>
> For some things, which first became clear to me by the mechanical method, were after-
> wards proved geometrically, because their investigation by the said method does not furnish
> an actual demonstration. *For it is easier to supply the proof when we have previously
> acquired, by the method, some knowledge of the questions than it is to find it without any
> previous knowledge.* [My emphasis] (Livio 2009)

Think of the *Method* as an ur-precursor to today's interactive geometry software—with the caveat that, for example, *Cinderella* actually does provide certificates for much Euclidean geometry. As 2006 Abel Prize winner Leonard Carleson described in his 1966 ICM speech on his positive resolution of Luzin's 1913 conjecture, about the pointwise convergence of Fourier series for square-summable functions, after many years of seeking a counter-example he decided none could exist. The importance of this confidence he expressed as follows:

> The most important aspect in solving a mathematical problem is the conviction of what is
> the true result. Then it took 2 or 3 years using the techniques that had been developed during
> the past 20 years or so.

1.2 Digitally Mediated Mathematics

I shall now assume that all proofs discussed are "non-trivial" in some fashion appropriate to the level of the material since the issue of using inductive methods is really only of interest with this caveat. Armed with these terms, it remains to say that by *digital assistance* I intend the use of such *artefacts* as

- *Modern Mathematical Computer Packages*—be they Symbolic, Numeric, Geometric, or Graphical. I would capture all as "modern hybrid workspaces". One should also envisage much more use of stereo visualisation, *haptics*,[2] and auditory devices.
- *More Specialist Packages* or *General Purpose Languages*, such as Fortran, C++, CPLEX, GAP, PARI, SnapPea, Graffiti, and MAGMA. The story of the *SIAM 100-Digits Challenge* (Borwein 2005) illustrates the degree to which mathematicians now start computational work within a hybrid platform such as *Maple*, *Mathematica* or MATLAB and make only sparing recourse to more specialist packages when the hybrid work spaces prove too limited.
- *Web Applications*, such as Sloane's Online Encyclopedia of Integer Sequences, the Inverse Symbolic Calculator, Fractal Explorer, Jeff Weeks' Topological Games, or Euclid in Java.[3]
- *Web Databases*, including Google, MathSciNet, ArXiv, JSTOR, Wikipedia, MathWorld, Planet Math, Digital Library of Mathematical Functions (DLMF), MacTutor, Amazon, and many more sources that are not always viewed as part of the palette. Nor is necessary that one approve unreservedly, say of the historical reliability of MacTutor, to acknowledge that with appropriate discrimination in its use it is a very fine resource.

All the above entail *data-mining* in various forms Franklin (2005) argues that what Steinle has termed "exploratory experimentation" facilitated by "widening technology" as in pharmacology, astrophysics, and biotechnology, is leading to a reassessment of what is viewed as a legitimate experiment, in that a "local model" is not a prerequisite for a legitimate experiment. Henrik Sørensen (2010) cogently makes the case that *experimental mathematics*—as "defined" below—is following similar tracks:

> These aspects of exploratory experimentation and wide instrumentation originate from the philosophy of (natural) science and have not been much developed in the context of experimental mathematics. However, I claim that e.g. the importance of wide instrumentation for an exploratory approach to experiments that includes concept formation also pertain to mathematics.

Danny Hillis is quoted as saying recently that:

> Knowing things is very 20th century. You just need to be able to find things.[4]

about how *Google* has already changed how we think. This is clearly not yet true and will never be, yet it catches something of the changing nature of cognitive style

---

[2] With the growing realisation of the importance of gesture in mathematics "as the very texture of thinking," (Sfard 2009, p. 92) it is time to seriously explore tactile devices.

[3] A cross-section of such resources is available through http://ddrive.cs.dal.ca/~isc/portal/.

[4] In Achenblog http://blog.washingtonpost.com/achenblog/ of July 1 2008.

in the twenty-first century. Likewise, in a provocative article (2008), Chris Anderson, the Editor-in-Chief of *Wired*, recently wrote

> There's no reason to cling to our old ways. It's time to ask: What can science learn from Google?

In consequence, the boundaries between mathematics and the natural sciences and between inductive and deductive reasoning are blurred and getting blurrier. This is discussed at some length by Jeremy Avigad (2008). A very useful discussion of similar issues from a more explicitly pedagogical perspective is given by de Villiers (2004) who also provides a quite extensive bibliography.

1.3 Experimental Mathodology

We started *The Computer as Crucible* (Borwein and Devlin 2009) with then United States Supreme court Justice Potter Stewart's famous, if somewhat dangerous, 1964 Supreme Court judgement on pornography:

> I know it when I see it. (Borwein and Devlin 2009, p. 1)

I complete this subsection by reprising from Borwein and Bailey (2008) what somewhat less informally we mean by *experimental mathematics*. I say 'somewhat', since I do not take up the perhaps vexing philosophical question of whether a true *experiment* in mathematics is even possible—without adopting a fully realist philosophy of mathematics—or if we should rather refer to 'quasi-experiments'? Some of this is discussed in Bailey et al. (2007, Chap. 1) and Borwein and Bailey (2008, Chaps. 1,2, and 8), wherein we further limn the various ways in which the term 'experiment' is used and underline the need for mathematical experiments with predictive power.

1.3.1 What Is Experimental Mathematics?

1. Gaining insight and *intuition*.
   Despite my agreement with Pólya, I firmly believe that—in most important senses—intuition, far from being "knowledge or belief obtained neither by reason nor by perception," as the Collin's English Dictionary and Kant would have it, is acquired not innate. This is well captured by Lewis Wolpert's 2000 title *The Unnatural Nature of Science*, see also Gregory and Miller (1998).
2. *Discovering* new relationships
   I use "discover" in Giaquinto's terms as quoted above.
3. *Visualising* math principles.
   I intend the fourth Random House sense of "to make perceptible to the mind or imagination" not just Giaquinto's more direct meaning.

4. *Testing* and especially *falsifying* conjectures.

Karl Popper's "critical rationalism" asserts that induction can never lead to truth and hence that one can only falsify theories (Brown 2008). Whether one believes this is the slippery slope to *Post modernist interpretations of science* (Brown's term abbreviated *PIS*) or not is open to debate, but Mathematics, being based largely on deductive science, has little to fear and much to gain from more aggressive use of falsification.

5. *Exploring* a possible result to see if it *merits* formal proof.

"Merit" is context dependent. It may mean one thing in a classroom and quite another for a research mathematician.

6. *Suggesting* approaches for *formal proof*.

I refer to computer-assisted or computer-directed proof which is quite far from completely *Formal Proof*—the topic of a special issue of the *Notices of the AMS* in December 2008.

7. *Computing* replacing lengthy hand derivations.

Hales' recent solution of the *Kepler problem*, described in the 2008 *Notices* article, pushes the boundary on when "replacement" becomes qualitatively different from, say, factoring a very large prime. In the case of factorisation, we may well feel we understand the entire sequence of steps undertaken by the computer.

8. *Confirming* analytically derived results.

The *a posteriori* value of confirmation is huge, whether this be in checking answers while preparing a calculus class, or in confirming one's apprehension of a newly acquired fact.

Of these, the first five play a central role in the current context, and the sixth plays a significant one.

1.4 Cognitive Challenges

Let me touch upon the *Stroop effect*[5] illustrating *directed attention* or *interference*. This classic cognitive psychology test, discovered by John Ridley Stroop in 1935, is as follows.

Consider the picture in Fig. 4.1, in which various coloured words are coloured in one of the colours mentioned, but not necessarily in the same one to which the word refers. First, say the **colour** which the given **word mentions**.

Second, say the **colour** in which the **word is written**.

Most people find the second task harder. You may find yourself taking more time for each word, and may frequently say the word, rather than the colour in which the word appears. Proficient (young) multitaskers find it easy to suppress information and so perform the second task faster than traditionally. Indeed, Cliff Nass' work in the CHIME lab at Stanford suggests that neurological changes are taking place amongst

---

[5]http://www.snre.umich.edu/eplab/demos/st0/stroopdesc.html has a fine overview.

Fig. 4.1 An illustration of the Stroop test

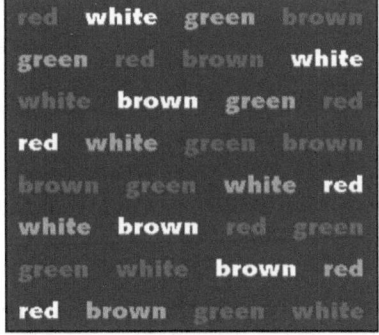

the 'born-digital.'[6] If such cognitive changes are taking place there is even more reason to ensure that epistemology, pedagogy, and cognitive science are in concert.

1.5   Paradigm Shifts

Old ideas give way slowly; for they are more than abstract logical forms and categories. They are habits, predispositions, deeply engrained attitudes of aversion and preference. Moreover, the conviction persists-though history shows it to be a hallucination that all the questions that the human mind has asked are questions that can be answered in terms of the alternatives that the questions themselves present. But in fact intellectual progress usually occurs through sheer abandonment of questions together with both of the alternatives they assume an abandonment that results from their decreasing vitality and a change of urgent interest. We do not solve them: we get over them.

Old questions are solved by disappearing, evaporating, while new questions corresponding to the changed attitude of endeavor and preference take their place. Doubtless the greatest dissolvent in contemporary thought of old questions, the greatest precipitant of new methods, new intentions, new problems, is the one effected by the scientific revolution that found its climax in the "Origin of Species".—John Dewey (1859–1952)[7]

Thomas Kuhn (1922–1996) has noted that a true *paradigm shift*—as opposed to the cliché—is "a conversion experience."[8] You (and enough others) either have one or you don't. Oliver Heaviside (1850–1925) said in defending his operator calculus before it could be properly justified:"Why should I refuse a good dinner simply because I don't understand the digestive processes involved?"

---

[6] See http://www.snre.umich.edu/eplab/demos/st0/stroop_program/stroopgraphicnonshockwave.gif.

[7] In Dewey's introduction to his book (1910). Dewey, a leading pragmatist (or instrumentalist) philosopher and educational thinker of his period, is also largely responsible for the Trotsky archives being at Harvard, through his activities on the *Dewey Commission*.

[8] This was said in an interview in Regis (1986), not only in Kuhn's 1962 *The Structure of Scientific Revolutions*, which Brown notes is "the single most influential work in the history of science in the twentieth century." In Brown's accounting (2008) Kuhn bears more responsibility for the slide into PIS than either Dewey or Popper. An unpremeditated example of digitally assisted research is that—as I type—I am listening to *The Structure of Scientific Revolutions*, having last read it 35 years ago.

But please always remember as Arturo Rosenblueth and Norbert Wiener wrote: "The price of metaphor is eternal vigilance."[9] I may not convince you to reevaluate your view of Mathematics as an entirely deductive science—if so indeed you view it—but in the next section I will give it my best shot.

2   Mathematical Examples

I continue with various explicit examples. I leave it to the reader to decide how much or how frequently he or she wishes to exploit the processes I advertise. Nonetheless they all controvert Picasso's "Computers are useless they can only give answers."[10] and confirm Hamming's "The purpose of computing is insight not numbers."[11] As a warm-up illustration, consider Fig. 4.2. The lower function in both graphs is $x \mapsto x - x^2 \log x$. The left-hand graph compares $x \mapsto x - x^2$ while the right-hand graph compares $x \mapsto x^2 - x^4$ each on $0 \ 0 \leq x \leq 1.$.

Before the advent of plotting calculators if asked a question like *"Is $- x^2 \log x$ less than $x - x^2$ on the open interval between zero and one?"* one immediately had recourse to the calculus. Now that would be silly, clearly they cross. In the other case, if there is a problem it is at the right-hand end point. 'Zooming' will probably persuade you that $-x^2 \log x \leq x^2 - x^4$ on $0 \leq x \leq 1$ and may even guide a calculus proof if a proof is needed.

The examples below contain material on sequences, generating functions, special functions, continued fractions, partial fractions, definite and indefinite integrals, finite and infinite sums, combinatorics and algebra, matrix theory, dynamic geometry and recursions, differential equations, and mathematical physics, among other things. So they capture the three main divisions of pure mathematical thinking: algebraic-symbolic, analytic, and topologic-geometric, while making contact with more applied issues in computation, numerical analysis and the like.

Example I: What Did the Computer Do?

> This computer, although assigned to me, was being used on board the International Space Station. I was informed that it was tossed overboard to be burned up in the atmosphere when it failed.—anonymous NASA employee[12]

In my own work, computer experimentation and digitally-mediated research now invariably play a crucial part. Even in many seemingly non-computational areas of

---

[9]Quoted by R. C. Leowontin, in *Science* p. 1264, Feb 16, 2001 (the *Human Genome Issue*).

[10]Michael Moncur's (Cynical) Quotations #255 http://www.quotationspage.com/collections.html

[11]Richard Hamming's philosophy of scientific computing appears as preface to his influential 1962 book (1962).

[12]*Science*, August 3, 2007, p. 579: "documenting equipment losses of more than $94 million over the past 10 years by the agency."

Fig. 4.2 Try visualisation or calculus first?

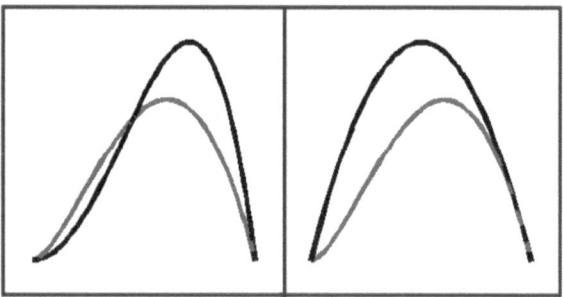

functional analysis and the like, there is frequently a computable consequence whose verification provides confidence in the result under development. Moreover, the process of specifying my questions enough to program with them invariably enhances my understanding and sometimes renders the actual computer nearly superfluous. For example, in a recent study of expectation or "box integrals" (Bailey et al. 2009) we were able to evaluate a quantity which had defeated us for years, namely

$$K_1 := \int_3^4 \frac{\arcsec(x)}{\sqrt{x^2 - 4x + 3}} dx$$

in *closed-form* as

$$K_1 = \mathrm{Cl}_2(\theta) - \mathrm{Cl}_2\left(\theta + \frac{\pi}{3}\right) - \mathrm{Cl}_2\left(\theta - \frac{\pi}{2}\right) + \mathrm{Cl}_2\left(\theta - \frac{\pi}{6}\right)$$
$$- \mathrm{Cl}_2\left(3\theta + \frac{\pi}{3}\right) + \mathrm{Cl}_2\left(3\theta + \frac{2\pi}{3}\right) - \mathrm{Cl}_2\left(3\theta - \frac{5\pi}{6}\right) + \mathrm{Cl}_2\left(3\theta + \frac{5\pi}{6}\right)$$
$$+ \left(6\theta - \frac{5\pi}{2}\right)\log\left(2 - \sqrt{3}\right). \tag{1}$$

where $\mathrm{Cl}_2(\theta) := \sum_{n=1}^{\infty} \sin(n\theta)/n^2$ is the *Clausen function*, and

$$3\theta := \arctan\left(\frac{16 - 3\sqrt{15}}{11}\right) + \pi.$$

Along the way to the evaluation above, after exploiting some insightful work by George Lamb, there were several stages of symbolic computation, at times involving an expression for K_1 with over 28,000 characters (perhaps 25 standard book pages). It may well be that the closed form in (1) can be further simplified. In any event, the very satisfying process of distilling the computer's 28,000 character discovery, required a mixture of art and technology and I would be hard pressed to assert categorically whether it constituted a conventional proof. Nonetheless, it is correct and has been checked numerically to over a thousand-digit decimal precision. ◁

I turn next to a mathematical example which I hope will reinforce my assertion that there is already an enormous amount to be mined mathematically on the internet. And this is before any mathematical character recognition tools have been made generally available and when it is still very hard to search mathematics on the web.

Example II: What Is That Number?

The dictum that everything that people do is 'cultural' … licenses the idea that every cultural critic can meaningfully analyze even the most intricate accomplishments of art and science. … It is distinctly weird to listen to pronouncements on the nature of mathematics from the lips of someone who cannot tell you what a complex number is!—Norman Levitt[13]

In 1995 or so Andrew Granville emailed me the number

$$\alpha := 1.4331274267223\ldots \tag{2}$$

and challenged me to identify it; I think this was a test I could have failed. I asked *Maple* for its continued fraction. In the conventional concise notation I was rewarded with

$$\alpha = [1, 2, 3, 4, 5, 6, 7, 8, 9, 10, 11, \ldots]. \tag{3}$$

Even if you are unfamiliar with continued fractions, you will agree that the changed representation in (3) has exposed structure not apparent from (2)! I reached for a good book on continued fractions and found the answer

$$\alpha = \frac{I_1(2)}{I_0(2)} \tag{4}$$

where I_0 and I_1 are *Bessel functions* of the first kind. Actually I remembered that all arithmetic continued fractions arise in such fashion, but as we shall see one now does not need to.

In 2009 there are at least three "zero-knowledge" strategies:

1. Given (3), type "arithmetic progression", "continued fraction" into *Google*.
2. Type "$1, 4, 3, 3, 1, 2, 7, 4, 2$" into *Sloane's Encyclopedia of Integer Sequences*.[14]
3. Type the decimal digits of α into the *Inverse Symbolic Calculator*.[15]

---

[13] In *The flight From Science and Reason*. See *Science*, Oct. 11, 1996, p. 183.

[14] See http://www.research.att.com/~njas/sequences/.

[15] The online *Inverse Symbolic Calculator* http://isc.carma.newcastle.edu.au/ was newly web-accessible in the same year, 1995.

Continued Fraction Constant -- from Wolfram MathWorld

- 3 visits - 14/09/07 Perron (1954-57) discusses _continued fractions_ having terms even more general than the _arithmetic progression_ and relates them to various special functions. ...
  mathworld.wolfram.com/**ContinuedFraction**Constant.html - 31k

HAKMEM -- _CONTINUED FRACTIONS_ -- DRAFT, NOT YET PROOFED

The value of a _continued fraction_ with partial quotients increasing in _arithmetic progression_ is I (2/D) A/D [A+D, A+2D, A+3D, ...
www.inwap.com/pdp10/hbaker/hakmem/cf.html - 25k -

On simple _continued fractions_ with partial quotients in _arithmetic_ ...

0. This means that the sequence of partial quotients of the _continued fractions_ under. investigation consists of finitely many _arithmetic progressions_ (with ...
www.springerlink.com/index/C0VXH713662G1815.pdf - by P Bundschuh – 1998

Moreover the MathWorld entry includes

$$[A + D, A + 2D, A + 3D, ...] = \frac{I_{A/D}\left(\frac{2}{D}\right)}{I_{1+A/D}\left(\frac{2}{D}\right)}$$

(Schroeppel 1972) for real A and $D \neq 0$.

Fig. 4.3 What Google and MathWorld offer

I illustrate the results of each strategy.

1. On October 15, 2008, on typing "arithmetic progression", "continued fraction" into _Google_, the first three hits were those shown in Fig. 4.3. Moreover, the MathWorld entry tells us that any arithmetic continued fraction is of a ratio of Bessel functions, as shown in the inset to Fig. 4.3, which also refers to the second hit in Fig. 4.3. The reader may wish to see what other natural search terms uncover (4)—perhaps in the newly unveiled _Wolfram Alpha_.

2. Typing the first few digits into Sloane's interface results in the response shown in Fig. 4.4. In this case we are even told what the series representations of the requisite Bessel functions are, we are given sample code (in this case in _Mathematica_), and we are lead to many links and references. Moreover, the site is carefully moderated and continues to grow. Note also that this strategy only became viable after May 14th 2001 when the sequence was added to the database which now contains in excess of 158,000 entries.

3. If one types the decimal representation of α into the Inverse Symbolic Calculator (ISC) it returns

```
Best guess: BesI(0,2)/BesI(1,2)
```

Most of the functionality of the ISC is built into the "identify" function in versions of _Maple_ starting with version 9.5. For example, "<identify(4.45033263602792)>" returns $\sqrt{3} + e$. As always, the experienced user will be able to extract more from this tool than the novice for whom the ISC will often produce more. ◁

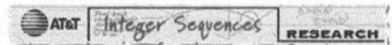

Greetings from The On-Line Encyclopedia of Integer Sequences!

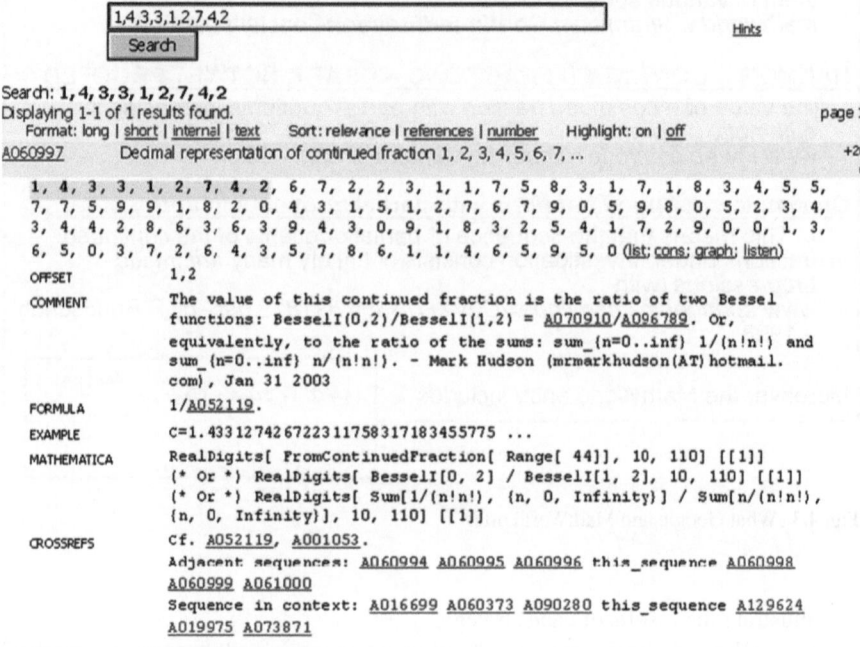

Search: **1, 4, 3, 3, 1, 2, 7, 4, 2**
Displaying 1-1 of 1 results found.   page 1
    Format: long | short | internal | text   Sort: relevance | references | number   Highlight: on | off
A060997   Decimal representation of continued fraction 1, 2, 3, 4, 5, 6, 7, ...   +2(

1, 4, 3, 3, 1, 2, 7, 4, 2, 6, 7, 2, 2, 3, 1, 1, 7, 5, 8, 3, 1, 7, 1, 8, 3, 4, 5, 5,
7, 7, 5, 9, 9, 1, 8, 2, 0, 4, 3, 1, 5, 1, 2, 7, 6, 7, 9, 0, 5, 9, 8, 0, 5, 2, 3, 4,
3, 4, 4, 2, 8, 6, 3, 6, 3, 9, 4, 3, 0, 9, 1, 8, 3, 2, 5, 4, 1, 7, 2, 9, 0, 0, 1, 3,
6, 5, 0, 3, 7, 2, 6, 4, 3, 5, 7, 8, 6, 1, 1, 4, 6, 5, 9, 5, 0 (list; cons; graph; listen)

OFFSET   1,2

COMMENT   The value of this continued fraction is the ratio of two Bessel
                  functions: BesselI(0,2)/BesselI(1,2) = A070910/A096789. Or,
                  equivalently, to the ratio of the sums: sum_{n=0..inf} 1/(n!n!) and
                  sum_{n=0..inf} n/(n!n!). - Mark Hudson (mrmarkhudson(AT)hotmail.
                  com), Jan 31 2003

FORMULA   1/A052119.

EXAMPLE   C=1.433 12742 67223 11758 31718 3455775 ...

MATHEMATICA   RealDigits[FromContinuedFraction[Range[44]], 10, 110] [[1]]
                  (* Or *) RealDigits[BesselI[0, 2] / BesselI[1, 2], 10, 110] [[1]]
                  (* Or *) RealDigits[Sum[1/(n!n!), {n, 0, Infinity}] / Sum[n/(n!n!),
                  {n, 0, Infinity}], 10, 110] [[1]]

CROSSREFS   Cf. A052119, A001053.
                  Adjacent sequences: A060994 A060995 A060996 this_sequence A060998
                  A060999 A061000
                  Sequence in context: A016699 A060373 A090280 this_sequence A129624
                  A019975 A073871

KEYWORD   cons,easy,nonn

AUTHOR   Robert G. Wilson v (rgwv(AT)rgwv.com), May 14 2001

Fig. 4.4 What *Sloane's Encyclopedia* offers

Example III: From Discovery to Proof

> Besides it is an error to believe that rigour in the proof is the enemy of simplicity.—David Hilbert[16]

The following integral was made popular in a 1971 *Eureka*[17] article

$$0 < \int_0^1 \frac{(1-x)^4 x^4}{1+x^2}\, dx = \frac{22}{7} - \pi \tag{5}$$

as described in Borwein and Bailey (2008). As the integrand is positive on $(0, 1)$ the integral yields an area and hence $\pi < 22/7$. This problem was set on a 1960 Sydney

---

[16]In his *23 Mathematische Probleme* lecture to the Paris International Congress, 1900 (Yandell 2002).

[17]*Eureka* was an undergraduate Cambridge University journal.

honours mathematics final exam (5) and perhaps originated in 1941 with the author of the 1971 article—Dalzeil who chose not reference his earlier self! Why should we trust this discovery? Well *Maple* and *Mathematica* both 'do it'. But this is *proof by appeal to authority* less imposing than, say, von Neumann (Inglis and Mejia-Ramos 2009) and a better answer is to ask *Maple* for the indefinite integral

$$\int_0^t \frac{(1-x)^4 x^4}{1+x^2} dx = ?$$

The computer algebra system (CAS) will return

$$\int_0^t \frac{x^4 (1-x)^4}{1+x^2} dx = \frac{1}{7}t^7 - \frac{2}{3}t^6 + t^5 - \frac{4}{3}t^3 + 4t - 4\arctan(t), \tag{6}$$

and now differentiation and the *Fundamental theorem of calculus* proves the result.

This is probably not the proof one would find by hand, but it is a totally rigorous one, and represents an "instrumental use" of the computer. The fact that a CAS will quite possibly be able to evaluate an indefinite integral or a finite sum whenever it can evaluate the corresponding definite integral or infinite sum frequently allows one to provide a certificate for such a discovery. In the case of a sum, the certificate often takes the form of a mathematical induction (deductive version). Another interesting feature of this example is that it appears to be quite irrelevant that 22/7 is an early, and the most famous, continued-fraction approximation to π (Lucas 2009). Not every discovery is part of a hoped-for pattern. ◁

Example IV: From Concrete to Abstract

The plural of 'anecdote' is not 'evidence'.—Alan L. Leshner[18]

We take heed of Leshner's caution but still celebrate accidental discovery.

1. In April 1993, Enrico Au-Yeung, then an undergraduate at the University of Waterloo, brought to my attention the result

$$\sum_{k=1}^{\infty} \left(1 + \frac{1}{2} + \cdots + \frac{1}{k}\right)^2 k^{-2} = 4.59987\ldots \approx \frac{17}{4}\zeta(4) = \frac{17\pi^4}{360}$$

He had spotted from six place accuracy that $0.047222\ldots = 17/360$. I was very skeptical, but Parseval's identity computations affirmed this to high precision. This is effectively a special case of the following class

---

[18]Leshner, the publisher of *Science*, was speaking at the Canadian Federal Science & Technology Forum, October 2, 2002.

$$\zeta(s_1, s_2, \cdots, s_k) \;=\; \sum_{n_1 > n_2 > \cdots > n_k > 0} \prod_{j=1}^{k} n_j^{-|s_j|} \sigma_j^{-n_j},$$

where s_j are integers and $\sigma_j = \mathrm{signum} s_j$. These can be rapidly computed as implemented at http://www.cecm.sfu.ca/projects/ezface+ (Borwein et al. 2004). In the past 20 years they have become of more and more interest in number theory, combinatorics, knot theory and mathematical physics. A marvellous example is Zagier's conjecture, found experimentally and now proven in Borwein et al. (2004), viz;

$$\zeta\left(\overset{n}{\overline{3,1,3,1,\cdots,3,1}} \right) = \frac{2\pi^{4n}}{(4n+2)!}. \tag{7}$$

Along the way to finding the proof we convinced ourselves that (7) held for many values including $n = 163$ which required summing a slowly convergent 326-dimensional sum to $1,000$ places with our fast summation method. Equation 7 is a remarkable non-commutative counterpart of the classical formula for $\zeta(2n)$ (Borwein et al. 2004, Chap. 3).

2. In the course of proving empirically-discovered conjectures about such multiple zeta values (Borwein and Bailey 2008) we needed to obtain the coefficients in the *partial fraction* expansion for

$$\frac{1}{x^s (1-x)^t} = \sum_{j \geq 0} \frac{a_j^{s,t}}{x^j} + \sum_{j \geq 0} \frac{b_j^{s,t}}{(1-x)^j}. \tag{8}$$

It transpires that

$$a_j^{s,t} = \binom{s+t-j-1}{s-j}$$

with a symmetric expression for $b_j^{s,t}$. This was known to Euler and once known is fairly easily proved by induction. But it can certainly be discovered in a CAS by considering various rows or diagonals in the matrix of coefficients—and either spotting the pattern or failing that by asking Sloane's Encyclopedia. Partial fractions like continued fractions and Gaussian elimination are the sort of task that *once mastered* are much better performed by computer while one focusses on the more conceptual issues they expose.

3. We also needed to show that $M := A + B - C$ was invertible where the $n \times n$ matrices A, B, C respectively had entries

$$(-1)^{k+1}\binom{2n-j}{2n-k}, \quad (-1)^{k+1}\binom{2n-j}{k-1}, \quad (-1)^{k+1}\binom{j-1}{k-1}. \tag{9}$$

Thus, A and C are triangular while B is full. For example, in nine dimensions M is displayed below

$$
\begin{bmatrix}
1 & -34 & 272 & -1360 & 4760 & -12376 & 24752 & -38896 & 48620 \\
0 & -16 & 136 & -680 & 2380 & -6188 & 12376 & -19448 & 24310 \\
0 & -13 & 105 & -470 & 1470 & -3458 & 6370 & -9438 & 11440 \\
0 & -11 & 88 & -364 & 1015 & -2093 & 3367 & -4433 & 5005 \\
0 & -9 & 72 & -282 & 715 & -1300 & 1794 & -2002 & 2002 \\
0 & -7 & 56 & -210 & 490 & -792 & 936 & -858 & 715 \\
0 & -5 & 40 & -145 & 315 & -456 & 462 & -341 & 220 \\
0 & -3 & 24 & -85 & 175 & -231 & 203 & -120 & 55 \\
0 & -1 & 8 & -28 & 56 & -70 & 56 & -28 & 9
\end{bmatrix}
$$

After messing around futilely with lots of cases in an attempt to spot a pattern, it occurred to me to ask *Maple* for the *minimal polynomial* of M.
```
> linalg[minpoly](M(12),t);
```
returns $-2 + t + t^2$. Emboldened I tried
```
> linalg[minpoly](B(20),t);
> linalg[minpoly](A(20),t);
> linalg[minpoly](C(20),t);
```

and was rewarded with $-1+t^3, -1+t^2, -1+t^2$. Since a typical matrix has a full degree minimal polynomial, we are quite assured that A, B, C really are roots of unity. Armed with this discovery we are lead to try to prove

$$A^2 = I, \quad BC = A, \quad C^2 = I, \quad CA = B^2 \tag{10}$$

which is a nice combinatorial exercise (by hand or computer). Clearly then we obtain also

$$B^3 = B \cdot B^2 = B(CA) = (BC)A = A^2 = I \tag{11}$$

and the requisite formula

$$M^{-1} = \frac{M+I}{2}$$

is again a fun exercise in formal algebra; in fact, we have

$$M^2 = AA + AB - AC + BA + BB - BC - CA - CB + CC$$
$$= I + C - B - A + I$$
$$= 2I - M$$

It is also worth confirming that we have discovered an amusing presentation of the symmetric group S_3. ◁

Characteristic or minimal polynomials, entirely abstract for me as a student, now become members of a rapidly growing box of concrete symbolic tools, as do many matrix decomposition results, the use of Groebner bases, Robert Risch's 1968 decision algorithm for when an elementary function has an elementary indefinite integral, and so on.

Example V: A Dynamic Discovery and Partial Proof

> Considerable obstacles generally present themselves to the beginner, in studying the elements of Solid Geometry, from the practice which has hitherto uniformly prevailed in this country, of never submitting to the eye of the student, the figures on whose properties he is reasoning, but of drawing perspective representations of them upon a plane. ... I hope that I shall never be obliged to have recourse to a perspective drawing of any figure whose parts are not in the same plane.—Augustus De Morgan (1806–1871) (Rice 1999, p. 540)

In a wide variety of problems (protein folding, 3SAT, spin glasses, giant Sudoku, etc.) we wish to find a point in the intersection of two sets A and B where B is non-convex but "divide and concur" works better than theory can explain. Let $P_A(x)$ and $R_A(x) := 2P_A(x) - x$ denote respectively the *projector* and *reflector* on a set A, as shown in Fig. 4.5, where A is the boundary of the shaded ellipse. Then "divide and concur" is the natural geometric iteration "reflect-reflect-average":

$$x_{n+1} = \rightarrow \frac{x_n + R_A\left(R_B(x_n)\right)}{2}. \tag{12}$$

Fig. 4.5 Reflector (*interior*) and projector (*boundary*) of a point external to an ellipse

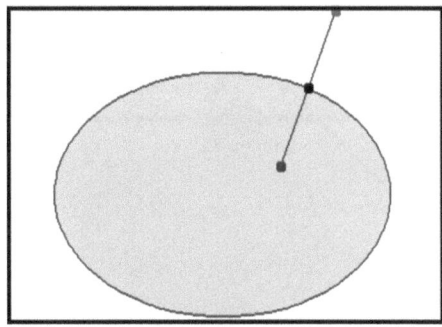

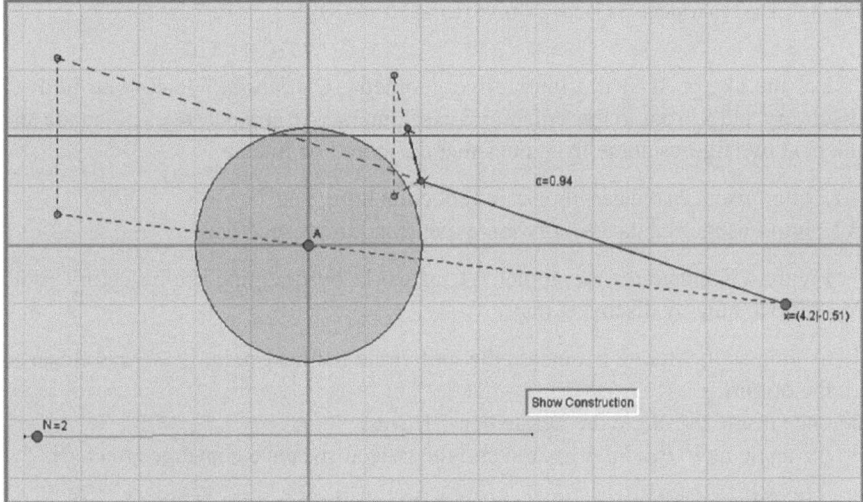

Fig. 4.6 The first three iterates of (13) in *Cinderella*

Consider the simplest case of a line A of height α (all lines may be assumed horizontal) and the unit circle B. With $z_n := (x_n, y_n)$ we obtain the explicit iteration

$$x_{n+1} := \cos\theta_n, y_{n+1} := y_n + \alpha - \sin\theta_n, \quad (\theta_n := \arg z_n). \tag{13}$$

For the infeasible case with $\alpha > 1$ it is easy to see the iterates go to infinity vertically. For the tangent $\alpha = 1$ we provably converge to an infeasible point. For $0 < \alpha < 1$ the pictures are lovely but proofs escape me and my collaborators. Spiraling is ubiquitous in this case. The iteration is illustrated in Fig. 4.6 starting at $(4.2, -0.51)$ with $\alpha = 0.94$. Two representative *Maple* pictures follow in Fig. 4.7.

For $\alpha = 0$ we can prove convergence to one of the two points in $A \cap B$ if and only if we do not start on the vertical axis, where we provably have *chaos*.

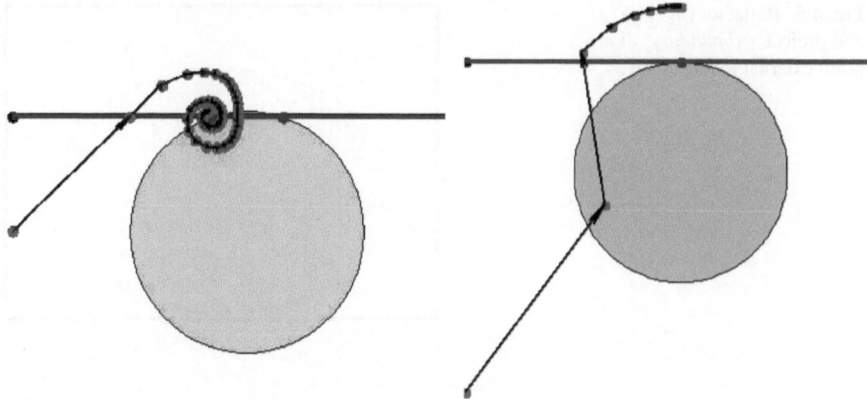

Fig. 4.7 The behaviour of (13) for $\alpha=0.95$ (L) and $\alpha=1$ (R)

Let me sketch how the interactive geometry *Cinderella*[19] leads one both to discovery and a proof in this equatorial case. Interactive applets are easily made and the next two figures come from ones that are stored on line at

A1. http://users.cs.dal.ca/~jborwein/reflection.html; and
A2. http://users.cs.dal.ca/~jborwein/expansion.html respectively.

Figure 4.8 illustrates the applet **A1**. at work: by dragging the trajectory (with $N=28$) one quickly discovers that

1. as long as the iterate is outside the unit circle the next point is *always* closer to the origin;
2. once inside the circle the iterate *never* leaves;
3. the angle now *oscillates* to zero and the trajectory hence converges to $(1,0)$.

All of this is quite easily made algebraic in the language of (13).

Figure 4.9 illustrates the applet **A2**. which takes up to $10,000$ starting points in the rectangle $\{(x,y):0 \le x \le 1, |y-\alpha| \le 1\}$ coloured by distance from the vertical axis with red on the axis and violet at $x=1$, and produces the first 100 iterations in gestalt. Thus, we see clearly but I cannot yet prove, that all points not on the y-axis are swept into the feasible point $(\sqrt{1-\alpha^2},\alpha)$. It also shows that to accurately record the behaviour *Cinderella*'s double precision is inadequate and hence provides a fine if unexpected starting point for a discussion of numerical analysis and instability.

Here we have a fine counter-example to an old mathematical bugaboo:

> A heavy warning used to be given [by lecturers] that pictures are not rigorous; this has never had its bluff called and has permanently frightened its victims into playing for safety. Some pictures, of course, are not rigorous, but I should say most are (and I use them whenever possible myself).—J. E. Littlewood, (1885–1977)[20]

---

[19] Available at http://www.cinderella.de.
[20] From p. 53 of the 1953 edition of Littlewood's *Miscellany* and so said long before the current fine graphic, geometric, and other visualisation tools were available; also quoted in Inglis and Mejia-Ramos (2009).

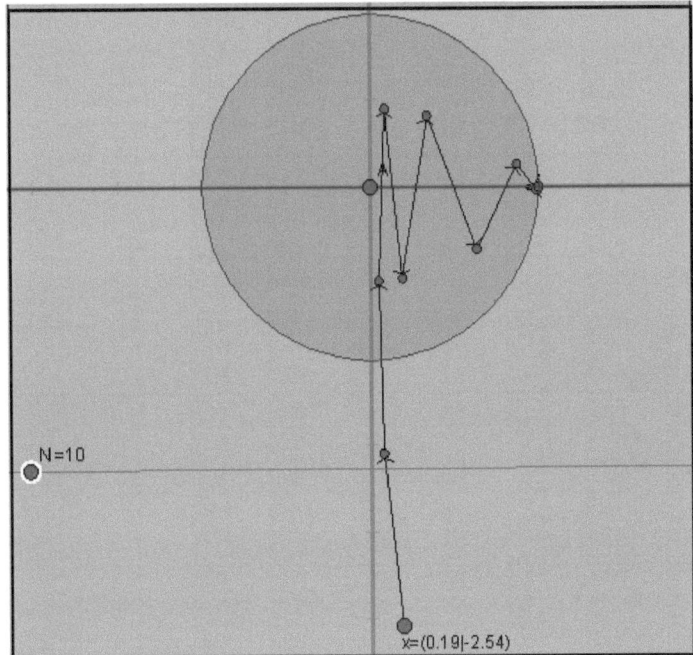

Fig. 4.8 Discovery of the proof with $\alpha=0$

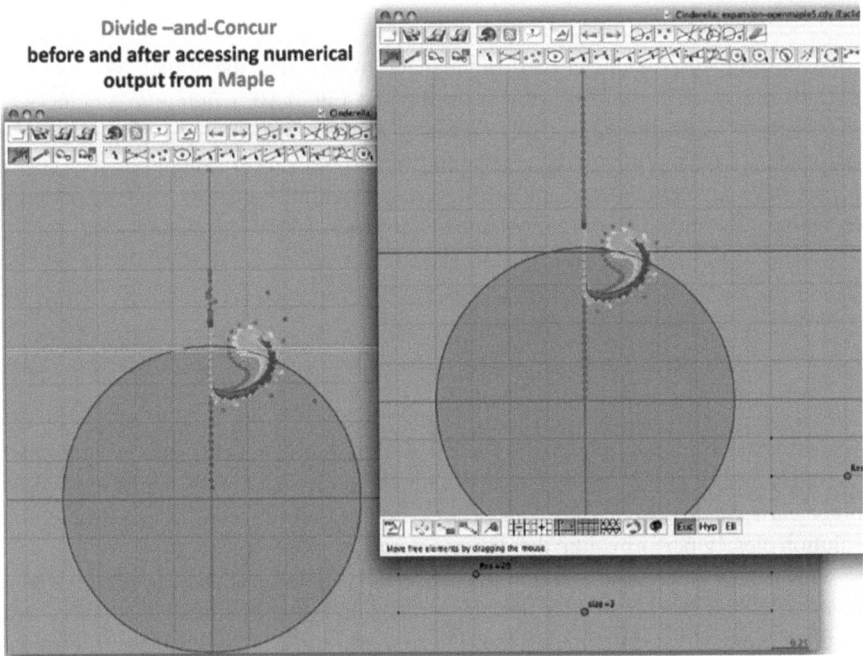

Fig. 4.9 Gestalt of 400 third steps in *Cinderella* without (L) and with *Maple* data (R)

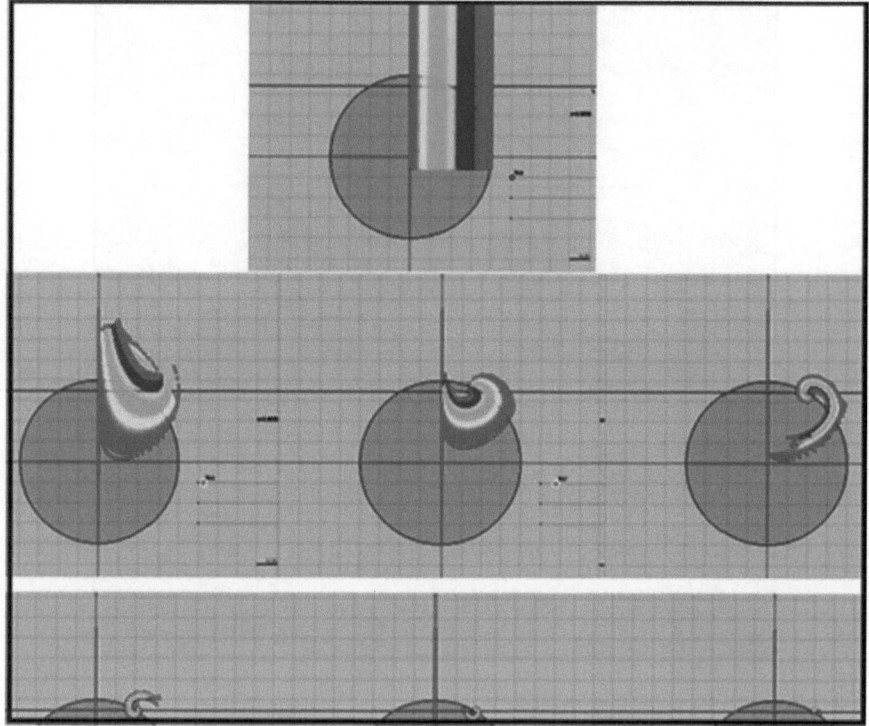

Fig. 4.10 Snapshots of 10,000 points after 0, 2, 7, 13, 16, 21, and 27 steps in *Cinderella*

Á la Littlewood, I find it hard to persuade myself that the applet **A2.** does not constitute a *generic proof* of what it displays in Fig. 4.10. *Cinderella*'s numerical instability is washed away in this profusion of accurate data. For all intents and purposes, we have now run the algorithm from all relevant starting points.

We have also considered the analogous differential equation since asymptotic techniques for such differential equations are better developed. We decided

$$x'(t) = \frac{x(t)}{r(t)} - x(t)r(t) := \sqrt{[x(t)^2 + y(t)^2]}$$

$$y'(t) = \alpha - \frac{y(t)}{r(t)} \tag{14}$$

was a reasonable counterpart to the Cartesian formulation of (13)—we have replaced the difference $x_{n+1} - x_n$ by $x'(t)$, etc.—as shown in Fig. 4.11. Now we have a whole other class of discoveries without proofs. For example, the differential equation solution clearly performs like the discrete iteration solution.

This is also an ideal problem to introduce early under-graduates to research as it involves only school geometry notions and has many accessible extensions in two or three dimensions. Much can be discovered and most of it will be both original

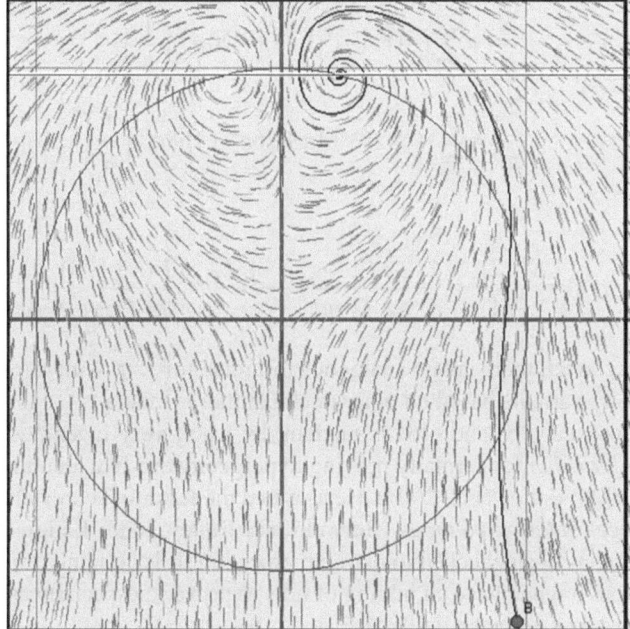

Fig. 4.11 ODE solution and vector field for (14) with α=0.97 in *Cinderella*

and unproven. Consider what happens when B is a line segment or a finite set rather than a line or when A is a more general conic section.

Corresponding algorithms, like "project-project-average", are representative of what was used to correct the Hubble telescope's early optical abberation problems. ◁

Example VI: Knowledge Without Proof

> All physicists and a good many quite respectable mathematicians are contemptuous about proof.—G. H. Hardy (1877–1947)[21]

A few years ago Guillera found various Ramanujan-like identities for π, including three most basic ones:

$$\frac{128}{\pi^2} = \sum_{n=0}^{\infty} (-1)^n r(n)^5 (13 + 180n + 820n^2) \left(\frac{1}{32}\right)^{2n} \tag{15}$$

---

[21] In his famous *Mathematician's Apology* of 1940. I can not resist noting that modern digital assistance often makes more careful referencing unnecessary and sometimes even unhelpful!

$$\frac{8}{\pi^2} = \sum_{n=0}^{\infty} (-1)^n r(n)^5 (1+8n+20n^2) \left(\frac{1}{2}\right)^{2n} \qquad (16)$$

$$\frac{32}{\pi^3} \stackrel{?}{=} \sum_{n=0}^{\infty} r(n)^7 (1+14n+76n^2+168n^3) \left(\frac{1}{8}\right)^{2n}. \qquad (17)$$

where

$$r(n) = \frac{(1/2)_n}{n!} = \frac{1/2 \cdot 3/2 \cdots (2n-1)/2}{n!} = \frac{\Gamma(n+1/2)}{\sqrt{\pi}\,\Gamma(n+1)}$$

As far as we can tell there are no analogous formulae for $1/\pi^N$ with $N \geq 4$. There are, however, many variants based on other Pochhammer symbols.

Guillera proved (15) and (16) in tandem, by using very ingeniously the *Wilf–Zeilberger algorithm* for formally proving hypergeometric-like identities (Borwein and Bailey 2008; Bailey et al. 2007). He ascribed the third to Gourevich, who found it using *integer relation methods* (Borwein and Bailey 2008; Bailey et al. 2007). Formula (17) has been checked to extreme precision. It is certainly true but has no proof, nor does anyone have an inkling of how to prove it, especially as experiment suggests that it has no mate, unlike (15) and (16).

My by-now-sophisticated intuition on the matter tells me that if a proof exists it is most probably more a verification than an explication and so I for one have stopped looking. I am happy just to know the beautiful identity is true. It may be so for no good reason. It might conceivably have no proof and be a very concrete Gödel statement. ◁

Example VII. A Mathematical Physics Limit

Anyone who is not shocked by quantum theory has not understood a single word.—Niels Bohr[22]

The following N-dimensional integrals arise independently in mathematical physics, indirectly in statistical mechanics of the *Ising Model* and as we discovered later more directly in *Quantum Field Theory*:

$$C_N = \frac{4}{N!} \int_0^{\infty} \cdots \int_0^{\infty} \frac{1}{\left(\sum_{j=1}^{N} (u_j + 1/u_j)\right)^2} \frac{du_1}{u_1} \cdots \frac{du_N}{u_N}. \qquad (18)$$

---

[22] As quoted in Barad (2007, p. 54) with a footnote citing *The Philosophical Writings of Niels Bohr* (1998). (1885–1962).

We first showed that C_N can be transformed to a 1-D integral:

$$C_N = \frac{2^N}{N!} \int_0^\infty t K_0^N(t)\, dt \tag{19}$$

where K_0 is a *modified Bessel function*—Bessel functions which we met in Example I are pervasive in analysis.

We then computed 400-digit numerical values. This is impossible for $n \geq 4$ from (18) but accessible from (19) and a good algorithm for K_0. Thence, we found the following, now proven, results (Bailey et al. 2008):

$$C_3 = L_{-3}(2) := \sum_{n \geq 0} \left\{ \frac{1}{(3n+1)^2} - \frac{1}{(3n+2)^2} \right\}$$

$$C_4 = 14\zeta(3)$$

We also observed that

$$C_{1024} \quad = \quad 0.6304735033743867961220401927108789043545 87\ldots$$

and that the limit as $N \to \infty$ was the same to many digits. We then used the Inverse Symbolic Calculator, the aforementioned online numerical constant recognition facility, at http://isc.carma.newcastle.edu.au/ which returned

```
Output:  Mixed  constants,  2  with  elementary  trans-
forms..6304735033743867 = sr(2)^2/exp(gamma)^2
```

from which we discovered that

$$C_{1024} \approx \lim_{n \to \infty} C_n = 2e^{-2\gamma}$$

Here $\gamma = 0.57721566490153\ldots$ is *Euler's constant* and is perhaps the most basic constant which is not yet proven irrational (Havel 2003). The limit discovery showed the Bessel function representation to be fundamental. Likewise $\zeta(3) = \sum_{n=1}^\infty 1/n^3$ the value of the Riemann zeta-function at 3, also called Apéry's constant, was only proven irrational in 1978 and the irrationality of $\zeta(5)$ remains unproven.

The discovery of the limit value, and its appearance in the literature of Bessel functions, persuaded us the Bessel function representation (19) was fundamental—not just technically useful—and indeed this is the form in which C_N, for odd N appears in quantum field theory (Bailey et al. 2008). ◁

Example VIII: Apéry's Formula

> Another thing I must point out is that you cannot prove a vague theory wrong. ... Also, if the process of computing the consequences is indefinite, then with a little skill any experimental result can be made to look like the expected consequences.—Richard Feynman (1918–1988)

Margo Kondratieva found the following identity in the 1890 papers of Markov (Bailey et al. 2007):

$$\sum_{n=0}^{\infty} \frac{1}{(n+a)^3} = \frac{1}{4}\sum_{n=0}^{\infty} \frac{(-1)^n (n!)^6 \left(5(n+1)^2 + 6(a-1)(n+1) + 2(a-1)^2\right)}{(2n+1)! \prod_{k=0}^{n} (a+k)^4}. \quad (20)$$

Apéry's 1978 formula

$$\zeta(3) = \frac{5}{2}\sum_{k=1}^{\infty} \frac{(-1)^{k+1}}{k^3 \binom{2k}{k}}, \quad (21)$$

which played a key role in his celebrated proof of the irrationality of $\zeta(3)$, is the case with $a=0$.

Luckily, by adopting Giaquinto's accounting of discovery we are still entitled to say that Apéry discovered the formula (21) which now bears his name.

We observe that *Maple* 'establishes' identity (20) in the hypergeometric formula

$$-\frac{1}{2}\Psi(2,a) = -\frac{1}{2}\Psi(2,a) - \zeta(3) + \frac{5}{4}\,{}_4F_3\left(\begin{array}{c}1,1,1,1\\2,2,\frac{3}{2}\end{array}\middle| -\frac{1}{4}\right),$$

that is, it has reduced it to a form of (21). ◁

Like much of mathematics, this last example leads to something whose computational consequences are very far from indefinite. Indeed, it is the rigidity of much algorithmic mathematics that makes it so frequently the way hardware or software errors, such as the 'Pentium Bug', are first uncovered.

Example IX: When Is Easy Bad?

Many algorithmic components of CAS are today extraordinarily effective when two decades ago they were more like 'toys'. This is equally true of extreme-precision calculation—a prerequisite for much of my own work (Baillie et al. 2008; Bailey et al. 2009) and others (Borwein 2005)—or in combinatorics.

Consider the *generating function* of the number of *additive partitions*, $p(n)$ of a natural number where we ignore order and zeroes. Thus,

$$5 = 4+1 = 3+2 = 3+1+1 = 2+2+1 = 2+1+1+1 = 1+1+1+1+1$$

and so $p(5)=7$. The *ordinary generating function* (22) discovered by Euler is

$$\sum_{n=0}^{\infty} p(n)q^n = \frac{1}{\prod_{k=1}^{\infty} (1-q^k)}. \quad (22)$$

This is easily obtained by using the geometric formula for each $1/(1-q^k)$ and observing how many powers of q^n are obtained. The famous and laborious computation by MacMahon of $p(200) = 3972999029388$ early last century, if done *symbolically and entirely naively* from (22) on a reasonable laptop took 20 min in 1991, and about 0.17 sec today, while

$$p(2000) = 4720819175619413888601432406799959512200344166$$

took about 2 min in 2009.

Moreover, Richard Crandall was able, in December 2008, to calculate $p(10^9)$ in 3 s on his laptop, using Hardy-Ramanujan and Rademacher's 'finite' series along with FFT methods. The current ease of computation of $p(500)$ directly from (22) raises the question of what interesting mathematical discoveries does easy computation obviate?

Likewise, the record for computation of π has gone from under 30 million decimal digits in 1986 to over 5 trillion places this year (Borwein and Bailey 2008).

3 Concluding Remarks

We [Kaplansky and Halmos] share a philosophy about linear algebra: we think basis-free, we write basis-free, but when the chips are down we close the office door and compute with matrices like fury.—Paul Halmos (1916–2006) (Ewing and Gehring 1991)

Theory and practice should be better comported!

The students of today live, as we do, in an information-rich, judgement-poor world in which the explosion of information, and of tools, is not going to diminish. So we have to teach judgement (not just concern with plagiarism) when it comes to using what is already possible digitally. This means mastering the sorts of tools I have illustrated. Additionally, it seems to me critical that we mesh our software design—and our teaching style more generally—with our growing understanding of our cognitive strengths and limitations as a species (as touched upon in Sect. 1). Judith Grabiner, in her contribution to this volume, has noted that a large impetus for the development of modern rigour in mathematics came with the Napoleonic introduction of regular courses: Lectures and text books force a precision and a codification that apprenticeship obviates.

As Dave Bailey noted to me recently in email:

Moreover, there is a growing consensus that human minds are fundamentally not very good at mathematics, and must be trained as Ifrah points out (Ifrah 2000). Given this fact, the computer can be seen as a perfect complement to humans—we can intuit but not reliably calculate or manipulate; computers are not yet very good at intuition, but are great at calculations and manipulations.

We also have to acknowledge that most of our classes will contain students with a very broad variety of skills and interests (and relatively few future mathematicians).

Properly balanced, discovery and proof, assisted by good software, can live side-by-side and allow for the ordinary and the talented to flourish in their own fashion. Impediments to the assimilation of the tools I have illustrated are myriad as I am only too aware from my own recent teaching experiences. These impediments include our own inertia and organisational and technical bottlenecks (this is often from poor IT design—not so much from too few dollars). The impediments certainly include under-prepared or mis-prepared colleagues and the dearth of good material from which to teach a modern syllabus.

Finally, it will never be the case that quasi-inductive mathematics supplants proof. We need to find a new equilibrium. Consider the following empirically-discovered identity

$$\sum_{n=-\infty}^{\infty} \text{sinc}(n)\text{sinc}(n/3)\text{sinc}(n/5)\cdots\text{sinc}(n/23)\text{sinc}(n/29)$$

$$= \int_{-\infty}^{\infty} \text{sinc}(x)\text{sinc}(x/3)\text{sinc}(x/5)\cdots\text{sinc}(x/23)\text{sinc}(x/29)dx \qquad (23)$$

where the denumerators range over the primes.

Provably, the following is true: The analogous "sum equals integral" identity remains valid for more than the first $10,176$ primes but stops holding after some larger prime, and thereafter the "sum minus integral" is positive but *much less than one part in a googolplex* (Baillie et al. 2008). It is hard to imagine that inductive mathematics alone will ever be able to handle such behaviour. Nor, for that matter, is it clear to me what it means psychologically to digest equations which are false by a near infinitesimal amount.

That said, we are only beginning to scratch the surface of a very exciting set of tools for the enrichment of mathematics, not to mention the growing power of formal proof engines. I conclude with one of my favourite quotes from George Pólya and Jacques Hadamard:

> This "quasi-experimental" approach to proof can help to de-emphasise a focus on rigor and formality for its own sake, and to instead support the view expressed by Hadamard when he stated "The object of mathematical rigor is to sanction and legitimise the conquests of intuition, and there was never any other object for it." (Pólya 1981, (2) p. 127)

Unlike Frank Quinn (Jaffe and Quinn 1991) perhaps, I believe that in the most complex modern cases certainty, in any reasonable sense, is unattainable through proof. I do believe that even then quasi-inductive methods and experimentation can help us improve our level of certainty. Like Reuben Hersh (1997), I am happy to at least entertain some "non-traditional forms of proof." Never before have we had such a cornucopia of fine tools to help us develop and improve our intuition. The challenge is to learn how to harness them, how to develop and how to transmit the necessary theory and practice.

Acknowledgements I owe many people thanks for helping refine my thoughts on this subject over many years. Four I must mention by name: my long-standing collaborators Brailey Sims, Richard Crandall and David Bailey, and my business partner Ron Fitzgerald from *MathResources*, who has taught me a lot about balancing pragmatism and idealism in educational technology—among other things. I also thank Henrik Sørensen whose thought-provoking analysis gave birth to the title and the thrust of the paper, and my student Chris Maitland who built most of the *Cinderella* applets.

References

Anderson, C. (2008). The end of theory. *Wired*, http://www.wired.com/science/discoveries/magazine/16-07/pb_theory.

Avigad, J. (2008). Computers in mathematical inquiry. In P. Mancuso (Ed.), *The philosophy of mathematical practice* (pp. 302–316). Oxford: Oxford University Press.

Bailey D. H., & Borwein, J. M. (2011, November). Exploratory experimentation and computation. *Notices of the AMS*. November 1410–1419.

Bailey, D., Borwein, J., Calkin, N, Girgensohn, R., Luke, R., & Moll, V. (2007). *Experimental mathematics in action*. Wellesley: A K Peters.

Bailey, D. H., Borwein, J. M., Broadhurst, L. M., & Glasser, L. (2008). Elliptic integral representation of Bessel moments. *Journal of Physics A: Mathematics & Theory, 41*, 5203–5231.

Bailey, D. H., Borwein, J. M., & Crandall, R. E. (2010, March). Advances in the theory of box integrals. *Mathematics of Computation, 79*, 1839–1866.

Baillie, R., Borwein, D., & Borwein, J. (2008). Some sinc sums and integrals. *American Mathematical Monthly, 115*(10), 888–901.

Barad, K. M. (2007). *Meeting the universe halfway*. Durham: Duke University Press.

Borwein, J. M. (2005). The SIAM 100 digits challenge. Extended review in the *Mathematical Intelligencer, 27*, 40–48. [D-drive preprint 285].

Borwein, J. M., & Bailey, D. H. (2008). *Mathematics by experiment: Plausible reasoning in the 21st century* (2nd ed.). Wellesley: A K Peters.

Borwein, J. M., & Devlin, K. (2009). *The computer as crucible*. Wellesley: A K Peters.

Borwein, J. M., Bailey D. H., & Girgensohn, R. (2004). *Experimentation in mathematics: Computational pahts to discovery*. Wellesley: A K Peters.

Brown, R. D. (2008). *Are science and mathematics socially constructed? A mathematician encounters postmodern interpretations of science*. Singapore: World Scientific.

de Villiers, M. (2004). The role and function of quasi-empirical methods in Mathematics. *Canadian Journal of Science, Mathematics and Technology Education, 4*, 397–418.

Dewey, J. (1910). *The influence of darwin on philosophy and other essays*. New York: Henry Holt.

Ewing, J. H., & Gehring, F. W. (1991). *Paul Halmos. Celebrating 50 years of mathematics*. New York: Springer.

Franklin, L. R. (2005). Exploratory experiments. *Philosophy of Science, 72*, 888–899.

Giaquinto, M. (2007). *Visual thinking in mathematics. An epistemological study*. Oxford: Oxford University Press.

Gregory, J., & Miller, S. (1998). *Science in public, communication, culture and credibility*. Cambridge: Basic Books.

Hamming, R. W. (1962). *Numerical methods for scientists and engineers*. New York: McGraw-Hill.

Havel, J. (2003). *Gamma: Exploring Euler's constant*. Princeton: Princeton University Press.

Hersh, R. (1997). *What is mathematics, really?* New York: Oxford University Press.

Ifrah, G. (2000). *The universal history of numbers*. New York: Wiley.

Inglis, M., & Mejia-Ramos, J. P. (2009). The effect of authority on the persuasiveness of mathematical arguments. preprint.

Jaffe, A., & Quinn, F. (1991). "Theoretical mathematics": Toward a cultural synthesis of mathematics and theoretical physics. *Bulletin of the American Mathematical Society, 29*(1), 1–13.

Livio, M. (2009). *Is God a mathematician?*. New York:Simon and Schuster.

Lucas, S. K. (2009). Approximations to π derived from integrals with nonnegative integrands. *American Mathematical Monthly, 116*(10), 166–172.

Pinker, S. (2007). *The stuff of thought: Language as a window into human nature*. London: Allen Lane.

Pólya, G. (1981). *Mathematical discovery: On understanding, learning, and teaching problem solving* (Combined ed.). New York: Wiley.

Regis, E. (1986). *Who got Einstein's office?* Reading: Addison-Wesley.

Rice A. (1999). What makes a great mathematics teacher? *American Mathematical Monthly, 106*(6), 534–552.

Sfard, A. (2009). What's all the fuss about gestures: A commmentary. Special issue on gestures and multimodality in the construction of mathematical meaning. *Educational Studies in Mathematics, 70*, 191–200.

Smail, D. L. (2008). *On deep history and the brain*. Berkeley: Caravan Books/University of California Press.

Sørensen, H. K. (2010). Exploratory experimentation in experimental mathematics: A glimpse at the PSLQ algorithm. In B. Löwe & T. Müller (Eds.), *PhiMSAMP. Philosophy of mathematics: Sociological aspects and mathematical practice. Texts in Philosophy* (Vol. 11, pp. 341–360). London: College Publications.

Yandell, B. (2002) *The honors class*. Natick: A K Peters.

Chapter 5
Experimental Approaches to Theoretical Thinking: Artefacts and Proofs

Ferdinando Arzarello, Maria Giuseppina Bartolini Bussi,
Allen Yuk Lun Leung, Maria Alessandra Mariotti, and Ian Stevenson

1 Introduction

From straight-edge and compass to a variety of computational and drawing tools, throughout history instruments have been deeply intertwined with the genesis and development of abstract concepts and ideas in mathematics. Their use introduces an "experimental" dimension into mathematics, as well as a dynamic tension between the *empirical nature* of activities with them, which encompasses perceptual and operational components– and the *deductive nature* of the discipline, which entails rigorous and sophisticated formalisation. As Pierce writes of this peculiarity:

> (It) has long been a puzzle how it could be that, on the one hand, mathematics is purely deductive in its nature, and draws its conclusions apodictically, while on the other hand, it presents as rich and apparently unending a series of surprising discoveries as any observational science.
>
> (Peirce, C.P., 3.363: quoted in Dörfler 2005, p. 57)

F. Arzarello (✉)
Department of Mathematics, University of Torino, Torino, Italy
e-mail: ferdinando.arzarello@unito.it

M.G. Bartolini Bussi
Department of Mathematics, University of Modena and Reggio Emilia (UNIMORE),
Modena, Italy
e-mail: bartolini@unimore.it

A.Y.L. Leung
Department of Education Studies, Hong Kong Baptist University, Kowloon Tong, Hong Kong
e-mail: aylleung@hkbu.edu.hk

M.A. Mariotti
Department of Mathematics and Computer Science, University of Siena, Siena, Italy
e-mail: mariotti21@unisi.it

I. Stevenson
Department of Education and Professional Studies, King's College, London, UK
e-mail: ian.stevenson@kcl.ac.uk

© The Author(s) 2021
G. Hanna and M. de Villiers (eds.), *Proof and Proving in Mathematics Education*,
New ICMI Study Series, https://doi.org/10.1007/978-94-007-2129-6_5

The main goal of our chapter centres on the dynamic tension between the empirical and the theoretical nature of mathematics. Our purpose is to underline the elements of historical continuity in the stream of thought today called experimental mathematics, and show the concrete possibilities it offers to today's teachers for pursuing the learning of proof in the classroom, especially through the use of their computer tools.

Specifically, we examine how this dynamic tension regulates the actions of students who are asked to solve mathematical problems by first making explorations with technological tools, then formulating suitable conjectures and finally proving them.

The latest developments in computer and video technology have provided a multiplicity of computational and symbolic tools that have rejuvenated mathematics and mathematics education. Two important examples of this revitalisation are *experimental mathematics* and *visual theorems*:

> Experimental mathematics is the use of a computer to run computations – sometimes no more than trial-and-error tests – to look for patterns, to identify particular numbers and sequences, to gather evidence in support of specific mathematical assertions that may themselves arise by computational means, including search. Like contemporary chemists – and before them the alchemists of old – who mix various substances together in a crucible and heat them to a high temperature to see what happens, today's experimental mathematicians put a hopefully potent mix of numbers, formulas, and algorithms into a computer in the hope that something of interest emerges.

> (Borwein and Devlin 2009, p. 1)

> Briefly, a visual theorem is the graphical or visual output from a computer program – usually one of a family of such outputs – which the eye organizes into a coherent, identifiable whole and which is able to inspire mathematical questions of a traditional nature or which contributes in some way to our understanding or enrichment of some mathematical or real world situation.

> (Davis 1993, p. 333)

Such developments throw a fresh light on mathematical epistemology and on the processes of mathematical discovery; consequently, we must also rethink the nature of mathematical learning processes. In particular, the new epistemological and cognitive viewpoints have challenged and reconsidered the phenomenology of learning proof (cf. Balacheff 1988, 1999; Boero 2007; de Villiers 2010; Chap. 3). These recent writers have scrutinised and revealed not only *deductive* but also *abductive* and *inductive* processes crucial in all mathematical activities, emphasising the importance of experimental components in teaching proofs. The related didactical phenomena become particularly interesting when instructors plan proving activities in a technological environment (Arzarello and Paola 2007; Jones et al. 2000), where they can carefully design their interventions. By "technological environment", we do not mean just digital technologies but any environment where instruments are used to learn mathematics (for a non-computer technology, see Bartolini Bussi 2010). We discuss this issue from different linked perspectives: historical, epistemological, didactical and pedagogical.

In Part 1, we consider some emblematic events from the history of Western mathematics where instruments have played a crucial role in generating mathematical concepts.

Next, Part 2 analyses some didactical episodes from classroom life, where the use of instruments in proving activities makes the dynamic tension palpable.

We carefully analyse students' procedures whilst using tools and derive some theoretical frameworks that explain how that tension can be used to design suitable didactical situations. Within these, students can learn practices with the tools that help them pass from the empirical to the theoretical side of mathematics. In particular, we discuss the complex interactions between inductive, abductive and deductive modalities in that transition. By analysing the roles for technologies within our framework, we show that instructors can and should make the history and cultural aspects of experimental mathematics visible to students.

Last, in Part 3 we show how a general pedagogical framework (Activity Theory) makes sense of the previous microanalyses within a general, unitary educational standpoint.

2 Part 1: From Straight-Edge and Compass to Dynamic Geometry Software

2.1 Classical European Geometry

Since antiquity, geometrical constructions have had a fundamental theoretical importance in the Greek and later Western traditions (Heath 1956, p. 124); indeed, construction problems were central to Euclid's work. This centrality is clearly illustrated by the later history of the classic 'impossible' problems, which so puzzled Euclid and other Greek geometers (Henry 1993). Despite their apparent practical objective, geometrical constructions (like drawings produced on papyrus or parchment) do have a theoretical meaning. In Euclid's masterpiece, the *Elements*, no real, material tools are envisaged; rather their use is objectified into the geo-metrical objects defined by definitions and axioms. However, Arsac (1987) shows that the observational, empirical component was also present in the *Elements*. Euclid was aware of the dialectic between the decontextualised aspects of pure geometry and the phenomenology of our perception of objects in space and our representations of them in the plane. In his *Optics* (Euclide 1996) masterpiece, he gives a rationale for this tension. Giusti writes: "the mathematical objects are not generated through abstraction from real objects [...] but they formalize human operations"[1](Giusti 1999). In addition, they are shaped by the tools with which people perform such operations.

Consequently, the tools and the rules for their use have a counterpart in the axi-oms and theorems of a theoretical system, so that we may conceive of any construc-tion as a theoretical problem stated inside a specific theoretical system. The solution of a problem is correct; therefore, insofar as we can validate it within such a

---

[1] 'gli oggetti matematici provengono non dall'astrazione da oggetti reali [...] ma formalizzano l'operare umano'.

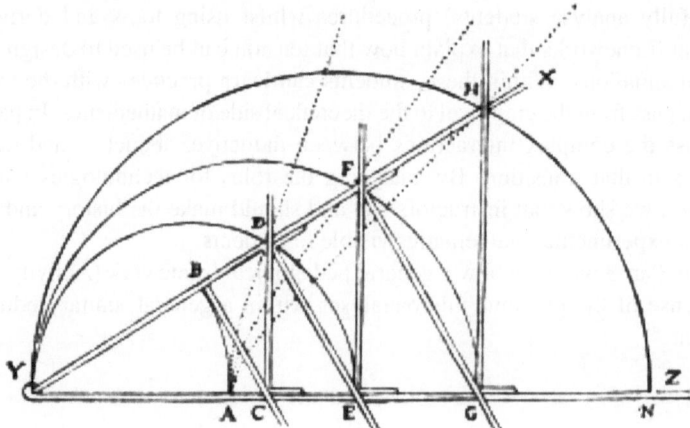

Fig. 5.1 Descartes compass

theoretical system, any successful construction corresponds to a specific theorem, and *validates* the specific relationships between the properties of the geometrical figure represented by the drawing, obtained after the construction.

From the perspective of classical geometry, drawing tools, despite their empirical manifestation, may also be conceived as theoretical tools defining a particular geometry. Hence, the tradition of calling classic Euclidean geometry "straight-edge and compass geometry" refers to both the origins and limitations of its objects.

2.2 The Modern Age in Europe

In the Western Euclidean tradition, interesting European developments about the theoretical status of drawing tools and geometric objects have accumulated since the seventeenth century. For example, in the *'Géométrie'*, Descartes clearly states two methods of representing curves: (a) by a continuous motion and (b) by an equation (Bos 1981). Descartes invented a tool, his compass (Fig. 5.1), to make evident what he meant: moving the different t-squares YBC, DCE, EDF, ... moves a point like H and generates a curve with the features described under (a) and (b).

In contrast, in the Hippias trisectrix (Fig. 5.2), the points of the curve APQT are generated through a continuous synchronous motion of the ray DP (which rotates uniformly around D like the hand of a clock) and of the horizontal ray MP (which moves uniformly downwards, so that whilst the ray rotates from DA to DC the ray MP moves from AB to DC). In fact, it is impossible to empirically obtain the position of all the points of the curve APQT since one can only imagine them through the description of the movement, not concretely depict them as in the case of a figure drawn using straight-edge and compass.

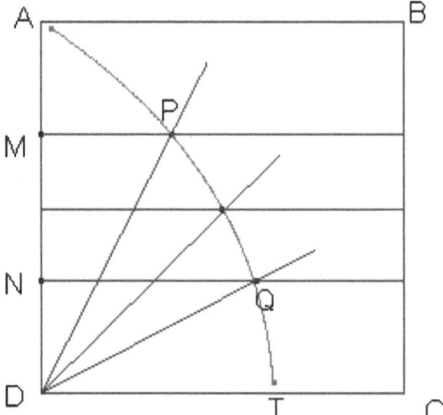

Fig. 5.2 Hippias trisectrix

As Lebesgue (1950) claims, a curve traced pointwise is obtained by approximation; it is only a graphic solution. However, if one designs a tracing instrument, the graphic solution becomes a mechanical solution. The seventeenth century mathematicians found the mechanical solution acceptable because it refers to one of the basic intuitions about the continuum: namely, the movement of an object. Descartes did not confront the question of whether the two given criteria – the mechanical and the algebraic – are equivalent or not. This problem requires more advanced algebraic tools and, more important, changing the status of the new drawing instruments from tools for solving geometric problems to objects of a theory.

In the classical age and in the seventeenth century, changing the drawing tools would clearly have changed the set of solvable problems. So, if one accepts only the straight-edge and compass (i.e., only straight lines and circles), one cannot rigorously solve the problems of cube duplication and angle trisection. If, on the contrary, other tools are admitted (e.g., the Nicomedes compass that draws a conchoid; see Heath 1956), these problems can be solved rigorously.

The previous examples highlighted a crucial dialectical relationship between practical and theoretical problems. The core of this relationship resides in the notion of *construction* as related to the specific tools available. Therefore, the practical realisation of any graphical element has a counterpart in a theoretical element, in either an axiom that states how to use a tool or a theorem that validates the construction procedure according to the stated axioms. In these terms, we can consider a geometrical construction archetypal for a theoretical approach to geometry.

However, in spite of their long tradition, geometrical constructions have recently lost their centrality and almost disappeared from the Geometry curriculum, at least in the Western world. One can rarely find any reference to 'drawing tools' when geometrical axioms are stated, and geometrical constructions no longer belong to the set of problems commonly proposed in the textbooks. This disappearance began as nineteenth century mathematicians from Pasch to Peano to Hilbert tried to eliminate the observational "intuitive" hidden hypotheses from Geometry.

Similarly, the school of Weierstrass eliminated any reference to space or motion in the geometric definition of limits through the epsilon-delta machinery (Lakoff and Núñez 2000): in fact, the new definition entails only purely logical relationships ("for all epsilon, there is a delta...") and any reference to motion and time (e.g. "whilst x approaches x_0, $f(x)$ approaches l") is eliminated. Until recently, observational components had apparently been completely banished from the geometric scene. However, philosophical criticisms by many recent scholars (for a survey, cf. Tymoczko 1998) and the development of computational techniques have produced a fresh approach to mathematical learning and discovery. They have revived epistemological stances which underlie the observational, experimental and empirical aspects of mathematical inquiry, including the use of geometric constructions (see Lovasz 2006).

Indeed, geometric constructions are rich in meaning and perfectly suitable for implementation in today's classrooms, even though the relationship between a geometrical construction and the theorem which validates it is very complex and certainly not immediate for students, as Schoenfeld (1985) discussed. As he explained, "many of the counterproductive behaviors we see in students are learned as unintended by-products of their mathematics instruction" (p. 374). Apparently, the very nature of the construction problem may make it difficult to take a theoretical perspective (cf. Mariotti 1996).

Nevertheless, the analysis above allows us to state a specific hypothesis, namely, that *geometrical construction* can serve as a *key to accessing* the meaning of proof. Different research groups have undertaken to test or apply this hypothesis, in different directions with different tools and different mathematical theories.

2.3 Constructions with Straight-Edge and Compass in the Mathematics Classroom

A recent teaching experiment in Italy has shown the potential of straight-edge and compass for developing an experimental approach with theoretical aims (Bartolini Bussi et al. in print). The project involved a group of 80 mathematics teachers (only six from primary school, the others equally divided between junior secondary and high school; see Martignone 2010) and nearly 2,000 students (scattered all over a large region of Northern Italy). Straight-edge and compass problems were set in the larger context of mathematical "machines" (Bartolini Bussi 2000, p. 343), tools that force a point to follow a trajectory or to be transformed according to a given law. A common theoretical framework (see below; also Bartolini Bussi and Mariotti 2008) structured the exploration of the tools and of the functions they served in the solution of geometrical problems by construction. Similar learning processes were implemented with the participants. First, the teachers received an in-service course of six meetings; then they instructed their students. A total of 79 teaching experiments, with detailed documentations, were collected; 25% of them concerned straight-edge and compass.

The general structure of the approach comprised:

A. Exploration and analysis of the tool (shorter for teachers; longer for students, in order to make them aware of the relationship between the physical structure of the compass and Euclid's definition of a circle).
B. Production of very simple constructions of geometrical figures (e.g., "draw an equilateral triangle with a given side") in open form, in order to allow a variety of constructions based on different known properties.
C. Comparison of the different constructions in large group discussions, to show that the "same" drawing may be based on very different processes, each drawing on either implicit or explicit assumptions and on the technical features of the tool.
D. Production of proofs of the constructions exploiting each times the underlying assumptions.

These stages were structured around three key questions concerning the compass as a tool:

1. How is it made?
2. What does it do?
3. Why does it do that?

The third question, dependent on the others, aimed at connecting the tool's practical use to the theoretical content. In fact, the justification of a construction draws on the geometrical properties of the compass, as is clearly shown in the proof of Proposition 1, Book 1 of Euclid's *Elements* (Heath 1956, p. 241), with the construction of an equilateral triangle.

2.4 Constructions in a DGS

The interest in constructions has been renewed in particular by the appearance of Dynamic Geometry Systems (DGS), where the basic role played by construction has been reinforced by the use of graphic tools available in a dynamic system, like *Cabri-géomètre*, *Sketchpad*, *Geogebra*, etc. Any DGS figure is the result of a construction process, since it is obtained after the repeated use of tools chosen from those available in the "tool bar". However, what makes DGS so interesting compared to the classic world of paper and pencil figures is not only the construction facility but also the direct manipulation of its figures, conceived in terms of the embedded logic system (Laborde and Straesser 1990; Straesser 2001) of Euclidean geometry. DGS figures possess an intrinsic logic, as a result of their construction, placing the elements of a figure in a hierarchy of relationships that corresponds to the procedure of construction according to the chosen tools and in a hierarchy of properties, and this hierarchy corresponds to a relationship of logical conditionality. This relationship is made evident in the "dragging" mode, where what cannot be dragged by varying the basic points (elements) of a built figure constitutes the results of the construction. The dynamics of the DGS figures preserves its intrinsic logic; that is, the logic of its construction. The DGS figure is the complex of these elements, incorporating various relationships which can be differently referred to the definitions and theorems of geometry.

The presence of the dragging mode introduces in the DGS environment a specific criterion of validation for the solution of the construction problems: A solution is valid if and only if the figure on the screen is stable under the dragging test. However, the system of DGS figures embodies a system of relationships consistent with the broad system of geometrical theory. Thus, solving construction problems in DGS means not only accepting all the facilities of the software but also accepting a logic system within which to make sense of them.

The DGS's intrinsic relation to Euclidean geometry makes it possible to interpret the control 'by dragging' as corresponding to theoretical control 'by proof and definition' within the system of Euclidean Geometry, or of another geometry that allows recourse to a larger set of tools. In other words, there is a correspondence between the world of DGS constructions and the theoretical world of Euclidean Geometry.

2.5 DGS Constructions in the Classroom

Mariotti (2000, 2001) carried out teaching experiments with grade 10 students attending first year in a science-oriented school (Liceo Scientifico). The design of the teaching sequence was based on the development of the field of experience (Boero et al. 1995) of geometrical constructions in a DGS (*Cabri-Géomètre*). The educational aim was to introduce students to a theoretical perspective; its achievement relied on the potential correspondence between DGS constructions and geometric theorems.

The activity started by revisiting drawings and concrete artefacts which the pupils had already experienced: for example, the compass. The students were more or less familiar with the artefacts' constraints, which determine possible actions and expected results; for instance, a compass's intrinsic properties directly affect the properties of the graphic trace it produces. Revisitation involved transferring the drawing activity into the *Cabri* environment, thus moving the external context from the physical world of straight-edge and compass to the virtual world of DGS figures and commands.

In a DGS environment, the new 'objects' available are Evocative Computational Objects (Hoyles 1993; Hoyles and Noss 1996, p. 68), characterised by their computational nature and their power to evoke geometrical knowledge. For *Cabri*, they comprise:

1. The *Cabri*-figures realising geometrical figures;
2. The *Cabri*-commands (primitives and macros), realising the geometrical relationships which characterise geometrical figures;
3. The dragging function, which provides a perceptual control of the construction's correctness, corresponding to a theoretical control consistent with geometric theory.

The development of the field of experience occurred through activities in *Cabri*'s world, such as construction tasks, interpretation and prediction tasks and mathematical discussions. However, that development also involved making straight-edge and compass constructions, which became both concrete referents and signs of the *Cabri* figures. Relating the drawings on paper and the *Cabri* figures gave the students a unique experience with a 'double face', one physical and the other virtual.

In the DGS environment, a construction activity, such as drawing figures through the commands on the menu, is integrated with the dragging function. Thus, a construction task is accomplished if the figure on the screen passes the dragging test.

In Mariotti's (2000, 2001) research, the necessity of justifying the solution came from the need to validate one's own construction, in order to explain why it worked and/or to foresee that it would work. Although dragging the figure might suffice to display the correctness of the solution, the second component of the teaching/learning activities came into play at this point. Namely, construction problems become part of a social interchange, where the students reported and compared their different solutions. This represented a crucial element of the experience.

2.6 Experiments and Proofs with the Computer

Typically, current experimental mathematics involves making computations with a computer. Crucially, validating numerical solutions, which may have already been found, requires producing suitable proofs (cf. Borwein and Devlin 2009, for example). We illustrate with an example precisely how a so called CAS (Computer Algebra System: it processes not only numerical values but also algebraic expressions with letters and infinite-precision rational numbers) can be used as a tool for promoting the production of proofs for found numerical solutions.

Arzarello (2009) researched Grade 9 students, attending first year in a science-oriented higher secondary school (Liceo Scientifico), who were studying functions through tables of differences. The students had already learnt that for first-degree functions, the first differences are constant. The teacher asked them to make conjectures on which functions have the first differences that change linearly and to arrange a spreadsheet as in Fig. 5.3a, where they utilise:

1. Columns A, B, C, D to indicate respectively the values of the variable x, of the function $f(x)$ (in B_i there is the value of $f(A_i)$) and of its related first and second differences (namely in C_i there is the value $f(A_{i+1}) - f(A_i)$ and D_j there is the value $C_{j+1} - C_j$);
2. Variable numbers in cells E2, F2, …,I2 to indicate respectively: the values x_0 (the first value for the variable x to put in A2); a, b, c for the coefficients of the second degree function $ax^2 + bx + c$; the step h of which the variable in column A is incremented each time for passing to A_i to A_{i+1}.

By modifying the values of E2, F2, …, I2, the students could easily do their explorations. This practice gradually became shared in the classroom, through

a

	A	B	C	D	E	F	G	H	I
1	x	f(x)	df(x)	ddf(x)	x0	a	b	c	h
2	0	3	-1	-4	0	-2	1	3	1
3	1	2	-5	-4					
4	2	-3	-9	-4					
5	3	-12	-13	-4					
6	4	-25	-17	-4					
7	5	-42	-21	-4					
8	6	-63	-25	-4					
9	7	-88	-29	-4					
10	8	-117	-33	-4					
11	9	-150	-37	-4					
12	10	-187	-41	-4					
13	11	-228	-45	-4					
14	12	-273	-49	-4					
15	13	-322	-53	-4					
16	14	-375	-57						
17	15	-432							

b

	A	B	C	D	E	F	G	H
1	x	f(x)	df(x)	ddf(x)	x0	a	b	c	h
2	x0	a*x0^2+b*x0..	a*(h^2+2*h*x0)+..	2*a*h^2	x0	a	b	c	h
3	h+x0	a*(h+x0)^2+..	a*(3*h^2+2*h*x0..	2*a*h^2				
4	2*h+x0	a*(2*h+x0)^2..	a*(5*h^2+2*h*x0..	2*a*h^2				
5	3*h+x0	a*(3*h+x0)^2..	a*(7*h^2+2*h*x0..	2*a*h^2				
6	4*h+x0	a*(4*h+x0)^2..	a*(9*h^2+2*h*x0..	2*a*h^2				
7	5*h+x0	a*(5*h+x0)^2..	a*(11*h^2+2*h*x..	2*a*h^2				
8	6*h+x0	a*(6*h+x0)^2..	a*(13*h^2+2*h*x..	2*a*h^2				
9	7*h+x0	a*(7*h+x0)^2..	a*(15*h^2+2*h*x..	2*a*h^2				
10	8*h+x0	a*(8*h+x0)^2..	a*(17*h^2+2*h*x..	2*a*h^2				
11	9*h+x0	a*(9*h+x0)^2..	a*(19*h^2+2*h*x..	2*a*h^2				
12	10*h..	a*(10*h+x0)^..	a*(21*h^2+2*h*x..	2*a*h^2				
13	11*h..	a*(11*h+x0)^..	a*(23*h^2+2*h*x..	2*a*h^2				
14	12*h..	a*(12*h+x0)^..	a*(25*h^2+2*h*x..	2*a*h^2				
15	13*h..	a*(13*h+x0)^..	a*(27*h^2+2*h*x..	2*a*h^2				
16	14*h..	a*(14*h+x0)^..	a*(29*h^2+2*h*x..					
17	15*h..	a*(15*h+x0)^..						

Fig. 5.3 (**a**) Numerical finite differences. (**b**) Algebraic finite differences

the intervention of the teacher. In fact the class teacher stressed its value as an instrumented action (Rabardel 2002), to support explorations in the numerical environment. Students realised that:

1. If they changed only the value of c, column B changed, whilst columns C and D of the first and second differences did not change; hence, they argued, the way in which a function increased/decreased did not depend on the coefficient c;
2. If they changed the coefficient b, then columns B and C changed but column D did not; many students conjectured that the coefficient b determines whether a function increases or decreases, but not its concavity;
3. If they changed the coefficient a, then columns B, C and D changed; hence the coefficient a was responsible for the concavity of the function.

Here, it is difficult to understand why such relationships hold and to produce at least an argument or even a proof of such conjectures. The tables of numbers do not suggest any justification. Now, the symbolic power of the spreadsheet became useful.[2] The students' very interesting instrumented actions consisted in substituting letters for the numbers (Fig. 5.3b); in most cases, the teacher had suggested this practice, but a couple of students used it autonomously. The resulting spreadsheet shows clearly that the value of the second difference is $2ah^2$. The letters condense the symbolic meaning of the numerical explorations, so proofs can be produced (with teacher's help) because of the spreadsheet's symbolic support. In the subsequent lesson, the teacher stressed the power of the symbolic spreadsheet; a fresh practice had entered the classroom.

Finally, a typical algebraic proof, where the main steps are computations – like the proof produced through use of the spreadsheet – apparently differs from the more discursive proofs produced in elementary geometry. Such algebraic proofs

---

[2] They were using the TI-Nspire software of Texas Instruments.

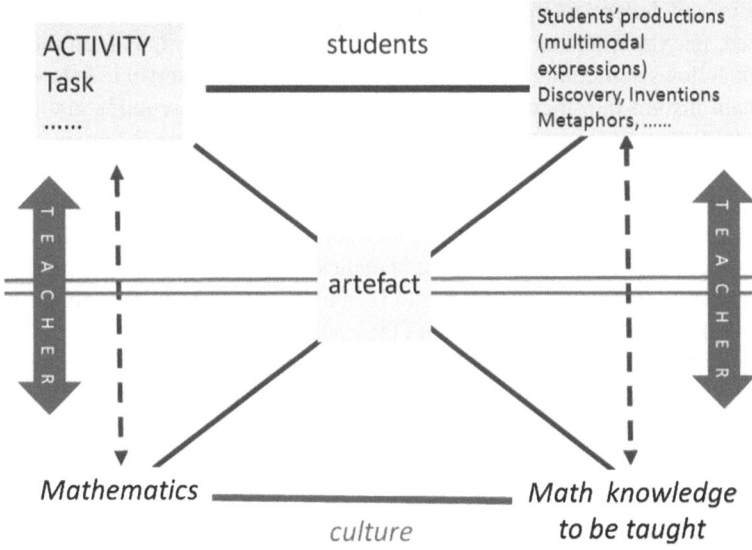

Fig. 5.4 Relationships between teachers, students, mathematics and artefacts in didactical activities

result from the "algebraisation" of geometry, which started with Descartes and improved further in the succeeding development of mathematics (e.g., the Erlangen programme by F. Klein 1872; see also Baez 2011). Consequently, so-called synthetic proofs have been surrogated by computations developed in linear algebra environments. Though students find big obstacles in learning algebra as a meaningful topic (e.g. Dorier 2000), CAS environments can support students in conceiving and producing such computational proofs, which are typically not so easy to reach in paper and pencil environments.

2.7 Implementation in Mathematics Classrooms

In all the cases above (straight-edge and compass as well as DGS or CAS), the teacher's role is crucial. The teacher not only selects suitable tasks to be solved through constructions and visual, numerical or symbolic explorations, but also orchestrates the complex transition from practical actions to theoretical argumentations. Students' argumentations rest on their experimental experiences (drawing, dragging, computing, etc.), so the transition to a validation within a theoretical system requires delicate mediation by an expert (see diagram, Fig. 5.4).

The upper part of Fig. 5.4 represents the student's space. The students are given a task (left upper vertex) to be solved with an artefact or set of artefacts. The presence of the artefact(s) calls into play experimental activities: for example, drawing with straight-edge and compass; creating DGS-figures with DGS-tools; or using numbers and letters in the symbolic spreadsheet. An observer, the teacher for instance, may monitor the process: students gesticulate, point, and tell themselves or their fellows something about their actions; from this observable behaviour one may gain insight into their cognitive processes. If the task requires giving a final report (either oral or written), traces of the experience are likely to remain in the text produced. Such reports may thus differ from the decontextualised texts typical of mathematics; nevertheless, they can evoke specific mathematical meanings.

The lower part of Fig. 5.4 represents the mathematical counterpart of the students' experience. There is the activity of mathematics in general as a cultural product, and there is the mathematical knowledge to be taught according to curricula. The link between the students' productions and the mathematics to be taught is the responsibility of the teacher, who has to construct a suitable process that connects the students' personal productions with the statements and proofs expected in the mathematics to be taught.

Hence, Fig. 5.4 highlights two important responsibilities for the teacher:

1. Choosing suitable tasks (left side);
2. Monitoring and managing of the process from students' productions to mathematical statements and proofs (right side)

The second point constitutes the core of the semiotic mediation process, in which the teacher is expected to foster and guide the students' evolution towards recognisable mathematics. The teacher acts both at the cognitive and the metacognitive levels, by fostering the evolution of meanings and guiding the pupils to awareness of their mathematical status (see the idea of mathematical norms, Cobb et al. 1993; see also chapter 5 in this volume). From a sociocultural perspective, one may interpret these actions as the process of relating students' "personal senses" (Leont'ev 1964/1976, pp. 244 ff.) to mathematical meanings, or of relating "spontaneous" to "scientific" concepts (Vygotsky 1978/1990, p. 286 ff.). The teacher, as an expert representative of mathematical culture, participates in the classroom discourse to help it proceed towards sense-making within mathematics.

Within this perspective, several investigations have focused on the teacher's contribution to the development of a mathematical discourse in the classroom, specifically in the case of classroom activities centred on using an artefact (Bartolini Bussi et al. 2005; Mariotti 2001; Mariotti and Bartolini Bussi 1998). The researchers aimed at identifying specific "semiotic games" (Arzarello and Paola 2007; Mariotti and Bartolini Bussi 1998) played by the teacher, when intervening in the discourse, in order to make the students' personal senses emerge from their common experience with the artefact and develop towards shared meanings consistent with the target mathematical meanings. Analysis of the data highlighted a recurrent pattern of interventions encompassing a sequence of different types of operations (Bartolini Bussi and Mariotti 2008; for further discussion, see Mariotti 2009; Mariotti and Maracci 2010).

Thus, artefacts have historically been fruitful in generating the idea of proof and consequently can provide strong didactical support for teaching proofs, specifically, if the teacher acts as a semiotic mediator. In the next section, we illustrate this issue from the point of view of students.

3 Part 2: A Student-Centred Analysis

Suitably designed technology can help students to face and possibly to overcome the obstacles between their empirical mathematical tasks and the discipline's theoretical nature. When integrated in the teaching of proofs, artefacts trigger a network of interactive activities amongst different components categorisable at two different epistemological levels:

1. The convincing linguistic logical arguments that explain WHY according to the specific theory of reference;
2. The artefact-dependent convincing arguments that explain WHY according to the mathematical experimentation facilitated by an artefact.

Approaching proof in school consists in promoting a *network of interactive activities* in order to connect these different components. For example, as we discuss below, abductive processes can support interactions between (1) and (2) above. Other interactive activities concern students' multimodal behaviour[3] whilst interacting within technological environments. Such activities feature in the transition to proof within experimental mathematics, a transition with novel and specific features compared to the transition to proof within more traditional approaches. Here, we scrutinise when and how the distance between arguments and formal proofs (Balacheff 1999; Pedemonte 2007) produced by students can diminish because of the use of technologies within a precise pedagogical design.

To focus the didactical and epistemological aspects of this claim, we recall four theoretical constructs taken from the current literature:

1. *Almost-empiricism* and experimental mathematics;
2. *Abductive* vs. *deductive* activities in mathematics learning;
3. *Cognitive unity* between arguments and proofs;
4. *Negation* from a mathematical and cognitive point of view.

Using these theoretical constructs, we scrutinise some studies of students asked to explore different mathematical situations with different artefacts and

---

[3] The notion of *multimodality* has evolved within the paradigm of *embodiment*, which has been developed in recent years (Wilson 2002). Embodiment is a movement in cognitive science that grants the body a central role in shaping the mind. It concerns different disciplines, e.g. cognitive science and neuroscience, interested with how the body is involved in thinking and learning. It emphasises sensory and motor functions, as well as their importance for successful interaction with the environment, particularly palpable in human-computer interactions. A major consequence is that the boundaries among perception, action and cognition become *porous* (Seitz 2000). Concepts are so analysed not on the basis of 'formal abstract models, totally unrelated to the life of the body, and of the brain regions governing the body's functioning in the world' (Gallese and Lakoff, 2005, p.455), but considering the *multimodality* of our cognitive performances. We shall give an example of multimodal behaviours of students when discussing the multivariate language of students who work in DGE. For a more elaborate discussion, see Arzarello and Robutti (2008).

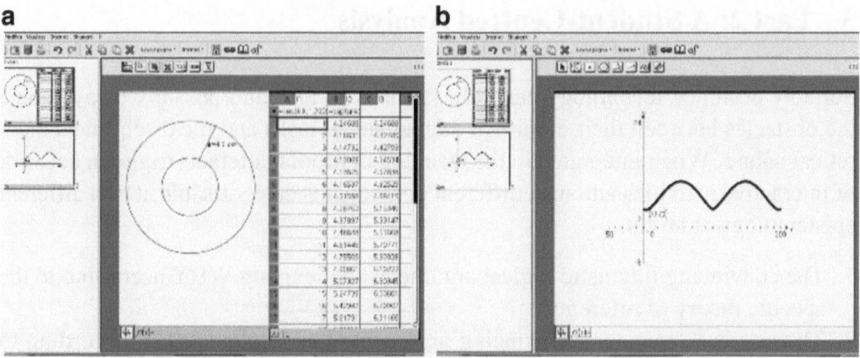

Fig. 5.5 (**a**) Solving a problem with data capture in TI-nspire. (**b**) Graph of the solution

within different pedagogical designs. Specifically, we show that a suitable use of technologies may improve the *almost-empirical* aspects in students' mathematical activities through a specific production of *abductive arguments*, which generate a *cognitive unity* in the transition from arguments to proofs. We also focus on some reasons why such a unity may not be achieved, particularly in the case of arguments and proofs by contradiction, where the *logic of negation* typically presents a major difficulty for students.

3.1   Almost-Empiricism and Experimental Mathematics

The notion of *almost-empirical actions*, introduced by Arzarello (2009), describes some instrumented actions[4] within DGS and CAS environments. It refines the usual epistemic/pragmatic dyadic structure of the instrumental approach. We provide a brief emblematic example.

In Arzarello's (2009) study, students of the 10th grade faced a simple problem, originated by the PISA test:

> *The students A and B attend the same school, which is 3 Km far from A's home and 6 Km far from B's home. What are the possible distances between the two houses?*

They produced a solution by using TI-Nspire software as illustrated in Fig. 5.5. They drew two circles, whose centre is the school and which represent the

---

[4] The so called *instrumentation approach* has been described by Vérillon & Rabardel (1995) and others (Rabardel 2002; Rabardel and Samurçay 2001; Trouche 2005). In our case particular ways of using an artefact, e.g. specific dragging practices in DGS or data capture in TI-Nspire, may be considered an *artefact* that is used to solve a particular *task* (e.g. for formulating a conjecture). When the user has developed particular *utilisation schemes* for the artefact, we say that it has become an *instrument* for the user.

possible positions of the two houses with respect to the school. They then created two points, say *a* and *b*, moving on each circle, constructed the segment *ab* and measured it using a software command. Successively they created a sequence of the natural numbers in column A of the spreadsheet (Fig. 5.5a) and through two animations (moving *a* and *b* respectively) they collected the corresponding lengths of *ab* in columns B and C. In the end, they built the "scattered plot" A vs. B and A vs. C (Fig. 5.5b), and drew their conclusions about the possible distances of A's and B's houses by considering the regularities of the scatter graph and discussing why it is so.

The variable points and the ways they are manipulated in the example are typical of the software, which allows a collection of data similar to those accomplished in empirical sciences. One first picks out the variables involved, then through sequence A one gets a device to reckon the time in the animation conventionally; namely, the time variable is made explicit. The instrumented actions of TI-Nspire software naturally induce students to do so. The scattered plot thus combines the time variable A versus the length variable B or C, because the TI-Nspire software enables making the time variable explicit within mathematics itself.[5]

Given a mathematical problem like that above, one can "do an experiment" very similar to those made in empirical sciences. One picks out the important variables and makes a concrete experiment using them (e.g., collecting the data in a spreadsheet through the data-capture command). On can study mutual inter-relationships between variables (e.g., using the scatter plot) and conjecture and validate a mathematical model, possibly by new experiments. In the end, one can investigate why such a model is obtained and produce a proof of a mathematical statement. All these steps follow a precise protocol: pick out variables, design the experiment, collect data, produce the mathematical model, and validate it. The protocol is made palpable by different specific commands in the (TI-Nspire) software, such as naming variables, animation or dragging, data capture, and producing a scatter plot.

Such practices within TI-Nspire are as crucial as the dragging practices within DGS. Both incorporate almost-empirical features that can support the transition from the empirical to the deductive side of mathematics. Baccaglini-Frank (in print) has suggested how this can happen when the students are able to internalise such practices and to use them as *psychological tools* (Kozulin 1998; Vygotsky 1978, p. 52 ff) for solving conjecture-generation problems.

In this sense, the practices with the software introduce new methods in mathematics. Of course, the teacher must be aware of these potentialities of the software and integrated them into a careful didactical design. Such practices consist not only in the possibility of making explorations but also in the precise protocols that students learn to follow according to the teacher's design. Similarly, external data concerning certain quantities are passed to a computer through the use of probes. In our

---

[5] This procedure is very similar to the way Newton introduced his idea of scientific time as a quantitative variable, distinguishing it from the fuzzy idea of time about which hundreds of philosophers had (and would have) speculated (Newton, CW, III, p. 72).

case, the measures are collected through the "data capture" from the "internal experiment" made in the TI-Nspire mathematical world by connecting the three environments illustrated in Fig. 5.5a, b (Geometrical, Numerical, Cartesian) through suitable software commands. From one side, these methods are empirical, but from the other side they concern mathematical objects and computations or simulations with the computer, not physical quantities and experiments. Hence, the term *almost-empirical* (Arzarello 2009), which recalls the vocabulary used by some earlier scholars; for example, Lakatos (1976) and Putnam (1998) claimed that mathematics has a *quasi-empirical* status (cf. Tymoczko 1998). However, "almost-empirical" stresses a different meaning: The main feature of almost-empirical methods is the precise protocol that the users follow to make their experiments, in the same way that experimental scientists follow their own precise protocols in using machines.

Almost-empirical methods also apply within DGS environments; in fact, there are strong similarities between instrumented actions produced in TI-Ns and DGS environments. In addition, almost-empirical actions made by students in either environment are not exclusively pragmatic but also have an epistemic nature. As we discuss below, they can support the production of abductions and, hence, the transition from an inductive, empirical modality to a deductive, more formal one.

3.2 Abductions in Mathematics Learning

Abduction is a way of reasoning pointed out by Peirce, who observed that abductive reasoning is essential for every human inquiry, because it is intertwined both with perception and with the general process of invention: "It [abduction] is the only logical operation which introduces any new ideas" (C.P. 5.171).[6] In short, abduction becomes part of the *process of inquiry* along with induction and deduction.

Peirce gave different definitions of abduction, two of which are particularly fruitful for mathematical education (Antonini and Mariotti 2009; Arzarello 1998; Arzarello and Sabena in print; Baccaglini-Frank 2010a), particularly when technological tools are considered:

1. The so-called *syllogistic abduction* (C.P. 2.623), according to which a *Case* is drawn from a *Rule* and a *Result*. There is a well-known Peirce example about beans:

   Rule: All the beans from this bag are white
   Result: These beans are white
   Case: These beans are from this bag

   Such an abduction is different from a *Deduction* that would have the form: the *Result* is drawn from the *Rule* and the *Case*, and it is obviously different from an *Induction*, which has the form: from a *Case* and many *Results* a *Rule* is drawn.

---

[6] Peirce's work is usually referred to in the form C.P. n.m., with the following meaning. C.P. = Collected Papers; n = number of volume; m = number of paragraph.

Of course the conclusion of an abduction holds only with a certain probability. (In fact Pólya 1968, called this abductive argument an *heuristic syllogism*.)
2. Abduction as "the process of forming an *explanatory hypothesis*" (Peirce, CP 5.171; our emphasis).

Along this stream of thought, Magnani (2001, pp. 17–18) proposed the following conception of abduction: the process of inferring certain facts and/or laws and hypotheses that render some sentences plausible, that explain or discover some (eventually new) phenomenon or observation. As such it is the process of reasoning in which explanatory hypotheses are formed and evaluated. A typical example is when a logical or causal dependence of two observed properties is captured during the exploration of a situation. The dependence is by all means an "explanatory hypothesis" developed to explain a situation as a whole.

As pointed out by Baccaglini-Frank (2010a, pp. 46–50), the two types of abduction correspond to two different logics of producing a hypothesis: the logic of *selecting a hypothesis* from amongst many possible ones (first type) versus the logic of *constructing a hypothesis* (second type). According to Peirce (C.P. 5.14-212), an abduction in either form should be *explanatory*, *testable*, and *economic*. It is an *explanation* if it accounts for the facts, but remains a suggestion until it is verified, which explains the need for *testability*. The motivation for the *economic* criterion is twofold: it is a response to the practical problem of having innumerable explanatory hypotheses to test, and it satisfies the need for a criterion to select the best explanation amongst the testable ones.

Abductions can be produced within DGS environments, and can bridge the gap between perceptual facts and their theoretical transposition through supporting a *structural cognitive unity* (see below) between the explorative and the proving phase, provided there is a suitable didactic design.

For example, Arzarello (2000) gave the following problem to students of ages 17–18 (Grade 11–12) who knew Cabri-géomètre very well and had already had a course in Euclidean geometry. Moreover, the students knew how to explore situations when presented with open problems (see Arsac et al. 1992) and could construct the main geometrical figures. The students were already beyond the third van Hiele level and were entering the fourth or fifth one. (For the use of van Hiele levels in DGS environments, see Govender and de Villiers 2002.) The problem read:

Let ABCD be a quadrangle. Consider the perpendicular bisectors of its sides and their intersection points H, K, L, M of pairwise consecutive bisectors. Drag ABCD, considering all its different configurations: What happens to the quadrangle HKLM? What kind of figure does it become?

Many pairs of expert[7] students typically solved the problem in five "phases":

1. The students start to shape ABCD into standard figures (parallelogram, rectangle, trapezium) and check what kind of figures they get for HKLM. In some cases they see that all the bisectors pass through the same point.

---

[7] Students who have acquired a sufficient instrumented knowledge of dragging practices according to a precise didactical design. The word is taken from Baccaglini-Frank (2010a).

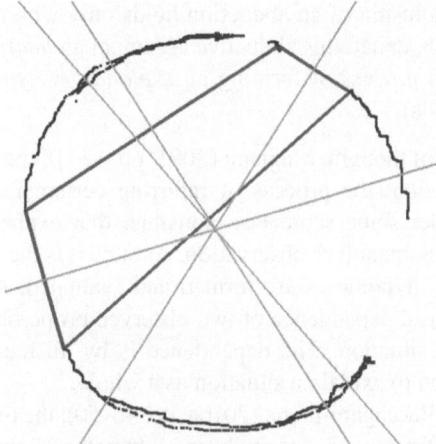

Fig. 5.6 Dragging with trace: generating a conjecture

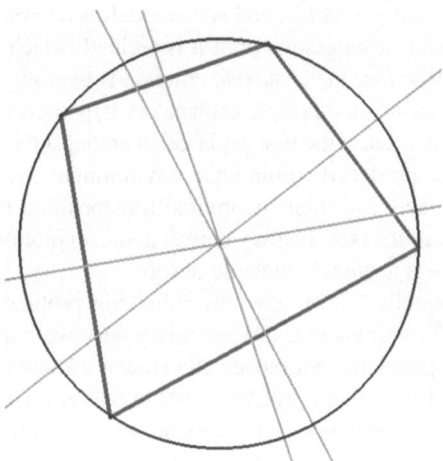

Fig. 5.7 Checking the conjecture with a construction

2. As soon as they see that HKLM becomes a point when ABCD is a square, they consider this interesting; therefore they drag a vertex of ABCD (starting from ABCD as a square) so that H, K, L, M keep on being coincident.
3. They realise that this kind of configuration is also true with quadrilaterals that apparently have no special property. Using the trace command, they find that whilst dragging a vertex along a curve that resembles a circle they can keep the four points together (Fig. 5.6). Hence they formulate the conjecture: *If the quadrilateral ABCD can be inscribed in a circle, then its perpendicular bisectors meet in one point, centre of the circle.*
4. They validate their conjecture by constructing a circle, a quadrilateral inscribed in this circle and its perpendicular bisectors, and observing that all of them meet in the same point (Fig. 5.7).

5. They write a proof of the conjecture. This process mainly consists in transforming (or eliminating) parts of the discussion held in the previous phases into a linear discourse, which is essentially developed according to the formal rules of proof.

Two major phenomena characterise the development above and are emblematic of these types of open tasks:

1. The production of an abduction: it typically marks a crucial understanding point in the process of solution;
2. The structural continuity between the conjecturing phases 1–4 and the transforming-eliminating activities of the last phase.

In producing an abduction, students first see a perceptual invariant, namely the coincidence of the four points in some cases (phases 1 & 2). So they start an exploration in order to see what conditions make the four points H, K, L, M coincide (phases 2 & 3). A particular kind of dragging (*maintaining dragging*: Baccaglini-Frank 2010b) supports this exploration: Using the trace command they carefully move the vertexes of ABCD so that the other four points remain together; finally, they realise they have thus produced a curve that resembles a circle (phase 3), namely a second invariant. At this point, they conjecture a link between the two invariants and see the second as a possible "cause" of the first; namely they produce an abduction in the form of an "explanatory hypothesis" (phase 4).

In producing a proof, (phase 5) the students write a proof that exhibits a strong continuity with their discussion during their previous explorations; more precisely, they write it through linguistic eliminations and transformations of those aforementioned utterances.

3.3 Maintaining Dragging as an Acquired Instrumented Action

The results discussed above were acquired through suitably designed teaching interventions, carefully considering the instrumented practices with the software, aimed at students' interiorising those practices as psychological tools (cf. Vygotsky 1978) they can use to solve mathematical problems. One approach to the instrumentation of dragging in DGS accords with this aim: the configuration of the '*Maintaining Dragging* Conjecturing Model' for describing a specific process of conjecture-generation, as developed by A. Baccaglini-Frank (Baccaglini-Frank 2010a) and by Baccaglini-Frank and Mariotti (2010). Enhancing Arzarello et al.'s (2002) analysis of dragging modalities (and of the consequent abductive processes), Baccaglini-Frank developed a finer analysis of dragging. She has also advanced hypotheses on the potential of dragging practices, introduced in the classroom, becoming a psychological tool, not only a list of automatic practices learnt by rote.

According to the literature (Olivero 2002), spontaneous use of some typologies of dragging does not seem to occur frequently. Consequently, Baccaglini-Frank first explicitly introduced the students to some dragging modalities, elaborated from

Arzarello et al.'s (2002) classification,[8] then asked the students to solve open tasks like the problem on quadrilaterals in the previous section. In such problem-solving activities, a specific modality of dragging appeared particularly useful to students: *maintaining dragging* (MD). Maintaining dragging consists of trying to drag a base point whilst also maintaining some interesting property observed. In the example above, the solvers noticed that the quadrilateral HKLM, part of the *Cabri*-figure, could "become" a single point; thus, they could attempt to drag a base point whilst trying to keep the four points together. In other words, MD involves both the recognition of a particular configuration as interesting and the attempt to induce the particular property to remain invariant during dragging. Healy's (2000) terminology would denote such an invariant as a soft invariant as opposed to a robust invariant, which derives directly from the construction steps. Maintaining dragging is an elaboration of *dummy locus dragging* but differs slightly: dummy locus dragging can be described as "wandering dragging that has found its path," a dummy locus that is not yet visible to the subject (Arzarello et al. 2002, p. 68), whilst MD is "the mode in which a base point is dragged, not necessarily along a pre-conceived path, with the specific intention of the user to maintain a particular property." (Baccaglini-Frank and Mariotti 2010).

In the example above, MD happened in phases 2 and 3, when the students dragged the vertices of the quadrilateral in order to keep together the four points H, K, L, M. As in phases 3–4 of the example, when MD is possible, the invariant observed during dragging may automatically become "the regular movement of the dragged-base-point along the curve" recognised through the trace mark; this can be interpreted geometrically as the property "dragged-base-point belongs to the curve" (Baccaglini-Frank in print). As pointed out by Baccaglini-Frank (2010b), the *expert* solvers proceed smoothly through the perception of the invariants and immediately interpret them appropriately as conclusion and premise in the final conjecture. However, becoming expert is not immediate, since it requires a careful didactical design that pushes the students towards a suitable instrumented use of the MD-artefact. In fact, "from the perspective of the instrumental approach, MD practices may be considered a utilization scheme for expert users of the *MD- artefact* thus making MD an *instrument* (the *MD-instrument*) for the solver with respect to the task for producing a conjecture" (Baccaglini-Frank, *ibid.*).

----

[8] Arzarello and his collaborators distinguish between the following typologies of dragging:

– *Wandering dragging*: moving the basic points on the screen randomly, without a plan, in order to discover interesting configurations or regularities in the figures.

– *Dummy locus dragging*: moving a basic point so that the figure keeps a discovered property; that means you are following a hidden path even without being aware of it.

– *Line dragging*: moving a basic point along a fixed line (e.g. a geometrical curve seen during the dummy locus dragging).

– *Dragging test*: moving draggable or semi-draggable points in order to see whether the figure keeps the initial properties. If so, then the figure passes the test; if not, then the figure was not constructed according to the desired geometric properties.

Baccaglini-Frank has organised this development of instrumented maintaining dragging (MD) in a Model (Baccaglini-Frank 2010a) that serves as a precise protocol for students, who follow it in order to produce suitable conjectures when asked to tackle open problems (Arsac 1999). This protocol structurally resembles that illustrated above for data-capture with TI-Nspire software. It is divided into three main parts:

1. Determine a configuration to be explored by inducing it as a (soft) invariant. Through wandering dragging the solver can look for interesting configurations and conceive them as potential invariants to be intentionally induced. (See phases 1–2 in our DGS example).
2. Searching for a Condition through MD: students look for a condition that makes the intentionally induced invariant be visually verified through maintaining dragging from path to the geometric interpretation of the path. Genesis of a Conditional Link through the production of an abduction. (See phases 2–3).
3. Checking the Conditional Link between the Invariants and verifying it through the dragging test. (See phases 3–4).

After a conjecture has been generated through this process, the students (try to) prove their conjecture (see phase 5).

The MD-conjecturing Model relates dragging and the perception of invariants with the developing a conjecture, especially with the emergence of the premise and the conclusion. This apparently common process is well-illustrated by the MD-conjecturing protocol as a sequence of tasks a solver can engage in. Baccaglini-Frank's model allows us to "unravel" the abductive process that supports both the formulation of a conjecture and the transition from an explorative phase to one in which the conjecture is checked.[9] The *path* (in our DGS example the circle created by the students in phase 4, Fig. 5.7) plays a central role by incorporating an answer to the solver's "search for a cause" for the intentionally induced invariant (phase 3, Fig. 5.6), and thus leading to the premise of a potential conjecture. Its figure-specific component (the actual curve that can be represented on the screen) contains geometrical properties that may be used as a bridge to proof.

---

[9] Arzarello et al. (1998a,b, 2000, 2002) showed that the transition from the inductive to the deductive level is generally marked by an *abduction*, accompanied by a cognitive shift from *ascending to descending* epistemological modalities (see Saada-Robert 1989), according to which the figures on the screen are looked at. The modality is ascending (from the environment to the subject) when the user explores the situation, e.g., a graph on the screen, with an open mind and to see if the situation itself can show her/him something interesting (like in phases 1, 2, 3 of our example); the situation is descending (from the subject to the environment) when the user explores the situation with a conjecture in mind (as in phase 4 of our example). In the first case the instrumented actions have an explorative nature (to see if something happen); in the second case they have a checking nature (to see if the conjecture is corroborated or refuted). Epistemologically, the cognitive shift is marked by the production of an *abduction*, which also determines the transition from an inductive to a deductive approach.

When MD is used expertly, abduction seems to reside at a meta-level with respect to the dynamic exploration. However, abduction at the level of the dynamic explorations only seems to occur when MD is used as a psychological tool (Kozulin 1998; Vygotsky 1978, p. 52 ff). According to Baccaglini-Frank analysis, it seems that:

> if solvers who have appropriated the MD- instrument *also internalize it* transforming it into a psychological tool, or a fruitful "mathematical habit of mind" (Cuoco 2008) that may be exploited in various mathematical explorations leading to the generation of conjectures, a greater cognitive unity (Pedemonte 2007) might be fostered. In other words, it may be the case that when the MD instrument is used as a psychological tool the conjecturing phase is characterized by the emergence of arguments that the solver can set in chain in a deductive way when constructing a proof (Boero et al. 1996).

(Baccaglini-Frank in print)

Something similar pertains to the protocol of data-capture with TI-Nspire software, which also involves almost-empirical actions (discussed above). Such almost-empirical methods seem fruitful for supporting the transition to the theoretical side of mathematics, provided their instrumentation can produce their internalisation as psychological tools and foster cognitive unity. On the contrary, when such protocols are merely used "automatically" they tend to lead to conjectures with no theoretical elements to bridge the gap between the premise and the conclusion of the conditional link; in other words, they do not encourage cognitive unity.

Since it is crucial in the transition from arguments to proofs, from the empirical to the theoretical, in the next section we discuss cognitive unity as the latest research has elaborated it.

3.4 Cognitive Unity

Boero has defined *cognitive unity* as the continuity that may exist between the argumentation of producing a conjecture and the construction of its proof (Boero et al. 1996). He hypothesises that, in some cases, "this argumentation can be exploited by the student in the construction of a proof by organizing some of the previously produced arguments into a logical chain" (Boero et al. 2010, p. 183). Pedemonte (2007) has further refined this concept, introducing the notion of *structural continuity* between argumentation and proof; that is, when inferences in argumentation and proof are connected through the same structure (abduction, induction, or deduction). For example, there is structural continuity between argumentation and proof if some abductive steps used in the argumentation are also present in the proof, as was the case in the problem of the distances of the houses from the school (see Fig. 5.5 and Boero et al. 2010).

Recently, Boero and his collaborators (Boero et al. 2010,) have integrated their analysis of *cognitive unity* with Habermas' elaboration (Habermas 2003) of *rational behaviour in discursive practices*. They have adapted Habermas' three components of rational behaviour (teleologic, epistemic, communicative) to the discursive practice of proving and have identified:

(A) An *epistemic aspect,* consisting in the conscious validation of statements according to shared premises and legitimate ways of reasoning...;
(B) A *teleological aspect,* inherent in the problem-solving character of proving, and the conscious choices to be made in order to obtain the desired product;
(C) A *communicative aspect,* consisting in the conscious adhering to rules that ensure both the possibility of communicating steps of reasoning and the conformity of the products (proofs) to standards in a given mathematical culture.

(Boero et al. 2010, pp. 188)

In this model, the expert's behaviour in proving processes can be described in terms of (more or less) conscious constraints upon the three components of rationality: "constraints of epistemic validity, efficiency related to the goal to achieve, and communication according to shared rules" (*ibid.*, p. 192). As the authors point out, such constraints result in *two levels of argumentation*:

− a level (that we call *ground level*) inherent in the specific nature of the three components of rational behaviour in proving;
− a *meta-level,* "inherent in the awareness of the constraints on the three components"

(*ibid.*, p. 192).

The two notions − cognitive unity and levels of argumentations − are important for analysing students' thought processes in the transition from argumentations to proofs within technological environments (especially DGS) and in particular very useful for analysing indirect proofs.

3.5   *Indirect Proofs*

Antonini and Mariotti (2008) have developed a careful analysis of indirect proofs and related argumentations from both a mathematical and a cognitive point of view, and have elaborated a model appropriate for interpreting students' difficulties with such proofs. Essentially, the model splits any indirect proof of a sentence S (principal statement) into a pair (s, m), where s is a direct proof (within a theory T, for example Euclidean Geometry) of a secondary statement S* and m is a meta-proof (within a meta-theory MT, generally coinciding with classical logic) of the statement S* $\rightarrow$ S. However, this meta-proof m does *not* coincide with Boero et al.'s (2010) meta-level considered above; rather, it is at the meta-mathematical level. As an example, they consider the (principal) statement S: "Let a and b be two real numbers. If ab=0 then a=0 or b=0" and the following indirect proof: "Assume that ab=0, a≠0, and b≠0. Since a≠0 and b≠0 one can divide both sides of the equality ab=0 by a and by b, obtaining 1=0". In this proof, the secondary statement S* is: "let a and b be two real numbers; if ab = 0, a ≠ 0, and b ≠ 0 then 1 = 0". A direct proof is given. The hypothesis of this new statement is the negation of the original statement and the thesis is a false proposition ("1 = 0").

Antonini and Mariotti (2008) use their model to point out that the main difficulty for students facing indirect proof consists in switching from s to m. Yet the difficulty seems less strong for statements that require a proof by contrapositive; that is, to prove B' → A' (secondary statement) in order to prove A → B (principal statement). Integrating the two models, we can say that switching from s to m requires a well-established epistemic and teleological rationality in the students and in this respect does need the activation of Boero et al.'s (2010) meta-level of argumentation.

The distinction between this meta-level and the ground level in Boero et al.'s (2010) model may be very useful in investigating the argumentation and proving processes related to indirect proof. Based on this distinction, we introduce the notion of *meta-cognitive unity*: a cognitive unity between the two levels of argumentation described above, specifically between the *teleological* component at the *meta-level* and the *epistemic* component at the *ground level*.

Different from structural and referential cognitive unity (Garuti et al. 1996; Pedemonte 2007), meta-cognitive unity is not concerned with two diachronic stages in students' discursive activities (namely argumentation and proving, which are produced sequentially), rather it refers to a synchronic integration between the two levels of argumentation. We hypothesise that the existence of such a meta-cognitive unity is an important condition for producing indirect proofs. In other words, lacking the integration between the two levels of argumentation can block students' proving processes or produce cognitive breaks like those described in the literature on indirect proofs. Meta-cognitive unity may also entail structural cognitive unity at the ground level and may develop through what we call '*the logic of not*' (see Arzarello and Sabena, in print).

3.6   The Logic of Not

The '*logic of not*' is an interesting epistemological and cognitive aspect of argumentation that sometimes is produced by students who tackle a problem where a direct argument is revealed as not viable.

Their strategy is similar to that of a chemist, who in the laboratory has to detect the nature of some substance. For example, knowing that the substance must belong to one of three different categories (a, b, c), the chemist uses suitable reagents to test: if the substance reacts in a certain way to a certain reagent it may be of type a or b but *not* c, and so on. In such practices, abductive processes are usually used: if, as a *Rule*, the substance S makes blue the reagent r and if the *Result* of the experiment shows that the unknown substance X makes blue the reagent r, then the chemist reasons that X = S (*Case* of the abduction).

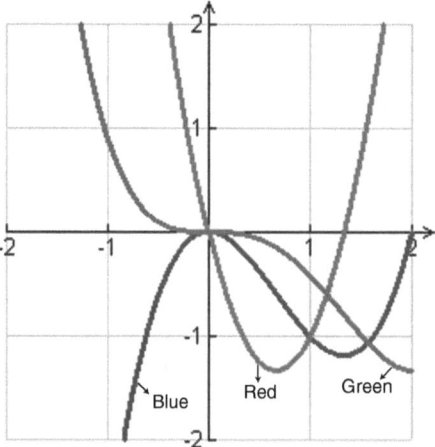

Fig. 5.8 The given task

For example, we summarise the case of a student comprehensively discussed in Arzarello and Sabena (in print). The student S (grade 9 in a science-oriented school) is solving the following task:

> *The drawing* (Fig. 5.8) *shows the graphs of: a function f, its derivative, one of its anti-derivatives. Identify the graph of each function, and justify your answer.*

The functions are differently coloured: the parabola red (indicated with R); the cubic (with a maximum point in the origin) is blue (B); the last (a quartic with an inflection point in the origin) is green (G). S does not know the analytic representations of the functions but has only their graphs. As such, he refers to the functions only by their colours.

In the first part of his protocol S checks which of the three functions can be *f*. He does this by looking for possible abductions, which involve the features of the given graphs.

For example, he starts supposing that *f* is the red function, probably because it is the simplest graph, and wonders whether he can apply an abductive argument with the following form to conclude that its derivative possibly is the green function:

1. Rule: "any derivative of a decreasing function is negative"
2. Result: "the green function is negative";   (ARG. 1)
3. Case: "the green function is the derivative of *f*"

Like the metaphorical chemist, S is able first to find a 'reagent' that discriminates between the substances (functions) he is analysing and then to validate his hypothesis with a further discriminating experiment, using his learnt practices with the graphs of functions. Arzarello and Sabena (in print) argue that some of S's argumentations are *teleological* and at the *meta-level*: they address S's own successive actions and his control of what is happening. The teleological component at the meta-level intertwines with the epistemological component at the ground level in a deep unity. This complex unity allows S to produce a proof by contraposition (the reasoning that logicians call "*modus tollens*": from "A implies B" to "not B implies not A"). Through this transition to a new epistemological status for his statements, S can lighten the cognitive load of the task using Arzarello and Sabena's 'logic of not', as written in part of S's protocol:

> Then I compared the "red" with the "green" function: but, the "green" function cannot be a derivative of the "red" one, because in the first part, when the "red" function is decreasing, its derivative should have a negative sign, but the "green" function has a positive sign.

Here the structure of the sentence is more complex than before: S is thinking towards a possible argument in the following form:

1. "any derivative of a decreasing function is negative"
4. "the "green" function has a positive sign"   (ARG. 2)
5. "the "green" function cannot be a derivative of an increasing function"

Unlike the possible abduction ARG. 1 above, ARG. 2 has the form: (1) and not (2); hence not (3). Crucially, the refutation of the usual Deduction (Rule, Case; hence Result) has the same structure, because of the converse of an implication ("A implies B" is equivalent to "not B implies not A"). In other words, the refutation of an argument by abduction coincides with the refutation of an argument by deduction. Whilst abductions and deductions are structurally and cognitively different, their refutations are identical formally. So S can produce a form of deductive argument "naturally" within an abductive modality – though remarkably from an epistemological and cognitive point of view, because the apparently "natural" abductive approach of students in the conjecturing phases (Arzarello et al. 1998) often does not lead to the deductive approach of the proving phase (Pedemonte 2007). The transition from an abductive to a deductive modality requires a sort of "somersault", an inversion in the functions and structure of the argument (the Case and Result functions are exchanged) which may cognitively load the students. However, this inversion is not necessary in either the refutation of an abduction or the refutation of a deduction. An "impossible" abductive argument already has the structure of a deduction; namely, it is an argument by contraposition. Of course greater cognitive effort is required to manage the refutation of an abduction than to develop a simple direct abduction. But the coincidence between abduction and deduction in cases of refutation allows avoiding the "somersault".

De Villiers (who does not use this terminology) has pointed out another possible use of the 'logic of not'. He observes that in DGS environments it is important:

> ...to sensitize students to the fact that although *Sketchpad* is very accurate and extremely useful for exploring the validity of conjectures, one could still make false conjectures with it if one is not very careful. Generally, even if one is measuring and calculating to 3 decimal accuracy, which is the maximum capacity of *Sketchpad 3*, one cannot have absolute certainty that there are no changes to the fourth, fifth or sixth decimals (or the 100th decimal!) that are just not displayed when rounding off to three decimals. This is why a logical explanation/proof, even in such a convincing environment as *Sketchpad*, is necessary for absolute certainty.

(de Villiers 2002, p. 9)

One way of promoting students' sensibility is to create some cognitive conflict to counteract students' natural inclination to just accept the empirical evidence that the software provides. For example, one can use an activity where students are led to make a false conjecture; though they are convinced it is true, it turns out false: In such cases the logic of not can drive them to produce a proof.

Thus, DGS has the potential of introducing students to indirect arguments and proofs; specifically, the use of "maintaining dragging" (MD) supports producing abductions. This can be fruitfully analysed in terms of the "logic of not".

3.7   Indirect Proof Within DGS

Theorem acquisition and justification in a DGS environment is a "cognitive-visual dual process potent with structured conjecture-forming activities, in which dynamic visual explorations through different dragging modalities are applied on geometrical entities" (Leung and Lopez-Real, 2002, p. 149). In this duality, visualisation plays a pivotal role in the development of epistemic behaviour like the Maintaining Dragging Model (Baccaglini-Frank and Mariotti 2010). On the cognitive side, DGS facilitates experimental identification of geometrical invariants through functions of variation induced by dragging modalities which serve as cognitive-visual tools to conceptualise conjectures and DGS-situated argumentative discourse (Leung 2008). With respect to indirect proof within DGS, Leung proposed a visualisation scheme to "see a proof by contradiction" in a DGS environment (Leung and Lopez-Real 2002). The scheme's key elements were the DGS constructs of pseudo-object and locus of validity; together, they serve as the main cognitive-visual bridge to connect the semiotic controls and the theoretical controls in the argumentation process. This scheme developed out of a *Cabri* problem-solving workshop conducted for a group of Grade 9 and Grade 10 students in Hong Kong. The researcher gave the following problem to students to explore in the *Cabri* environment:

> Let ABCD be a quadrilateral such that each pair of interior opposite angles adds up to
> 180°. Find a way to prove that ABCD must be a cyclic quadrilateral.

3.7.1   The Proof

After exploration, a pair of students wrote down the following "*Cabri*-proof" (Fig. 5.9).

The labelling of the angles in their diagram was not part of the actual *Cabri* figure. The key idea in the proof was the construction of an impossible quadrilateral EBFD. However, the written proof did not reflect the dynamic variation of the impossible quadrilateral in the *Cabri* environment that promoted the argumentation. An in-depth interview with the two students on how they used *Cabri* to arrive at the proof led to the construction of the cognitive-visual scheme.

PROOF:

Assume that for a quadrilateral with each pair of interior opposite angles adding up to 180°, the four vertices can be on different circles.

From the diagram we see that it has a contradiction as the sum of the opposite angles of the blue quadrilateral (EBFD) is 360°, which is impossible.

Therefore, for a quadrilateral with each pair of interior opposite angles adding up to 180°, the four vertices must be on the same circle.

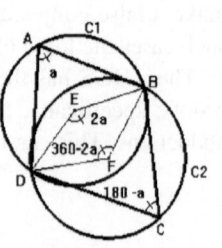

Fig. 5.9 The proof of the students

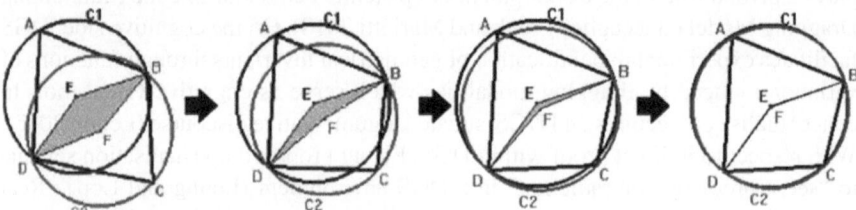

Fig. 5.10 Dragging the pseudo-quadrilateral EBFD

3.7.2 The Argumentation

The impossible quadrilateral EBFD, henceforth called a *pseudo-quadrilateral*, in Fig. 5.9 plays a critical role in organising the cognitive-visual process that leads to the construction and justification of a theorem. EBFD is a visual object that measures the degree of anomaly of a biased *Cabri* world with respect to the different positions of the vertices A, B, C and D. There are positions where the pseudo-quadrilateral EBFD vanishes when a vertex of ABCD is being dragged. Figure 5.10 depicts a sequence of snapshots in a dragging episode when C is being dragged until EBFD vanishes.

The last picture in the sequence shows that when C lies on the circumcircle C1 of the quadrilateral ABCD, E and F coincide, and at this instance, $\angle DEB + \angle DFB = 360°$ (a contradiction arising from the pseudo-quadrilateral EBFD). Furthermore, this condition holds only when C lies on C1; that is, when A, B, C and D are concyclic. The pseudo-quadrilateral EBFD and the circumcircle C1 play a dual role in an argumentation process. First, they restrict the quadrilateral ABCD to a special configuration that leads to the discovery that ABCD possesses

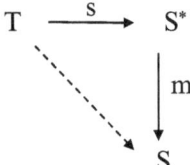

Fig. 5.11 Indirect proofs scheme

certain properties (through abductive inference). Second, they generate a convincing argument that collapses onto a Reductio ad Absurdum proof (Fig. 5.9) acceptable in Euclidean Geometry.

3.7.3   The Scheme

Suppose A is a figure (quadrilateral ABCD in the *Cabri*-proof) in a DGS environment. Assume that A satisfies a certain condition C(A) (interior opposite angles are supplementary) and impose it on all figures of type A in the DGS environment. This *forced presupposition* evokes a 'mental labelling' (the arbitrary labelling of $\angle DAB = 2a$ and $\angle DCB = 180\text{-}2a$ in the *Cabri*-proof) which leads one to *act cognitively* on the DGS environment. Thus C(A) makes an object of type A biased with extra meaning that might not necessarily be true in the actual DGS environment. This *biased DGS environment* exists as a kind of hybrid state between the *visual-true* DGS (a virtual representation of the Euclidean world) and a *pseudo-true* interpretation, C(A), insisted on by the user. In this pseudo world, the user can construct an object associated with A which inherits a local property that is not necessarily consistent with the Euclidean world because of C(A) (e.g., the impossible quadrilateral EBFD in the *Cabri*-proof): We call such an object associated a *pseudo object* and denote it by O(A). When part of A (the point C) is being dragged to different positions, O(A) might vanish (or degenerate; i.e., a plane figure to a line, a line to a point). The path or locus on which this happens gives a constraint (both semiotic and theoretical) under which the forced presupposition C(A) is "Euclidean valid"; that is, where the biased microworld is being realised in the Euclidean world. This path is called the *locus of validity* of C(A) associated with O(A) (the circle C1).

In the Indirect Proof context (Antonini and Mariotti, 2008), one can interpret this scheme as follows: S is the principle statement "If the interior opposite angles of a quadrilateral add up to 180°, then it is a cyclic quadrilateral"; S* is the secondary statement "If the interior opposite angles of a quadrilateral add up to 180° and its vertices can lie on two circles, then there exists a quadrilateral with the property that a pair of interior opposite angles add up to 360°." (Fig. 5.11)

In this scheme, T is Euclidean Geometry; s is a direct Euclidean proof. In a DGS environment, the meta-proof m could be a kind of dragging-based visual logic. In the case discussed above, when a pseudo object and a locus of validity arise, m could be a drag-to-vanish MD visual logic. Thus the *composite proof*

$$m \circ s : \quad T \dashrightarrow S$$

is an indirect proof that is both theoretical and DGS-mediated.

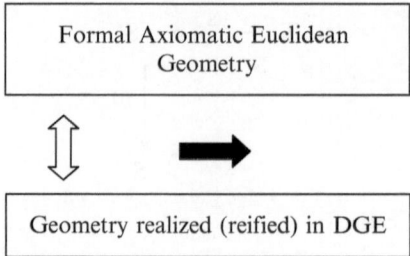

Fig. 5.12 From Lopez-Real and Leung (2006)

In relation to the Boero-Habermas model (Boero et al. 2010), the theoretical part (s) is the epistemological component (theoretical control) at the ground level where the existence of the pseudo-quadrilateral EBFD was deduced. The DGS-mediated part (m) is the teleological component (semiotic control) at the meta-level where the dragging-based argumentation took place. Hence the intertwined composite proof (m o s) can be seen as a meta-cognitive unity in which argumentation crystallises into a Reductio ad Absurdum proof.

Within the "logic of not", the DGS-mediated part (m) allows a link from the abductive modality to the deductive modality. In the previous example of S, the distance between the two modalities was annihilated because of the coincidence between the negations of the abduction and of the deduction; here the distance is shortened through m: in both cases, the cognitive effort required is reduced.

Lopez-Real and Leung (2006) suggested that Formal Axiomatic Euclidean Geometry (FAEG) and Dynamic Geometry Environment (DGE) are 'parallel' systems that are "situated in different semiotic phenomena" instead of two systems having a hierarchical relationship (Fig. 5.12).

> The vertical two-way arrow denotes the connection (networking) that enables an exchange of meaning between the systems. The horizontal arrow stands for a concurrent mediation process that signifies some kind of mathematical reality. This perspective embraces the idea that dragging in DGE is a semiotic tool (or a conceptual tool) that helps learners to form mathematical concepts, rather than just a tool for experimentation and conjecture making that doesn't seem to match the 'logical rigour' in FAEG. (Lopez-Real and Leung 2006, p.667)

In this connection, the MD dragging scheme – together with the construct of pseudo-object and locus of validity, and with the associated reasoning carried out, on the one hand, in the context of the DGS and, on the other hand, in T by the solver – may serve as channels to enable an exchange of meaning between the two systems (Fig. 5.13).

The idea of composite proof in DGS environment could possibly be expanded to a wider scope where there is a hybrid of Euclidean and DGS registers. Leung (2009) presented such a case where a student produced a written proof that intertwined Euclidean and DGS registers. The first results of Leung's analysis are promising, opening new perspectives of investigation.

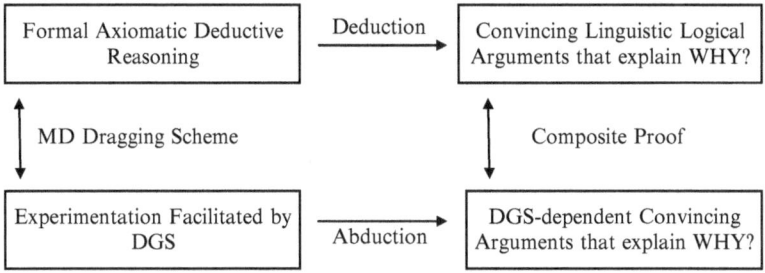

Fig. 5.13 The two semiotic systems: FAEG and DGE

4 Part 3: Towards a Framework for Understanding the Role of Technologies in Geometrical Proof

As discussed in Part 1, geometry may be split into what Einstein (1921) called "practical geometry", obtained from physical experiment and experience, and "purely axiomatic geometry" containing its logical structure. Central to learning geometry is an understanding of the relationship between the technologies of geometry and its epistemology. Technology in this context is the range of artefacts (objects created by humans) and the associated techniques which together are needed to achieve a desired outcome. Part 2 has set out in detail how the process of coordinating technologies with the development of geometric reasoning combines artefactual "know-how" with cognitive issues. In Part 3, we first provide a model, using Activity Theory, that highlights the role of technology in the process. Second, this part discusses the mediational role of digital technology in learning geometry, and the implications for developing proofs.

4.1 Modelling Proof in a Technological Context

To analyse proof in a technological context, it is useful to consider a framework derived from Activity Theory, shown in Fig. 5.14 (Stevenson 2008). The framework provides a way of describing the use of artefacts, (e.g., digital devices, straight-edges, etc.) in processes of proving. Activity Theory, a framework for analysing artefact-based social activity, is a "theory" in the sense that it claims that such activity can be described as a system using the categories shown in Fig. 5.14. In this section, we "flesh out" the epistemological, cultural and psychological dimensions of the system in relation to technology and proof.

In Fig. 5.14, the "object" of the system is the formulation of a problem which motivates and drives the proof process, with the "outcome" being the proof created from this system. "Artefacts" are any material objects used in the process, which in our case includes straight-edges, compasses and DGS. A "subject" is a person who

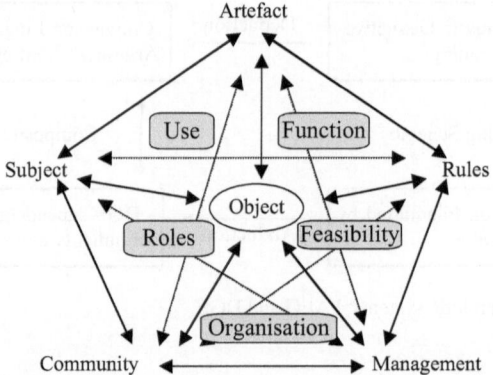

Fig. 5.14 Proving in a technological context: a framework from activity theory

is part of a "community" in which the activity is set; the community is "managed" through power structures that assign roles and status within a given context (e.g. classroom, professional mathematics community). "Rules" comprise the relationships that define different aspects of the system.

A "technology" in this system consists of the artefacts used, objects to which they can be applied, persons permitted to use the artefacts, and the sets of rules appropriate to each of the relationships which make up the system. Figure 5.14 identifies five aspects of the interactions with technology useful for the analysis. 'Function' covers the set of techniques applicable with the artefacts to a specific object, within a particular setting and social grouping. 'Use' relates to the ways in which individuals or groups actually behave with the artefacts within the social context, governed by the norms of community organisation. 'Roles' indicates the types of linguistic interactions adopted by the participants, and 'Organisation' refers to the groupings of those participating in the technology-based activities. Finally, 'Feasibility' relates to the practical constraints placed on an activity by the physical and temporal location.

Figure 5.4 (Part 1) and its associated description highlight the cultural dimension of a mathematics classroom engaging in proof activities with technology (straightedge and compass or DGS) by linking together the five aspects of the model in Fig. 5.14 (Use, Function, Roles, Feasibility and Organisation, cf. Stevenson 2008). The model brings together both the selection of tasks related to the objectives and outcomes of activities and the teacher's use of artefacts. In particular, it expresses how teachers tailor their use of artefacts to their specific classrooms in order to mediate ideas about proof and its forms to their students. As a result, the model expresses how specific forms of activity and styles of linguistic interaction between pupils and teacher in a given physical location provide the context for studying proof.

Epistemologically, the system is defined by the rules governing the formal object of mathematical knowledge that is the context for the proof process (e.g., geometry). For "standard" proof, the technology consists of the artefacts needed to create the proof (e.g., paper and pencil) and the rules of inference that govern how statements are organised. Their rules of inference describe the syntactical dimension of the process. A major claim for using proof as a method for obtaining knowledge is that, if "applied correctly", the rules of inference preserve the semantic integrity of the argument forming the proof. Semantics deals with the meanings given to geometric statements and is concerned with the truth and knowledge claims that a proof contains. Technology, therefore, plays a key role in the relationship between the semantics and syntactical structures of a proof, raising the issue of whether syntax and semantics are separable or inextricably intertwined with the technology.

As the section "Abductions in mathematics learning" (Part 2) proposes, conjectures become proofs by applying the technology of logical inference. Abduction is, epistemologically, a non-linear process which develops over a period of time and involves iterations between facts and the conjectured rules that gradually come to explain those facts. Constructing a proof involves restructuring a conjecture to suit the linear form of logical inference so that the technique can be applied to organise the argument "on the page". Such linearisation re-interprets (or removes) the diachronic aspect of abduction as an epistemological structure. In the process of translating conjecture into a proof, constructions, false starts, and strings of informal calculations are removed. References to sensorimotor processes in geometry are suppressed by talking about "ideal" points and lines, with the paper surface acting as a kind of window on the "real" geometry. (Livingstone 2006). In terms of the model in Fig. 5.14, the role of the teacher is crucial in helping learners make this linguistic transition. The extracts of dialogue related to this process of linearisation in Part 2 indicate how the cues and leads given by the teacher aid the learner in filtering the conjecture so that the techniques of inference can be used to organise it appropriately.

Adding DGS to this situation does not change the essential dynamic tensions resulting from the need to translate from one technological setting (straight-edge and compass or DGS) to another (rules of inference). The discussions in Part 2 of the "logic of not" and "indirect proof" (the 'Use' aspects of the model in Fig. 5.14) imply that abduction arises as a strategy to deal with those dynamic tensions.

4.2 Digital Technology as a Mediational Artefact

Learning how to use geometrical equipment, whether physical or digital, is part of the instrumentation of geometry (Verillion and Rabardel 1995), the interplay between facility with artefacts and the development of psychological concepts. Physically, one has the experience of using a straight edge to draw a "straight line" and compasses to make a "curve". In Lakoff's (1988) framework, such actions can be interpreted as developing a "prototype". The motor-sensory action of using a straight-edge and pencil, combined with the word "straight" and the Gestalt perception of the resulting mark on a surface, embodies the concept of "straightness".

Fig. 5.15 A genesis of the Poincaré disc for hyperbolic geometry

This account raises the question of how far technology, in general, mediates the understanding of geometrical concepts, and how that mediation relates to proof in the formal sense. For example, rather than simply motivating proofs of results, does/ can/should DGS play an integral part in forming the conceptual structures that constitute geometry? Much of the interest in DGSs lies in the representation of Euclidean geometry, but as Part 2 implies, DGSs provide a different kind of geometry from that obtained by paper and pencil construction, or by the axiomatic version of Euclidean geometry. Consequently, different versions of geometry emerge with these different technologies: "pencil geometry", "digital geometry", and "axiomatic geometry". The question is not whether technology mediates knowledge, but how different technologies mediate different kinds of knowledge. Proof, as a means of establishing knowledge claims, should therefore take account of the mediational role that artefacts play in epistemology.

Learning non-Euclidean geometry, for example, has a number of complexities when compared to the Euclidean case. On the one hand, spherical geometry is relatively straightforward, since learners may have everyday opportunities to develop visual intuitions. Being "smaller" than Euclidean space, spherical surfaces are both closed and bounded, and allow both physical and digital manipulation. On the other hand, hyperbolic space poses a different problem; it is "larger" than Euclidean space, so learners have difficulty defining a complete physical surface to manipulate (Coxeter 1969). The learning process also lacks opportunities for visual intuition and suffers from difficulties in finding appropriate artefacts to support instrumentation. However, digital technologies offer possibilities for engaging with hyperbolic geometries that cannot be found otherwise (Jones et al. 2010). Figure 5.15 shows how a two-dimensional Euclidean model can be obtained by projecting a hyperbolic surface, and illustrates the geometry associated with the projection.

Imagine that an Escher tessellation is spread isometrically across the hyperboloid on the left-hand side of Fig. 5.15. Viewing it as a projection onto a flat disc gives the image in the centre of Fig. 5.15. The grid for the tessellation is shown on the right of Fig. 5.15, together with the basic hyperbolic triangle OAB used to tessellate the disc. Triangle OAB shows one of the key differences between hyperbolic geometry and Euclidean: the angle sum of the hyperbolic triangles is less than 180°. The edge of the circle represents infinity (it is called the "horizon"), which can be

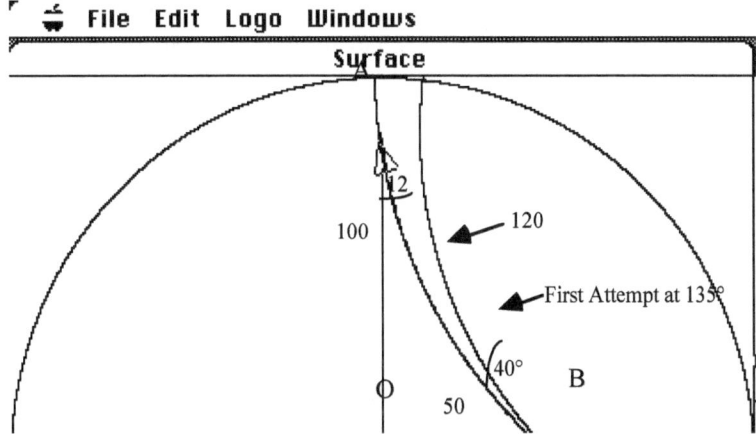

Fig. 5.16 The turtle within the hyperbolic world

approached but never reached, as indicated by the "bunching" of the tessellations at the circumference of the central image and of the grid lines on the right. Further, "straight" lines in the hyperbolic world can be either Euclidean straight lines (in fact diameters of the horizon) or circular arcs (orthogonal to the horizon).

Using Turtle geometry, it is possible to animate the two-dimensional projective model of the hyperbolic surface to provide an artefact for exploring the geometry (Jones et al., *ibid*). Taken from Stevenson (2000), the following snippet shows the work of two adults (S and P) using the non-Euclidean Turtle microworld to illustrate the role of artefacts in mediating understanding of the geometry. In Fig. 5.16, S and P first draw the lines OA and OB; then they attempt to find the line (AB) to close the triangle. Starting with the Turtle at B, pointing to the right of the screen, they turn it left through 135°, and use a built-in procedure called "Path" which indicates how the Turtle would travel if moved along that heading. They see that line does not close the triangle, so they turn the Turtle by a further 5° to the left. This time the path goes through A, and they reflect on the screen results (Fig. 5.16).

S picks up a hyperbolic surface provided for them and reminds himself about the projection process (the left-hand side of Fig. 5.15). S comments on the diagram in Fig. 5.16:

> S: We haven't got 180, but it's walking a straight-line path (reflectively).
> (S here refers to the metaphor that, in order to trace a straight line on a curved surface – in this case the hyperboloid, the Turtle must take equal strides, hence "walking a straight-line path".)
> P: Yeah, you've probably got to turn.
> (Instinctively, P thinks that a Turtle walking on the curved surface must turn to compensate for the curvature. S is clear that this is not happening.)
> S: No, you don't have to turn. It's actually drawing a triangle on the surface.
> (Looking at the hyperboloid, S imagines the Turtle marking out the triangle on the surface.)
> S: The projection defies Pythagoras. No! Hang on, walking on the surface is defying it, isn't it! Because we walk straight lines on the surface; we just see them as curves on the projection.

For S, the projection clearly preserves the geometric properties of the Turtle's path on the surface. S believes that what they see is a fact about the geometry, not a result of the software or the projection. Two points are significant: First, S's insight would be impossible without both the material artefacts (physical surface and digital application) and the metaphor that turtles walk straight lines on curved surfaces by taking "equal-strides without turning" (Abelson and diSessa 1980, p. 204). Second, S is convinced by what he sees, and provides an explanation about why the image preserves something about the geometry of the hyperbolic surface.

There remains the problem of what might constitute a proof that the angle sum of any hyperbolic triangle is always less than 180°. Given that the artefacts mediate the object of study (hyperbolic geometry), one can convert S's insight into the technology of logical inference with its associated linearisation and the re-interpretation of the diachronic aspect. Proofs of the angle sum do exist, but they rely on the axiomatic approach described by Einstein and on the technology of inference (e.g. Coxeter 1969, p. 296 ff.). However, the proofs are abstract and lack visual intuition; the digital and physical artefacts described here offer learners a more concrete and visual situation.

In the not too distant future, learners may use electronic media, rather than paper, to develop their work, which would enable them to embed digital applications. Effectively, this process will separate the technology of inference from the need to lay out arguments on paper as some kind of final statement. As for the model presented in Fig. 5.15, its value lies in being able to provide the cultural and pedagogic context for these activities; it embeds technologies in social relationships and human motivation. It also shows how dynamic tensions arise in reasoning due to conflicts between technologies. Coupled with Balacheff's analysis of proof types (2008), the model identifies how the assumptions and expectations of those engaging in proof generate contradictions in their practices (Stevenson 2011).

In closing, in this chapter we have discussed some strands of experimental mathematics from both an epistemological and a didactical point of view. We have introduced some past and recent historical examples in Western culture in order to illustrate how the use of tools has driven the genesis of many abstract mathematical concepts.

The intertwining between concrete tools and abstract ideas introduces both an "experimental" dimension in mathematics and a dynamic tension between the *empirical nature* of the activities with the tools –which encompass perceptual and operational components– and the *deductive nature* of the discipline –which entails a rigorous and sophisticated formalisation. This *almost empirical* aspect of mathematics was hidden in the second half of the nineteenth and the first half of the twentieth century because of a prevailing formalistic attitude. More recently, the perceptual and empirical aspects of the discipline have come again on the scene. This is mainly due to the heavy use of the new technology, which is deeply and quickly changing both research and teaching in mathematics (Lovasz 2006).

We have illustrated the roles both perception and empiricism now play in proving activities within the classroom and have introduced some theoretical frameworks which highlight the dynamics of students' cognitive processes whilst working in CAS and DGS environments. The learners use those technologies to explore problematic situations, to formulate conjectures and finally to produce proofs. We have pointed out the complex interplay between inductive, abductive, and deductive modalities in the delicate transition from the empirical to the theoretical side in the production of proofs. This dynamic can be strongly supported by a suitable use of technologies, provided the students learn some practices in their use, for example the *maintaining dragging* scheme in DGS. We have also shown how the induced instrumental genesis can help learners in producing indirect proofs.

Finally, we have used Activity Theory to model the dynamic tension between empiricism and deduction as a consequence of *translating between different technologies*, understood in the broadest sense as something that can mediate between different ontologies and epistemologies.

Acknowledgements We thank all the participants in Working Group 3 in Taipei, Ferdinando Arzarello, Maria Giuseppina Bartolini Bussi, Jon Borwein, Liping Ding, Allen Leung, Giora Mann, Maria Alessandra Mariotti, Víctor Larios-Osorio, Ian Stevenson, and Nurit Zehavi, for the useful discussions we had there. Special thanks go to Anna Baccaglini-Frank for the fruitful discussions that at least four of the authors had with her during and after the preparation of her dissertation and for her contribution to this chapter. Many thanks also to the referees for their helpful suggestions. Finally, we wish to express our deepest appreciation to John Holt and to Sarah-Jane Patterson for doing a remarkable editing job.

References

Abelson, H., & di Sessa, A. (1980). *Turtle geometry*. Boston: MIT Press.

Antonini, S., & Mariotti, M. A. (2008). Indirect proof: What is specific to this way of proving? *ZDM Mathematics Education, 40*, 401–412.

Antonini, S., & Mariotti, M. A. (2009). Abduction and the explanation of anomalies: The case of proof by contradiction. In: Durand-Guerrier, V., Soury-Lavergne, S., & Arzarello, F. (Eds.), *Proceedings of the 6th ERME Conference*, Lyon, France.

Arsac, G. (1987). L'origine de la démonstration: Essai d'épistémologie didactique. *Recherches en didactique des mathématiques, 8*(3), 267–312.

Arsac, G. (1999). Variations et variables de la démonstration géométrique. *Recherches en didactique des mathématiques, 19*(3), 357–390.

Arsac, G., Chapiron, G., Colonna, A., Germain, G., Guichard, Y., & Mante, M. (1992). *Initiation au raisonnement déductif au collège*. Lyon: Presses Universitaires de Lyon.

Arzarello, F. (2000, 23–27 July). Inside and outside: Spaces, times and language in proof production. In T. Nakahara & M. Koyama (Eds.), *Proceedings of PME XXIV* (Vol. 1, pp. 23–38), Hiroshima University.

Arzarello, F. (2009). New technologies in the classroom: Towards a semiotics analysis. In B. Sriraman & S. Goodchild (Eds.), *Relatively and philosophically earnest: Festschrift in honor of Paul Ernest's 65th birthday* (pp. 235–255). Charlotte: Information Age Publishing, Inc.

Arzarello, F., & Paola, D. (2007). Semiotic games: The role of the teacher. In J. H. Woo, H. C. Lew, K. S. Park, & D. Y. Seo (Eds.), *Proceedings of PME XXXI* (Vol. 2, pp. 17–24), University of Seoul, South Korea.

Arzarello, F., & Robutti, O. (2008). Framing the embodied mind approach within a multimodal paradigm. In L. English (Ed.), *Handbook of international research in mathematics education* (pp. 720–749). New York: Taylor & Francis.

Arzarello, F., & Sabena, C. (2011). Semiotic and theoretic control in argumentation and proof activities. *Educational Studies in Mathematics, 77*(2–3), 189–206.

Arzarello, F., Micheletti, C., Olivero, F., Paola, D. & Robutti, O. (1998). A model for analysing the transition to formal proof in geometry. In A. Olivier & K. Newstead (Eds.), *Proceedings. of PME-XXII* (Vol. 2, pp. 24–31), Stellenbosch, South Africa.

Arzarello, F., Olivero, F., Paola, D., & Robutti, O. (2002). A cognitive analysis of dragging practices in Cabri environments. *Zentralblatt fur Didaktik der Mathematik/International Reviews on Mathematical Education, 34*(3), 66–72.

Baccaglini-Frank, A. (2010a). *Conjecturing in dynamic geometry: A model for conjecture-generation through maintaining dragging.* Doctoral dissertation, University of New Hampshire, Durham, NH). Published by ProQuest ISBN: 9781124301969.

Baccaglini-Frank, A. (2010b). The maintaining dragging scheme and the notion of instrumented abduction. In P. Brosnan, D. Erchick, & L. Flevares (Eds.), *Proceedings of the 10th Conference of the PMENA* (Vol. 6, 607–615), Columbus, OH.

Baccaglini-Frank, A. (in print). Abduction in generating conjectures in dynamic through maintaining dragging. In B. Maj, E. Swoboda, & T. Rowland (Eds.), *Proceedings of the 7th Conference on European Research in Mathematics Education*, February 2011, Rzeszow, Poland.

Baccaglini-Frank, A., & Mariotti, M. A. (2010). Generating conjectures through dragging in a DGS: The MD-conjecturing model. *International Journal of Computers for Mathematical Learning, 15*(3), 225–253.

Baez (2011). J. Felix Klein's Erlangen program. Retrieved April 13, 2011, from http://math.ucr.edu/home/baez/erlangen/

Balacheff, N. (1988). *Une étude des processus de preuve en mathématique chez des élèves de collège.* Thèse d'état, Univ. J. Fourier, Grenoble.

Balacheff, N. (1999). Is argumentation an obstacle? Invitation to a debate. *Newsletter on proof.* Retrieved April 13, 2011, from http://www.lettredelapreuve.it/OldPreuve/Newsletter/990506.html

Balacheff, N. (2008). The role of the researcher's epistemology in mathematics education: An essay on the case of proof. *ZDM Mathematics Education, 40*, 501–512.

Bartolini Bussi, M. G. (2000). Ancient instruments in the mathematics classroom. In J. Fauvel & J. Van Maanen (Eds.), *History in mathematics education: An ICMI study* (pp. 343–351). Dordrecht: Kluwer Academic.

Bartolini Bussi, M. G. (2010). Historical artefacts, semiotic mediation and teaching proof. In G. Hanna, H. N. Jahnke, & H. Pulte (Eds.), *Explanation and proof in mathematics: Philosophical and educational perspectives* (pp. 151–168). Berlin: Springer.

Bartolini Bussi, M. G., & Mariotti, M. A. (2008). Semiotic mediation in the mathematics classroom: Artefacts and signs after a vygotskian perspective. In L. English, M. Bartolini, G. Jones, R. Lesh, B. Sriraman, & D. Tirosh (Eds.), *Handbook of international research in mathematics education* (2nd ed., pp. 746–783). New York: Routledge/Taylor & Francis Group.

Bartolini Bussi, M. G., Mariotti, M. A., & Ferri, F. (2005). Semiotic mediation in the primary school: Du¨rer glass. In M. H. G. Hoffmann, J. Lenhard, & F. Seeger (Eds.), *Activity and sign – grounding mathematics education: Festschrift for Michael otte* (pp. 77–90). New York: Springer.

Bartolini Bussi, M. G., Garuti, R., Martignone, F., Maschietto, M. (in print). Tasks for teachers in the MMLAB-ER Project. In *Proceedings PME XXXV*, Ankara.

Boero, P. (2007). *Theorems in schools: From history, epistemology and cognition to classroom practice.* Rotterdam: Sense.

Boero, P., Dapueto, C., Ferrari, P., Ferrero, E., Garuti, R., Lemut, E., Parenti, L., & Scali, E. (1995). Aspects of the mathematics-culture relationship in mathematics teaching-learning in compulsory school. In L. Meira & D. Carraher (Eds.), *Proceedings of PME-XIX.* Brazil: Recife.

Boero P., Garuti R., Lemut E. & Mariotti, M. A. (1996). Challenging the traditional school approach to theorems: A hypothesis about the cognitive unity of theorems. In L. Puig & A. Gutierrez (Eds.), *Proceedings of 20th PME Conference* (Vol. 2, pp. 113–120), Valencia, Spain.

Boero, P., Douek, N., Morselli, F., & Pedemonte, B. (2010). Argumentation and proof: A contribution to theoretical perspectives and their classroom implementation. In M. M. F. Pinto & T. F. Kawasaki (Eds.), *Proceedings of the 34th Conference of the International Group for the Psychology of Mathematics Education* (Vol. 1, pp. 179–204), Belo Horizonte, Brazil.

Borwein, J. M., & Devlin, K. (2009). *The computer as crucible: An introduction to experimental mathematics*. Wellesley: A K Peters.

Bos, H. J. M. (1981). On the representation of curves in Descartes' géométrie. *Archive for the History of Exact Sciences, 24*(4), 295–338.

Cobb, P., Wood, T., & Yackel, E. (1993). Discourse, mathematical thinking and classroom practice. In E. A. Forman, N. Minick, & C. A. Stone (Eds.), *Contexts for learning: Socio cultural dynamics in children's development*. New York: Oxford University Press.

Coxeter, H. S. M. (1969). *Introduction to geometry* (2nd ed.). New York: Wiley.

Cuoco, A. (2008). Introducing extensible tools in elementary algebra. In C. E. Greens & R. N. Rubinstein (Eds.), *Algebra and algebraic thinking in school mathematics, 2008 yearbook*. Reston: NCTM.

Davis, P. J. (1993). Visual theorems. *Educational Studies in Mathematics, 24*(4), 333–344.

de Villiers, M. (2002). *Developing understanding for different roles of proof in dynamic geometry*. Paper presented at ProfMat 2002, Visue, Portugal, 2–4 October 2002. Retrieved April 13, 2011, from http://mzone.mweb.co.za/residents/profmd/profmat.pdf

de Villiers, M. (2010). Experimentation and proof in mathematics. In G. Hanna, H. N. Jahnke, & H. Pulte (Eds.), *Explanation and proof in mathematics: Philosophical and educational perspectives* (pp. 205–221). New York: Springer.

Dörfler, W. (2005). Diagrammatic thinking. Affordances and constraints. In Hoffmann et al. (Eds.), *Activity and Sign. Grounding Mathematics Education* (pp. 57–66). New York: Springer.

Dorier, J. L. (2000). *On the teaching of linear algebra*. Dordrecht: Kluwer Academic Publishers.

Einstein, E. (1921). *Geometry and experience*. Address to the Prussian Academy of Sciences in Berlin on January 27th, 1921. Retrieved April 13, 2011, from http://pascal.iseg.utl.pt/~ncrato/Math/Einstein.htm

Euclide, S. (1996). *Ottica, a cura di F.* Incardona: Di Renzo Editore.

Gallese, V., & Lakoff, G. (2005). The brain's concepts: The role of the sensory-motor system in conceptual knowledge. *Cognitive Neuropsychology, 21*, 1–25.

Garuti, R., Boero, P., Lemut, E., & Mariotti, M. A. (1996). Challenging the traditional school approach to theorems. *Proceedings of PME-XX* (Vol. 2, pp. 113–120). *Valencia*.

Giusti, E. (1999). *Ipotesi Sulla natura degli oggetti matematici*. Torino: Bollati Boringhieri.

Govender, R. & de Villiers, M. (2002). *Constructive evaluation of definitions in a Sketchpad context*. Paper presented at AMESA 2002, 1–5 July 2002, Univ. Natal, Durban, South Africa. Retrieved April 13, 2011, from http://mzone.mweb.co.za/residents/profmd/rajen.pdf

Habermas, J. (2003). *Truth and justification*. (B. Fulmer, Ed. & Trans.) Cambridge: MIT Press.

Healy, L. (2000). Identifying and explaining geometric relationship: Interactions with robust and soft Cabri constructions. In T. Nakahara & M. Koyama (Eds.), *Proceedings of PME XXIV* (Vol. 1, pp. 103–117), Hiroshima, Japan.

Heath, T. L. (1956). *The thirteen books of Euclid's elements translated from the text of Heiberg with introduction and commentary* (pp. 101–122). Cambridge: Cambridge University Press.

Henry, P. (1993). Mathematical machines. In H. Haken, A. Karlqvist, & U. Svedin (Eds.), *The machine as mehaphor and tool* (pp. 101–122). Berlin: Springer.

Hoyles, C. (1993). Microworlds/schoolworlds: The transformation of an innovation. In C. Keitel & K. Ruthven (Eds.), *Learning from computers: Mathematics education and technology* (NATO ASI, series F: Computer and systems sciences, Vol. 121, pp. 1–17). Berlin: Springer.

Hoyles, C., & Noss, R. (1996). *Windows on mathematical meanings*. Dordrecht: Kluwer Academic Press.

Jones, K., Gutierrez, A., & Mariotti, M.A. (Guest Eds). (2000). Proof in dynamic geometry environments. *Educational Studies in Mathematics, 44*(1–3), 1–170. A PME special issue.

Jones, K., Mackrell, K., & Stevenson, I. (2010). Designing digital technologies and learning activities for different geometries. In C. Hoyles & J. B. Lagrange (Eds.), *Mathematics education and technology-rethinking the terrain New ICMI study series* (Vol. 13, pp. 47–60). Dordrecht: Springer.

Klein, F. (1872). Vergleichende Betrachtungen über neuere geometrische Forschungen [A comparative review of recent researches in geometry], Mathematische Annalen, 43 (1893), 63–100. (Also: Gesammelte Abh. Vol. 1, Springer, 1921, pp. 460–497). An English translation by Mellen Haskell appeared in *Bull. N. Y. Math. Soc* 2 (1892–1893): 215–249.

Kozulin, A. (1998). *Psychological tools a sociocultural approach to education*. Cambridge: Harvard University Press.

Laborde, J. M., & Strässer, R. (1990). Cabri-géomètre: A microworld of geometry for guided discovery learning. *Zentralblatt für Didaktik der Mathematik, 90*(5), 171–177.

Lakatos, I. (1976). *Proofs and refutations*. Cambridge: Cambridge University Press.

Lakoff, G. (1988). *Women, fire and dangerous things*. Chicago: University of Chicago Press.

Lakoff, G., & Núñez, R. (2000). *Where mathematics comes from?* New York: Basic Books.

Lebesgue, H. (1950). *Leçons sur les constructions géométriques*. Paris: Gauthier-Villars.

Leont'ev, A. N. (1976/1964). *Problemi dello sviluppo psichico*, Riuniti & Mir. (Eds.) (Problems of psychic development).

Leung, A. (2008). Dragging in a dynamic geometry environment through the lens of variation. *International Journal of Computers for Mathematical Learning, 13*, 135–157.

Leung, A. (2009). Written proof in dynamic geometry environment: Inspiration from a student's work. In F.-L. Lin, F.-J. Hsieh, G. Hanna & M. de Villiers (Eds.), *Proceedings of the ICMI 19 Study Conference: Proof and proving in mathematics education* (Vol. 2, pp. 15–20), Taipei, Taiwan.

Leung, A., & Lopez-Real, F. (2002). Theorem justification and acquisition in dynamic geometry: A case of proof by contradiction. *International Journal of Computers for Mathematical Learning, 7*, 145–165.

Livingstone, E. (2006). The context of proving. *Social Studies of Science, 36*, 39–68.

Lopez-Real, F., & Leung, A. (2006). Dragging as a conceptual tool in dynamic geometry. *International Journal of Mathematical Education in Science and Technology, 37*(6), 665–679.

Lovász, L. (2006). *Trends in mathematics: How they could change education?* Paper delivered at the Future of Mathematics Education in Europe, Lisbon.

Magnani, L. (2001). *Abduction, reason, and science. Processes of Discovery and Explanation*. Dordrecht: Kluwer Academic/Plenum.

Mariotti, M. A. (1996). Costruzioni in geometria. *L'insegnamento della Matematica e delle Scienze Integrate, 19B*(3), 261–288.

Mariotti, M. A. (2000). Introduction to proof: The mediation of a dynamic software environment. *Educational Studies in Mathematics (Special issue), 44*(1&2), 25–53.

Mariotti, M. A. (2001). Justifying and proving in the Cabri environment. *International Journal of Computers for Mathematical Learning, 6*(3), 257–281.

Mariotti, M. A. (2009). Artefacts and signs after a Vygotskian perspective: The role of the teacher. *ZDM Mathematics Education, 41*, 427–440.

Mariotti, M. A. & Bartolini Bussi M. G. (1998). From drawing to construction: Teachers mediation within the Cabri environment. In *Proceedings of the 22nd PME Conference* (Vol. 3, pp. 247–254), Stellenbosh, South Africa.

Mariotti, M. A., & Maracci, M. (2010). Un artefact comme outil de médiation sémiotique: une ressource pour l'enseignant. In G. Gueudet & L. Trouche (Eds.), *Ressources vives. Le travail documentaire des professeurs en mathématiques* (pp. 91–107). Rennes: Presses Universitaires de Rennes et INRP.

Martignone F. (Ed.) (2010). MMLab-ER - Laboratori delle macchine matematiche per l'Emilia-Romagna (Azione 1). In USR E-R, ANSAS ex IRRE E-R, Regione Emilia Romagna, Scienze e Tecnologie in Emilia-Romagna – Un nuovo approccio per lo sviluppo della cultura scientifica e tecnologica nella Regione Emilia-Romagna. Napoli: Tecnodid, 15–208. Retrieved April 13, 2011, from http://www.mmlab.unimore.it/on-line/Home/ProgettoRegionaleEmiliaRomagna/RisultatidelProgetto/LibroProgettoregional/documento10016366.html

Olivero, F. (2002). *The proving process within a dynamic geometry environment.* Unpublished PhD Thesis, University of Bristol, Bristol, UK.

Pedemonte, B. (2007). How can the relationship between argumentation and proof be analysed? *Educational Studies in Mathematics, 66*(1), 23–41.

Pólya, G. (1968). *Mathematics and plausible reasoning, Vol. 2: Patterns of plausible inference, (2nd edn.).* Princeton (NJ): Princeton University Press.

Putnam, H. (1998). What is mathematical truth? In T. Tymoczko (Ed.), *New directions in the philosophy of mathematics* (2nd ed., pp. 49–65). Boston: Birkhäuser.

Rabardel, P. (2002). *People and technology — A cognitive approach to contemporary instruments* (English translation of Les hommes et les technologies : une approche cognitive des instruments contemporains). Paris: Amand Colin.

Rabardel, P., & Samurçay, R. (2001, March 21–23). *Artefact mediation in learning: New challenges to research on learning.* Paper presented at the International Symposium Organized by the Center for Activity Theory and Developmental Work Research, University of Helsinki, Helsinki, Finland.

Saada-Robert, M. (1989). La microgénèse de la représentation d'un problème. *Psychologie Française, 34*(2/3), 193–206.

Schoenfeld, A. (1985). *Mathematical problem solving.* New York: Academic Press.

Seitz, J. A. (2000). The bodily basis of thought: New ideas in psychology. *An International Journal of Innovative Theory in Psychology, 18*(1), 23–40.

Stevenson, I. (2000). Modelling hyperbolic geometry: Designing a computational context for learning non-Euclidean geometry. *International Journal for Computers in Mathematics Learning, 5*(2), 143–167.

Stevenson, I. J. (2008). Tool, tutor, environment or resource: Exploring metaphors for digital technology and pedagogy using activity theory. *Computers in Education, 51*, 836–853.

Stevenson, I. J. (2011). An Activity Theory approach to analysing the role of digital technology in geometric proof. Invited paper for Symposium on *Activity theoretic approaches to technology enhanced mathematics learning orchestration.* Laboratoire André Revuz. Paris.

Straesser, R. (2001). Cabri-géomètre: Does dynamic geometry software (DGS) change geometry and its teaching and learning? *International Journal of Computers for Mathematical Learning, 6*, 319–333.

Trouche, L. (2005). Construction et conduite des instruments dans les apprentissages mathématiques: Nécessité des orchestrations. *Recherche en Did. des Math, 25*(1), 91–138.

Tymoczko, T. (1998). *New directions in the philosophy of mathematics* (2nd ed.). Boston: Birkhäuser.

Verillion, P., & Rabardel, P. (1995). Cognition and artefacts: A contribution to the study of thought in relation to instrumented activity. *European Journal of Psychology of Education, 10*(1), 77–101.

Vygotsky, L. S. (1978). *Mind in society: The development of higher psychological processes.* Cambridge: Harvard University Press.

Wilson, M. (2002). Six views of embodied cognition. *Psychonomic Bulletin and Review, 9*(4), 625–636.

5   Response to "Experimental Approaches to Theoretical Thinking: Artefacts and Proofs"

Jonathan M. Borwein, and Judy-anne Osborn

An overview of the chapter. The material we review focuses on the teaching of proof, in the light of the empirical and deductive aspects of mathematics. There is emphasis on the role of technology, not just as a pragmatic tool but also as a shaper of concepts. Technology is taken to include ancient as well as modern tools with their uses and users. Examples, both from teaching studies and historical, are presented and analysed. Language is introduced enabling elucidation of mutual relations between tool-use, human reasoning and formal proof. The article concludes by attempting to situate the material in a more general psychological theory.

We particularly enjoyed instances of the 'student voice' coming through, and would have welcomed the addition of the 'teacher's voice' as this would have further contextualised the many descriptions the authors give of the importance of the role of the teacher. We are impressed by the accessibility of the low-tech examples, which include uses of straight-edge and compass technology, and commonly available software such as spreadsheets. Other more high-tech examples of computer geometry systems were instanced and it would of interest to know how widely available these technologies are to schools in various countries (examples in the text were primarily Italian with one school from Hong Kong) and how much time-investment is called upon by teachers to learn the tool before teaching with it. The general principles explicated by the authors apply equally to their low-tech and high-tech examples, and are thus applicable to a broad range of environments including both low and high-resourced schools.

The main theoretical content of the chapter is in the discussion of why and how tool-use can lighten cognitive load, making the transition from exploring to proving easier. On the one hand tool-use is discussed as it relates to the discovery of concepts, both in the practical sense of students coming to a personal understanding, and in the historical sense of how concepts make sense in the context of the existence of a given tool. On the other hand the kinds of reasoning used in the practice of mathematics are made explicit – deduction, induction and a third called 'abduction' – with their roles in the stages of mathematical discovery, as well as how tool-use can facilitate these kinds of reasoning and translation between them.

Frequent use of the term 'artefact' is made in the writing, thus it is pertinent to note that the word has different and opposing meanings in the educational and the science-research literature. In the educational context the word means a useful purposely human-created tool, so that a straight-edge with compass or computer-software is an artefact in the sense used within the article. In the science-research literature, an artefact is an accidental consequence of experimental design which is misleading until identified, so for instance a part of a graph which a computer gets wrong due to some internal rounding-error is an artefact in this opposite sense.

The structure of the chapter. There is an introduction: essentially a reminder that mathematics has its empirical side as well as the deductive face which we see in formal proofs. Then there are three parts. Part 1 begins with a discussion of the history of mathematics, with reference to the sometimes less-acknowledged aspect of empiricism. In its second half, Part 1 segues from history into modern teaching examples. Part 2 is the heart of the article. It deals with the kind of reasoning natural to conjecture-forming, 'abduction', the concept of 'instrumentation' and cognitive issues relating to 'indirect proofs'; all through detailed examples and theory. Part 3 reads as though the authors are trying to express a large and fledgling theory in a small space. A general psychological paradigm called 'Activity Theory', is introduced, which deals with human activities and artefacts. An indication of digital technology as a means of translating between different ways of thinking is given in this context. We now discuss each Part in more detail.

Part 1. The historical half of Part 1 deals with geometric construction as a paradigmatic example, with the authors showing that since Euclid, tools have shaped concepts. For example, the straight-edge and compass is not just a practical technology, but helps define what a solution to a construction problem means. For instance cube duplication and angle trisection are impossible with straight-edge and compass (i.e. straight lines and circles) alone, but become solvable if the Nicomedes compass which draws a conchoid are admitted. A merely approximate graphic solution becomes a mechanical solution with the new tool. The moral is that changing the set of drawing tools changes the set of theoretically solvable problems, so that practical tools become theoretical tools. Another theme is the ambiguity noted in Descartes' two methods of representing a curve, by either a continuous motion or an equation; and subsequent historical developments coming with Pasch, Peano, Hilbert and Weierstrass, in which the intuition of continuous motion is suppressed in favour of purely logical relations. The authors perceive that historical suppression as having a *cost* which is only beginning to be counted, and rejoice that the increased use of computers is accompanying a revived intuitive geometric perspective.

This revival also offers the prospect of teachers who better understand mathematics in its historical context. Ideally, their students will gain a better appreciation of the lustrous history of mathematics. It is not unreasonable that students find hard concepts which took the best minds in Europe decades or centuries to understand and capture.

Part 1 is completed by examples from three educational studies followed by a discussion of the importance of the role of the teacher. Each example uses a tool to explore some mathematical phenomenon, with the teaching aim being that students develop a theoretical perspective. The first study involved over 2,000 students in various year-groups and 80 teachers, setting straight-edge and compass in the wider context of mathematical "machines". The second study, of Year 10 students, sits in the context of a particular DGS (Dynamical Geometry System), specifically the software called *'Cabri'*. Students first revised physical straight-edge and compass work, then worked in the virtual *Cabri* world, in which their drawings become what are termed 'Evocative Computational Objects' | no longer just shapes but shapes with associated *Cabri* commands and the capacity to be 'dragged' in interesting ways whose stability relates to the in-built hierarchical structure of the object. Interesting assertions made by the authors are that drag-ability relates to prove-ability, and that the original pencil drawings become signs for the richer *Cabri* objects. As in the previous study, a central aspect was student group discussion and comparison of solutions. The third study was of the use Year 9 students made of a CAS (Computer Algebra System) to explore the behaviour of functions. The students initially made numerical explorations, from which they formulated conjectures. Then, largely guided by a suggestion from their teacher, they substituted letters for numbers, at which point the path to a proof became evident.

The way in which the role of the teacher is crucial, in all three studies, is described with reference to a model expressed in Fig. 5.4. On the left of the diagram, activities and tasks chosen by the teacher sit above and relate to mathematics as a general entity within human culture. On the right side of the diagram, student's productions and discoveries from carrying out the tasks sit above and relate to the mathematical knowledge required by the school curriculum. The artefact (purpose-created tool) sits in the middle. Reading the picture clockwise in an arc from bottom left to bottom right neatly captures that teachers need to choose suitable tasks, students carry them out, and teachers help the students turn their discovered personal meanings into commonly understood mathematics. It is pointed out that as students discuss their use of artefacts, teachers get an insight into students' thought-processes.

Part 2. In Part 2, we get to the core of the article's discussion of proving as the mental process of transitioning between the exploratory phase of understanding a mathematical problem to the formal stage of writing down a deductive proof. The central claim is that this transition is assisted by tools such as Dynamic Geometry System (DGS) softwares and Computer Assisted Algebra (CAS) softwares, provided these tools are used within a careful educational design. The concept of abduction is central to the authors' conceptual framework. The term is used many times before it is defined – a forward reference to the definition in about the tenth paragraph of Part 2 would have been useful to us. It is worth quoting the definition (due to Peirce) verbatim:

> The so-called syllogistic abduction (C.P.2.623), according to which a Case is drawn from a Rule and a Result. There is a well-known Peirce example about beans:
>
> Rule: All beans from this bag are white
>
> Result: These beans are white Case:
>
> These beans are from this bag

Clearly this kind of reasoning is not deduction. The conclusion doesn't necessarily hold. But it might hold. It acts as a potentially useful conjecture. Nor, as the authors note, is this kind of reasoning induction, which requires one case and many results from which to suppose a rule.

The authors' naming and valuing of abduction sits within their broader recognition and valuing of the exploratory and conjecture-making aspects of mathematics which can be hidden in final-form deductive proofs. Their purpose is to show how appropriate abductive thinking arises in experimentation and leads to deductive proofs, when the process is appropriately supported.

The role of abductive reasoning in problem solving strikes these reviewers as a very useful thing to bring to educator's conscious attention. One of us personally recalls observing an academic chastise a student for reasoning which the academic saw as incorrect use of deduction, but which we now see as correct use of abduction in the early part of attempting to find a proof.

An example of 10th grade students faced with a problem about distances between houses and armed with a software called TI-Nspire is presented in detail in this section. The empirical aspect of mathematical discovery is described in an analogy with a protocol for an experiment in the natural sciences. We note that this example could be usefully adapted to a non-computerised environment. A point which the authors make, specific to the use of the computer in this context, is that the software encourages/requires useful behaviour such as variable-naming; which can then assist students in internalising these fundamental mathematical practices as psychological tools.

A teaching/learning example regarding a problem of finding and proving an observation about quadrangles, presented to 11th and 12th grade students is given. The authors give a summary of the steps most students used to solve the problem, and then interpret the steps in terms of the production of an abduction followed by a proof. The authors write

> In producing a proof, (Phase 5) the students write a proof that exhibits a strong continuity with their discussion during their previous explorations; more precisely, they write it through linguistic eliminations and transformations of those aforementioned utterances.

This statement is in the spirit of a claim at the start of Part 2 that empirical behaviour using software appropriately in mathematics leads to abductive arguments which supports cognitive unity in the transition to proofs.

The next main idea in Part 2 after 'abduction' is that of 'instrumentation'. The special kind of 'dragging' which has been referred to during discussions of DGS softwares is recognised as maintaining dragging (MD), where what is being maintained during the dragging is some kind of visible mathematical invariant. Furthermore the curve that is traced out during dragging is key to conjecture-formation and potentially proof.

The third main concept dealt with in Part 2 is that of 'indirect proof' and the difficulties that students often have with it. The authors usefully describe how software-mediated abductive reasoning may help, which they support with two plausible arguments. First, the authors note that indirect proofs can be broken up into direct proofs of a related claim (they use the term 'ground level') together with a proof of the relationship between the two claims (they use the term 'meta level'). Thus an argument for software-mediated reasoning says that use of software helps students keep track of the two levels of argument.

Second, the authors argue that abduction is useful to students partly because of what happens to formal (mathematical) claims when they are negated. The authors state that in some sense the cognitive distance between the conjecture and the proof is decreased in the negation step. They give two examples. The first is a study of a Year 9 student presented with a delightful problem about functions and their derivatives and anti-derivatives. In this case, refutation of an argument by abduction turns out to coincide with refutation of an argument by deduction. The second example is of a study of two students from Years 9 and 10 in Hong Kong given a problem about cyclic quadrilaterals in a *Cabri* environment. In this case, 'dragging' behaviour led the students to an argument which collapsed to a formal 'reductio ad absurdum'.

To summarise one stream of thought from Part 2 relating to the practice of learning and teaching: (a) tool-use facilitates exploration, especially visual exploration; (b) exploration (in a well-designed context) leads to conjecture-making; (c) practical tool-use forces certain helpful behaviours such as variable-naming; (d) this 'instrumentation' can lead to internalising tools psychologically; (e) for indirect proofs, the way negation works helps bridge the distance between kinds of reasoning used in conjecture-making and proof.

In the closing section of Part 2, the authors go beyond the claim that abduction supports proof and become more speculative. They quote Lopez-Real and Leung to claim that deduction and abduction are parallel processes in a pair of 'parallel systems', Formal Axiomatic Euclidean Geometry on the one hand, and Geometry realised in a Dynamic Geometry Environment (DGE) on the other hand, and that interaction between the two ways of knowing and storing information could be productive in ways not yet fully elucidated. It would be fascinating to see these ideas fleshed out.

Part 3. Part 3 introduces 'Activity Theory', a general framework concerned with the know-how that relates to artefacts, and attempts to situate the discussions of Parts 1 and 2 in this context, however as readers we found it difficult to gain insight from this formulation lacking as we do previous detailed knowledge of Activity Theory. The attempted translation between languages is scanty, although there are some illuminating examples, for instance an interesting use of 'turtle geometry' to explore hyperbolic geometry is presented in this section, where the turtle geometry is regarded as an artefact within Activity Theory. This part also expands upon the idea of instrumentation, linking ideas about concept-development, Gestalt perception and embodiment. There is much here which could be further developed; and which assuredly will be.

Conclusions. We first highlight a notion which is implicit throughout the chapter, which is the valuing of the teaching of proof in schools. Proof is a central component of mathematics however the valuing of the teaching of proof is not always taken for granted. For instance in the Australian context we know of instances of stark contrast, where the current state-based curricula does not emphasise proof (it is mentioned in the context of upper level advanced classes only), although we know of cases in which teacher-training does emphasise it. In short, we believe that there is often not enough teaching of proof in schools and that the chapter under review may help by providing a conceptual and practical bridge for students and their teachers between the activities of exploring mathematics and of creating and understanding proofs.

We also highlight the authors' own advisements about implementation in practice of the theory they have articulated. The authors emphasise that the role of the teacher is crucial both in lesson design and classroom interaction; as is neatly captured by Fig. 5.4 near the end of Part 1. They observe, for instance in their discussion of "maintaining dragging" in Part 2, that desired student understandings and behaviours often do not arise spontaneously. Further, they warn early in Part 1 (quoting Schoenfeld) that counterproductive student behaviour can arise as unintended by-products of teaching. At the end of Part 1 the authors give references to studies in which the kinds of *useful* interventions that teachers repeatedly make are analysed. It is helpful to have the centrality of the mathematics teacher made so clear. The importance of design and interaction are emphasised in quotations such as "The teacher not only *selects suitable tasks* to be solved through constructions and visual, numerical or symbolic explorations, but also *orchestrates* the complex transition from practical actions to theoretic arguments"; and "The teacher, as an expert representative of mathematical culture, *participates in the classroom discourse* to help it proceed towards sense-making in mathematics" (our emphasis).

In summary, this work repays the effort to read it. The historical perspective at the beginning brings the duality between empiricism and deductive reasoning usefully to mind. The examples, language and theory developed in Part 2 are likely to be clarifying and inspiring to both educators and theorists. The more speculative aspects at the end of Part 2 and in Part 3 call for further elucidation and development to which we look forward.

Part III
Historical and Educational
Perspectives of Proof

Chapter 6
Why Proof? A Historian's Perspective

Judith V. Grabiner

1 Introduction

If you ask a mathematician "why?" the mathematician will give you a proof. But when you ask a historian "why?" the historian will tell you a story. I'm a historian, and the title of this article asks why: "Why proof?" So I want to tell a story, or rather a series of stories, that address some questions about proof in the history of mathematics. These are: Why was logically based proof developed in the first place? Why were new types of proofs developed? And finally, what happens when standards of proof change?

These questions are important because every time we teach proof, we recapitulate a little of its history in the classroom. Reviewing the history of a topic lets us "remember" what it was like *not* to know about it. History, unlike logic or philosophy, shows what the effects of various ideas actually have been. In the classroom, if we can recreate the motivation, we can recreate the effect. Furthermore, proof did not develop in either a cultural or intellectual vacuum; we can recreate some of its historical context in the classroom. Finally, we can recreate the excitement that creating proof in mathematics historically has had. We need that excitement in the classroom too!

This essay traces the history of mathematical proof in the western tradition. It first addresses the birth of logical proof in Greek geometry and why the Greeks moved beyond visualisation to purely logical proof. Next, I look at the use of visual demonstration in Western mathematics after the Greeks. I then address two characteristics of more modern mathematics, abstraction and symbolism, and their power. Then follows a discussion of how and why standards of proof change, in particular the role of ideas imported from philosophy. Finally, I discuss how proof in mathematics interacts with the 'real world.'

J.V. Grabiner (✉)
Department of Mathematics, Pitzer College, Claremont, CA, USA
e-mail: jgrabiner@pitzer.edu

© The Author(s) 2021   147
G. Hanna and M. de Villiers (eds.), *Proof and Proving in Mathematics Education*,
New ICMI Study Series, https://doi.org/10.1007/978-94-007-2129-6_6

2   The Birth of Logical Proof in Greek Geometry

Demonstration at first was visualisation, the response to the question, "Is that so? *Show* me." In fact, the root of the Greek word "theorem" is a verb meaning "to look at". We see visual proofs in the mathematics of many cultures, including the Greek. Figure 6.1, from Plato's dialogue the *Meno*, shows how to prove, by counting triangles, that, given the square with area four in the lower-left-hand corner, the square with twice its area is the square on the diagonal (Fauvel and Gray 1987, pp. 61–67; Plato 2004, 81e–86c).

Even arithmetic uses visual proofs. Whether we use a tally system, Egyptian hieroglyphic numbers, an abacus, or add small numbers by counting on our fingers, the elementary truths of arithmetic are made visually apparent. The Greeks knew this too. There were visual proofs in Pythagorean number theory, where a dot represented a unit. For example, Fig. 6.2 presents the Pythagorean visual demonstration that the sum of the successive odd numbers is always a square.

Eventually, though, visual demonstration did not suffice for the Greeks. We know this because they made logical proof essential to their geometry. In the simplest sense, a logical proof deduces that something is a logical consequence of something else already believed to be true. Such proofs are necessary when what is being proved is not apparent. The Greeks felt that they needed to give an argument in such cases. But they did not limit proving to just a few statements; they built the entire subject so that it had a logical structure, assuming as its basis the smallest possible number of "already believed" results.

Scholars have suggested a wealth of historical explanations for why this particular culture, that of the ancient Greeks, was the one to give geometry this kind of logical structure and why the Greeks thought that doing this was so significant.

Fig. 6.1 The double square is the square on the diagonal

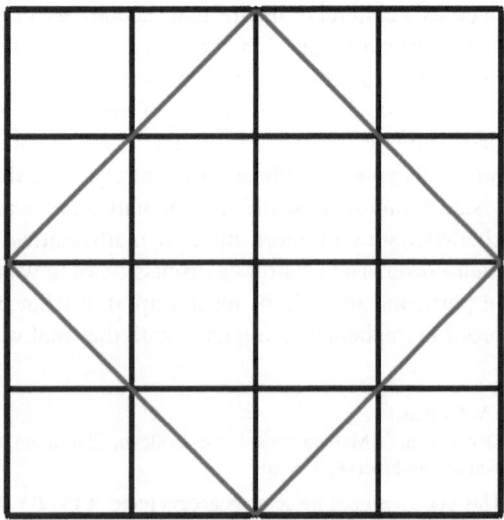

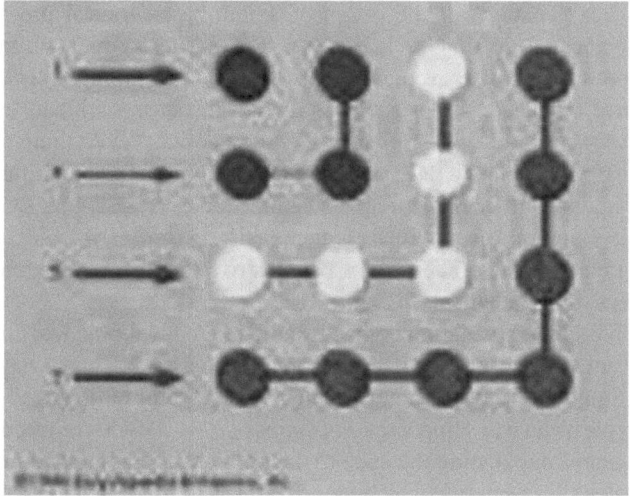

Fig. 6.2 The sum of the successive odd numbers is always a square

First, they argue, the Greek achievement did not come out of nowhere. Greek culture was heir to two previous mathematical traditions, the Egyptian and the Babylonian, each of which included a wealth of results (Imhausen 2007; Robson 2007). But these two traditions did not always agree. For instance, in finding the areas of circles, the Babylonians approximated the areas in ways equivalent to setting what we now call π first as 3, and later as 3 1/8,[1] in modern decimal notation 3.125 (Boyer and Merzbach 1989, p. 44). The best Egyptian area computations are equivalent to a value for π of 256/81, which is about 3.16 (Imhausen 2007, p. 31). Both values cannot be correct; the Greeks must have noticed this and asked what the true value is. Perhaps they reasoned that the way to avoid having multiple answers to the same question is to make only those assumptions about which nobody could disagree, like "all right angles are equal," and then deduce other things from those undoubtable assumptions to see how far they could get. As it turned out, they got pretty far.[2]

---

[1] Since the Babylonians used fractions with a base of 60, a practice which, incidentally, is the origin of our base-60 divisions of time into minutes and seconds, it would be more accurate to say that the area calculation given is equivalent to approximating π as $3 + 7/60 + 30/60^2$.

[2] None of the present discussion is meant to rule out the possibility that there were already steps towards forming axioms and logical proofs in Egyptian and Babylonian mathematics. No evidence of this is known to me at the present time, but scholarship on ancient mathematics continues, and one should keep an open mind. Even should such evidence be found, though, we would still need to explain why the Greeks chose to make logical proofs, using the smallest possible set of axioms, so central to their mathematics.

A second explanation for the origin of axiom-based logical proof in Greece comes from the nature of Greek science. The earliest Greek philosophers of nature tried to find a single explanatory principle that could make sense out of the entire universe. Thales (c. 624 – c. 547 BCE), for instance, said that "everything is water." Anaximenes (c. 585 – c. 528 BCE) said that "everything is air." The Pythagoreans (Sixth century BCE) said that "all is number." Democritus (c. 460– c. 370 BCE) said that "all is made of atoms." Empedocles (c. 492–432 BCE) said that everything is made up of the "four elements" fire, water, earth, and air. So, as in nature, so in mathematics: The Greeks would strive to reduce everything to simple first principles, to its "elements."

Another possible explanation comes from mathematical practice. Greek mathematicians wanted to solve problems (Knorr 1993). One effective way to solve a problem is to reduce it to a simpler problem whose solution is already known. A classic example is the way Hippocrates of Chios (470–410 BCE) reduced the famous Greek problem of duplicating the cube to the problem of finding two mean proportionals between two numbers, which in turn reduces to finding the intersection of a hyperbola and a parabola. How this reduction worked is clear if we look at the problem in modern notation. To find x such that $x^3 = 2a^3$, it suffices to find x and y such that $a/x = x/y = y/2a$. Equating the first two terms yields the equation $x^2 = y$, a parabola, and equating the first and third yields the equation $xy = 2a^2$, a hyperbola. The intersection of these two curves gives the required value for x (Katz 2009, pp. 40–41). Hippocrates also reduced finding the area of some "lunes" – areas bounded by two circular arcs – to finding the area of triangles.

When they reduced hard problems to simpler problems, and then reduced these to yet simpler problems, the Greek mathematicians were creating sets of linked ideas, from the complex to the simple. Now, suppose we run such a set of linked ideas in reverse order. This gives us a proof structure: simple things on which rest more complex things on which rest yet more complex things. The simplest things at the beginning are the "elements," and the intermediate ones are the fruitful, but not final, results we now call "lemmas." In fact, Hippocrates of Chios himself is credited with being the first to distinguish between those theorems interesting just for themselves and those theorems which lead to something else. And this explanation of the origin of proof – running reductions backwards – is supported by the fact that it was Hippocrates of Chios who wrote the best logically structured *Elements of Geometry* until Euclid wrote his own *Elements* 150 years afterwards. Much later, Pappus of Alexandria (Fourth century CE) gave the classic description of the relationship between these two processes. He called working backwards to find the simple things from which a problem can be solved "analysis," and subsequently proving that the solution follows from those simple things "synthesis" (Fauvel and Gray 1987, pp. 208–209).

A fourth, cultural explanation for logical proof in Greece is based on the nature of classical Greek society (Katz 2009, pp. 33, 39–40, 43; Lloyd 1996). In the sixth and fifth centuries BCE, Greece was made up of small city-states run by their citizens. Especially in Athens, the give and take of argument between disputing parties, from the law courts to the public assemblies, required, and therefore helped advance, logical skills. A good way to persuade people is to find their premises and then construct one's own argument by reasoning from their premises, and a good way to disprove other people's views is to find some logical consequence of their views that appears absurd. An ideal place to see these techniques illustrated in the Greek social context is in the dialogues of Plato. So the cultural emphasis on logical thinking would easily be incorporated into mathematics.

Finally, yet another explanation for "why logical proof in Greece?" comes from the influence that Greek philosophy had on mathematics. Greek philosophy was deeply argumentative. Contemporary thinkers disagreed; later thinkers began by trying to logically refute their predecessors. Zeno, for instance, presented his paradoxical arguments not to prove that motion is impossible but to use logic to challenge others' intuition and common-sense assumptions. The very fact that Plato presents his philosophy in dialogue form both illustrates and demonstrates that Greek philosophy was about logical argument as much as about conclusions. In this respect, Greek mathematics is like Greek philosophy, and not by accident. Greek philosophy was intimately linked to mathematics – not just for Plato, who made it the centre of his prescription for the education of the rulers of his ideal Republic, but also for his great rival and successor Aristotle. Aristotle wanted every science to start, like geometry, with explicitly stated elementary first principles, and then to logically deduce the rest of the subject (Fauvel and Gray 1987, pp. 93–94; Katz 2009, p. 44; McKirahan 1992). So Greek philosophy issued marching orders to mathematicians, and Euclid followed these orders.

All those factors worked together to promote the Greek invention of logical proof in mathematics. Now, though, let us focus on one particular type of proof: indirect proof, or proof by contradiction. The argument form, "If you accept that, why, then, you must also accept so-and-so,…but this contradicts such-and-such," was part and parcel of the educated Greek's weapons of refutation. But proof by contradiction is not merely destructive. It also allows us to rigorously test conjectures that cannot be tested directly and, if they are true, to demonstrate them. Take, for example, two supremely important historical instances of this process from ancient Greek mathematics.

First, Euclid defined parallel lines as lines that never meet. But we can never show directly that two lines never meet. One can only propose that the two lines *do* in fact meet and then prove that this leads to a contradiction. That exact move made Euclid's theory of parallels possible (Euclid 1956, *e.g.*, Book I, prop. 27).

For the second example, consider the nature of $\sqrt{2}$. There must be a square root of 2, because of the Pythagorean theorem for isosceles right triangles. One literally sees this in the diagram from Plato's *Meno* (Fig. 6.1). But is the square root of 2 the ratio of two whole numbers? Because there are infinitely many rational numbers, one cannot prove the irrationality of the square root of 2 by squaring every possible rational number to see if the square equals 2. Furthermore, no picture of an isosceles right triangle, no matter how carefully drawn, can possibly distinguish a rational length from an irrational length. However, if we assume that there is a rational number whose square is two, logic then leads us to a contradiction and so forces us to conclude that there can be no such rational number. The Pythagoreans thus proved that $\sqrt{2}$ was irrational. By these means, the Greeks created a whole new set of mathematical objects: magnitudes that they could prove were not rational numbers.

To sum up what the Greeks had now introduced into mathematics: Logic lets us reason about things that are beyond experience and intuition, about things that absolutely cannot be observed. The Greek proofs by contradiction changed the way later mathematicians thought about the subject-matter of mathematics.[3] Mathematics now had come to include objects whose existence cannot be visualised and which cannot be physically realised. Mathematics had become the study of objects neither visible nor tangible, objects transcending material reality, objects visible only to the eye of the intellect. Logical proof created these new objects, and these developments reinforced the role of proof as the heart of Western mathematics.

Thus, many factors contributed to the central role played by proof in Greek mathematics: disagreement between older results, the desire to establish elementary first principles, the logical structure produced when problems are solved by reduction to simpler problems, the role of argument in Greek society, the central importance of philosophical argument in Greek thought, and the major contributions to mathematics resulting from using proof by contradiction. For present purposes, one can use all of these ideas in the classroom to motivate the teaching of proof.

Once the Greeks had established this ideal of logical proof in mathematics, the ideal took on a life of its own. Even before Euclid, Aristotle was advocating the ideal of a science based on demonstration. Since mathematicians apparently had achieved truth by means of proof, practitioners of other areas of Western thought wanted to do the same. So thinkers, in theology, politics, philosophy, and science tried to imitate the mathematicians' method. Here, I discuss three of many important historical examples.

In 1675, the rationalist philosopher Baruch Spinoza wrote *Ethics Demonstrated in Geometrical Order* (Spinoza 1953). In that book, he first defined his terms – terms like "God" and "eternity." He then laid down axioms about things like existence and causality. On the basis of these definitions and axioms, he proved his philosophical conclusions. In particular, he gave an proof – an indirect proof – for the existence of God, demonstrating that "God, or substance consisting of infinite attributes, each one of which expresses eternal and infinite essence, necessarily exists" (Spinoza 1953, Prop. XI).

Even more important in the seventeenth century, Isaac Newton wrote his great work the *Principia*, not like a modern physics book but with the same definition-axiom-theorem structure as Euclid's *Elements*. In fact, Newton called his famous three laws "*Axioms*, or Laws of Motion" (Fauvel and Gray 1987, pp. 389–390;

---

[3] I have used the phrase 'Greek proofs by contradiction'. In philosophy, proof by contradiction exists in cultures independent of the Greek, notably in China (Leslie 1964; Siu 2009). But as far as I know, only the Greeks and their mathematical heirs used it within mathematics; Professor Siu (2009) says that he does not know of an example from China before the coming of the Jesuits in the seventeenth century.

Newton 1995, emphasis added). From these axioms, Newton deduced many propositions, such as: If a body moves in an ellipse, the force directed towards a focus of the ellipse varies inversely as the square of the distance between the body and that focus (Newton 1995, Book I, Proposition XI). Newton even deduced his law of gravity in the form of two theorems (Newton 1995, Book III, Propositions VII and VIII).

Third, the American Declaration of Independence also pays homage to the Euclidean proof method. The principal author, Thomas Jefferson, was well versed in contemporary mathematics (Cohen 1995). Jefferson began with postulates, saying "We hold these truths to be self-evident", including "that all men are created equal…" and, that if a government does not protect human rights, "it is the right of the people to alter or abolish it, and set up new government." The Declaration then says that it will "prove" – Jefferson's word – that King George's government does not protect human rights. Once Jefferson has proved this to his satisfaction, the Declaration of Independence concludes: "We *therefore* … publish and declare that these United Colonies are, and of right ought to be free and independent states" (Jefferson 1776, emphasis added). Indeed, Jefferson could have ended his argument, as had Spinoza and Newton, with the geometer's "QED."

All these examples suggest how important it is for all of liberal education, not just for mathematics, that students understand logical proofs. I am not repeating the general and often quoted statement that "mathematics trains the mind to think." I mean something more specific and more important. In the systems of secondary education in most countries, if mathematics teachers do not teach logic, logic doesn't get taught. Now, just as in ancient Athens, citizens of democracies need to be able to reason, to tell good arguments from bad. In the words of the American cultural historian Jacques Barzun, "The ability to feel the force of an argument apart from the substance it deals with is the strongest possible weapon against prejudice" (Barzun 1945, p. 121). The historical function of proof in geometry has been not just to prove theorems in geometry, though of course it did that, but also to exemplify and teach logical argument in every field from philosophy to politics to religion. We should be proud to be the guardians of this tradition and to pass it on to our students.

3 Visual Demonstration Revisited

However, logical proof, deduction from self-evident first principles, is not the only kind of proof in mathematics. The visual proof remains, and the visual frequently not only is more convincing psychologically but also is a means of further discovery. So some successors of Greek geometry creatively linked visual arguments to logical proof.

Figure 6.3 presents a well-known visual argument, found in Babylonian, Chinese, and Indian mathematics (Dauben 2007, pp. 222–226; Pfloker 2007, p. 392; Robson 2007, pp. 102–110), that can convince the modern student of algebra that $(a + b)^2 = a^2 + 2ab + b^2$.

Fig. 6.3 $(a+b)^2 = a^2 + 2ab + b^2$
visually

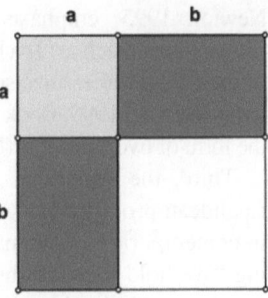

If the teacher is lucky, the student says, "Great! Now I understand, I see why it's not $a^2 + b^2$ but $a^2 + 2ab + b^2$." So the ancient visualisation helps explain algebraic computation.

However, Euclid brought the Greek logical attitude to testing this "obvious" visual result. He logically proved it in Book II, Proposition IV of his *Elements of Geometry* (Euclid 1956, vol. I, pp. 379–380; Fauvel and Gray 1987, pp. 118–119). His proof required seven different theorems from Book I of the *Elements*: theorems about parallel lines and angles, about parallelograms, and about the construction of squares.

Later, mathematicians in the medieval Islamic world combined visual demonstrations with the proof tradition from the Greeks and with computational traditions from the Eastern mathematical sources. This produced something new. For example, take a famous problem from the ninth-century *Algebra* of Muhammad ibn Musa al-Khwarizmi (Fauvel and Gray 1987, pp. 228–231; Struik 1969, pp. 58–60).

Al-Khwarizmi (and many later writers in the Islamic world) used the example which we would write $x^2 + 10 \ x = 39$. Al-Khwarizmi stated it as a computational problem to be solved: The root multiplied by itself, added to ten times the root, gives 39; but he could also represent the equation geometrically. (Figure 6.4 shows the equation and al-Khwarizmi's geometric solution.)

There is no obvious way to solve the diagrammed problem in the original form "L-shaped thing equals a number." However, it is easy to solve a problem with the form "a square equals a number." So al-Khwarizmi added something to the L-shaped figure to turn it into a square; that is, he completed the square. He filled the 'gap' in the L with a square with sides of 5.

He now had a larger square with sides of $x + 5$. But adding the square, with area 25, to the L-shaped left-hand side of the equation requires adding 25 to the right side of the equation as well. The right-hand side now becomes a square with area 64 ($39 + 25$). The side of that square must be 8 ($\sqrt{64}$), so $x + 5$ is 8 (he considered positive roots only); therefore x is 3.

How did al-Khwarizmi and his successors know that their method worked correctly in general, not just for that easily checked example? They knew because the geometrical representation of the method is guaranteed by Euclid's theorem II, 4 (Euclid 1956, vol. I, pp. 379–380). So al-Khwarizmi had both a discovery and a proof.

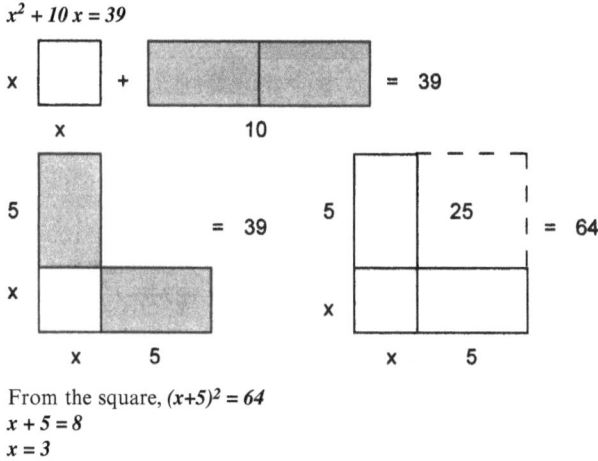

From the square, $(x+5)^2 = 64$
$x + 5 = 8$
$x = 3$

Fig. 6.4 How al-Khwarizmi completes the square

This example and many more show that algebra in the Islamic world brought together the activities of numerical problem-solving and logically based proof (Katz 2009, Chapter 9, Berggren 2007). This linkage was both fruitful and influential. In addition, this episode exemplifies how different people, and different cultures, may prefer one way of thinking about mathematical proof over another. Synthesising such diverse perspectives on what constitutes a proof has, historically, helped drive mathematical progress.

4   Abstraction, Symbolism, and Their Power

Historic changes in the style of proof also took place in Europe in the seventeenth and eighteenth centuries. The ruling proof paradigm changed from the geometric to the algebraic.

This shift began when François Viète (Franciscus Vieta) first introduced general symbolism in algebra in 1591 (Struik 1969, pp. 74–81; Viète 1968). Elementary school children learn that for every unequal pair of numbers like 9 and 7, not only does $9 + 7 = 16$, so does $7 + 9$. There are infinitely many such pairs of numbers. But Vieta's general symbolism lets us write down the infinite number of such facts all at once:

$$B + C = C + B.$$

So Vieta, by inventing general algebraic symbolism, began the transformation of algebra from a set of ways to solve individual problems to the general study of mathematical structures. A century after Vieta, Isaac Newton summed up the power and generality of this symbolic revolution by calling algebra "a universal arithmetic" (Newton 1967, title of the work) Newton meant that we could prove many facts if we just took for granted the universal validity of those symbolic manipulations that obey the laws of ordinary arithmetic.

For example, there are many different ways to arrange the square, linear, and constant terms to make up a quadratic. We have seen how earlier algebraists solved equations like $x^2 + 10\ x = 39$. But solving equations with forms like $x^2 + 21 = 10x$ required a somewhat different approach (Berggren 1986, p. 104). Since Vieta introduced the idea of general notation for constants, though, there has been really only one quadratic equation. In notation slightly more modern than Vieta's,[4] one can write $ax^2 + bx + c = 0$ for *every* quadratic equation and then derive the general solution for every quadratic by completing the square.

The notation has even more power than this. Saying that "3" solves al-Khwarizmi's equation gives no information about how that answer was obtained. But here, as Vieta's English follower William Oughtred put it in 1647, the operations we have performed to get the answer leave their footsteps behind (Fauvel and Gray 1987, p. 302). We know exactly how the answer was obtained from the coefficients in the equation. Furthermore, we have proved that this is the answer, because the truth of the generalised rules of arithmetical manipulation demonstrates it.

Another historically important seventeenth-century example shows symbolism itself discovering and proving results. Following Vieta, in the 1630s René Descartes (Struik 1969, pp. 89–93) and Thomas Harriott (Stedall 2003), independently, proved many results in what we now call the elementary theory of equations. For instance, they would consider an equation with two roots, a and b. Since $x = a$ and $x = b$, it follows that $x - a = 0$, $x - b = 0$, and therefore that $(x - a)(x - b) = 0$. Multiplying this product out yields:

$$x^2 - (a + b)x + ab = 0.$$

By this means, Descartes and Harriott demonstrated the root-coefficient relations of every quadratic equation. This process shows that a quadratic can have no more than two roots and makes clear why this is so. It shows also that the constant term has to be the product of the roots, and that the coefficient of the linear term must be the sum of the roots with the sign changed. This same approach let Descartes and Harriott discover and demonstrate similar relations for polynomial equations of any degree. For instance, for the first time, mathematicians knew for sure that an *nth* degree equation can have no more than n roots.

The power of symbolism and abstraction led to even greater progress. In 1799, Joseph-Louis Lagrange argued that the idea of algebra as universal arithmetic was

---

[4] Vieta, and many mathematicians in the century following him, thought of expressions like x^3 as volumes, x^2 as areas, and x as lines, and so they would not write an expression like $ax^2 + bx + c$. Vieta also designated his unknowns by upper-case vowels and his constants by upper-case consonants; the use of the lower-case letters x and y for the principal unknowns was introduced later by Descartes. But these details do not affect the main point here.

not sufficiently general. Algebra, he said, is more than a universal arithmetic; it is the general study of systems of operations (Lagrange 1808; compare Fauvel and Gray 1987, pp. 562–563). This further abstraction, the idea of studying systems of operations, helped lead Lagrange to study the permutations of the roots of algebraic equations, work that led eventually to the theory of groups as well as to the proof that there is no general solution in radicals of algebraic equations of degrees higher than four.

Furthermore, in the eighteenth century mathematicians turned away from the visual almost entirely, giving primacy to algebraic symbolism. One reason was that algebraists like Lagrange thought that algebra was "pure" and that intuition, especially geometric intuition, could lead us astray. This belief led mathematicians to seek algebraic proofs even for things that seemed obvious from pictures, like the intermediate-value theorem for continuous functions. In fact Lagrange was the first to try to prove the intermediate-value theorem (Grabiner 1981, p. 73). His proof was deficient, but the theorem was proved successfully in the early nineteenth century by Cauchy and by Bolzano (modulo implicitly assuming the completeness of the real numbers; Bolzano 1817; Cauchy 1821, pp. 378–380, reprinted by Fauvel and Gray 1987, pp. 570–571, excerpted in Fauvel and Gray 1987, pp. 564–566). So this new algebraic paradigm advanced many fields of mathematics considerably.

In the nineteenth century, mathematicians, beginning with William Rowan Hamilton, even deliberately investigated systems of operations that disobey the laws of arithmetic, non-commutative (Hankins 1976) or even non-associative systems. As a result, people began to think of mathematics itself in a new way, as the study of formal systems.

But there was a cost in the classroom – not in rigour, to be sure, but in losing geometric intuition. For example, there are algebraic proofs of the quadratic formula in textbooks. But when I ask my university students if they have ever seen the geometric diagrams for completing the square, they say they have not. The history of al-Khwarizmi's proof is therefore well worth recapturing. Seeing his proof helps understand why the quadratic formula, and its symbolic derivation, are true.

Furthermore, the cost of downgrading the visual was not only in the classroom. Over-reliance on formal reasoning could also lead mathematicians astray. Eighteenth-century mathematicians often manipulated formal symbols and paid little attention to what they stood for. For instance, Euler happily used the root-coefficient relations between polynomials and their factors to go back and forth between infinite series and infinite products. Often this led to valid new results (Dunham 1999; Euler 1748; Katz 2009, pp. 617–625), but sometimes it did not.

Consider this eighteenth-century example of algebra extended to the infinite that raises more problems than it solves. Algebraic long division gives us this formal relationship:

$$1/(1+r) = 1 - r + r^2 - r^3 + r^4 - \dots.$$

Is this equation true in general? How about if $r = 1$? In 1703, Guido Grandi said that if $r = 1$, yielding ½ for the left-hand side, then the infinite sum on the right-hand side was also ½ (Boyer and Merzbach 1989, p. 487–8). Leibniz agreed with him. It took until 1821 to rein in such excesses, when Augustin-Louis Cauchy, who both introduced and proved the standard tests for convergence of series, warned against the common assumption that algebra was completely general. "Most algebraic formulas," Cauchy wrote, "hold true only under certain conditions, and for certain values of the quantities they contain" (Cauchy 1821, p. iii), and that quotation from Cauchy is a good lead-in to our next topic: changing standards of proof.

5 When Standards of Proof Change

Not just styles of proof, but standards of proof, change. For example, consider the history of the calculus. In the late seventeenth century, Newton and Leibniz independently invented the algorithms and basic concepts of the calculus. The problem-solving power of this new subject was tremendous, and the calculus was applied to Newtonian physics with amazing success. Eighteenth-century mathematicians therefore did not worry too much about the logical basis of the calculus. The calculus was advanced by means of plausible arguments and heuristic derivations without what modern mathematicians would consider adequate justification. There was, in particular, a great deal of questionable reasoning about infinites and infinitesimals, like arguing that a curve is made up of infinitely many infinitesimal straight lines. And Newtonians used the idea of velocity, which some eighteenth-century critics said was part of physics, to justify mathematical conclusions about derivatives and integrals.

People noticed these logical defects. In 1734, George Berkeley, Bishop of Cloyne, attacked the logical rigour of eighteenth-century calculus. (Berkeley 1951; Fauvel and Gray 1987, pp. 556–558; Struik 1969, pp. 333–338). Berkeley had theological motives for attacking eighteenth-century science, but that didn't make his arguments wrong. Berkeley said that all his contemporaries' computations of what we now call "derivatives" were logically inconsistent. For example, consider finding the instantaneous rate of change of x^2. A typical solution began: Let x become $x + h$. Then the ratio of the differences of x^2 and x is $[(x + h)^2 - x^2] / h$. After multiplying this out and simplifying it, the ratio becomes $2x + h$. Then, the practitioners of the calculus said, as h vanishes, this expression becomes $2x$, and that *equals* the instantaneous rate of change.

Berkeley's criticism of such arguments pointed out that either h is zero or it is not. If h were zero, the top and bottom of the original ratio would both be zero, so there would be no ratio to set up. So h cannot be zero. But if h is not zero, what justifies finally discarding it? Even good students in modern calculus courses have trouble refuting this attack.

Several of Berkeley's mathematical contemporaries were disturbed by his arguments. Some of them tried to repair the foundations of the calculus. For instance, Colin Maclaurin used indirect proofs, based on inequalities between constant and varying velocities. (Grabiner 1997; Maclaurin 1742; Sageng 2005). Lagrange used symbolic reasoning about infinite series (Grabiner 1990; Lagrange 1797). However, these attempted rigorisations suffered from their own logical problems, and, unsatisfactorily, they did not really embody the intuitions that underlie the calculus.

This history demonstrates how difficult the rigorisation of the calculus was: too hard even for Newton, Maclaurin, Leibniz, and Lagrange. The eventual rigorisation of analysis in the nineteenth century came about by using delta-epsilon definitions and proofs. It began with Cauchy 150 years after the invention of the calculus, and was only concluded, later yet, by Weierstrass and his school in Berlin in the 1860s and beyond (Katz 2009, pp. 766–7). That it took over a century and a half is a good sign that the rigorous basis for the calculus is hard.

This difficulty deserves closer examination. What Cauchy and Weierstrass did to the limit concept was counterintuitive. In a common example of the intuitive idea of limit, the circle was said to be the limit of the inscribed regular polygons (d'Alembert and de la Chapelle 1789). One can see that, and that the polygons can be drawn as close as one likes to the circle. The intuition is that one eventually reaches the limit, so polygonal approximations actually give the area of the circle. More precisely: if a variable approaches a fixed value so that it eventually differs from that fixed value by less than any given quantity, the variable and that fixed value become, in Newton's words, "ultimately equal" (Newton 1995, Book I, Section I, Lemma I; Fauvel and Gray 1987, p. 391). Similarly, the slope of the secant ultimately becomes equal to the slope of the tangent. For Newton, the calculus produces precise, exact results, for example that the rate of change of x^2 is equal to $2x$.

But the nineteenth-century delta-epsilon definition of limit changed the meaning of "equal" when dealing with limits. Thus, logically, when one says that a fixed value is equal to the limit of the value of a variable, one is describing an infinite set of *inequalities*. That is, given every possible positive number epsilon, one can find a corresponding number delta such that if the absolute value of h is less than delta, the absolute value of the difference between, say, the ratio of the finite differences and the derivative is less than epsilon. That is what it means to say that something is equal to the limit of some process. That infinite set of inequalities – given any epsilon, one can find the appropriate delta – is sufficient to prove theorems about the value of limits. So now mathematicians can construct rigorous proofs about the concepts of the calculus. But to support rigorous proofs, the new definition has reinterpreted the intuitive idea of limit to define precise equality in terms of inequalities. That reinterpretation is part of what made rigorisation hard and makes the calculus a much deeper subject than it sometimes appears.

In fact, Berkeley's logical criticism that a secant can never become a tangent, that there is always a difference, never got an intuitive answer. "You can get as close as you like, and that's all you need for proofs," may be true, but it is not wholly satisfying. Such difficulties, which not only troubled Maclaurin and Lagrange but which trouble our students, are profound and have no simple answer.

I want to draw three more insights from this episode in the history of the calculus. First, there have been times in history when demanding proof, rather than plausibility arguments, was premature. After all, the great progress of the calculus in the eighteenth century did not require the full delta-epsilon machinery to justify the true results and to rule out the false ones. Things other than rigorous proof can keep mathematics honest. In the eighteenth century, intuition, numerical checking, and successful applications to the natural world kept mathematics honest. There are times, the eighteenth century being one of them, when settling for less than complete rigour may be the correct way. Another such time may be in introductory calculus courses.

Second, the first incentive for the new rigour was not, as one might think, the correction of errors; instead, it was the need to teach (Grabiner 1981, p. 25). For instance, Lagrange's foundations of calculus were delivered in 1797 as a series of lectures at the *Ecole polytechnique* in Paris, and Lagrange said that he had first thought about the foundations of the calculus when lecturing at the military school in Turin (Lagrange 1759). Cauchy's "foundations" were part of his great *Cours d'analyse* (Cauchy 1821) at the *Ecole polytechnique* and of his subsequent lectures there on the calculus (Cauchy 1823). Weierstrass's work on foundations of analysis came from his lectures at Berlin (Birkhoff 1973, pp. 71–92; Katz 2009, pp. 786–787). Dedekind said that he began thinking about the nature of the natural numbers as a result of lecturing in Zurich in 1858 (Dedekind 1963, p. 1; Fauvel and Gray 1987, p. 573). The need to teach first became urgent at the end of the eighteenth century, because, earlier in the seventeenth and eighteenth centuries, mathematicians often had been attached to royal courts or on the payroll of noblemen. But starting with the *Ecole polytechnique*, founded in the wake of the French Revolution, professional mathematicians by and large became teachers. And the need to teach a class – to present a subject systematically to people not already working with the concepts – focused mathematicians' attention on the nature of the basic concepts and their most essential properties. It still does.

Third, although they were long not really needed for the progress of the calculus, eventually delta-epsilon proofs became absolutely crucial. There comes a point, even in the applications of the calculus, where failure to make the key distinctions, to have unexceptionable definitions, does produce mistakes. For instance, in the nineteenth century, Cauchy found distinct functions that had the same Taylor series expansions (Cauchy 1823, pp. 230). Abel gave counterexamples to Cauchy's supposed theorem that an infinite series of continuous functions is itself continuous (Birkhoff 1973, p. 70). Riemann showed that a function need not be continuous in order to have a definite integral (Birkhoff 1973, pp. 16–23). Bolzano and others found functions that are everywhere continuous and nowhere differentiable (Boyer 1959, p. 282).

The distinctions made as a result of these discoveries – for example, the distinction between pointwise and uniform convergence – gave rise, as had the Greek discovery of irrationals, to a whole new set of mathematical objects, which lie even farther beyond intuition than do the concepts of the calculus. For example, Cantor's theory of the infinite arose out of trying to specify the structure of the sets of real numbers on which Fourier series converge (Dauben 1978, Chapter 2). Cantor's problem could not even have been formulated in eighteenth-century mathematical language, but it produced the whole theory of transfinite numbers. There are times when informal arguments may suffice, but there are also times when stronger proofs are absolutely necessary for any further progress in a subject. This holds true in the classroom as well as in the history of mathematics; it is an important pedagogical task to appreciate the difference.

6 How Ideas from Philosophy Shape Proofs

I return now to the history of proof to investigate what factors outside of mathematics have driven changes in proof. I would argue that philosophy has played a causal role in the evolution of proofs. The most important philosophical principle involved is what Leibniz called "the principle of sufficient reason": "Nothing happens unless there is a reason why it happens that way and not otherwise" (Russell 1937, p. 31). This has also been called the "principle of symmetry" and the "principle of indifference." Only by its use can one prove that a lever with equal weights at equal distances from the fulcrum must balance, or justify the fact that a geometric proof works equally well with a mirror image of its original diagram. This principle also operates in elementary probability theory, my first example.

In elementary probability theory, the probability of an event E is defined as the ratio $P(E) = n/N$, where n is the number of equally likely outcomes that make up the event E and N is the total number of equally likely outcomes. For instance, this definition tells us that the probability of getting an even number when we throw a single fair die is $3/6$.

But we have defined probability by using the term "equally likely outcomes," which can only mean "outcomes with the same probability." That would apparently make the definition of probability circular.

In fact, though, the notion of "equally probable" is simpler than that of probability in general, because of the philosophical principle of indifference. When speaking of a fair die, a seventeenth-century probability theorist was assuming that every one of the six sides is symmetrically placed and otherwise identical. In that case, there is no reason that the die should come up one way rather than another; each of the six possible outcomes is equally probable, equally likely. So Pascal and Fermat were able to invent probability theory in the 1650s (Katz 2009, pp. 490–495). Given the definition of probability in terms of equally likely outcomes, they could use simple combinatorics to prove all the basic properties of probability.

Being able to prove things about probabilities, historically, was no minor matter. Aristotle had said it couldn't be done. More precisely, he said, "There is no knowledge by demonstration of chance conjunctions; for chance conjunctions exist neither by necessity or as general connections" (Aristotle 1994, I.30.87b19); so a theory of probability could not meet the criteria of generality and necessity he had laid down for mathematical proofs.

What Aristotle said sounds compelling. It is one reason that it took 2,000 years after Aristotle before Pascal and Fermat invented probability theory. The explicit recognition of the philosophical principle of indifference in the seventeenth century was necessary for proofs about probability to meet Aristotle's criteria – really, the criteria for all mathematical proofs – of generality and necessity.

My second example concerns eighteenth-century ideas about space. The principle of indifference justified the eighteenth-century view that 'real' space had to be the same in all directions, and the principle of symmetry justified the view that it had to be flat. Similarly, real space had to be infinite because there would be no reason for space to stop at any given point (Grabiner 2009, pp. 6–7; Koyré 1957). Such explicit philosophical arguments helped shape the eighteenth-century consensus that space not only is, but must necessarily be, Euclidean.

Thus, the discovery of non-Euclidean geometry had to overcome powerful philosophical forces embedded in eighteenth-century thinking about space by physicists, philosophers, and geometers. Nevertheless, the discovery occurred, and did not even require imaginative or crazy people, hostile to mathematics, speculating about alternative realities. It was another triumph of human reason and logic over intuition and experience. Like the discovery of irrational numbers, the invention of non-Euclidean geometry came from indirect proof.

Non-Euclidean geometry ultimately grew out of attempts to prove Euclid's Fifth Postulate, the so-called parallel postulate. Such proofs were undertaken because the postulate, as stated by Euclid,[5] seemed considerably less self-evident than his others. From antiquity onwards, people felt that it ought to be a theorem rather than an assumption; so many eminent mathematicians tried to prove Euclid's fifth postulate from his others.

However, in the eighteenth century, mathematicians tried something new. People like Gerolamo Saccheri, Johann Heinrich Lambert, and Lagrange all tried to prove the fifth postulate indirectly; that is, they assumed it to be false, and deduced what appeared to be absurd consequences from that assumption. In fact, they were proving theorems in an alternative geometry, but what they thought was that some of the conclusions were contradictory or absurd. The 'absurdity' of many of the logical implications of denying Euclid's Fifth Postulate – for instance, parallel lines are not everywhere equidistant, a quadrilateral with three right angles need not have a fourth right angle, parallels are not unique – stems from their contradicting our deep intuitive sense of symmetry (Grabiner 2009, pp. 6–14). But in the nineteenth century, Gauss, Bolyai, and Lobachevsky, each independently, realised that these conclusions were not absurd at all, but were perfectly valid theorems in an alternative geometry (Gray 1989, Chapter 10; for excerpts from many of the original sources, see Fauvel and Gray 1987, Chapter 16).

These innovators described this alternative geometry in different terms. Lobachevsky, by analogy with imaginary numbers, called the subject "imaginary geometry" (Gray 1989, p. 118). Bolyai more theologically called it "a new world created out of nothing" (Gray 1989, p. 107). But Gauss, acknowledging the logical move that made this new subject possible, called it non-Euclidean geometry (Letter to Schumacher, 1831, quoted by Bonola 1955, p. 67).

---

[5] 'If a straight line falling on two straight lines makes the interior angles on the same side less than two right angles, then the two straight lines, if produced indefinitely, meet on that side where the angles are less than two right angles' (Euclid 1956, Postulate 5, p. 155; Fauvel and Gray 1987, p. 101).

Geometry needed its historical commitment to logical proof for mathematicians to be able to overcome their intuitive and psychological and philosophical commitment to Euclidean symmetry. Again, in order to understand the properties of the non-visible, the non-intuitive, or the counterintuitive, mathematicians need logic; we need proof. Indeed, non-Euclidean geometry is the ultimate triumph of the Euclidean method of proof.

Learning that there is such a thing as non-Euclidean geometry, an alternative system just as internally consistent as Euclid's geometry, is a great way to enable students to learn easily what nineteenth-century philosophers laboured to conclude: that the essence of mathematics (as opposed to physics) is its freedom – a freedom to choose any consistent set of axioms that meets our sense of what is important, beautiful, and fruitful, as long as we get the logic right. And then, as eventually happened in the case of non-Euclidean geometry and general relativity, such a freely chosen system can apply to the real world after all (Gray 1989, Chapter 20).

7 How Proof in Mathematics Interacts with the Real World

As to the 'real world': Many philosophers have addressed the relationship between proof in mathematics and our understanding of the world of nature and society. Aristotle and Spinoza, for instance, argued that the relationship between premises and conclusions is like the relationship between cause and effect. The "if" part explains the "then" part. What makes the base angles equal? The fact that it's an isosceles triangle.

Figure 6.5 presents a more sophisticated geometric example: the parallel lines marking parking spaces.

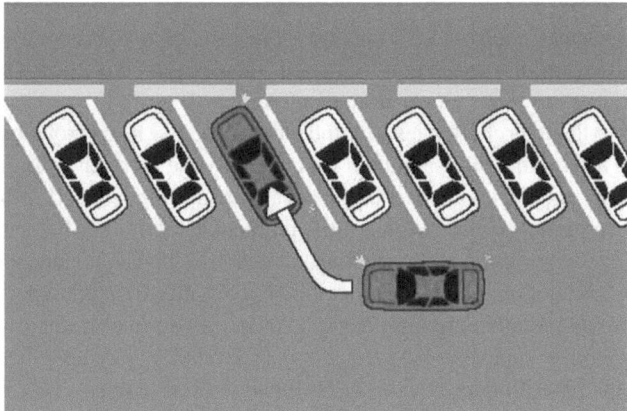

Fig. 6.5 What causes these lines to be parallel?

How do the painters make the lines between the parking spaces parallel? Not by extending the lines for a mile or so to make sure they do not intersect. They have a device that paints the same angle (to the curb) over and over again. They are using a theorem in Euclid (Euclid 1956, Book I, Proposition 28): If two lines are cut by a third so that the corresponding angles formed are equal, then the two lines are parallel. Hence for the painters of parking spaces, making the corresponding angles equal causes the lines to be parallel. Premises are to conclusions as causes are to effects.

Thus logical proofs escape from the hermetically sealed world of pure thought into the real world. The logical structure of mathematics, imported into the natural and social sciences, produces explanations in all fields. A law or observation is explained by being deduced from an accepted general principle. Reasoning in scientific fields, from antiquity to the nineteenth century, has historically followed the Euclidean/Aristotelian model. (Hence, statistical mechanics and, later, quantum mechanics challenged the prevailing philosophies of science, because their "explanations" were not causal.)

8   Conclusion

My colleagues at this study conference (ICMI-19) have catalogued a number of functions for proof in the classroom: explanation, verification, discovery, systematisation, and intellectual challenge. All these functions are exemplified in the historical record discussed above.

To start with, proofs provide explanations by convincing us that various results make sense, for example, the visual demonstration of the value of the algebraic product $(a + b)^2$. Proofs also perform the act of verification when they help us distinguish between the true and the merely plausible. For instance, only a proof can definitively tell us whether $\sqrt{2}$ or π are rational or irrational. Proofs accomplish discovery where, for example, algebraic symbolism shows that an *nth* degree equation can have no more than n roots. As for proofs creating systematisation, one good example is the way Cauchy's delta-epsilon foundation of analysis determined which arguments about infinite series and limits are true and which are not. The historical record, of course, provides many more examples.

Finally, intellectual challenges abound in proofs. We learn from mathematical proof how to reason logically, and then we can apply the methods of logical demonstration to other subjects, as the examples of Newton, Spinoza, and Jefferson make clear. We learn to see the limitations of the mathematical models used in other subjects, as when we realise that physical space need not be Euclidean. We learn that we can study and understand things that are beyond intuition and experience, ranging from irrational numbers to infinite sets. We should learn also not always to rank perfect logic above imperfect but real progress; Berkeley may have been logically correct, but the practitioners of calculus still had the last laugh.

I chose the examples of proof in this paper because of their historical importance. What happens historically must of course be psychologically possible, but what happens historically is not necessarily psychologically or pedagogically optimal. Nevertheless, when what happens historically does coincide with what educational research says works best in the classroom, it means something. It does not mean that we should simply replicate the history in the classroom. But it does mean that the history of mathematics is a source of instructive examples and of inspiration, worthy of study.[6]

This paper has also illustrated another pair of generalisations: First, proof takes place always in historical, philosophical, cultural and social context. Second, proofs require ideas, not just going through the motions. Recapturing the ideas, the context, and the effects can help us, as well as help our students, recapture the original excitement. History, then, not only answers the question, "Why do we have proof?" History also helps us do a better job of teaching proof and proving.

References

Aristotle (1994). *Posterior analytics* [4th century BCE] (J. Barnes, Trans.). Oxford: Clarendon Press.

Barzun, J. (1945). *Teacher in America*. London: Little-Brown.

Berggren, J. L. (1986). *Episodes in the mathematics of medieval Islam*. New York: Springer.

Berggren, J. L. (2007). Mathematics in medieval Islam. In V. J. Katz (Ed.), *The mathematics of Egypt, Mesopotamia, China, India, and Islam: A sourcebook* (pp. 515–675). Princeton: Princeton University Press.

Berkeley, G. (1951). The analyst, or a discourse addressed to an infidel mathematician [1734]. In A. A. Luce & T. R. Jessop (Eds.), *The works of George Berkeley* (Vol. 4, pp. 65–102). London: T. Nelson.

Birkhoff, G. (Ed.). (1973). *A source book in classical analysis. With the assistance of Uta Merzbach*. Cambridge: Harvard University Press.

Bolzano, B. (1817). *Rein analytischer Beweis des Lehrsatzes dass zwischen je zwey [sic] Werthen, die ein entgegengesetztes Resultat gewaehren, wenigstens eine reele Wurzel der Gleichung liege*. Prague: Gottlieb Haase. English translation in Russ, S. B. (1980). A translation of Bolzano's paper on the intermediate value theorem. Historia *Mathematica 7* (pp. 156–185). Also in Russ, S. B. (Ed.). (2004) *The mathematical works of Bernard Bolzano* (pp. 251–278). Oxford: Oxford University Press.

---

[6] If one wants to read only one history of mathematics, I recommend Katz (2009). For further readings, Katz's bibliography will take one as far as one would like. Another good scholarly general history is Boyer and Merzbach (1989).

For a collection of original sources in the history of mathematics from the period between 1200 and 1800, with excellent commentary, I recommend Struik (1969). Another fine collection of original sources, with shorter excerpts and commentary but covering the entire period from antiquity to the twentieth century, is Fauvel and Gray (1987). Original sources from Egypt, Babylon, China, India and the Islamic world may be found in Katz (2007), and, in classical analysis, in Birkhoff (1973). In this paper, I have given references to these four source books whenever relevant materials are accessible there. Finally, I highly recommend the Mathematical Association of America's online 'magazine' *Convergence* (http://mathdl.maa.org/mathDL/46/) dedicated to the history of mathematics and its use in teaching. Besides articles, *Convergence* includes book reviews, translations of original sources, quotations about mathematics, portraits, 'mathematics in the news', and a great deal more.

Bonola, R. (1955). *Non-Euclidean geometry: A critical and historical study of its development. With a Supplement containing "the theory of parallels" by Nicholas Lobachevski and "the science of absolute space" by John Bolyai.* [1912] (H. S. Carslow, Trans.). New York: Dover.

Boyer, C. (1959). *The history of the calculus and its conceptual development.* New York: Dover.

Boyer, C., & Merzbach, U. (1989). *A history of mathematics* (2nd ed.). New York: Wiley.

Cauchy, A.-L. (1821). Cours d'analyse de l'école royale polytechnique. 1re partie: Analyse algébrique [all that was published]. Paris: Imprimerie royale. In A.-L. Cauchy, Oeuvres (Series 2, Vol. 3)

Cauchy, A.-L. (1823). Résumé des leçons données a l'école royale polytechnique sur le calcul infinitésimal. In A.-L. Cauchy (Ed.), *Oeuvres* (Series 2, Vol. 4, pp. 5–261). Paris: Imprimérie royale.

Cohen, I. B. (1995). *Science and the founding fathers.* New York: W. W. Norton.

Convergence (n. d.). [The Mathematical Association of America's online magazine on the history of mathematics and its uses in the classroom.]. http://mathdl.maa.org/mathDL/46/

Dauben, J. (1978). *Georg Cantor.* Cambridge: Harvard University Press.

Dauben, J. (2007). Chinese mathematics. In V. J. Katz (Ed.), *The mathematics of Egypt, Mesopotamia, China, India, and Islam: A sourcebook* (pp. 187–385). Princeton: Princeton University Press.

D'Alembert, J., & de la Chapelle, J.-B. (1789). Limite. In J. d'Alembert, C. Bossut, & J. J. de Lalande (Eds.), *Dictionnaire encyclopédique des mathématiques.* Paris: Hotel de Thou.

Dedekind, R. (1963). *Essays on the theory of numbers.* New York: Dover.

Dunham, W. (1999). *Euler: The master of us all.* Washington, DC: Mathematical Association of America.

Euclid (1956). In T. L. Heath (Ed.), *The thirteen books of Euclid's elements.* New York: Dover.

Euler, L. (1748). Introductio in analysin infinitorum, Lausanne: Bousquet. In Euler *Opera omnia* (Series 1, Vols. 8–9). [English translation by John Blanton, as Leonhard Euler, Introduction to the Analysis of the Infinite in 2 Vols.]. New York et al: Springer, 1988–1990.

Fauvel, J., & Gray, J. (Eds.). (1987). *The history of mathematics: A reader.* London: Macmillan, in Association with the Open University.

Grabiner, J. V. (1981). *The origins of Cauchy's rigorous calculus.* Cambridge: MIT Press. Reprinted New York: Dover, 2005.

Grabiner, J. V. (1990). *The Calculus as Algebra: J.-L. Lagrange, 1736–1813.* New York and London: Garland Publishing, Inc.

Grabiner, J. V. (1997). Was Newton's calculus a dead end? The continental influence of Maclaurin's *Treatise of Fluxions. The American Mathematical Monthly, 104*, 393–410.

Grabiner, J. V. (2009). Why did Lagrange "prove" the parallel postulate? *The American Mathematical Monthly, 116*, 3–18.

Gray, J. (1989). *Ideas of space: Euclidean, Non-Euclidean and relativistic* (2nd ed.). Oxford: Clarendon.

Hankins, T. L. (1976). Algebra as pure time: William Rowan Hamilton and the foundations of algebra. In P. Machamer & R. Turnbull (Eds.), *Motion and time, space and matter* (pp. 327–359). Columbus: Ohio State University Press.

Imhausen, A. (2007). Egyptian mathematics. In V. J. Katz (Ed.), *The mathematics of Egypt, Mesopotamia, China, India, and Islam: A sourcebook* (pp. 7–56). Princeton: Princeton University Press.

Jefferson, T. (1776). *The declaration of independence.* Often reprinted, *e.g.,* T. Jefferson, *The declaration of independence.* Introduction by M. Hardt; additional material by G. Kindervater. London: Verso, 2007.

Katz, V. J. (Ed.). (2007). *The mathematics of Egypt, Mesopotamia, China, India, and Islam: A sourcebook.* Princeton: Princeton University Press.

Katz, V. J. (2009). *A history of mathematics: An introduction* (3rd ed.). Boston: Addison-Wesley.

Knorr, W. (1993). *The ancient tradition of geometric problems.* New York: Dover.

Koyré, A. (1957). *From the closed world to the infinite universe.* Baltimore: Johns Hopkins Press.

Lagrange, J.-L. (1759). Letter to Leonhard Euler, 24 November 1759, reprinted in Lagrange (1973), vol. XIV, pp. 170–174.

Lagrange, J.-L. (1770). Réflexions sur la resolution algébrique des équations. *Nouveaux mémoires de l'Académie des sciences de Berlin*. Reprinted in J.-L. Lagrange, *Oeuvres* (Vol. 3, pp. 205–424).

Lagrange, J.-L. (1797). *Théorie des fonctions analytiques*. Paris: Imprimerie de la République. An V [1797]. 2nd ed. Paris: Courcier, 1813. Reprinted as J.-L. Lagrange, *Oeuvres* (Vol. 9).

Lagrange, J.-L. (1808). *Traité de la résolution des équations numériques de tous les degrés, avec des notes sur plusieurs points de la théorie des équations algébriques*; Paris: Courcier. Reprinted in *Oeuvres de Lagrange*, vol. VIII (Paris: Gauthier-Villars, 1879).

Leslie, D. (1964). *Argument by contradiction in Pre-Buddhist Chinese reasoning*. Canberra: Australian National University.

Lloyd, G. E. R. (1996). *Adversaries and authorities: Investigations into ancient Greek and Chinese science*. Cambridge: Cambridge University Press.

Maclaurin, C. (1742). *A treatise of fluxions in two books*. Edinburgh: T. Ruddimans.

McKirahan, R. D., Jr. (1992). *Principles and proofs: Aristotle's theory of demonstrative science*. Princeton: Princeton University Press.

Newton, I. (1967). Universal Arithmetick [1728]. In D. T. Whiteside (Ed.), *Mathematical works of Newton* (Vol. 2, pp. 3–134). New York: Johnson.

Newton, I. (1995). *The principia: Mathematical principles of natural philosophy* [1687] (I. B. Cohen & A. Whitman, Trans.). Berkeley: University of California Press.

Pfloker, K. (2007). Mathematics in India. In V. J. Katz (Ed.), *The mathematics of Egypt, Mesopotamia, China, India, and Islam: A sourcebook* (pp. 385–514). Princeton: Princeton University Press.

Plato (2004). *Plato's Meno* [4th century BCE] (G. Anastaplo & L. Berns, Trans.). Newburyport: Focus Publishing.

Robson, E. (2007). Mesopotamian mathematics. In V. J. Katz (Ed.), *The mathematics of Egypt, Mesopotamia, China, India, and Islam: A sourcebook* (pp. 56–186). Princeton: Princeton University Press.

Russell, B. (1937). *A critical exposition of the philosophy of Leibniz*. London: Allen and Unwin.

Sageng, E. (2005). 1742: Colin MacLaurin, a treatise of fluxions. In I. Grattan-Guinness (Ed.), *Landmark writings in western mathematics, 1640–1940* (pp. 143–158). Amsterdam: Elsevier.

Siu, M. K. (2009, May 15). *Plenary Panel presentation given at ICMI-19*, Taiwan Normal University, Taipei.

Spinoza, B. (1953). *Ethics* [1675]. J. Gutmann (Ed.). New York: Hafner.

Stedall, J. A. (Ed.). (2003). *The greate invention of algebra: Thomas Harriot's treatise on equations*. Oxford: Oxford University Press.

Struik, D. J. (Ed.). (1969). *A source book in mathematics, 1200–1800*. Cambridge: Harvard University Press.

Viète, F. (Vieta) (1968). Introduction to the analytic art (J. W. Smith, Trans.). In J. Klein (Ed.), *Greek mathematical thought and the origin of Algebra*. Cambridge, MA: MIT Press.

Chapter 7
Conceptions of Proof – In Research and Teaching

Richard Cabassut, AnnaMarie Conner, Filyet Aslı İşçimen, Fulvia Furinghetti, Hans Niels Jahnke, and Francesca Morselli

1 Conceptualisations of Proof

1.1 Conceptions by Mathematicians

The education of professional mathematicians very successfully transmits a practically precise conception of proof. Mathematically educated persons who specialise in and know a certain domain of mathematics will generally agree that a given piece of mathematical text is an adequate proof of a given statement. Nevertheless, no explicit general definition of a proof is shared by the entire mathematical community. Consequently, in attempting such a conceptualisation it

R. Cabassut
LDAR Laboratoire de Didactique André Revuz, Paris 7 University, Paris, France

IUFM Institut Universitaire de Formation des Maîtres, Strasbourg University, Strasbourg, France
e-mail: richard.cabassut@unistra.fr

A. Conner
Department of Mathematics & Science Education, University of Georgia, Athens, GA, USA
e-mail: aconner@uga.edu

F.A. İşçimen
Department of Mathematics and Statistics, Kennesaw State University, Kennesaw, GA, USA
e-mail: ersozas@yahoo.com

F. Furinghetti • F. Morselli
Dipartimento di Matematica, Università di Genova, Genoa, Italy
e-mail: furinghe@dima.unige.it; morselli@dima.unige.it

H.N. Jahnke(✉)
Fakultät für Mathematik, Universität Duisburg-Essen, Essen, Germany
e-mail: njahnke@uni-due.de

© The Author(s) 2021
G. Hanna and M. de Villiers (eds.), *Proof and Proving in Mathematics Education*,
New ICMI Study Series, https://doi.org/10.1007/978-94-007-2129-6_7

is wise to resort first to mathematical logic for a definition of proof and to look afterwards at how working mathematicians comment on such a definition.

According to Rav (1999, p. 11), in a formalised theory **T** a linear derivation is a finite sequence of formulas in the language of **T**, each member of which is a logical axiom, or an axiom of **T**, or the result of applying one of the finitely many explicitly stated rules of inference to previous formulas in the sequence. A tree derivation can be similarly defined. A formula of **T** is said to be derivable if it is the end-formula of a linear or tree derivation.

Obviously, the structure characterised in this formal definition echoes the axiomatic method of Euclid's *Elements* (c. 300 B.C.). Hence, we can consider the notion of proof as some combination of the axiomatic method and formalism, the latter called 'rigour' since the time of Cauchy.

That definition may be considered as a 'projection' of the real practice of mathematical proof onto the skeleton of formal logic. A projection inherits some properties of the original, but is as a rule, poorer. Consequently, Rav distinguishes the formal idea of proof from that of a 'conceptual proof', by which he means an informal proof "of customary mathematical discourse, having an irreducible semantic content" (Rav 1999, p. 11; see also Hanna and Barbeau 2009, p. 86). As a rule, working mathematicians insist on the informal and semantic components of proof. As Rav stresses, beyond establishing the truth of a statement, proof contributes to getting new mathematical insights and to establishing new contextual links and new methods for solving problems. (Functions of proof beyond that of verification are also discussed by Bell (1976), de Villiers (1990), and many others.)

Working mathematicians also stress the *social process* of checking the validity of a proof. As Manin put it: "A proof only becomes a proof after the social act of 'accepting it as a proof'. This is true for mathematics as it is for physics, linguistics, and biology" (Manin 1977, p. 48). By studying the comments of working mathematicians Hanna came to the conclusion that the public process of accepting a proof not only involves a check of deductive validity, but is also determined by factors like 'fit to the existing knowledge', 'significance of the theorem', the 'reputation of the author' and 'existence of a convincing argument' (Hanna 1983, p. 70; see also Neubrand 1989). Bell (1976) also stressed the essentially public character of proof.

All in all, formal definitions of proof cover the meaning of the notion only incompletely, whereas mathematicians are convinced that, in practice, they know precisely what a proof is. This situation is difficult to handle in the teaching of mathematics at schools, since there exist no easy explanations of what proof and proving are that teachers could provide to their pupils. Proof is not a "stand-alone concept", as Balacheff nicely puts it (2009, p. 118), and is aligned to the concept of a "theory" (see also Jahnke 2009b, p. 30).

1.2 Conceptions by Mathematics Educators

Genuinely *didactical* conceptions of proof are determined by two clearly distinguishable sets of motives. One line of thought tries to devise genetic ideas of proof.

These ideas are pedagogically motivated in that they try to devise a learning path from a cognitive state in which an individual learner is able to construct argumentations with some deductive components to a state in which the learner manages to understand and develop mathematical proofs in their proper sense. The other line of thought builds conceptions of proof with the intention of doing empirical research. Both lines of thought define categories that allow one to classify individuals' argumentative behaviours and strategies observed in classrooms or in interview situations. In both, it is essential to distinguish between arguments which are not yet proofs from mathematical proofs proper. No wonder different researchers come to different conclusions about the demarcation between the two modes of reasoning. Balacheff (1988) and Duval (1991), for example, draw a sharp line of demarcation, whereas other authors stress the continuity between argumentation and proof – thereby embedding proof in a general theory of argumentation. In mathematics education, in recent years it has become customary to use the term 'argumentation' for reasoning which is 'not yet' proof and the term 'proof' for mathematical proof proper.

1.2.1   Genetic Ideas of Proof

Genetic conceptions distinguish between different stages in the development of proof essentially along three lines. First, the *type of warrant* for a general statement is at stake. Pupils might infer from some special cases a general rule or statement. In such cases problems of proof and generalisation are intermingled. Generalisation is an important scientific activity but, of course, different from proof. Consequently, it is an important step in these pupils' cognitive development to understand that a general statement can only be derived from other general statements. Second, if pupils argue on the level of general statements the *type of principle* they refer to differs according to their proximity to established mathematical principles and norms. For example, in comparing the lengths of different paths between two points, pupils might apply a physical argument like the stretching of an elastic band as a warrant. Whether this is accepted as a valid argument is, in principle, a matter of classroom convention. Nevertheless, some authors would not classify this as a mathematical proof, because the principle does not belong to the accepted principles of mathematics proper. Third, the *mode of representing* an argument might also be a distinguishing feature between different stages in the acquisition of proof. Though, in principle, it is extrinsic to the very heart of proof whether it is displayed in verbal or symbolic form, growth in the ability to handle the symbolic language of mathematics is an indispensable condition for growth of a learner's ability to understand and develop mathematical proofs; the complexity of most mathematical relations is such that they can hardly be expressed without recourse to symbolic representation.

A famous paper by van Dormolen (1977) exemplifies genetic conceptions of proof. In regard to different proof situations, van Dormolen gives possible solutions which reflect different stages in the development of argumentative skills. One task asks for a proof of the statement that the diagonals in an isosceles trapezium are equal (Fig. 7.1). A beginning pupil might react by measuring the diagonals and finding them equal. On a second level, the pupil might mentally cut out the trapezium,

Fig. 7.1 Van Dormolen's
trapezium problem

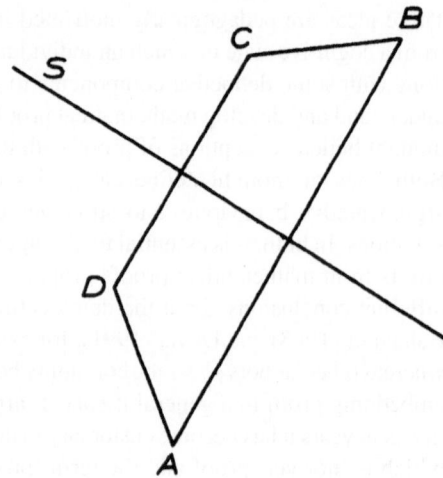

turn it about and put it back into its hole, with the result that each diagonal is now in the position of the other. On a third level, a pupil might formally argue from the symmetry of the figure and apply a reflection, so that one diagonal is mapped on the other.

Van Dormolen considers these three stages in the framework of van Hiele's theory of levels of thinking. On a first level the student is bound to special objects. On a second level the student can think about properties of classes of objects, and on the third level the student is able to logically organise an argument.

As a further example of such a pedagogically motivated genetic conception, Kirsch (1979) distinguished between 'pre-mathematical' and 'mathematical' proofs, referring to the earlier paper (Semadeni n.d.). Later, Kirsch changed his terminology and spoke about 'pre-formal' versus 'formal' proofs (Blum and Kirsch 1991).

1.2.2 Conceptions from Empirical Research

In discussing conceptions of proof motivated by empirical research, Balacheff (1988) states explicitly that his view rests on an experimental approach and that he is interested in studying pupils' practices. For this purpose, he distinguishes between 'pragmatic' and 'conceptual' proofs. Pragmatic proofs have recourse to real actions, whereas conceptual proofs deal with properties and the relations between them and, consequently, do not involve actions. However, there can and should be bridges between the two types, insofar as pragmatic proofs might have a generic quality and be a step towards conceptual proofs. Though Balacheff's primary interest lies in the classification of pupils' practices, these distinctions interfere with genetic ideas, since pragmatic proofs operate on a lower, earlier level than conceptual proofs. Balacheff explicitly says that his categories "form a hierarchy ... Where a particular type of proof falls in this hierarchy depends on how much the demands of generality and the conceptualisation of knowledge are involved," (Balacheff 1988, p. 218). It is also plausible to see in the distinction between pragmatic and conceptual proofs an influence of Piaget's cognitive psychology which is based on the idea that thoughts are preceded by actions.

Balacheff has further refined the distinction between pragmatic and conceptual proofs into sub-categories. Pragmatic proofs split into 'naïve empiricism' and 'crucial experiment': 'Naïve empiricism' refers to asserting the truth of a result by verifying special cases. 'Crucial experiment' means that a pupil considers a special example and argues that the proposition in question must be true if it is true for this (extreme) example. Argumentative behaviour of this kind shows some attention to the problem of a statement's generality, but is still bound to the special case.

The 'generic example', Balacheff's first sub-category of a conceptual proof, is different. The generic example makes the reasons for the truth of an assertion explicit by operations or transformations on an object that is a characteristic representative of its class. The difference from the 'crucial experiment' is subtle but decisive. The final category, 'thought experiment', invokes internalised actions and is detached from particular representations. If we understand Balacheff, this level still falls short of a professional approach to proof but represents the best students can attain in their school lives.

Harel and Sowder (1998) have proposed another influential system of categories. They give priority to the function of proof as a *convincing argument* and derive from this idea categories for classifying individuals' argumentative behaviours. Their starting point is a pedagogical motive. "The goal is to help students refine their own conception of what constitutes justification in mathematics: from a conception that is largely dominated by surface perceptions, symbol manipulation, and proof rituals, to a conception that is based on intuition, internal conviction, and necessity" (1998, p. 237). Consequently, their categories reflect students' ideas about what a convincing argument might be. These categories they call 'proof schemes': Individuals' ways of thinking associated with the proving act and consisting of "what constitutes ascertaining and persuading for that person" (1998, p. 244).

Harel and Sowder's (1998) whole system of categories rests on a number of empirical studies and splits into three large domains: 'External conviction proof schemes', 'Empirical proof schemes' and 'Analytical proof schemes'. In the domain of 'External conviction proof schemes' the authors distinguish between 'Authoritarian', 'Ritual' and 'Symbolic' proof schemes. Authoritarian means that students refer to an authority – be it their teacher, a book or whatever – to convince themselves of the truth of a statement. Ritual and Symbolic proof schemes reflect the fact that many students come to the conclusion that ritual and form constitute mathematical justification. The Symbolic schemes describe a behaviour of approaching a solution without first comprehending the meaning of the symbols involved.

In the domain of Empirical proof schemes, the authors distinguish between 'inductive' and 'perceptual' proof schemes (Harel and Sowder 1998). 'Inductive' means obtaining a general statement from some special examples or measurements, whereas 'perceptual' designates uncritically taking for granted some visual property of an object or a configuration.

In general, Harel and Sowder define an Analytical proof scheme as one "that validates conjectures by means of logical deductions" (1998, p. 258). At first glance, it seems surprising that this scheme splits into a 'Transformational' and an 'Axiomatic' proof scheme. The 'Axiomatic' can be considered as what an educated mathematician might mean when speaking about proof. The 'Transformational' refers to arguments which identify an invariant by systematically changing a configuration. For example, consider angles inscribed in a circle over a fixed chord. According to a well-known theorem these angles are equal. If the vertex of an angle is moved on the circumference to one of the end-points of the chord, then in the limiting position one of the legs coincides with the tangent and the other with the chord. Consequently, the inscribed angle becomes the angle between chord and tangent and equal to the latter. Of course, this is also a well-known theorem in elementary geometry. Obviously, such a transformational argument is not a rigorous proof; yet it operates on a general level and can rightly be considered as belonging to an analytical domain.

Both Balacheff's (1988) and Harel and Sowder's (1998) systems of categories have been applied with some modifications in a number of empirical studies. Both systems share the problem that it is not always straightforward to relate an argumentative behaviour to a certain category; rather, it may require a considerable amount of interpretation. What distinguishes the two systems is the fact that Balacheff's system involves a genetic sequence whereas Harel and Sowder do not seem to be interested in such a sequence; their point of view is more the co-existence or even competition between conceptions of proof and argumentation in various fields of human life (everyday life and other sciences). This difference might be due to the fact that their subjects are college students, whereas Balacheff is mostly interested in grade-school teaching.

2   Proving and Beliefs of Teachers and Students

2.1   The System of Beliefs

This sub-chapter examines the epistemological and pedagogical beliefs about the practice of proof in the classroom:

- beliefs about the nature and role of proof in mathematics.
- beliefs about the role of proof in school mathematics.
- beliefs about difficulties in proving.
- beliefs about how proof should be taught in school.
- beliefs about oneself as mathematical thinker in the context of proof.

Moreover, due to the central role of proof in mathematical activity, beliefs involved in the practice of proof are not confined to the subject of proof but also include beliefs about mathematics, about mathematics teaching and learning, about oneself in relation to mathematics, and so on.

Furinghetti and Pehkonen (2002) have shown to what extent beliefs and related concepts are controversial issues. They also have pointed out the different uses of the term 'conception' and the mutual relationship of beliefs, conceptions and knowledge. Following their recommendations, below we clarify our assumptions about the meanings of the terms we use and their mutual relationships.

As regards conceptions and beliefs, we follow Philipp (2007, p. 259) who describes conception as "a general notion or mental structure encompassing beliefs, meanings, concepts, propositions, rules, mental images, and preferences". As regards beliefs and knowledge, our position is expressed by the following passage by Leatham:

> Of all the things we believe, there are some things that we "just believe" and other things we "more than believe – we know". Those things we "more than believe" we refer to as knowledge and those things we "just believe" we refer to as beliefs. Thus beliefs and knowledge can profitably be viewed as complementary subsets of the things we believe. (2006, p. 92)

For us, "things we know" are those that rely on a social agreement inside a given community (for mathematics, the community of mathematicians).

In the same vein, Philipp describes knowledge as "beliefs held with certainty or justified true belief. What is knowledge for one person may be belief for another, depending upon whether one holds the conception as beyond question." (2007, p. 259). Philipp goes on to describe beliefs as:

> Psychologically held understandings, premises, or propositions about the world that are thought to be true. Beliefs are more cognitive, are felt less intensely, and are harder to change than attitudes. Beliefs might be thought of as lenses that affect one's view of some aspect of the world or as dispositions toward action. Beliefs, unlike knowledge, may be held with varying degrees of conviction and are not consensual. Beliefs are more cognitive than emotions and attitudes. (I do not indent this definition under affect because, although beliefs are considered a component of *affect* by those studying affect, they are not seen in this way by most who study teachers' beliefs.) (ibid., p. 259)

We mainly use the term 'belief'; we use 'conception' (in the sense explained above) when referring to a set of beliefs.

2.2 Teachers' Epistemological and Pedagogical Beliefs

Researchers have investigated the beliefs about proof of pre-service and in-service elementary and secondary school teachers. We have organised their findings around four major themes: teachers' knowledge of proof, teachers' beliefs about the nature and role of proof in mathematics, teachers' beliefs about the role of proof in school mathematics and teachers' beliefs about themselves as mathematical thinkers in the context of proof.

2.2.1 Teachers' Knowledge of Proof

The majority of researchers who have investigated teachers' knowledge of proof have focused on teachers' acceptance of empirical versus deductive arguments as valid proofs. Knuth (2002a), who investigated what constitutes proof for 16 in-service secondary school mathematics teachers, and Martin and Harel (1989), who assessed the notions of proof held by 101 pre-service elementary school teachers, gave their participants statements accompanied by predetermined arguments and asked them to rate these in terms of their validity. Whereas Martin and Harel asked for written responses only, Knuth conducted in-depth interviews with his participants.

Both Knuth (2002a) and Martin and Harel (1989) concluded that although most teachers correctly identify a valid argument, they also wrongly accept invalid arguments as proofs. Several pre-service elementary teachers accepted empirically based arguments as proofs (Martin and Harel 1989; see also Morselli 2006; Simon and Blume 1996). What secondary school teachers find convincing in an argument – inclusion of a concrete feature, specific examples and visual reference – (Knuth 2002a) might also explain why elementary teachers accepted empirical arguments as proofs. (On visualisation, see Biza et al. 2009)

The criteria teachers used to evaluate an argument differed widely but there were some commonalities. Several teachers adopted Symbolic or Ritual proof schemes (Harel and Sowder 1998; see Sect. 1.2.2 above). For example, some teachers focused on the correctness of the algebraic manipulations or on the form of an argument as opposed to its nature, (Knuth 2002a), whilst others accepted false proofs based on their ritualistic aspects (Martin and Harel 1989). Although they rated correct deductive arguments as valid proofs, teachers still did not find them convincing (Knuth 2002a). Treating the proof of a particular case as the proof for the general case was also common amongst most teachers (Knuth 2002a; Martin and Harel 1989).

Employing a different method, Jones (2000) asked recent mathematics graduates enrolled in a 1-year course to become secondary school teachers to construct concept maps reflecting their conceptions of mathematical proof. Analysis of the concept maps revealed that participants who had barely received pass degrees in mathematics courses needed "considerable support in developing a secure knowledge base of mathematics" (2000, p. 57). On the other hand, Jones reported, technical fluency in writing proofs did not necessarily imply richly connected knowledge of proof.

2.2.2 Teachers' Beliefs About the Nature and Role of Proof in Mathematics

Chazan stated that "many teachers do not seem to understand why mathematicians place such a premium on proof" (1993, p. 359). However, all of the in-service secondary school teachers in Knuth's (2002a) study indicated that the role of proof in mathematics was to establish the truth of a statement. They also suggested various other roles: explaining why something is true with a procedural focus rather than promoting understanding; the communicative role of proof (social interaction, communicating and convincing others); and the creation of knowledge and systematisation of results.

In a survey study with 30 pre-service elementary school teachers and 21 students majoring in mathematics with an emphasis in secondary education, Mingus and

Grassl (1999) asked the participants what constitutes a proof and asked about the role of proof in mathematics. In their definitions of proof, the secondary-education majors emphasised explanatory power, whereas the elementary-education majors focused on verification. The majority of the participants also pointed out the importance of proofs in helping "students understand the mathematics they are doing" (1999, p. 441). Furthermore, the secondary-education majors "also considered the role of *proof* for maintaining and advancing the structure of mathematics" (ibid., p. 441).

Although the teachers in Knuth's (2002a) study could identify the roles of proof in mathematics and the explanatory power of proofs was mentioned by the secondary-education majors in the Mingus and Grassl (1999) study, Harel and Sowder later concluded from a review of the literature that teachers "do not seem to understand other important roles of proof, most noticeably its explanatory role" (2007, p. 48). Also, one important question is how teachers' beliefs about proof relate to other aspects of their classroom practice. In a small qualitative study, Conner (2007) found that three student teachers' conceptions of proof (particularly their beliefs about the purpose and role of proof in mathematics) aligned closely with how they supported argumentation (not proof in particular, but asking for and providing data and warrants for claims) in secondary classrooms.

2.2.3   Teachers' Beliefs About the Role of Proof in School Mathematics

The roles of proof in school mathematics that the secondary teachers in Knuth's (2002b) study talked about included all the roles they mentioned for proof in mathematics in general (Knuth 2002a) except for systematising statements into an axiomatic system. In the subsequent report (Knuth 2002b), the secondary teacher participants added some new roles for proof when they discussed it in the context of school mathematics: developing logical-thinking skills and displaying student thinking.

Although the roles that the teachers attached to proof in secondary-school mathematics seemed promising, their beliefs about the centrality of proof were limited (Knuth 2002a, b). Several teachers did not think that proof should be a central idea throughout secondary school mathematics, but only for advanced mathematics classes and students studying mathematics-related fields. On the other hand, all the teachers considered that *informal* proof should be a central idea *throughout* secondary-school mathematics. This is consistent with Healy and Hoyles' reporting: "For many teachers it was more important that the argument was clear and uncomplicated than that it included any algebra" (2000, p. 413).

The majority of the teachers in Knuth's research (2002b) viewed Euclidean geometry or upper-level mathematics classes as appropriate places to introduce proof to students. All of them said that they would accept an empirically based argument as a valid argument from students in a lower-level math class. Two of them, however, explained that they would discuss its limitations. Probably these beliefs were shaped by the teachers' own experiences with proofs, since high-school

Euclidean geometry is the "usual locus" (Sowder and Harel 2003, p. 15) for introducing proof in U.S. curricula. Furthermore, "the only substantial treatment of proof in the secondary mathematics curriculum occurs" (Moore 1994, p. 249) in this (usually 1-year) geometry course. Knuth's findings also point out that teachers view proof as a subject to be taught separately rather than as a learning tool that can be integrated throughout mathematics.

On the other hand, the majority (69%) of the pre-service teachers in Mingus and Grassl's (1999) study advocated the introduction of proof before 10th-grade geometry classes. Furthermore, the participants who had taken college-level mathematics courses believed that proofs needed to be introduced earlier, in the elementary grades, in contrast to the participants who had only experienced proofs at the high-school level. Mingus and Grassl argued that the former group "may have recognized that a lack of exposure to formal reasoning in their middle and high school backgrounds affected their ability to learn how to read and construct *proofs*" (1999, p. 440).

Other studies reveal different beliefs about the role of proof. Furinghetti and Morselli (2009a) investigated how secondary teachers treat proofs, and which factors (especially beliefs) affect that treatment, in a qualitative study of ten cases via individual, semi-structured interviews. Nine of the ten teachers declared that they teach proof in the classroom. The other said that she does not because Euclidean geometry is not in her school curriculum. The other teachers also referred mainly to Euclidean geometry as the most suitable domain for teaching proof. Sowder and Harel (2003) already pointed out beliefs about geometry being the ideal domain for the teaching of proof or, even more, the teaching of proof being confined to geometry. Concerning the way proof is treated in the classroom, Furinghetti and Morselli (2009a) identified two tendencies: teaching theorems versus teaching via the proof. The first sees proof as a means for convincing and systematising mathematical facts, whilst the second uses proof mainly to promote mathematical understanding. The first focuses on proof as a product, the second as a process.

2.2.4 Teachers' Beliefs About Proof and Themselves as Mathematical Thinkers

Teacher attitudes towards using mathematical reasoning, their abilities in constructing proofs, and their abilities to deal with novel ideas are especially important, because "ideas that surprise and challenge teachers are likely to emerge during instruction" (Fernandez 2005, p. 267). In such situations teachers should be able to "reason, not just reach into their repertoire of strategies and answers" (Ball 1999, p. 27). However, the U.S. teachers in Ma's (1999) study were not mathematically confident to deal with a novel idea and investigate it. Like students, these teachers relied on some authority – a book or another teacher – to be confident about the truth of a statement.

Although it was not their main purpose, Simon and Blume (1996) also found evidence that prospective elementary teachers appealed to authority. Their study differed from others in the sense that they investigated pre-service elementary-school teachers' conceptions of proof in the context of a mathematics course "which was run as a whole class constructivist teaching experiment" (1996, p. 3). The participants had previously experienced mathematics only in traditional classrooms where the authority was the teacher; the goal of the instructor in the study was to shift the "authority for verification and validation of mathematical ideas from teacher and textbook to the mathematical community (the class as a whole)" (ibid., p. 4). The authors argued that this shift was significant because it "can result in the students' sense that they are capable of creating mathematics and determining its validity" (ibid., p. 4).

Simon and Blume's (1996) findings illustrate how pre-service teachers' prior experiences with proofs (or the lack thereof) and views about mathematics influence how they initially respond to situations where proof is necessary. At the beginning of the semester when the study's instructor asked them to justify mathematical ideas, the participants referred to their previous mathematics courses or provided empirical reasons. They also did not necessarily make sense of the others' general explanations if they were not operating at the same level of reasoning. However, Simon and Blume claimed that "norms were established over the course of the semester, that ideas expressed by community members were expected to be justified and that those listening to the justification presented would be involved in evaluating them" (1996, p. 29).

More recently, Smith (2006) compared the perceptions of, and approaches to, mathematical proof by undergraduates enrolled in lecture and problem-based "transition to proof" courses, the latter using the "modified Moore method," (2006, p. 74). Their key finding: "while the students in the lecture-based course demonstrated conceptions of proof that reflect those reported in the research literature as insufficient and typical of undergraduates, the students in the problem-based course were found to hold conceptions of, and approach the construction of, proofs in ways that demonstrated efforts to make sense of mathematical ideas" (ibid., p. 73). These promising results "suggest that such a problem-based course may provide opportunities for students to develop conceptions of proof that are more meaningful and robust than does a traditional lecture-based course" (ibid., p. 73).

2.3   Students' Beliefs About Proof

Many studies deal with students' approaches to proof. They mainly focus on students' difficulties; only rarely do they address the issue of beliefs directly and explicitly. Their most prevalent findings on students' beliefs about proof are that students find giving proofs difficult and that their views of the purpose and role of proof are very limited (Chazan 1993; Harel and Sowder 1998; Healy and Hoyles 2000). Some students are ignorant of the need to give a mathematical proof to verify a statement; others appeal to an authority – a teacher, a book or a theorem – to establish a truth (Carpenter et al. 2003).

According to Ball and Bass (2000), third-grade students "did not have the mathematical disposition to ask themselves about the completeness of their results when working on a problem with finitely many solutions" (2000, p. 910) early in the year. Similarly, Bell (1976) found that 70% of the 11- to 13-year-olds in his study could recognise and describe patterns or relationships but showed no attempt to justify or deduce them. Even some students at the university level may believe that "proof is only a formal exercise for the teacher; there is no deep necessity for it" (Alibert 1988, p. 31).

A common research finding is that students accept empirical arguments as proofs. They believe that checking a few cases is sufficient (Bell 1976). Healy and Hoyles (2000) found that 24% of 14- and 15-year-old algebra students, assigned a familiar mathematical problem accompanied by different arguments, indicated that the empirical argument would be the most similar to their own approach (39% for an unfamiliar problem). Chazan (1993), in his study of high-school geometry students' preferences between empirical and deductive reasoning, documented similar results.

Some students are aware that checking a few cases is not tantamount to proof, but believe that checking more varied and/or randomly selected examples *is* proof. They try to minimise the limitations of checking a few examples in a number of ways, including use of a pattern, extreme cases or special cases. In Chazan's (1993) study, some students believed that if they tried different kinds of triangles – acute, obtuse, right, equilateral, and isosceles – they would verify a given statement about triangles. One of Ball's third graders gave as a reason for accepting the truth of the statement that the sum of two odd numbers is even that she tried "almost 18 of them and even some special cases" (Ball and Bass 2003, p. 35).

On the other hand, some students are aware of the fact that checking a few examples is not enough and are also not satisfied with trying different cases. Interviews in Healy and Hoyles's (2000) study revealed that some students who chose empirical arguments as closest to their own were not really satisfied with these arguments but believed that they could not make better ones. There are also students who realise that some problems contain infinitely many numbers, so one cannot try them all; consequently, rather than considering a general proof, these students believe that no proof is possible (Ball and Bass 2003).

Bell (1976) found that students, although unable to give complete proofs, showed different levels of deductive reasoning, ranging from weak to strong. Some students can follow a deductive argument but believe that "deductive proof is simply evidence" (Chazan 1993, p. 362). Fischbein (1982) found that although 81.5% of them believed that a given proof was fully correct, just 68.5% of a student population accepted the theorem. Some students, given a statement claiming the same result for a subset of elements from an already generally proven category, think that another specific proof is necessary (Healy and Hoyles 2000). Some students also think that either further examples are necessary or that a deductive argument is subject to counterexamples. In Fischbein's (1982) study, only 24.5% of the students accepted the correctness of a given proof and *"at the same time"* (1982, p. 16) thought that they did not need additional checks.

Student-constructed proofs may take various forms. Even students who can produce valid mathematical proofs do not tend to give formal arguments using symbols. Healy and Hoyles found that students who went beyond a pragmatic approach were "more likely to give arguments expressed informally in a narrative style than to use algebra formally" (2000, p. 408). In Bell's (1976) study, none of the students used algebra in their proofs. Porteous (1990) also reported that it was "disappointing to find an almost total absence of algebra" (1990, p. 595). According to Healy and Hoyles, students did not use algebraic arguments because "it offered them little in the way of explanation … and [they] found them hard to follow" (2000, p. 415).

All the aforementioned studies mainly refer to pre-secondary school. Studies carried out at the more sophisticated high-school and college levels have had to consider further elements and difficulties.

As Moore claims, "the ability to read abstract mathematics and do proofs depends on a complex constellation of beliefs, knowledge, and cognitive skills" (1994, p. 250). Furinghetti and Morselli's study (2007, 2009b) instantiates how beliefs may intervene in this constellation and, at the same time, hints at the interpretative difficulties linked to this kind of investigation. Their analyses show the weight and role of beliefs as driving forces throughout the proving process as well as their mutual relationship with cognitive factors. For instance, in one case study (Furinghetti and Morselli 2009b) the student's choice of the algebraic representation, and the revision of such a choice after a difficulty is met, are hindered by the student's beliefs about self (low self-confidence), about mathematical activity as an automatic activity, and about the role of algebra as a proving tool. In another case (Furinghetti and Morselli 2007) the proving process is supported by the student's self-confidence and his belief about proof as a process aimed not only at proving but also at explaining. In this latter case, the authors underline the positive role of beliefs in supporting the construction of a proving process as well the final systematisation of the product.

3 Metaknowledge About Proof

3.1 Metaknowledge

Literally, metaknowledge is knowledge about knowledge. We use the term 'metaknowledge about proof' to designate the knowledge needed to reflect about, teach and learn proof. We distinguish metaknowledge from beliefs by placing beliefs closer to individuals' opinions, emotions and attitudes, whereas "metaknowledge" refers to consensually held ideas. Metaknowledge about proof includes concepts which refer to:

- the *structure of mathematical theories*, like axiom/hypothesis, definition, theorem;
- formal *logic*, like truth, conditional, connectives, quantifiers;

- *modes of representation*, like symbolic, pictorial and verbal reasoning; and
- relationships between proof in mathematics and related processes of argumentation in *other fields*, especially the empirical sciences.

Above, in Sects. 2.2 and 2.3, we discussed research findings about problematic dimensions of teachers' and students' beliefs about proof – above all, the preponderance of empirical ways of justification. These findings are internationally valid; consequently, we have to consider the problematic beliefs as outcomes of the usual way of teaching proof at school and university. Here, we expound our *thesis* that these shortcomings of teaching can be successfully overcome only when metaknowledge about proof is made a theme of mathematics teaching, beginning with the explicit introduction of the very notion of proof. The opinions of many educators, for example Healy and Hoyles (2000) and Hemmi (2008) support this view.

3.2   Place of Metaknowledge in the Curriculum

The role of proof in the curriculum varies across different countries. Of course, there is a broad international consensus that learning mathematical argumentation should start with the very beginning of mathematics in the primary grades (e.g. Ball and Bass 2000; Bartolini 2009; Wittmann 2009). However, the situation is different with regard to the explicit introduction of the notion of proof *per se*. In some countries, such as France, Germany and Japan, proof is seen as something to be explicitly taught. Cabassut (2005) notes that the introduction of proof in France and Germany takes place mostly in grade 8, which is also the situation in Japan (Miyazaki and Yumoto 2009). In these countries, the official syllabus makes explicit what should be taught about proof and/or textbooks contain chapters about proof (Cabassut 2009; Fujita, Jones, and Kunimune 2009). In other countries, such as Italy (Furinghetti and Morselli 2009a) and the United States (National Council of Teachers of Mathematics [NCTM] 2000), proof remains a more informal concept but is nevertheless made a theme by individual teachers. Furinghetti and Morselli report that most of the Italian teachers they have interviewed respond that they treat proof in their classes. In the United States, 'reasoning and proof' is identified as a process standard (NCTM 2000) to be integrated across content and grade levels rather than taught explicitly as one object of study. Regardless of whether proof is explicitly treated, it is important that mathematics teachers have well-founded metaknowledge about proof in order to communicate an adequate image of mathematics to their students.

3.3   Basic Components of Metaknowledge About Proof

In the practice of teaching, the attitude seems frequently to prevail that metaknowledge about proof emerges spontaneously from examples. Only a few ideas are available about how to provide metaknowledge about proof explicitly to pupils or teachers. For example, Arsac et al. (1992) give explicit 'rules' for discussions with pupils of the lower secondary level (11–15 years old): a mathematical assertion is either true or false; a counter-example is sufficient for rejection of an assertion; in mathematics people agree on clearly formulated definitions and properties as warrant of the debate; in mathematics one cannot decide that an assertion is true merely because a majority of persons agree with it; in mathematics numerous examples confirming an assertion are not sufficient to prove it; in mathematics an observation on a drawing is not sufficient to prove a geometrical assertion. However, listing such rules is not sufficient to develop metaknowledge, because the latter is broader and includes relationships to other fields.

Here, in identifying basic components of metaknowledge about proof, we confine ourselves to metaknowledge which should be made a theme already in the lower secondary grades and which (*a fortiori*) should be provided to future teachers of mathematics. We leave aside metaknowledge related to formal logic, since this topic is treated in other chapters of this volume and is appropriate only for a more advanced level. (We also exclude 'modes of representation'; see Cabassut 2005, 2009).

When introducing proof in the mathematics classroom, teachers usually say two things to their students: first, that proofs produce *certain knowledge*, "We think that this statement might be true, but to be sure we have to prove it"; second, that proof establishes *generally valid statements* – statements true not only for special cases but for all members of a class (e.g., all natural numbers or all triangles). Teachers all over the world thus try to explain proof to their students; we take those two messages as basic components of the necessary metaknowledge about proof. However, many teachers and educators are unaware that the two messages are incomprehensible by themselves and need further qualifications. One reason is the difference between these statements and statements made in science courses. Conner and Kittleson (2009) point out that students encounter similar problem situations in mathematics and science, but the ways in which results are established differ between these disciplines. In mathematics, a proof is required to establish a result; in science, results depend on a preponderance of evidence (not accepted as valid in mathematics).

3.3.1   The Certainty of Mathematics

It is important to convey to students the idea that proofs do not establish facts but 'if-then-statements'. We do not prove a 'fact' B but an implication 'If A then B'. Boero, Garuti and Lemut (2007, p. 249 *ff*) rightly speak about the conditionality of mathematical theorems. For example, we do not prove the 'fact' that all triangles have an angle sum of 180°; rather we prove that in a certain theory this consequence can be derived. The angle-sum theorem is an 'if-then-statement' whose 'if' part consists ultimately of the axioms/hypotheses of Euclidean geometry. Thus the *absolute certainty* of mathematics resides not in the facts but in the logical inferences, which are often implicit.

Whether mathematicians believe in the 'facts' of a theory is dependent on their confidence in the truth of the hypotheses/axioms. This confidence is the result of a more or less conscious *process of assessment*. Mathematicians find the axioms of arithmetic highly reliable and therefore can believe that there are infinitely many prime numbers. The situation is different in geometry; with 'medium-sized' objects Euclidean geometry is the best available theory, but in cosmological dimensions Riemannian geometry is taken as the appropriate model. The situation is even more complicated with applied theories in physics and other sciences.

The issues of the potential certainty of mathematical proof and of the conditionality of the theorems have to be made frequent themes in mathematics education, beginning at the secondary level. Teachers should discuss them with students in various situations if they expect the students to get an adequate understanding of mathematical proof. In particular, they should make students aware of the necessary process of assessing the reliability of a theory.

3.3.2 Universally Valid Statements

To an educated mathematician, it seems nearly unimaginable that the phrase "for all objects x with a certain property the statement A is true" should present any difficulty of understanding to a learner. Many practical experiences and some recent empirical studies show, however, that it does exactly that. Lee and Smith, in a recent study (2008, 2009) of college students, found that some of their participants held the notion that "true rules could always allow exceptions" or that "true means mostly true" or that there might be an "unknown exception to the rule." (Lee and Smith 2009, pp. 2–24). This is consistent with the experiences of students frequently not understanding that *one* counterexample suffices for rejection of a theorem. Galbraith (1981) found, for example, that one third of his 13- to 15-year-old students did not understand the role of counterexamples in refuting general statements (see also Harel and Sowder 1998).

Frequently, students do not think that the set to which a general statement refers has a definite extension but assume tacitly that under special circumstances an exception might occur. From the point of view of classical mathematics this is a 'misconception'; however considering general statements outside of classical mathematics one finds that concepts are generally seen as having indefinite extensions. Both everyday knowledge and the empirical sciences consider general statements which under certain conditions might include exceptions. To cover this phenomenon, one can distinguish between *open* and *closed* general statements, having respectively indefinite and definite domains of validity (Durand-Guerrier 2008; Jahnke 2007, 2008). In principle, closed general statements can occur only in mathematics, whereas disciplines outside of mathematics operate with open general statements with the tacit assumption that under certain conditions exceptions might occur. At the turn from the eighteenth to the nineteenth centuries even mathematicians spoke of "theorems which might admit exceptions" (on this issue see Sørensen 2005). Also, intuitionistic mathematics does not consider the concept of the set of all subsets (e.g., of the natural numbers) as a totality of definite extension. We well know that in cases where the domain for which a statement is valid does not have a definite extension the usual logical rules, especially the rule of the excluded middle, are no longer valid.

All in all, the seemingly simple phrase "for all" used in the formulation of mathematical theorems is not an obvious concept for the beginner. Rather, it is a sophisticated theoretical construct whose elaboration has taken time in history and needs time in individuals' cognitive development. Durand-Guerrier (2008, pp. 379–80) provides a beautiful didactic example about how to work on this concept with younger pupils.

3.3.3 Definitions

The theories which mathematicians construct by way of proof are hypothetical and consist of 'if-then-statements'. This fact implies that mathematical argumentation requires and presupposes *rigour*. Hypotheses/axioms and definitions have to be understood and applied in their exact meanings. This requirement sharply contrasts with everyday discourse, which does not commonly use definitions, at least in the mathematical sense. Consequently, the development of a conscious use of definitions is an important component of proof competence.

Most students at the end of their school careers do not understand the importance and meaning of mathematical definitions, even many university students (Lay 2009). Explicit efforts in teaching are required to develop a habit of using definitions correctly in argumentations. Since such a habit does not emerge spontaneously, students need metaknowledge about definitions. They should know that definitions are conventions but are not arbitrary; in general, a definition is constructed the way it is for good reasons.

Beginning university students of mathematics encounter an impressive example of the importance and meaning of definitions when they first operate with infinite sets: namely, how can one determine the 'size' of an infinite set? If one compares sets by way of the relation '$\subset$', then the set N of natural numbers is a proper subset of the set Q of rational numbers: $N \subset Q$, and N is 'smaller' than Q. If, however one compares sets by means of bijective mappings, a fundamental theorem of Cantorian set theory says that N and Q have the same cardinality. Hence, the outcome of a comparison of two sets depends on the definition of 'size'. Numerous further examples occur in analysis: for instance, whether an infinite series is convergent depends on the definition of convergence.

Not many examples of this type arise in secondary teaching. One instance of the importance and relevance of alternative definitions is the definition of a trapezoid (trapezium) as having at least two parallel sides versus having exactly two parallel sides. Asking whether a rectangle is a trapezium requires a student to look past the standard figures depicting the two types of quadrangle. If they apply a particular set of definitions, they conclude that answer is affirmative. Proof and deduction enter the game when the student realises that consequently the formulae for the perimeter and area of a trapezium must also give the perimeter and area of a rectangle. (For further ideas about teaching the construction of definitions, see Ouvrier-Buffet 2004 and 2006.)

3.3.4 Mini-theories as a Means to Elaborate Metaknowledge About Proof

We have suggested three basic components of metaknowledge about proof which naturally emerge in the teaching of proof and which should be more deeply elaborated in teaching: the *certainty* (*conditionality*) of mathematical theorems, the *generality* of the theorems and the conscious use of *definitions*. One possible method to further this learning is to develop *mini-theories* accessible to learners and sufficiently substantial to discuss meta-issues. The idea of such mini-theories, not completely new, resembles Freudenthal's (1973) concept of 'local ordering' or the use of a finite geometry as a surveyable example of an axiomatic theory. However, the study of finite geometries is not feasible in secondary teaching. Besides, our idea of a mini-theory differs in two aspects from Freudenthal's concept of local ordering. First, we would include in the teaching of a mini-theory phases of explicit reflection about the structure of axiomatic theories, the conditionality of mathematical theorems and the set of objects to which a theorem applies. Second, we would also take into account 'small theories' from physics, like Galileo's law of free fall and its consequences, and other examples of mathematised empirical science (Jahnke 2007).

Treating a mathematised empirical theory would provide new opportunities to make students aware of the process of assessing the truth of a theory (see above on certainty; Conner and Kittleson 2009; Jahnke 2009a, b). Usually, teachers only tell students that the axioms are intuitively true and that therefore all the theorems which can be derived from them are true; however, this is a one-sided image. In many other cases, one believes in the truth of a theory because its consequences agree with empirical evidence or because it explains what one wants to explain. For example, teachers generally treat Euclidean geometry in the latter way, at least at pre-tertiary levels. In the philosophy of science, this way of constructing and justifying a theory is called the 'hypothetico-deductive method'.

Barrier et al. (2009) developed a related idea for teaching the metaknowledge of proof. They discuss a dialectic between an 'indoor game' and an 'outdoor game'. The indoor game refers to the proper process of deduction, whereas the outdoor game deals with "the truth of a statement inside an interpretation domain" (2009, p. 78).

4 Conclusion

Our discussion in Sect. 1 has shown how strongly conceptualisations of proof are dependent of the professional background and aims of the respective researcher. Practising mathematicians, whilst agreeing on the acceptance of certain arguments as proof, stray from a formal definition of proof when explaining what one is.

Consequently, it is difficult to explain precisely what a proof is, especially to one who is a novice at proving (such as a child in school). Mathematics educators also differ in their distinctions between argumentation and proof (or inclusion of one in the other). Regardless of the classification scheme of the researchers, research reveals that students and teachers often classify arguments as proofs differently from the classifications accepted in the field of mathematics.

Existing research on beliefs about proof has focused on investigating proof conceptions of prospective and practising elementary and secondary school teachers (Sect. 2). Their beliefs about proving are wrapped around two main issues: what counts as proof in the classroom and whether the focus of teaching proof is on the product or on the process. Research has clearly hinted at the fact that quite a number of teachers tend to accept empirical arguments as proofs and have limited views about the role of proof in school mathematics. Given the influence of beliefs on the teaching and learning process at all levels of schooling continued research on beliefs about proof that focuses not only on detecting beliefs but also on understanding their origins seems highly necessary.

Research strongly suggests that beliefs about proof should be addressed more intensely in undergraduate mathematics and mathematics education courses and during professional development programmes in order to overcome the shortcomings which have been identified in the beliefs about proof. Consequently, we discuss in the last section of the chapter (Sect. 3) which type of metaknowledge about proof should be provided to students and how this can be done. We identify three components of metaknowledge about proof which should be made a theme in teacher training as well as in school teaching. These are the certainty of mathematics, universally valid statements and the role of definitions in mathematical theories. The elaboration of teaching units which allow an honest discussion of metaknowledge about proof seems an urgent desideratum of future work. Mini-theories could be one possible way of achieving this, and further research is necessary both to examine the feasibility of the use of mini-theories and to develop other ways of developing metaknowledge about proof.

References*

Alibert, D. (1988). Towards new customs in the classroom. *For the Learning of Mathematics, 8*(2), 31–35.

Arsac, G., Chapiron, G., Colonna, A., Germain, G., Guichard, Y., & Mante, M. (1992). *Initiation au raisonnement déductif au collège*. Lyon: Presses Universitaires de Lyon.

Balacheff, N. (1988). Aspects of proof in pupils' practice of school mathematics. In D. Pimm (Ed.), *Mathematics, teachers and children* (pp. 216–238). London: Hodder & Stoughton.

---

*NB: References marked with * are in F. L. Lin, F. J. Hsieh, G. Hanna, & M. de Villiers (Eds.) (2009). *ICMI Study 19: Proof and proving in mathematics education*. Taipei, Taiwan: The Department of Mathematics, National Taiwan Normal University.

Balacheff, N. (2009). Bridging knowing and proving in mathematics: A didactical perspective. In G. Hanna, H. N. Jahnke, & H. Pulte (Eds.), *Explanation and proof in mathematics. Philosophical and educational perspectives* (pp. 115–135). New York: Springer.

Ball, D. L. (1999). Crossing boundaries to examine the mathematics entailed in elementary teaching. In T. Lam (Ed.), *Contemporary mathematics* (pp. 15–36). Providence: American Mathematical Society.

Ball, D., & Bass, H. (2000). Making believe: The collective construction of public mathematical knowledge in the elementary classroom. In D. Phillips (Ed.), *Constructivism in education* (pp. 193–224). Chicago: University of Chicago Press.

Ball, D., & Bass, H. (2003). Making mathematics reasonable in school. In J. Kilpatrick, W. G. Martin, & D. Schifter (Eds.), *A research companion to principles and standards for school mathematics* (pp. 27–44). Reston: NCTM.

Barrier, T., Durand-Guerrier V., & Blossier T. (2009). Semantic and game-theoretical insight into argumentation and proof (Vol. 1, pp. 77–88).*

Bartolini Bussi, M. (2009). Proof and proving in primary school: An experimental approach (Vol. 1, pp. 53–58).*

Bell, A. (1976). A study of pupils' proof-explanations in mathematical situations. *Educational Studies in Mathematics, 7*(23–40).

Biza, I., Nardi, E., & Zachariades, T. (2009). Teacher beliefs and the didactic contract in visualization. *For the Learning of Mathematics, 29*(3), 31–36.

Blum, W., & Kirsch, A. (1991). Preformal proving: Examples and reflections. *Educational Studies in Mathematics, 22*(2), 183–203.

Boero, P., Garuti, R., & Lemut, E. (2007). Approaching theorems in grade VIII. In P. Boero (Ed.), *Theorems in schools: From history, epistemology and cognition to classroom practice* (pp. 249–264). Rotterdam/Taipei: Sense.

Cabassut, R. (2005). *Démonstration, raisonnement et validation dans l'enseignement secondaire des mathématiques en France et en Allemagne.* IREM Université Paris 7. Downloadable on http://tel.ccsd.cnrs.fr/documents/archives0/00/00/97/16/index.html

Cabassut, R. (2009). The double transposition in proving (Vol. 1, pp. 112–117).*

Carpenter, T., Franke, M., & Levi, L. (2003). *Thinking mathematically: Integrating arithmetic & algebra in elementary school.* Portsmouth: Heinemann.

Chazan, D. (1993). High school geometry students' justification for their views of empirical evidence and mathematical proof. *Educational Studies in Mathematics, 24*, 359–387.

Conner, A. (2007). Student teachers' conceptions of proof and facilitation of argumentation in secondary mathematics classrooms (Doctoral dissertation, The Pennsylvania State University, 2007). *Dissertations Abstracts International, 68*/05, Nov. 2007 (UMI No. AAT 3266090).

Conner, A., & Kittleson, J. (2009). Epistemic understandings in mathematics and science: Implications for learning (Vol. 1, pp. 106–111).*

de Villiers, M. (1990). The role and function of proof in mathematics. *Pythagoras, 24*, 17–24.

Dormolen, Jv. (1977). Learning to understand what giving a proof really means. *Educational Studies in Mathematics, 8*, 27–34.

Durand-Guerrier, V. (2008). Truth versus validity in mathematical proof. *ZDM – The International Journal on Mathematics Education, 40*(3), 373–384.

Duval, R. (1991). Structure du raisonnement déductif et apprentissage de la démonstration. *Educational Studies in Mathematics, 22*(3), 233–262.

Fernandez, C. (2005). Lesson study: A means for elementary teachers to develop the knowledge of mathematics needed for reform minded teaching? *Mathematical Thinking and Learning, 7*(4), 265–289.

Fischbein, E. (1982). Intuition and proof. *For the Learning of Mathematics, 3*(2), 9–24.

Freudenthal, H. (1973). *Mathematics as an educational task.* Dordrecht: Reidel.

Fujita T., Jones K., & Kunimune S. (2009). The design of textbooks and their influence on students' understanding of 'proof' in lower secondary school (Vol. 1, pp. 172–177).*

Furinghetti, F., & Morselli, F. (2007). For whom the frog jumps: The case of a good problem solver. *For the Learning of Mathematics, 27*(2), 22–27.

Furinghetti, F., & Morselli, F. (2009a). Leading beliefs in the teaching of proof. In W. Schlöglmann & J. Maaß (Eds.), *Beliefs and attitudes in mathematics education: New research results* (pp. 59–74). Rotterdam/Taipei: Sense.

Furinghetti, F., & Morselli, F. (2009b). Every unsuccessful solver is unsuccessful in his/her own way: Affective and cognitive factors in proving. *Educational Studies in Mathematics, 70*, 71–90.

Furinghetti, F., & Pehkonen, E. (2002). Rethinking characterizations of beliefs. In G. Leder, E. Pehkonen, & G. Toerner (Eds.), *Beliefs: A hidden variable in mathematics education?* (pp. 39–57). Dordrecht/Boston/London: Kluwer.

Galbraith, P. L. (1981). Aspects of proving: A clinical investigation of process. *Educational Studies in Mathematics, 12*, 1–29.

Hanna, G. (1983). *Rigorous proof in mathematics education*. Toronto: OISE Press.

Hanna, G., & Barbeau, E. (2009). Proofs as bearers of mathematical knowledge. In G. Hanna, H. N. Jahnke, & H. Pulte (Eds.), *Explanation and proof in mathematics. Philosophical and educational perspectives* (pp. 85–100). New York: Springer.

Harel, G., & Sowder, L. (1998). Students' proof schemes: Results from exploratory studies. In A. H. Schoenfeld, J. Kaput, & E. Dubisnky (Eds.), *Issues in mathematics education* (Research in collegiate mathematics education III, Vol. 7, pp. 234–283). Providence: American Mathematical Society.

Harel, G., & Sowder, L. (2007). Toward a comprehensive perspective on proof. In F. Lester (Ed.), *Second handbook of research on mathematics teaching and learning* (pp. 805–842). Charlotte: NCTM/Information Age Publishing.

Healy, L., & Hoyles, C. (2000). A study of proof conceptions in algebra. *Journal for Research in Mathematics Education, 31*(4), 396–428.

Hemmi, K. (2008). Students' encounter with proof: The condition of transparency. *The International Journal on Mathematics Education, 40*, 413–426.

Jahnke, H. N. (2007). Proofs and hypotheses. *ZDM – The International Journal on Mathematics Education, 39*(1–2), 79–86.

Jahnke, H. N. (2008). Theorems that admit exceptions, including a remark on Toulmin. *ZDM – The International Journal on Mathematics Education, 40*(3), 363–371.

Jahnke H. N. (2009a). Proof and the empirical sciences (Vol. 1, pp. 238–243).*

Jahnke, H. N. (2009b). The conjoint origin of proof and theoretical physics. In G. Hanna, H. N. Jahnke, & H. Pulte (Eds.), *Explanation and proof in mathematics. Philosophical and educational perspectives* (pp. 17–32). New York: Springer.

Jones, K. (2000). The student experience of mathematical proof at university level. *International Journal of Mathematical Education, 31*(1), 53–60.

Kirsch, A. (1979). Beispiele für prämathematische Beweise. In W. Dörfler & R. Fischer (Eds.), *Beweisen im Mathematikunterricht* (pp. 261–274). Hölder-Pichler-Tempspsky: Klagenfurt.

Knuth, E. J. (2002a). Secondary school mathematics teachers' conceptions of proof. *Journal for Research in Mathematics Education, 33*(5), 379–405.

Knuth, E. (2002b). Teachers' conceptions of proof in the context of secondary school mathematics. *Journal of Mathematics Teacher Education, 5*(1), 61–88.

Lay S. R. (2009) Good proofs depend on good definitions: Examples and counterexamples in arithmetic (Vol. 2, pp. 27–30).*

Leatham, K. (2006). Viewing mathematics teachers' beliefs as sensible systems. *Journal of Mathematics Teacher Education, 9*, 91–102.

Lee K., & Smith, J. P. III (2009). Cognitive and linguistic challenges in understanding proving (Vol. 2, pp. 21–26).*

Lee, K., & Smith, J. P. III. (2008). *Exploring the student's conception of mathematical truth in mathematical reasoning*. Paper presented at the Eleventh Conference on Research in Undergraduate Mathematics Education, San Diego.

Ma, L. (1999). *Knowing and teaching elementary mathematics*. Mahwah: Erlbaum Associates.

Manin, Y. (1977). *A course in mathematical logic*. New York: Springer.

Martin, W. G., & Harel, G. (1989). Proof frames of preservice elementary teachers. *Journal for Research in Mathematics Education, 20*(1), 41–51.

Mingus, T., & Grassl, R. (1999). Preservice teacher beliefs about proofs. *School Science and Mathematics, 99*, 438–444.

Miyazaki, M., & Yumoto, T. (2009). Teaching and learning a proof as an object in lower secondary school mathematics of Japan (Vol. 2, pp. 76–81).*

Moore, R. C. (1994). Making the transition to formal proof. *Educational Studies in Mathematics, 27*, 249–266.

Morselli, F. (2006). Use of examples in conjecturing and proving: An exploratory study. In J. Novotná, H. Moarová, M. Krátká, & N. Stelíchová (Eds.), *Proceedings of PME 30* (Vol. 4, pp. 185–192). Prague: Charles University.

National Council of Teachers of Mathematics (NCTM). (2000). *Principles and standards for school mathematics*. Reston: NCTM.

Neubrand, M. (1989). Remarks on the acceptance of proofs: The case of some recently tackled major theorems. *For the Learning of mathematics, 9*(3), 2–6.

Ouvrier-Buffet, C. (2004). Construction of mathematical definitions: An epistemological and didactical study. In *Proceedings of the 28th Conference of the International Group for the Psychology of Mathematics Education* (Vol. 3, pp. 473–480), Bergen, Norway.

Ouvrier-Buffet, C. (2006). Exploring mathematical definition construction processes. *Educational Studies in Mathematics, 63*, 259–282.

Philipp, R. A. (2007). Mathematics teachers' beliefs and affect. In F. K. Lester (Ed.), *Second handbook of research on mathematics teaching and learning* (pp. 257–315). Charlotte: NCTM/ Information Age Publishing.

Porteous, K. (1990). What do children really believe? *Educational Studies in Mathematics, 21*, 589–598.

Rav, Y. (1999). Why do we prove theorems? *Philosophia Mathematica, 7*(3), 5–41.

Semadeni, Z. (n.d.). *The concept of premathematics as a theoretical background for primary mathematics teaching*. Warsaw: Polish Academy of Sciences.

Simon, M., & Blume, G. (1996). Justification in the mathematics classroom: A study of prospective elementary teachers. *The Journal of Mathematical Behavior, 15*, 3–31.

Smith, J. C. (2006). A sense-making approach to proof: Strategies of students in traditional and problem-based number theory courses. *The Journal of Mathematical Behavior, 25*(1), 73–90.

Sørensen, H. K. (2005). Exceptions and counterexamples: Understanding Abel's comment on Cauchy's Theorem. *Historia Mathematica, 32*, 453–480.

Sowder, L., & Harel, G. (2003). Case studies of mathematics majors' proof understanding, production, and appreciation. *Canadian Journal of Science, Mathematics, and Technology Education, 3*, 251–267.

Wittmann, E. C. (2009). Operative proof in elementary mathematics (Vol. 2, pp. 251–256).*

Chapter 8
Forms of Proof and Proving in the Classroom

Tommy Dreyfus, Elena Nardi, and Roza Leikin

1 Introduction

In this chapter we draw on papers presented at the conference (Lin et al. 2009) in order to discuss forms of proof and proving in the learning and teaching of mathematics. By "forms of proof and proving" we refer to a variety of aspects that influence the appearance of proof and the manner in which these may be conceived by students and teachers trying to cope with understanding or producing proofs. These aspects include:

- different representations, including visual, verbal and dynamic, that may be used in the course of proof production;
- different ways of arguing mathematically, such as inductive example-based arguments, example-based generic arguments and general arguments, as well as individually versus socially produced arguments;
- different degrees of rigour and of detail in proving – including different degrees of pointing out assumptions, whether in terms of first principles or previously proven statements – and where and how these are used;
- multiple proofs; that is, different proofs for the same mathematical statement, which may be used in parallel or sequentially, by a single person or a group.

T. Dreyfus (✉)
Department of Mathematics, Science and Technology Education, School of Education,
Tel Aviv University, Tel Aviv, Israel
e-mail: tommyd@post.tau.ac.il

E. Nardi
School of Education, University of East Anglia, Norwich, UK
e-mail: e.nardi@uea.ac.uk

R. Leikin
Department of Mathematics Education, Faculty of Education, University of Haifa,
Haifa, Israel
e-mail: rozal@construct.haifa.ac.il

© The Author(s) 2021
G. Hanna and M. de Villiers (eds.), *Proof and Proving in Mathematics Education*,
New ICMI Study Series, https://doi.org/10.1007/978-94-007-2129-6_8

We read the complete conference proceedings and selected contributions that refer to forms of proof and proving in the above sense, grouping them appropriately and integrating them, to some extent, with other relevant work. All sections of the chapter draw heavily on these conference contributions. The majority of them are either empirical studies or thought pieces by colleagues with substantial pedagogical experience. No conference contributions explicitly offered or discussed a theoretical framework relating to our theme, but naturally this chapter reflects theoretical discussions in other chapters. Occasionally we highlight some of these resonances. Following careful reading of the conference proceedings – and given that the papers appear in the conference proceedings in their originally submitted version, reviewed but unrevised – we have selected only papers (or parts of papers) that we felt appropriate.

The chapter consists of four sections. In the section following this introduction we focus on external forms of proof (mostly visual, verbal and dynamic). We report work on student and teacher beliefs about visual aspects of proof and proving and discuss the importance of visibility and transparency of mathematical arguments, particularly with regard to the role of visualisation. We highlight the pedagogical potential of certain proving activities and also reflect briefly on whether all mathematical ideas are equally visualisable. We then briefly relate some of this discussion to work presented in the conference on verbal and dynamic proofs.

The next section discusses the importance of various mathematical, pedagogical, and cognitive aspects related to different forms of proof. This section also includes a discussion of multiple-proof tasks, which explicitly require different types of proofs for the same mathematical statement.

In the light of the variety of forms discussed up to this point, and the preponderance of empirical arguments and proof schemes (Harel and Sowder 1998, 2007) amongst students and sometimes even teachers, the penultimate section focuses on the question, which forms of proof might support students in making the transition from empirical arguments to general proofs. Part of this section draws lessons from the history of mathematics; most of the rest discusses the significant role that operative and generic proofs can play in the classroom.

We conclude with a section indicating some issues relating to forms of proof and proving that have not yet received sufficient research attention and where future research could improve the teaching and learning of proof. In this respect, we particularly highlight the need for more school-based studies, more longitudinal studies and more work on students' conceptualisations of the need for proof – as well as deep reflection and theory-building related to all these issues.

2   Visual and Other External Forms of Proof

In this section we examine issues related to visual and other forms of proof that were raised in a substantial number of the conference papers. These issues include: student and teacher beliefs about visual aspects of proof and proving (Biza et al. 2009a;

Nardi 2009; Whiteley 2009); the importance of the 'visibility' and transparency of mathematical arguments (Hemmi and Jaworski 2009; Raman et al. 2009); the pedagogical potential of 'geometrical sophisms' and other proving activities (Kondratieva 2009; Perry et al. 2009a, b); and some reflections on the non-universal visualisability of mathematical ideas (Iannone 2009; Mamona-Downs and Downs 2009). We close the section with brief notes on verbal aspects of proof and proving (Arzarello et al. 2009a; Tsamir et al. 2009) and on dynamic forms of proof (Arzarello et al. 2009b; Leung 2009; Stevenson 2009), issues that parallel some of our discussion of visual aspects of proof and proving in the main part of this section.

In recent years, the debate about the potential contribution of visual representations to mathematical proof has intensified (e.g., Mancosu et al. 2005) and the multidisciplinary community of diagrammatic reasoning (e.g., Stenning and Lemon 2001) has been steadily growing. Central to the debate is whether visual representations should be treated as adjuncts to proofs, as an integral part of proofs or as proofs themselves (e.g., Byers 2007; Giaquinto 2007; Hanna and Sidoli 2007). Within mathematics education, the body of work on the important pedagogical role of visualisation has also been expanding (see Presmeg 2006 for a substantial review). Overall, we still seem far from a consensus on the many roles visualisation can play in mathematical learning and teaching, as well as in both pre- and post- formal aspects of mathematical thinking in general.

Recently, the relationship between teachers' beliefs and pedagogical practice has attracted increasing attention from mathematics education researchers, (e.g., Cooney et al. 1998; Leatham 2006; Leder et al. 2002). In this section, we explore this relationship particularly with regard to teachers' beliefs about the role *visualisation* can play in mathematical reasoning.

Teachers' beliefs about the role of visualisation – pedagogical and epistemological – are complex and not always consistent. The study by Biza et al. (2009b) is a good case in point. Biza et al. consider two influences on secondary teachers' epistemological beliefs about the role of visualisation in mathematical proof: beliefs about the sufficiency and persuasiveness of a visual argument and personal concept images (in this case, tangent lines). The authors collected written and interview data collected through a task involving recognising a line as a tangent to a curve at an inflection point on a graph. About half of the teachers, 45 out of 91, appear to dwell on the geometric image of a circle tangent and incorrectly claim that the line is not a tangent, although it is. These teachers' prior extensive experiences with the image of the circle tangent in Euclidean Geometry were so strong that they did not feel compelled to explore the image offered to them (in a Calculus context) further and perhaps reshape their ideas about tangency accordingly. The two teachers cited in the paper rely on the Euclidean circle tangent image – in the words of Nogueira de Lima and Tall (2006) a perfectly valid "met-before" image – in order to reject (without questioning their definition of tangency and without checking algebraically) the line as a tangent. Therefore, a persistent image valid in one mathematical domain (e.g., Euclidean Geometry) supersedes the requirements of another mathematical domain (e.g., Calculus). Elsewhere, Biza et al. (2009b) discuss the need to construct an explicit didactical contract – through, for instance, discussion in the classroom – with regard to the role that such images can play in different mathematical domains.

Some studies focusing on university mathematicians also reported beliefs about the role of visualisation in mathematical reasoning. Many of these mathematicians – either in the form of self-reporting (Whiteley 2009) or in interviews (Iannone 2009; Nardi 2009) – made a relatively straightforward point: consider "what mathematicians often do" (Whiteley 2009, p. 258, also citing Brown 1997), how "mathematicians work" (Nardi 2009, p. 117), as one criterion for deciding pedagogical priorities for university mathematics teaching. In Iannone's (ibid.) words, "to study expert behaviour in doing mathematics, such as the behaviour of a research mathematician, can help [sic] understanding what skills students need to acquire to become experts themselves" (p. 224).

For example, Whiteley (2009) reflects on his own practices as a mathematician and highlights the significance of reasoning through examples and counter-examples, making conjectures, evaluating plausible ideas, working with sketched proofs, and writing. In assessing students and refereeing research papers, he draws on occasions to critically analyse presentations of reasoning and his own analytical processes. He typically challenges students to provide a counter-example to a false claim or an illustrative example for a true claim. Using his knowledge of mathematicians' professional practices and of students' typical difficulties with mathematical proof, he places particular emphasis on the pedagogical importance of exemplification. His support for visual, and also kinesthetic, arguments fits squarely within this emphasis: illustrations and gestures can be closer to the cognitive processes students need to carry out in order to develop understanding, sometimes even of the 'purest' mathematical idea. The usual critique against visual reasoning, that visuals 'are "merely" examples', 'too specific to be used in general proofs' should not deter us, he stresses: "Visuals are strong particularly because they are examples' and they can indeed 'carry general reasoning as symbols for the general case, provided the readers bring a range of variation to their cognition of the figure" (p. 260). Furthermore, not only is there nothing wrong with a 'partial' perception of a mathematical idea but also this very 'partiality' and any work students may do towards developing conventions and expressions for it can be instructive. Pedagogical practice that deprives students of these instructive opportunities is impoverished.

The mathematicians interviewed by Nardi (2009) elaborate Whiteley's "learning to see like a mathematician" (2004, p. 279) further. Part of the pedagogical role of the mathematician, they state, is to foster a fluent interplay between analytical rigour and (often visually based) intuitive insight. The need to foster this fluency is particularly poignant as students' relationship with visual reasoning is often turbulent. Even when students overcome resistance to employing visualisation, their reliance on it can be somewhat fraught: Pictures may appear unaccompanied by any explanation of how they came to be, or they may appear disconnected from the rest of the students' writing. In fact, students' reticence about employing visualisation has often been attributed (Nardi 2008) to what they perceive as the 'fuzzy' didactical contract (Brousseau 1997) of university mathematics: a contract that allows them to employ only previously proven statements but does not clarify which parts of their prior knowledge, or ways of knowing, count as proven or acceptable. Like Whiteley (2009), the mathematicians Nardi interviewed stress the potentially creative aspects of this 'fuzziness'. They argue that: a picture provides evidence, not proof; pictures are *natural, not obligatory* elements of mathematical thinking; pictures are "a third type of language" (Nardi 2009, p. 116). From these views emerges a clarified didactical contract in which students are allowed to use facts that have not been formally established; later, they are expected to establish those facts formally. The students are encouraged to make use of the power that visualisation allows them. However, they are required to do so in a sophisticated way – for example, through including articulate accounts of their thinking in their writing and through acknowledging the support (e.g., of a graphic calculator) that facilitated the emergence of an insight.

Raman et al. (2009) use videotaped proving incidents to offer a neat operationalisation of that didactical contract. Raman et al. aimed to highlight to students three crucial moments in the production of a proof that are familiar to mathematicians but not always articulated in teaching: "one that gives cause for believing the truth of a claim, one that indicates how a proof could be constructed, and one that formalizes the argument, logically connecting given information to the conclusion" (p. 154). Absent this articulation, students may see these three moments as more disconnected than they typically are. In the study, the students abandoned visual reasoning; "expecting discontinuity between a more intuitive argument and a more formal one, the students practically abandon their near-perfect proof for something that appears to them more acceptable as a formal proof" (p. 158). Clearly, students need more explicit explanation of what is acceptable and effective practice in mathematics.

Emerging from Raman et al.'s proposal of emphasis and explicitness are the questions of *what* and *how* these can be made available to students. Hemmi and Jaworski's (2009) discussion of the *condition of transparency* (Hemmi 2008) aptly addresses this question. Adopting Lave and Wenger's (1991) perspective – in which proof is an artefact in mathematical practice – they write about the "intrinsic balance in learning environments between the uses of artefacts on the one hand and the focusing on artefacts as such on the other hand" (p. 202). Artefacts "need to be seen (be visible) and to be used and seen through (be invisible)" to enhance students' understanding (ibid.). Combining these two characteristics creates the condition of transparency. Applying this theory to the artefact of proof, the authors define *visibility* as "referring to the different ways of focusing on various aspects of proof like logical structures, specific proof techniques, historical role and functions of proof in mathematics and meta-mathematical aspects connected to proof" (ibid.). *Invisibility* then "is the opposite: not focusing on particular aspects of proof, [...] not focusing on the process of proving but the products of proving like formulae and theorems and the justifying of solutions of problems" (ibid.). In relation to visualisation, in Hemmi and Jaworski's sense visual reasoning, as part of the proof artefact, often remains *invisible* in teaching.

The power of visualisation to generate insight, its *semantic potential*, is indubitable. Kondratieva's (2009) use of 'geometrical sophisms' – a paradoxical conclusion, which results from an impossible figure – is a telling illustration of this potential. Intrigued by the absurdity of a figure (or a statement, i.e., 'all triangles are isosceles', '64 = 65' etc.) students engage with what in essence is understanding the nature and purposes of proof. Searching for a flaw, they focus on the substance and details of the proving process, rather than on the truthfulness of a statement as perceived from either its substantive mathematical meaning or its fit to a physical model. The impossible figure provides an opportunity to shake learners' unconditional trust of images and becomes a trigger to "learn the art of deduction" (ibid.).

Analogous potential lies within the activities proposed by Perry et al. (2009a, b). The former focuses on plane geometry proof lessons with a pronounced three-level process of analysing definitions, enunciating propositions and proposing proofs for conjectures. The latter focuses on three sets of tasks used in proof in the Geometry section of a pre-service mathematics teacher education programme: the procedure of proving; proof within the framework of a reference axiomatic system; and issues of proving in geometry. The third set of tasks in Perry et al. (2009b) includes: *Obtain or use information that a graphic representation on paper or product of a dynamic geometry construction provides*. Identifying useful geometric relations, learning how to distinguish between legitimate and illegitimate extraction of information from a figure, identifying clues for a proof of a statement, making a geometric statement, and studying invariance – all become possible in a visually based environment.

But visualisation's semantic potential is not absolute. Iannone's (2009) discussion of the meta-mathematical skills that mathematicians view as important focuses on the ability to recognise when to proceed *semantically* and when to proceed *syntactically* (cf. Weber 2001; Weber and Alcock 2004) when constructing a proof, and on how that choice depends on the specific mathematical objects that appear in a given proof (cf. also Arzarello et al. 2009b; Pedemonte 2007). Iannone discusses *concept usage*, "the ways one operates with the concept in generating or using examples or doing proofs" (Moore 1994, p. 252). The mathematicians she quotes distinguish four types of concept usage: Two – "concepts for which syntactic knowledge can be used for proof production but only ineffectively" (p. 222) and "concepts for which syntactic knowledge alone cannot be used" (ibid.) – allow for the use of non-syntactic procedures (presumably including visualisation). The other two – "concepts without initial pictorial representation for which resorting to syntactic knowledge is the only suitable approach" (p. 221) and "concepts for which syntactic knowledge is an effective tool" (ibid.) – make up those for which visualisation is either unusable or less effective.

Of these latter two types, the former concerns "concepts which are very difficult to represent via mental images, for example via something that can be drawn on a number line or a Cartesian plane. These concepts can be used in proof production only via manipulation of the syntactic statement that defines them" (Iannone 2009, p. 221). For example, the interviewees offered the expression often used in Analysis, 'N arbitrarily large': The idea of arbitrary largeness is far more clearly expressed with symbols than with words or pictures. So are 'this function is not uniformly continuous' or 'this series does not converge'. The latter type are concepts for which "while it is possible to have a visual representation, a syntactic approach is more effective" (Iannone 2009, p. 222). One such example is the negation of the statement 'the sequence converges'.

A complicating factor in employing visualisation in teaching can be that images, as well as a preference for them over other ways of expressing mathematical meaning, can be "very private". Some simply feel "safer" in the syntactic world of symbolic manipulation; indeed, some may even be more effective in that world. As Mamona-Downs and Downs (2009) conclude, "a seemingly syntactic argument may have significant semantic overtones, and vice versa" (p. 99). For a further discussion of syntactic and semantic aspects of mathematical thinking we refer the reader to the chapter by Tall et al. (2011), where these aspects are discussed in relation to 'natural' aspects of mathematical thinking (based on embodied imagery) and 'formal' aspects (based on formal deduction).

In contrast, few presentations at the conference dealt with the verbal aspects of proof and proving. However, they raised issues similar to those we raised in this section in relation to visualisation. Tsamir et al. (2009) started from Healy and Hoyles' (2000) conclusion that students find verbal proofs appealing and tend to be better at deductive reasoning in verbal proofs than in other forms. Tsamir et al. investigated secondary teachers' reactions to verbal forms of proof and found that about half of the teachers tended to reject these proofs, often because they thought that the proofs were not general. Their findings resonate with those of other researchers about the impact teachers' views have on students: For example, Harel and Dreyfus (2009) found that secondary students held mixed opinions about the value of visual proofs – some wanted their teachers to use visual proofs or wished to use visual proofs themselves, whereas others did not – but they were uniformly sure that their teachers would not accept visual forms of proof.

Some of the issues around dynamic methods of proving and proof relate tangentially to those around both visualisation and verbal proofs. Leung (2009), Stevenson (2009), and Arzarello et al. (2009a) are all concerned with the potentially transformative relationship between proof in Dynamic Geometry (DG) environments and formal proofs (e.g., Euclidean proof). For example, Leung's work (2009) can contribute to a discussion of whether a Euclidean proof can be remodelled so as to become a DG proof. Stevenson's work (2009), closely related to Leung's, asks to what extent the notion of proof becomes different in a DG environment. Finally, Arzarello et al. (2009a) consider analogous questions, but in the context of proof in school Geometry and Calculus. They reports on a teaching experiment in Proof in Calculus, which offered students a multi-register encounter with the concept of functions in a variety of interactive and individual work. The students demonstrated a range of quasi-empirical, in Lakatos' sense (Tymoczko 1985), and quasi-theoretical, in Chevallard's sense (Bosch and Chevallard 1999), approaches. The former emerged under the influence also of the software used; the latter emerged in a largely symbolic-algebraic register throughout the activities. Even though they did not provide any complete, formal proofs, through these approaches, the students started to connect relevant pieces of knowledge in an increasingly sophisticated way.

Whilst neither purely inductive nor deductive (but rather abductive), their approaches indicate an emergence of theoretical understanding of proof in Calculus. Arzarello et al.'s findings thus resonate with those regarding effective proving activity in the classroom examined earlier in this section (e.g., Kondratieva 2009; Perry et al. 2009a, b).

In sum, despite the caveats we have outlined, the works reported here seem to be underlain by a robust appreciation of visual reasoning, particularly about images' capacity to encompass ideas that cannot be made overt in a formal definition or an algebraic computation. However, a concern that comes across, often intensely, from these works is that teachers' and, consequently, students' ambivalence towards visualisation may cause prejudice against it and a consequent loss of its unique benefits. Further research and pedagogical action are necessary to make sure that this is not (or ceases to be) the case.

3   Multiple Forms of Proof and Multiple Proof Tasks

Hanna et al. (2009) point out that distinctions amongst different forms of proofs can be mathematical, didactic, or cognitive (understanding-related). This section is structured in terms of these types of distinctions. It also includes a discussion of multiple proof tasks (MPTs), which explicitly require different types of proofs for the same mathematical statement.

Mathematical distinctions between proofs are based on the logical, structural, field-related, representation-related, and statement-related properties of the proofs. These distinctions are expressed in specific proof techniques and can be related to specific types of claims (Hanna et al. 2009). Some statements allow a variety of proofs in different forms. Based on this observation, Leikin (2009a) discusses MPTs as tasks that contain an explicit requirement for proving a statement in multiple ways. The differences between the proofs are *mathematical,* because they require: (a) different representations of a mathematical concept (e.g., proving the formula for the roots of a quadratic function using graphic representation, canonical symbolic representation, and polynomial symbolic representation); (b) different properties (definitions or theorems) of mathematical concepts related to a particular mathematical topic (e.g., Fig. 8.1, Task 1; Sun 2009; Sun and Chan 2009); (c) different mathematical tools and theorems from different branches of mathematics (e.g., Fig. 8.1, Task 2; Greer et al. 2009) or (d) different tools and theorems from various disciplines (not necessarily mathematics) that explicitly require proving a statement in different ways (Leikin 2009a; cf. Multiple Solution Tasks in Leikin 2007; Leikin and Levav-Waynberg 2007).

The most prominent MPT is Pythagoras' Theorem, which lends itself to numerous proofs of different types (for over 70 proofs of this theorem, see, e.g., http://www.cut-the-knot.org/pythagoras/index.shtml). Many specific, less familiar problems may also have multiple proofs. For example, Greer et al. (2009) focus their attention on an MPT (the ISIS problem) that asks: "Find which rectangles with sides of integral length (in

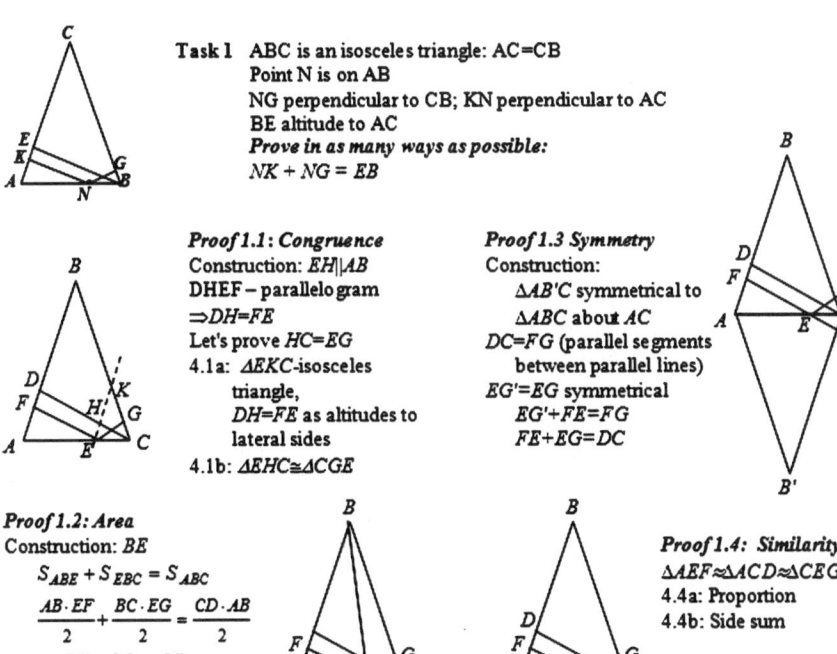

Task 1 ABC is an isosceles triangle: AC=CB
Point N is on AB
NG perpendicular to CB; KN perpendicular to AC
BE altitude to AC
Prove in as many ways as possible:
$NK + NG = EB$

Proof 1.1: Congruence
Construction: $EH\|AB$
DHEF – parallelogram
$\Rightarrow DH=FE$
Let's prove $HC=EG$
4.1a: $\triangle EKC$-isosceles
    triangle,
    $DH=FE$ as altitudes to
    lateral sides
4.1b: $\triangle EHC \cong \triangle CGE$

Proof 1.3 Symmetry
Construction:
  $\triangle AB'C$ symmetrical to
  $\triangle ABC$ about AC
$DC=FG$ (parallel segments
  between parallel lines)
$EG'=EG$ symmetrical
$EG'+FE=FG$
$FE+EG=DC$

Proof 1.2: Area
Construction: BE
$$S_{ABE} + S_{EBC} = S_{ABC}$$
$$\frac{AB \cdot EF}{2} + \frac{BC \cdot EG}{2} = \frac{CD \cdot AB}{2}$$
$$\Rightarrow EF + EG = CD$$

Proof 1.4: Similarity
$\triangle AEF \approx \triangle ACD \approx \triangle CEG$
4.4a: Proportion
4.4b: Side sum

Task 2: ***Prove in as many ways as possible:*** Of all the rectangles with a given perimeter P, the square has the maximal area.

Proof 2.1: Calculus
$P = 4p$; $f(x) = x \cdot (2p - x)$; $f'(x) = -2x + 2p$; $f'(x) = 0$; $f_{max}(x)$ $x = p$
$x = p$ is the side of the square with perimeter $P = 4p$

Proof 22: Algebra
$f(x) = x \cdot (2p - x)$ is a parabola with vertex (max) at $x = p$
2.a) according to the vertex formula;
2.b) according to symmetry of the parabola on the segment $]0; 2p[$

Proof 2.3: Geometry and algebraic manipulations

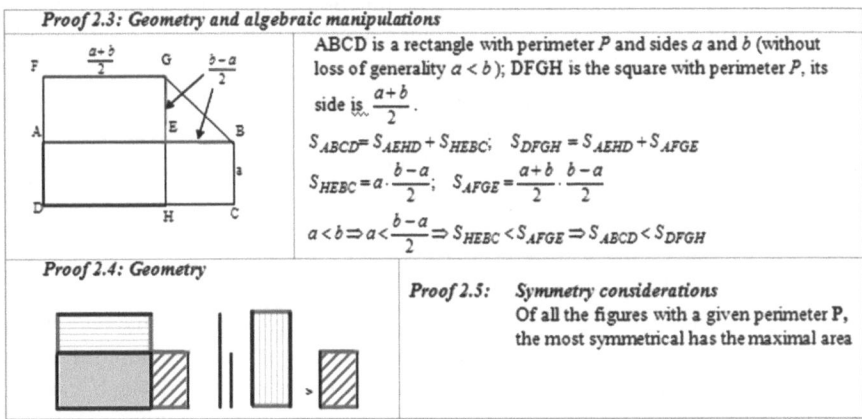

ABCD is a rectangle with perimeter P and sides a and b (without loss of generality $a < b$); DFGH is the square with perimeter P, its side is $\frac{a+b}{2}$.
$$S_{ABCD} = S_{AEHD} + S_{HEBC}; \quad S_{DFGH} = S_{AEHD} + S_{AFGE}$$
$$S_{HEBC} = a \cdot \frac{b-a}{2}; \quad S_{AFGE} = \frac{a+b}{2} \cdot \frac{b-a}{2}$$
$$a < b \Rightarrow a < \frac{b-a}{2} \Rightarrow S_{HEBC} < S_{AFGE} \Rightarrow S_{ABCD} < S_{DFGH}$$

Proof 2.4: Geometry

Proof 2.5: Symmetry considerations
Of all the figures with a given perimeter P, the most symmetrical has the maximal area

Fig. 8.1 Multiple proof tasks (From Leikin 2009a, 31)

some unit) have the area and the perimeter (numerically) equal, and prove the result." They characterise this problem as "notable for the variety of proofs (empirically grounded, algebraic, and geometrical) using different forms of argument, and their associated representations" (p. 184). Sun and Chan (2009) and Sun (2009) use the term "one problem, multiple solutions" when they consider student-generated proofs for the "mid-point theorem of triangles" and for the "area formula of a trapezoid." Sun and Chan (2009) present nine different proofs for the "mid-point theorem" and Sun (2009) demonstrates seven different proofs for the "area formula of a trapezoid."

Tabach et al. (2009) demonstrate that different types of mathematical statements can lead to different methods of proof. For example, a universal statement necessitates a general proof and a single counter-example is sufficient to refute the statement. By contrast, an existential statement can be proven by a single supportive example, and a general proof is necessary to refute it.

Mathematical differences between the proofs can also refer to different levels of mathematics (e.g., Fig. 8.1, Task 2, Proofs 2.1, 2.4). Morselli and Boero (2009), referring to Morselli (2007), argue that proofs at different mathematical levels can be seen when prospective teachers investigate the "properties of divisors of two consecutive numbers" by using divisibility, the properties of the remainder, or algebraic tools. Schwarz and Kaiser (2009), referring to Blum and Kirsch's (1991) distinction between pre-formal and formal proofs, agree that pre-formal proofs contribute to a deeper understanding of theorems by using application-oriented, experimental, and pictorial methods but argue that they are often incomplete (see the next section). By contrast, the authors claim the completeness of formal proofs often goes hand in hand with a degree of complexity that may impede students' understanding. Different levels of proofs can lead to different types of didactic situations (Brousseau 1997) for which proofs of different forms are appropriate. They can also result in different cognitive (understanding-related) processes and products, in addition to different curricular sequences and instructional designs.

Hanna et al. (2009) address distinctions between different types of proofs that are based on pedagogical properties and didactic functions. They view inductive proofs and generic (or transparent) proofs as "comprehensible to beginners" (p. xxiii) and of high didactic potential. Inductive proofs, however, may sometimes be invalid mathematically, they add. Amongst critical questions, these authors ask: "At which level, and in which situations, should the issue of the mathematical validity or lack of validity of inductive proofs be discussed, and how?" (p. xxiv) and "Is it important to introduce proof in a diversity of mathematical domains, and which proofs are more appropriate in which domains?" (p. xxiv). Various researchers have attempted a range of answers to these questions.

Some have used analysis of school mathematics textbooks to shed light on the didactic ideas implemented by teaching proofs. For example, Cabassut (2009) addressed the role of students' mathematical level and of corresponding didactic contracts (Brousseau 1997) in the use of different forms of proof through analysis of mathematics textbooks in France and Germany. Using as an example the "sign of quadratic function", he argues that in secondary education the didactic contracts can develop different techniques and functions of the proof for the same mathematical concept. At the same time, formulas of the perimeter and area of a circle are validated in French textbooks in the sixth and seventh grades through extra-mathematical warrants (visual, inductive, or pragmatic arguments) because mathematical warrants are not being used in these grades.

Meanwhile, after analysing textbooks commonly used for teaching students about proof in geometry in lower secondary schools in Japan, Fujita et al. (2009) argue that the proof and proving activities in these textbooks show geometry as a highly formal subject of study, which precludes discussion of the differences between a formal proof and experimental verification.

Sun and Chan (2009) and Sun (2009) argue that MPTs ("one problem, multiple solutions") are "a simple and powerful framework for guiding teaching and learning". However, students, teachers, textbook writers, and perhaps many researchers do not necessarily perceive them as such. When Sun (Sun 2009; Sun and Chan 2009) compared the use of "one problem, multiple solutions" approaches in Chinese and American textbooks, she found that the Chinese textbook uses MPTs better than its American counterpart. Sun and Chan (2009) and Sun (2009), following Sun (2007), claim that MPTs are widely used in China and that they appear in the Chinese didactical contract in Hong Kong schools as a constituent of a "spiral variation curriculum" (Sun 2007; Sun and Chan 2009; Wong 2007).

Spiral curriculum development and the connection-based approach to teaching mathematics gain expression in classroom activities centred around MPTs. However, Leikin (2009a) distinguishes between teacher-initiated and student-initiated multiple proofs. Frequently a teacher (or teacher educator, if the students are pre-service teachers) plans didactic situations that do not require each learner to provide multiple proofs. However, student-generated proofs can differ from each other; an a-didactic situation develops, contrary to the teacher's plans, led by the students (Leikin and Levav-Waynberg 2007; cf. Brousseau 1997). In other cases, the teacher may ask students to bring multiple proofs of a theorem (Leikin 2009a) or guide students in exploring their own methods of proof which differ from each other (Sun 2009). In these cases multiple proofs are teacher-initiated and are part of a didactic (i.e., the teacher carefully plans it) or a-didactic situation (i.e., the teacher initiates and monitors it, respecting students' responses; Leikin and Levav-Waynberg 2007; for types of didactical situations, Brousseau 1997).

Proofs and proving activities can differ also by the types of didactic settings in which they are incorporated. For example, games (e.g., Nim games) can be an effective form of teaching proofs, their purpose being to find and prove a winning strategy. Lin (2009) reports an experiment that used the game Hex to help students understand and use proofs by practising constructive proofs and proofs by contradiction. Winicki-Landman (2009) describes the design and implementation of a course for pre-service primary teachers based on mathematical games and analysis of a specific game. The analysis results in discovering various roles of mathematical proofs during engagement with the game.

As the suitability of various types of proof depends on the level of the students' mathematical knowledge, various ways of proving may be a factor in understanding proofs and in learning about proving (Hanna et al. 2009). First, different types of proof can be more or less explanatory or convincing. Nardi (2009) found that "students often have a turbulent relationship with visual means of mathematical expression" (p. 112). They also experience difficulties in connecting analytic with visual representations (see the previous section). Moreover, students express less positive attitudes about visual representations than they do about other forms. Nardi stresses "the importance of building bridges between the formal and the informal, in constant negotiation with the students" (p. 114).

However, such bridge-building may require a shift in teachers' attitudes. For example, Greer et al. (2009) describe a study of a group of 39 Flemish pre-service mathematics teachers presented with the Isis problem ("so called because of its connection with the Isis cult", Davis and Hersh 1981, p. 7). The task allowed empirical, algebraic and geometrical proofs. When they compared the proofs, the teachers preferred the empirically grounded and analytic-graphic proofs least, and the algebraic ones most. Tabach et al. (2009) asked teachers which correct and incorrect proofs their students might construct. Overall, the teachers expected their students to propose mainly correct formal proofs and mainly incorrect numeric proofs. At the same time, the teachers did not expect students to produce many verbal proofs. In addition, Chin et al. (2009) demonstrate that, although tending to favour narrative and empirical approaches, students think their teachers favour algebraic approaches to proofs. From the opposite point of view, Ersoz (2009) argues that students' learning proof in Euclidean geometry later creates obstacles when the students are learning to prove in other mathematical domains.

The use of MPTs could change both teachers' and students' views on various types of proofs (Leikin and Levav-Waynberg 2009). Linking mathematical ideas by using more than one approach to solving the same problem (e.g., proving the same statement) is an essential element in developing mathematical connections (NCTM 2000; Pólya 1945/1973; Schoenfeld 1985). Solving MPTs can develop students' divergent reasoning (Kwon et al. 2006), as well as their mental flexibility and fluency (Leikin 2009b; Silver 1997; Sriraman 2003). Using MPTs precludes fixating on a single mathematical idea and increases the chances of an original mathematical product (Kwon et al. 2006; Lithner 2008).

There is evidence that the teacher's role is indispensable in establishing didactic contracts associated with the use of different forms of proofs in the mathematics classroom (Furinghetti and Morselli 2009; Schwarz and Kaiser 2009; Stylianides and Stylianides 2009; Tabach et al. 2009). Using the example of the Pythagoras theorem, Furinghetti and Morselli (2009) argue that teachers should consider different proofs as equally valid, independently of whether they used them while they were students. Tabach et al. (2009) and Stylianides and Stylianides (2009) stress the importance of teachers' awareness of the roles and validity of different forms of proofs with respect both to mathematics as science and to students' mathematical level. Schwarz and Kaiser (2009) point out that teaching different types of proof places high demands on teachers and future teachers. They add that teachers should have university-level content knowledge of mathematics, including the abilities to identify different proof structures, to execute proofs on different levels, to know specific alternative mathematical proofs, and to recognise and establish connections between different topic areas.

4 The Function of Forms of Proof in the Classroom

Considering the variety and quality of forms of proof available, and given the preponderance of empirical proof schemes amongst students and even some teachers (see e.g., Fischbein 1982; Harel and Sowder 2007; Martin and Harel 1989), one may ask which forms of proof might help students make the transition from unsophisticated empirical arguments to general proofs, and how. This transition includes understanding that examples don't constitute proof, experiencing a need for general proof at least occasionally, and acquiring an ability to produce non-example-based proofs, again at least for some claims.

The history of mathematics might provide some guidance. Siu (2009) distinguishes between algorithmic and dialectic aspects of proof: Algorithmic mathematics is a problem-solving tool that invites action and generates results; dialectic mathematics, on the other hand, is a rigorously logical science, where statements are either true or false and where objects with specified properties either do or do not exist, which invites contemplation and generates insight. Siu shows that these two aspects intertwine in both ancient mathematics and students' thinking, and that one can learn from the link between them. One of Siu's examples is Al Khowarizmi's solution of quadratic equations. Grabiner (2011) shows that visual forms of proof have historically played an important role in the development of some proof ideas; she also points out that Al Khowarizmi used visual quadratic completion for developing the quadratic formula. Further, whilst Al Khowarizmi solved a specific equation, the same reasoning chain generalises to an entire class of equations; in this sense his proof is generic. Partly because of its algorithmic nature, Al Khowarizmi's proof satisfies Wittmann's (2009) criteria for operative proofs: It proceeds by actions on 'quasi-real' mathematical objects and is communicable in problem-oriented, non-technical language.

Generally, these forms of proof – algorithmic, visual, generic and operative – may be subsumed under the heading "preformal proofs", defined by Blum and Kirsch as "a chain of correct, but not formally represented conclusions which refer to valid, non-formal premises" (1991, p. 187). Preformal proofs have an illustrative style, and emphasise experimental and visual aspects of mathematics. Conceptions of pre-formal proofs are also closely related to forms of proof that arise from experimental mathematics, often via a computer-based activity. However, these proof forms have rather different epistemological and practical features from those discussed here and are dealt with elsewhere (e.g., Borwein 2011; Zehavi and Mann 2009). Several researchers have proposed using such proof forms to provide bridges for students on the path from empirical to general proof.

Proofs in which visual reasoning plays a central role have been a favourite playing field of mathematicians for some time. In particular, Nelsen (1993, 2000), and Alsina and Nelsen (2006) have published collections of such proofs, many of which are suitable for the high school level. Nevertheless, high school teachers and students accept visual reasoning in proofs, at best, with hesitation (Dreyfus 1994; Harel and Dreyfus 2009; see the section on visual forms of proof above).

Whilst described extensively in the German mathematics education literature since 1980, the idea of operative proofs is not easily accessible in the English literature (see, however, Selden 2005; Wittmann 1998, 2005, 2009). Characteristically, operative proofs arise from the exploration of a mathematical problem, are based on operations with "quasi-real" mathematical objects, and are communicable in problem-oriented language with little symbolism. Hence, they often use visual (rather than symbolic) representations for mathematical objects and, crucially, proceed via operations applicable to these objects. In other words, operative proofs usually have an algorithmic aspect, for which visual reasoning may well be central. Wittmann (2009) gives examples of operative proofs. First, students who have become familiar with odd and even numbers, represented by paired rows of counters with or without a leftover singleton, can operate on the representation to show that, say, the sum of two odd numbers is even. Or, students can deal with divisibility criteria (i.e., that a number is divisible by 9 if and only if the sum of its digits is) using operations on a visual representation of place value. Crucially, the students become familiar with the objects and their representations, discover properties and relationships between them, and the proof then becomes explanatory to them (Hanna 1989).

Generic proof has been treated more extensively in the literature, starting with Tall (1979) and including some empirical research studies. A generic proof aims to exhibit a complete chain of reasoning from assumptions to conclusion, just as in a general proof; however, as with operative proofs, a generic proof makes the chain of reasoning accessible to students by reducing its level of abstraction; it achieves this by examining an example that makes it possible to exhibit the complete chain of reasoning without the need to use a symbolism that the student might find incomprehensible. In other words, the generic proof, although using an (numerical) example, must not rely on any properties of this specific example. Consequently, many operative proofs are generic and vice versa.

The following generic proof shows that the sum of two even numbers is even. Dreyfus (2000) presented this proof to a group of 44 junior high school teachers, requesting that they comment on it as if it had been presented by one of their students.

Let's take two even numbers, say 14 and 32. We can split each, 14 and 32, into two equal parts:

$$
\begin{array}{rcccc}
14 & = & 7 & + & 7 \\
32 & = & 16 & + & 16 \\
\hline
46 & = & 23 & + & 23
\end{array}
$$

After summing the equal parts, 7 and 16, we still have two equal parts of 23; therefore their sum $23 + 23$ is even. It is always possible to do this with two even numbers.

Nine teachers made predominantly positive comments; 27 made predominantly negative ones. All but four stressed that an example or a special case does not constitute a proof. The remaining eight teachers did not comment. Only four teachers explicitly commented on the student's claim that the method always works. However, none of the four was fully convinced. Apparently, they appreciated the generic nature of the argument but were not completely satisfied by it. Ten other teachers made some comment about the student's correct way of thinking about the example. Dreyfus concluded that at best a third of the teachers showed some appreciation for the generic nature of the proof.

Mason and Pimm (1984) discuss the idea of *generic example* and its use. A generic example is an actual example but "presented in such a way as to bring out its intended role as the carrier of the general" (p. 287). Examples include finding the sign of a derivative on an interval by finding it at a single point in the interval, or more sweepingly, considering the graph of $y = x^2$ as representing any quadratic, which it does.

Mason and Pimm also raised many crucial issues and questions about generic proofs: How does the student come to understand that finding the derivative at a single point (or making an argument on x^2 only) is proof by means of a single example that is intended to stand for a more general case. It depends on how hidden or overt this intended role of generality is, and on how well the teacher can lead the students to investigate and become aware of the more general case which is being exemplified. In summary, "how can the necessary act of perception, of seeing the general in the particular, be fostered?" (p. 287). Mason and Pimm propose diagrams as a useful intermediary in this transition. However, diagrams are still only examples showing particular cases. Given the diagram of Al Khowarizmi's method for solving a quadratic (see above) how can a student know for which other equations this method does or does not work? Generally, how are students to learn seeing the general through any such particular diagram?

Balacheff (1987) stressed that the example used in a generic proof is generic in so far as it makes the reasons for the truth of the general assertion explicit. It is exactly that generic character of the situation which gives it the potential for passing from a pragmatic, example-based proof to an intellectual, general proof.

Meanwhile, Rowland (1998, 2001) argues that in order for a generic proofs to succeed, the generalisation made by the student needs to be of a structural nature: The deep structure of the argument rather than some surface features have to be put in evidence and generalised. Structural generalisations are based on the underlying meaning and achieve explanatory insight. Rowland proposes generic proofs as didactic devices that can be used to assist students to perceive and value that which is generic rather than particular in explanations and arguments. Here, generic proofs play a crucial role as transitional tools leading from inductive inference to deductive reasoning.

Like Rowland (2001), Malek and Movshowitz-Hadar (2009) refer to the use of generic proofs, which they call transparent pseudo-proofs, at the tertiary level with the intention that students can see through the presented generic argument or pseudo-proof to the formal proof. Continuing Rowland's formulation of guiding principles, they propose to look at different degrees of specialisation of general claims in terms of a 'generality level pyramid' and to analyse what considerations should be taken into account when constructing a generic proof. This approach appears relevant for complex statements such as are common at the tertiary level, with different parameters or variables that can be specialised by being given particular values. Malek and Movshowitz-Hadar particularly stress the importance of making sure that nothing specific to the example being proved enters the proof (i.e., that the presented proof is indeed generic). However, their more important contribution (Malek and Movshovitz-Hadar 2011) is a small scale (n = 10) empirical study, at the tertiary level, examining how the use of pseudo-proofs affects the acquisition of transferable cognitive proof-related structures. Their results are encouraging; they show that exposure to a generic proof supports students in articulating the main ideas of the proof, in writing a full proof for the theorem, and in proving a different theorem with a similar proof idea.

Finally, Leron and Zaslavksy (2009) add some further examples of generic proofs, again mostly suitable for the tertiary level; they also refine some of Mason and Pimm's (1984) statements. They point to the chosen example's degree of complexity as a critical attribute for making it sufficiently general (to allow the student to engage with the main ideas of the complete proof) but not too general (hence barring the student from following the argument). They also contribute further to characterising proofs that are amenable to be cast in generic form; they argue the importance of a proof's being constructive, since construction of a mathematical object or procedure can often be demonstrated by a generic example.

In summary, mathematics educators have invested quite a bit of thought in generic and other pre-formal forms of proof as steps in students' transition from empirical to deductive proof conceptions. Most of this work has been theoretical, analysing the nature and kinds of possible pre-formal proofs. A variety of examples have been proposed to suggest and support this theoretical work. Anecdotal case studies have provided some evidence that generic examples can provide "enlightenment" (Kidron and Dreyfus 2009; Rowland 1998; Tall 1979). The extant empirical research, however, is limited and relates almost exclusively to the tertiary level.

5 Towards a Research Agenda

Empirical research on students' conceptions of proof has been almost exclusively momentary, in the sense that it examined students' current conceptions; this research has led to several mutually compatible classifications of students' conceptions (e.g., Balacheff 1987; Harel and Sowder 1998) and hence to a rather solid theoretical base on which longer-term empirical studies could be built.

Researchers who have written about visual and generic proofs have started from the basic premise that these forms of proof can support students in the desirable transition from an empirical or pragmatic stance on proof to a deductive one. Research on students' conceptions of proof also shows that this is a key transition in mathematical education. In fact, the mathematics education community has been aware of the idea of generic proof for more than a quarter-century, and is in possession of a fair number of encouraging anecdotes and small-scale studies. However, no large-scale systematic research studies at the school level have been reported. This is surprising, given that mathematics education policy in many countries strongly supports the development of reasoning and proof. For example, in the USA the National Council of Teachers of Mathematics states that: "Mathematical reasoning and proof offer powerful ways of developing and expressing insights about a wide range of phenomena. Reasoning and proof should be a consistent part of students' mathematical experiences in pre-kindergarten through grade 12" (NCTM 2000, p. 56). Similar recommendations can be found in documents of the Australian Educational Council (1991), the Israeli Ministry of Education (1994) and others.

Empirical studies are necessary because we cannot on the basis of theoretical analyses predict in sufficiently fine-grained form what might happen empirically. For example, Malek and Movshovitz-Hadar's (2011) empirical results showed that the presumed positive effects of generic proofs empirically held for some generic proofs but not for others; specifically, they held only in cases where the proof required the employment of new (to the student) ideas or establishing new connections. Thus, the effectiveness of generic proofs might be related to new ideas that appear in a proof. On the other hand, the use of operative proofs involves carrying out computations via operations on the (often visual) representations that serve to generate the proof (Wittmann 2009). One might therefore expect such proofs to have an important role in supporting elementary-school students in the transition from empirical to general arguments. However, the differences between the elementary level, where Wittmann implemented operative proofs, and the tertiary level, where Malek and Movshowitz-Hadar did their research, remain considerable. Larger-scale empirical evidence is certainly needed: larger scale in age groups, in sub-domains of mathematics, in types of pre-formal proofs, even in types of proof, and in number of participants. In addition, different research studies might have different goals: for example, students' ability to construct a valid argument or even a general proof, students' insight into why an assertion is true or, at the most general level, students' transition from an empirical to a deductive stance with respect to what counts as proof. However, the latter research faces the difficulty that the transition from an empirical to a deductive proof stance may be a long-term process that depends heavily on the students' previous mathematics learning. Even the 3–5 years usually allotted to funded or doctoral research might be insufficient. Nevertheless, this must not be a reason for desisting from all research in the area.

In terms of the progressive teaching of proof, researchers might ask what, in addition to the availability of suitable generic or other pre-formal proofs is needed to achieve the desired result, whether the requirements are different for reading, reproducing, or constructing proofs (see Mejía-Ramos and Inglis 2009), what teacher actions and what classroom atmosphere might support or hinder achievement, and how all of these factors may vary with the age or level of the student, the type of mathematics, and the specificities of type of proof being used. For example, much of the existing work on generic proofs seems to deal with number theory; there is some work in algebra, but little or none in geometry, probability and calculus.

We need to know whether it is more difficult to find or construct appropriate generic proofs in these areas.

Finally, making generic, visual or pre-formal proofs accessible to students does not necessarily imply that the students see why they should prove or feel that they need to prove. Some researchers have addressed this need for proof (Buchbinder and Zaslavsky 2009; Dreyfus and Hadas 1996; Hadas et al. 2000; Kidron et al. 2010; Nardi 2008; Nardi and Iannone 2006). The issue of students' need for proof seems to us equally as important as the accessibility of proof. It should be taken into account; research programmes such as outlined above should investigate the influence of students' need for proof on their proof production.

In summary, we urgently and crucially need research studies on the long-term (several years) effects of specific approaches and interventions – in particular, the use of generic, operative or visual proof starting from elementary school onwards – on students' conceptions of proof and on their ability to produce and/or understand more formal proofs in later stages of learning. Long-term empirical studies on students' transition to deductive proof could provide the basis researchers need to build a coherent theory of learning proof and proving.

Acknowledgements This research was partly supported by the Israel Science Foundation under grants 843/09 and 891/03, as well as by an EU Erasmus Staff Mobility Bilateral Agreement between the University of East Anglia in the UK and the University of Athens in Greece.

References*

Alsina, C., & Nelsen, R. B. (2006). *Math made visual*. Washington, DC: The Mathematical Association of America.

Arzarello, F., Paola, D., & Sabena, C. (2009a). Proving in early calculus (Vol. 1, pp. 35–40).*

Arzarello, F., Paola, D., & Sabena, C. (2009b). Logical and semiotic levels in argumentation (Vol. 1, pp. 41–46).*

Australian Education Council. (1991). *A national statement on mathematics for Australian schools*. Melbourne: Curriculum Corporation.

Balacheff, N. (1987). Processus de preuve et situations de validation [Proof processes and situations of validation]. *Educational Studies in Mathematics, 18*(2), 147–176.

Biza, I., Nardi, E., & Zachariades, T. (2009a). Do images disprove but do not prove? Teachers' beliefs about visualization (Vol. 1, pp. 59–64).*

Biza, I., Nardi, E., & Zachariades, T. (2009b). Teacher beliefs and the didactic contract on visualization. *For the Learning of Mathematics, 29*(3), 31–36.

Blum, W., & Kirsch, A. (1991). Pre-formal proving: Examples and reflections. *Educational Studies in Mathematics, 22*(2), 183–203.

Borwein, J. (2011). Exploratory experimentation: Digitally-assisted discovery and proof. In G. Hanna & M. de Villiers (Eds.), *Proof and proving in mathematics education* (pp. XX–XX). Dordrecht: Springer.

---

*NB: References marked with * are in F. L. Lin, F. J. Hsieh, G. Hanna, & M. de Villiers (Eds.) (2009). *ICMI Study 19: Proof and proving in mathematics education*. Taipei, Taiwan: The Department of Mathematics, National Taiwan Normal University.

Bosch, M., & Chevallard, Y. (1999). Ostensifs et sensibilité aux ostensifs dans l'activité mathématique. *Recherches en didactique des mathématiques, 19*(1), 77–124.

Brousseau, G. (1997). *Theory of didactical situations in mathematics*. Dordrecht/Boston/London: Kluwer Academic Publishers.

Brown, J. (1997). Proofs and pictures. *The British Journal for the Philosophy of Science, 48*(2), 161–180.

Buchbinder, O., & Zaslavsky, O. (2009). *Uncertainty: a driving force in creating a need for proving*. Technion IIT: Technical report available from the authors. [Paper accepted by but withdrawn from the ICMI 19 Study conference.]

Byers, W. (2007). *How mathematicians think: Using ambiguity, contradiction, and paradox to create mathematics*. Princeton: Princeton University Press.

Cabassut, R. (2009). The double transposition in proving (Vol. 1, pp. 112–117).*

Chin, E-T., Liu, C-Y., & Lin, F-L. (2009). Taiwanese junior high school students' proof conceptions in algebra (Vol. 1, pp. 118–123).*

Cooney, J., Shealy, B. E., & Arvold, B. (1998). Conceptualizing belief structures of preservice secondary mathematics teachers. *Journal for Research in Mathematics Education, 29*(3), 306–333.

Davis, P. J., & Hersh, R. (1981). *The mathematical experience*. Boston: Birkhäuser.

Dreyfus, T. (1994). Imagery and reasoning in mathematics and mathematics education. In D. Robitaille, D. Wheeler, & C. Kieran (Eds.), *Selected lectures from the 7th International Congress on Mathematical Education* (pp. 107–122). Sainte-Foy: Les presses de l'université Laval.

Dreyfus, T. (2000). Some views on proofs by teachers and mathematicians. In A. Gagatsis (Ed.), *Proceedings of the 2nd Mediterranean Conference on Mathematics Education* (Vol. I, pp. 11–25). Nicosia: The University of Cyprus.

Dreyfus, T., & Hadas, N. (1996). Proof as answer to the question why. *Zentralblatt für Didaktik der Mathematik, 28*(1), 1–5.

Ersoz, F. A. (2009). Proof in different mathematical domains (Vol. 1, pp. 160–165).*

Fischbein, E. (1982). Intuition and proof. *For the Learning of Mathematics, 3*(2), 9–18.

Fujita, T., Jones, K., & Kunimune, S. (2009). The design of textbooks and their influence on students' understanding of 'proof' in lower secondary school (Vol. 1, pp. 172–177).*

Furinghetti, F., & Morselli, F. (2009). Teachers' beliefs and the teaching of proof (Vol. 1, pp. 166–171).*

Giaquinto, M. (2007). *Visual thinking in mathematics*. New York: Oxford University Press.

Grabiner, J. (2011). Why proof? A historian's perspective. In G. Hanna & M. de Villiers (Eds.), *Proof and proving in mathematics education* (pp. XX–XX). Dordrecht: Springer.

Greer, B., de Bock, D., & van Dooren, W. (2009). The ISIS problem and pre-service teachers' ideas about proof (Vol. 1, pp. 184–189).*

Hadas, N., Hershkowitz, R., & Schwarz, B. (2000). The role of contradiction and uncertainty in promoting the need to prove in dynamic geometry environments. *Educational Studies in Mathematics, 44*(1–2), 127–150.

Hanna, G. (1989). Proofs that prove and proofs that explain. In G. Vergnaud, J. Rogalski, & M. Artigue (Eds.), *Proceedings of the 13th Conference of the International Group for the Psychology of Mathematics Education* (Vol. 2, pp. 45–51). Paris: CNRS.

Hanna, G., & Sidoli, N. (2007). Visualisation and proof: A brief survey of philosophical perspectives. *ZDM – The International Journal on Mathematics Education, 39*, 73–78.

Hanna, G., de Villiers, M., Arzarello, F., Dreyfus, T., Durand-Guerrier, V., Jahnke, H. N., Lin, F.-L., Selden, A., Tall, D., & Yevdokimov, O. (2009). ICMI study 19: Proof and proving in mathematics education: Discussion document (Vol. 1, pp. xix-xxx).*

Harel, G., & Sowder, L. (1998). Students' proof schemes: Results from exploratory studies. In A. H. Schoenfeld, J. Kaput, & E. Dubinsky (Eds.), *Research in collegiate mathematics education III* (CBMS: Issues in mathematics education, Vol. 7, pp. 234–283). Providence: American Mathematical Society.

Harel, G., & Sowder, L. (2007). Toward a comprehensive perspective on proof. In F. Lester (Ed.), *Second handbook of research on mathematics teaching and learning* (pp. 805–842). Charlotte: Information Age Publishing.

Harel, R., & Dreyfus, T. (2009). Visual proofs: High school students' point of view. In M. Tzekaki, M. Kaldrimidou, & H. Sakonidis (Eds.), *Proceedings of the 33rd Conference of the International Group for the Psychology of Mathematics Education* (Vol. 1, p. 386). Thessaloniki: PME.

Healy, L., & Hoyles, C. (2000). A study of proof conceptions in algebra. *Journal for Research in Mathematics Education, 31*, 396–428.

Hemmi, K. (2008). Students' encounter with proof: The condition of transparency. *ZDM – The International Journal on Mathematics Education, 40*, 413–426.

Hemmi, K., & Jaworski, B. (2009). Transparency in a tutor-student .interaction concerning the converse of Lagrange's theorem (Vol. 1, pp. 202–207).*

Iannone, P. (2009). Concept usage in proof production: Mathematicians' perspectives (Vol. 1, pp. 220–225).*

Israeli Ministry of Education. (1994). *Tomorrow 98.* Jerusalem: Ministry of Education [in Hebrew].

Kidron, I., & Dreyfus, T. (2009). Justification, enlightenment and the explanatory nature of proof (Vol. 1, pp. 244–249).*

Kidron, I., Bikner-Ahsbahs, A., Cramer, J., Dreyfus, T., & Gilboa, N. (2010). Construction of knowledge: Need and interest. In M. M. F. Pinto & T. F. Kawasaki (Eds.), *Proceedings of the 34th Conference of the International Group for the Psychology of Mathematics Education* (Vol. 3, pp. 169–176). Belo Horizonte: PME.

Kondratieva, M. (2009). Geometrical sophisms and understanding of mathematical proofs (Vol. 2, pp. 3–8).*

Kwon, O. N., Park, J. S., & Park, J. H. (2006). Cultivating divergent thinking in mathematics through an open-ended approach. *Asia Pacific Education Review, 7*, 51–61.

Lave, J., & Wenger, E. (1991). *Situated learning: Legitimate peripheral participation.* Cambridge: Cambridge University Press.

Leatham, K. R. (2006). Viewing mathematics teachers' beliefs as sensible systems. *Journal of Mathematics Teacher Education, 9*, 91–102.

Leder, G. C., Pehkonen, E., & Törner, G. (Eds.). (2002). *Beliefs: A hidden variable in mathematics education?* Dordrecht: Kluwer.

Leikin, R. (2007). Habits of mind associated with advanced mathematical thinking and solution spaces of mathematical tasks. In D. Pitta-Pantazi & G. Philippou (Eds.), *Proceedings of the Fifth Conference of the European Society for Research in Mathematics Education – CERME-5* (pp. 2330–2339) (CD-ROM). Retrieved March 7, 2011, from http://www.erme.unito.it/CERME5b/

Leikin, R. (2009a). Multiple proof tasks: Teacher practice and teacher education (Vol. 2, pp. 31–36).*

Leikin, R. (2009b). Exploring mathematical creativity using multiple solution tasks. In R. Leikin, A. Berman, & B. Koichu (Eds.), *Creativity in mathematics and the education of gifted students* (pp. 129–145). Rotterdam: Sense Publishers.

Leikin, R., & Levav-Waynberg, A. (2007). Exploring mathematics teacher knowledge to explain the gap between theory-based recommendations and school practice in the use of connecting tasks. *Educational Studies in Mathematics, 66*, 349–371.

Leikin, R., & Levav-Waynberg, A. (2009). Development of teachers' conceptions through learning and teaching: Meaning and potential of multiple-solution tasks. *Canadian Journal of Science, Mathematics, and Technology Education, 9*(4), 203–223.

Leron, U., & Zaslavsky, O. (2009). Generic proving: Reflections on scope and method (Vol. 2, pp. 53–58).*

Leung, A. (2009). Written proof in dynamic geometry environment: Inspiration from a student's work (Vol. 2, pp. 15–20).*

Lin, C. C. (2009). How can the game of Hex be used to inspire students in learning mathematical reasoning? (Vol. 2, pp. 37–40).*

Lin, F.-L., Hsieh, F.-J., Hanna, G., & de Villiers, M. (Eds.). (2009). *Proceedings of the ICMI Study 19 Conference: Proof and Proving in Mathematics Education* (2 vols.). Taipei: Department of Mathematics, National Taiwan Normal University.

Lithner, J. (2008). A research framework for creative and imitative reasoning. *Educational Studies in Mathematics, 67*, 255–276.

Malek, A., & Movshovitz-Hadar, N. (2009). The art of constructing a transparent p-proof (Vol. 2, pp. 70–75).*

Malek, A., & Movshovitz-Hadar, N. (2011). The effect of using transparent pseudo-proofs in linear algebra. *Research in Mathematics Education, 13*(1), 33–58.

Mamona-Downs, J., & Downs, M. (2009). Proof status from a perspective of articulation (Vol. 2, pp. 94–99).*

Mancosu, P., Jorgensen, K. F., & Pedersen, S. A. (Eds.). (2005). *Visualization, explanation and reasoning styles in mathematics*. Dordrecht: Springer.

Martin, G., & Harel, G. (1989). Proof frames of preservice elementary teachers. *Journal for Research in Mathematics Education, 20*(1), 41–51.

Mason, J., & Pimm, D. (1984). Generic examples: Seeing the general in the particular. *Educational Studies in Mathematics, 15*, 277–289.

Mejía-Ramos, J. P., & Inglis, M. (2009). Argumentative and proving activities in mathematics education research (Vol. 2, pp. 88–93).*

Moore, R. C. (1994). Making the transition to formal proof. *Educational Studies in Mathematics, 27*(3), 249–266.

Morselli, F. (2007). *Sui fattori culturali nei processi di congettura e dimostrazione*. Unpublished Ph.D. dissertation. Università degli Studi di Torino, Turin.

Morselli, F., & Boero, P. (2009). Habermas' construct of rational behaviour as a comprehensive frame for research on the teaching and learning of proof (Vol. 2, pp. 100–105).*

Nardi, E. (2008). *Amongst mathematicians: Teaching and learning mathematics at university level*. New York: Springer.

Nardi, E. (2009). 'Because this is how mathematicians work!' 'Pictures' and the creative fuzziness of the didactical contract at university level (Vol. 2, pp. 112–117).*

Nardi, E., & Iannone, P. (2006). Conceptualising formal mathematical reasoning and the necessity for proof. In E. Nardi & P. Iannone (Eds.), *How to prove it: A brief guide for teaching proof to year 1 mathematics undergraduates* (pp. 5–16). Norwich: Higher Education Academy.

National Council of Teachers of Mathematics (NCTM). (2000). *Principles and standards for school mathematics*. Reston: National Council of Teachers of Mathematics.

Nelsen, R. B. (1993). *Proofs without words*. Washington, DC: The Mathematical Association of America.

Nelsen, R. B. (2000). *Proofs without words II*. Washington, DC: The Mathematical Association of America.

Nogueira de Lima, R., & Tall, D. (2006). The concept of equation: What have students met before? In J. Novotná, H. Moraová, M. Krátká, & N. Stehlíková (Eds.), *Proceedings of the 30th Conference of the International Group for the Psychology of Mathematics Education* (Vol. 4, pp. 233–241). Prague: PME.

Pedemonte, B. (2007). How can the relationship between argumentation and proof be analysed? *Educational Studies in Mathematics, 66*, 23–41.

Perry, P., Camargo, L., Samper, C. Echeverry, A., & Molina, Ó. (2009a). Assigning mathematics tasks versus providing pre-fabricated mathematics in order to support learning to prove (Vol. 2, pp. 130–135).*

Perry, P., Samper, C., Camargo, L., Molina, Ó., & Echeverry, A. (2009b). Learning to prove: Enculturation or…? (Vol. 2, pp. 124–129).*

Pólya, G. (1945/1973). *How to solve it*. Princeton: Princeton University.

Presmeg, N. C. (2006). Research on visualization in learning and teaching mathematics: Emergence from psychology. In A. Gutierrez & P. Boero (Eds.), *Handbook of research on the psychology of mathematics education* (pp. 205–235). Dordrecht: Sense Publishers.

Raman, M., Sandefur, J., Birky, G., Campbell, C., & Somers, K. (2009). "Is that a proof?": Using video to teach and learn how to prove at the university level (Vol. 2, pp. 154–159).*

Rowland, T. (1998). Conviction, explanation and generic examples. In A. Olivier & K. Newstead (Eds.), *Proceedings of the 22nd Conference of the International Group for the Psychology of Mathematics Education* (Vol. 4, pp. 65–72). Stellenbosch: PME.

Rowland, T. (2001). Generic proof in number theory. In S. Campbell & R. Zazkis (Eds.), *Learning and teaching number theory: Research in cognition and instruction.* Westport: Ablex.

Schoenfeld, A. H. (1985). *Mathematical problem solving.* New York: Academic.

Schwarz, B., & Kaiser, G. (2009). Professional competence of future mathematics teachers on argumentation and proof and how to evaluate it (Vol. 2, pp. 190–195).*

Selden, A. (2005). New developments and trends in tertiary mathematics education: Or, more of the same? *International Journal of Mathematical Education in Science and Technology, 36*(2/3), 131–147.

Silver, E. A. (1997). Fostering creativity through instruction rich in mathematical problem solving and problem posing. *ZDM – The International Journal on Mathematics Education, 3*, 75–80.

Siu, M-K. (2009). The algorithmic and dialectic aspects in proof and proving (Vol. 2, pp. 160–165).*

Sriraman, B. (2003). Mathematical giftedness, problem solving, and the ability to formulate generalizations. *Journal of Secondary Gifted Education, 14*, 151–165.

Stenning, K., & Lemon, O. (2001). Aligning logical and psychological perspectives on diagrammatic reasoning. *Artificial Intelligence Review, 15*(1–2), 29–62.

Stevenson, I. (2009). Dynamic geometry and proof: The cases of mechanics and non-Euclidean space (Vol. 2, pp. 184–189).*

Stylianides, G. J., & Stylianides, A. J. (2009). Ability to construct proofs and evaluate one's own constructions (Vol. 2, pp. 166–171).*

Sun, X. H. (2007). *Spiral variation (bianshi) curriculum design in mathematics: Theory and practice.* Unpublished doctoral dissertation, The Chinese University of Hong Kong, Hong Kong.

Sun, X. (2009). Renew the proving experiences: An experiment for enhancement trapezoid area formula proof constructions of student teachers by "one problem multiple solutions" (Vol. 2, pp. 178–183).*

Sun, X., & Chan, K. (2009). Regenerate the proving experiences: An attempt for improvement original theorem proof constructions of student teachers by using spiral variation curriculum (Vol. 2, pp. 172–177).*

Tabach, M., Levenson, E., Barkai, R., Tsamir, P., Tirosh, D., & Dreyfus, T. (2009). Teachers' knowledge of students' correct and incorrect proof constructions (Vol. 2, pp. 214–219).*

Tall, D. (1979). Cognitive aspects of proof, with special reference to the irrationality λf 2. In D. Tall (Ed.), *Proceedings of the 3rd Conference of the International Group for the Psychology of Mathematics Education* (pp. 206–207). Warwick: PME.

Tall, D., Yevdokimov, O., Koichu, B., Whiteley, W., Kondratieva, M., & Cheng, Y. (2011). Cognitive Development of Proof. In G. Hanna & M. de Villiers (Eds.), *Proof and proving in mathematics education* (pp. XX–XX). Dordrecht: Springer.

Tsamir, P., Tirosh, D., Dreyfus, T., Tabach, M., & Barkai, R. (2009). Is this verbal justification a proof? (Vol. 2, pp. 208–213).*

Tymoczko, T. (1985). *New directions in the philosophy of mathematics.* Basel: Birkhäuser.

Weber, K. (2001). Student difficulty in constructing proofs: The need for strategic knowledge. *Educational Studies in Mathematics, 48*, 101–119.

Weber, K., & Alcock, L. (2004). Semantic and syntactic proof productions. *Educational Studies in Mathematics, 56*, 209–234.

Whiteley, W. (2004). To see like a mathematician. In G. Malcolm (Ed.), *Multidisciplinary approaches to visual representations and interpretations* (Vol. 2, pp. 279–291). London: Elsevier.

Whiteley, W. (2009). Refutations: The role of counter-examples in developing proof (Vol. 2, pp. 257–262).*

Winicki-Landman, G. (2009). Mathematical games as proving seminal activities for elementary school teachers (Vol. 2, pp. 245–250).*

Wittmann, E. Ch. (1998). Operative proof in elementary school. *Mathematics in School, 27*(5).

Wittmann, E. Ch. (2005). The alpha and the omega of teacher education: Organizing mathematical activities. In D. Holton (Ed.), *The teaching and learning of mathematics at university level* (pp. 539–552). Dordrecht: Kluwer Academic Publishers.

Wittmann, E. Ch. (2009). Operative proof in elementary mathematics (Vol. 2, pp. 251–256).*

Wong, N. Y. (2007). Confucian heritage cultural learner's phenomenon: From "exploring the middle zone" to "constructing a bridge". Regular lecture, presented at *the Fourth ICMI-East Asia Regional Conference on Mathematical Education*. Penang.

Zehavi, N., & Mann, G. (2009). Proof and experimentation: Integrating elements of DGS and CAS (Vol. 2, pp. 286–291).*

Chapter 9
The Need for Proof and Proving: Mathematical and Pedagogical Perspectives

Orit Zaslavsky, Susan D. Nickerson, Andreas J. Stylianides, Ivy Kidron, and Greisy Winicki-Landman

We will organise our discussion of the need for proof and proving around three main questions: Why teach proof? What are (or may be) learners' needs for proof? How can teachers facilitate the need for proof?

First, we examine the close connection between the functions of proof within mathematics and the needs those evoke for teaching proof. We also examine the epistemology of proof in the history of mathematics in order to elucidate the needs that propelled the discipline to develop historically. The second section takes a learner's perspective on the need to prove. Learners face a problematic situation when they confront a problem that cannot be routinely solved by their current

O. Zaslavsky (✉)
Department of Teaching and Learning, New York University, New York, NY, USA

Department of Education in Technology and Science Technion – Israel
Institute of Technology, Haifa, Israel
e-mail: oritrath@gmail.com

S.D. Nickerson
Department of Mathematics and Statistics, San Diego State University,
San Diego, CA, USA
e-mail: snickers@sciences.sdsu.edu

A.J. Stylianides
Faculty of Education, University of Cambridge, Cambridge, UK
e-mail: as899@cam.ac.uk

I. Kidron
Applied Mathematics Department, Jerusalem College of Technology (JCT),
Jerusalem, Israel
e-mail: ivy@jct.ac.il

G. Winicki-Landman
Department of Mathematics and Statistics, California State Polytechnic University,
Pomona, CA, USA
e-mail: greisyw@csupomona.edu

© The Author(s) 2021
G. Hanna and M. de Villiers (eds.), *Proof and Proving in Mathematics Education*,
New ICMI Study Series, https://doi.org/10.1007/978-94-007-2129-6_9

knowledge. A person's *intellectual need* is a necessary condition for constructing new knowledge (Harel 1998). We examine categories of intellectual need that may drive learners to prove (i.e., for certitude, to understand, to quantify, to communicate, and for structure and connection; Harel in press). Interestingly, Harel's (in press) categories of learners' needs align with the roles that proof and proving play in mathematics. The third section addresses pedagogical issues involved in teachers' attempts to facilitate learners' need to prove. Uncertainty and cognitive conflict are the driving forces for proving. We discuss how a teacher can foster necessity-based learning that motivates the need to prove.

1   The Need for Teaching Proof

The role of proof and proving in mathematics directs us to the needs for teaching proof in schools. Here, we focus on proof's central roles within mathematics: validation, explanation, discovery, systemisation of results, incorporation into a framework, and conveying mathematical knowledge (de Villiers 1990; Hanna 2000; Rav 1999). A proof demonstrates that a mathematical assertion is true, assuming certain axioms. Yet, mathematicians look beyond establishing truth to seek insight into why; proofs can have explanatory power (Hanna 2000; Thurston 1994). Through the process of proof, mathematicians may discover new results. Proofs communicate mathematical knowledge and situate that knowledge systematically within a framework. Finally, Rav (1999) proposes that proofs are of primary importance in mathematics because they embody tools, methods, and strategies for solving problems.

It follows that one need for teaching proof in school is to support students' understanding of proof as practised in the discipline. Within the community of mathematicians, the truth of a mathematical assertion follows through valid deductive reasoning from established results; proof is a deductive "…demonstration that compels agreement by all who understand the concepts involved." (Hersh 2008, p. 100).

Hence, the mathematics curricula of countries worldwide have in common the goal of training students in deductive reasoning and logical inference (cf. Balacheff 1991; Healy and Hoyles 2000; Herbst 2002; Kunimune et al. 2008). In this respect, mathematics teachers play the role of a broker with membership status in both the mathematics community and the classroom community, bridging the two (Rasmussen et al. 2009; Yackel and Cobb 1996). Whether in secondary or elementary classrooms, teachers play an active role in judging and instructing on what arguments can establish validity or count as proof (Stylanides 2007).

Another role for proof in mathematics is explaining why a mathematical statement is true given certain assumptions. Often mathematicians undertake a proof because they are personally convinced of the truth or validity of a mathematical assertion, having explored it empirically and through its structure (de Villiers 1990, 2010; Weber 2008). Proofs communicate mathematical knowledge and its place within an organised structure. Therefore, another need for teaching proof is to 'shed light on' the origins and connections of mathematical knowledge.

If proof's only purpose were to establish validity, there would be no need to prove things in multiple ways (Dawson 2006; Hanna 2000; Siu 2008). Multiple perspectives provided by multiple proofs, coupled with reliance on examples provide a network of connections and deeper understanding of mathematical concepts. Mathematicians also generate multiple proofs of theorems "to demonstrate the power of different methodologies" or to discover new techniques (Dawson 2006; Rav 1999).

Hence, another reason to teach proof is the potential to teach methods of problem solving. For example, Hanna and Barbeau (2010) argued that mathematics teachers could 'leverage' proofs common in secondary curricula in order to explicitly introduce strategies, methods, and tools. They suggest, for instance, the derivation of the quadratic formula, which introduces students to the strategy of completing the square. They argue that students can thus learn a technique whose applicability extends beyond that situation as well as learning what can be discovered by reducing equations to a canonical form.

Proofs from geometry could be thus 'leveraged'; in fact, high school geometry is often where students first encounter proof. Geometry is a part of high school mathematics, which most obviously lends itself to the intellectual necessitation of rigorous mathematical structure. This could perhaps be enhanced by a programme that begins with *neutral geometry;* that is, geometry without the parallel postulate (Harel in press).

Obviously, students' experiences with proof and proving differ from mathematicians' experiences because their purposes differ; students in schools are not engaged in the activity of proving in order to discover new mathematical results (Herbst and Brach 2006; Hilbert et al. 2008). The reasons for teaching proof and proving in schools follow from the expectation that students have experiences in reasoning similar to those of mathematicians: learning a body of mathematical knowledge and gaining insight into why assertions are true. However, the proofs that students encounter in school are often presented complete in order to teach students the processes of logical thinking and communicating and, perhaps implicitly, problem solving by example. From the students' perspective, proving as merely an exercise in confirming assertions and learning theorems lacks intellectual purpose (Harel in press; Herbst 2002; Herbst and Brach 2006) because in this kind of situation students are not fully engaged in an attempt to search for a solution to a mathematical problem that they appreciate.

Historically, a proof is a rhetorical device for convincing someone else that a mathematical statement is true (Harel and Sowder 2007; Krantz 2007). In order to convince, a proof has to align with the norms (e. g., forms of reasoning, logical rules of inference, modes of argumentation and modes of argument representation) of the community to which it is presented (Harel and Sowder 2007; Siu 2008; Stylianides 2007). The standards that a proof must fulfil arise from an agreement amongst members of the community. A review of the history of the development of proof makes it clear that the rules and form have evolved over time and vary from culture to culture (cf. Arsac 2007; Hanna and Jahnke 1993; Kleiner 1991; Siu 1993).

Understanding the intellectual needs that might have brought about these changes in the development of proof historically does not directly lead to understanding a person's intellectual necessity for proof, though it may be instructive. In this brief account, we focus on the historical means of proving and the motives for change as they relate to the possible implications for pedagogy. For extensive descriptions of the history of the development of proof in mathematics, cf. Arsac 2007; Harel and Sowder 2007; Kleiner 1991; Krantz 2007; Siu 2008.

Historians characteristically discuss the historical-epistemological factors in three phases: pre-Greek, post-Greek, and modern. Prior to the Greeks' concept of deductive reasoning within an axiomatic system, Babylonians and Chinese provided justifications for mathematical assertions. Proofs, as explanations that convinced and enlightened, abounded in ancient texts (Siu 1993). Conjectures were proved by empirical evidence and commonly involved quantitative measures of actual physical entities. The subsequent evolution in mathematical practice from the pre-Greek to Greek phases encompassed a shift to abstract ideal entities and the development of a formal system.

The reasons for the shift encompass both internal and external factors; historians do not agree on which takes precedence. The needs in the mathematical community to generate changes were due to perturbations that needed resolution. The Greeks needed to alleviate inconsistencies found in their predecessors' mathematical work; they wished to create a system free of paradoxes (Harel and Sowder 2007). The Greeks were also driven by a need to resolve problems of incommensurability and irrationality (e. g., a square's diagonal is incommensurable with its side, 2 has no rational square root; Arsac 2007; Harel and Sowder 2007). Furthermore, the dialectic and intellectual milieu of the culture contributed to the promotion of the notion of deductive proof (Arsac 2007; Harel and Sowder 2007; Siu 2008).

In the post-Greek era, mathematics continued developing in Africa, India, China and the civilisations of Central and South America. The Arabs developed some of the seminal ideas in algebra. Symbolic algebra, beginning with Vieta's work, seems to have played a critical role in reconceptualisation of mathematics in general and proof in particular (Arsac 2007; Harel and Sowder 2007). The shift from Greek mathematics to modern mathematics entailed a shift from idealised physical realities to arbitrary entities not necessarily evident in "natural experience." In modern mathematics, a mathematical entity is dependent upon its connection to other entities within a structure. From the beginning of the 1900s, discussions about the foundations of mathematics resulted in insight into the need for axioms that can only be justified extrinsically, by virtue of their fruitfulness and explanatory power (Jahnke 2010). Axioms are viewed not as absolute truths but as agreed propositions. This shift, too, was born from a desire to create a system free of paradoxes. In addition, ongoing philosophical debate about the nature of understanding and Aristotlean causality, in particular, has played a role (Harel and Sowder 2007; Jahnke 2010).

The consequences of this history are borne out in the teaching of proof. Secondary school geometry in the Euclidean tradition is often the first (and possibly only) place where students learn to do proof. The limits of proof in this tradition include a lack of methods of discovery and the prominence of methods for proving without explanation (reductio ad absurdum, proof by exhaustion; Arsac 2007). Though modern mathematicians adopt axioms or hypotheses without perceiving them as evident or absolutely true, the Euclidean view of proof as taught in schools today communicates that proof establishes truth rather than validates assertions based on agreed axioms. The hypothetical nature of axioms "remains hidden from most pupils" (Jahnke 2010, p. 29). Furthermore, as Arsac (2007) notes, Euclidean mathematics is characterised by wanting to prove everything even when it is evident.

Mathematics educators need to understand *students'* perspectives on the need for proof and which situations, tasks and knowledge encourage that need in students. In the next section, we offer an overview of what research and practice tell us about learners' conceptions and beliefs regarding the need to prove.

2   The Need for Proof and Proving: A Learner's Perspective

Students can lack understanding of the functions of proof in mathematics. In studies in Japan of lower secondary school students, even though most performed well at proof writing, more than 60% of students did not understand why proofs are needed (Fujita and Jones 2003; Kunimune et al. 2008). Healy and Hoyles (2000) studied nearly 2,500 capable 14- and 15-year old British students, more than a quarter of them could not articulate the purpose or meaning of proof. About half of the students referred to verification as a purpose of proof; roughly one-third of them cited proof's function in explanation and communication. In follow-up interviews, many more students seemed to hold a view of proof as explanation. Twenty years earlier, Williams (1980) found that out of 255 11th-grade Canadian college preparation students in ten randomly selected classes from nine different senior high schools, about half expressed no need to prove a statement that they regarded as obvious. Fewer than 30% showed a grasp of the meaning of proof.

Coe and Ruthven's (1994) study illustrated how students may theoretically understand the function of proof yet not employ proof in their mathematics practice. The study was conducted with a cohort of advanced-level mathematics students towards the end of their first year college, who had followed a reformed secondary curriculum. The authors examined 60 pieces of student coursework and analysed the types of proof used. Seven students were interviewed. The students were aware that proof is required for mathematical knowledge; but only the best students said that proof convinced them of the truth of mathematical assertions.

Often, external requirements affect the need to prove. Some students produce proof because their teacher demands it, not because they recognise that proof is necessary in their practice (Balacheff 1988). Consequently, they may fail to understand reasons for the truth of a statement; they are convinced because their teacher asserts it. In the classification of proof schemes (means by which one obtains certainty), this authoritarian conception of proof belongs amongst the "external conviction" proof schemes (Harel and Sowder 1998, 2007).

Balacheff (1991) claims that the reason students do not engage in proving is not so much because they cannot but rather because they see no reasons or need for it.

If students lack understanding of proof's role and display external conviction proof schemes, a question arises: Is there only an external need for proving or can we find some inner need that may drive learners to prove?

To answer this question, we examine five categories of intellectual need: for certainty, for causality, for computation, for communication and for structure. These needs are linked and relate to functions of proof as practised in mathematics, as well as to proof schemes (Harel and Sowder 1998, 2007). The first two needs, for certainty and causality, are particularly salient to research on the learning and teaching of proof.

2.1   Need for Certainty

The first need, for certainty, is the human desire to verify an assertion. Though verification is one of the central roles of proof, students see no need for mathematical proof, because their need for certainty is personal, in the sense of requiring "personal" rather than "mathematical" convincing. For many students, empirical work is personally convincing. Observations of students' performance of proofs have produced some indirect evidence along these lines.

For example, Fischbein and Kedem (1982) studied 400 high school students who were presented with a mathematical theorem and its complete formal proof. The majority of the students affirmed that they were sure that the proof was complete and irreproachable but simultaneously claimed that analysis of examples would strengthen their confidence. In some ways this is not surprising, given the history of proof prior to the Greeks; in pre-Greek mathematics, concerned with actual physical entities, conjectures were proven by empirical evidence (Harel and Sowder 2007). Besides, this practice of looking for examples and counterexamples after reading a proof is one means mathematicians employ (Lakatos 1976). As Fischbein (1982) mentioned, the role of intuitive structures does not come to an end when analytical (formal) forms of thinking become possible.

However, when they declare that one might find contradictory evidence of a proven statement by examining further examples, students demonstrate that they do not understand the meaning of "mathematical proof." In addition to the conceptual logical schemes, the students need a feeling of agreement, a basis of belief expressed in this need for verification by further examples.

In Fischbein and Kedem's (1982) study, students needed verification by examples in addition to formal proof. In many other cases, exemplification replaced the formal proof. For example, Thompson (1991) studied advanced students in a course that emphasised reasoning and proof; nevertheless, a large number of the students 'proved' a statement by providing a specific example. This behaviour reveals an inductive proof scheme (Harel and Sowder 1998), one kind of empirical proof scheme. Several studies have observed students' adopting such empirical proof schemes. In fact, many school students, including advanced or high-attaining secondary students, (e.g., Coe and Ruthven 1994; Healy and Hoyles 2000; Knuth et al. 2002) and university students, including mathematics majors (e.g., Goetting 1995; Sowder and Harel 2003; Stylianides and Stylianides 2009a, b), consider empirical arguments to be proofs of mathematical generalisations. Although not universal (Iannone and Inglis 2010; Weber 2010), this pervasive misconception inhibits students' perceiving an intellectual need for proof as a mathematical construct.

In addition, a student who has this misconception, even if seeing a reason for developing a proof, will probably produce an empirical argument for a mathematical generalisation. Such students do not recognise the importance of producing a more general argument that meets the standard of proof (for a mathematician); for them, the empirical argument *is* the proof. A student thus satisfied with empirical arguments has little reason to bother to (learn how to) construct general arguments (notably proofs). Constructing general arguments is incomparably more complex than constructing empirical arguments. General arguments aim to cover appropriately the entire domain of a generalisation (which can have infinite cardinality; cf. Stylianides and Stylianides 2009a), whereas empirical arguments can be satisfied with the examination of only a subset of that domain. They may verify that subset, but the generalisation remains unverified and hence uncertain, though the student may not recognise this.

2.2   Need for Causality

Just as with certainty, humans have a desire to determine the cause of a phenomenon – to explain why an assertion is true. Causality relates to a proof's function as explanation (Harel in press). In fact, instructors can find it relatively easy to elicit students' curiosity about why a result is true.

Kidron and Dreyfus (2010) describe the need for causality in their case study of a teacher and a single learner's process of mathematical justification during the investigation of bifurcation points in dynamic systems. The learner's need to establish causality arose from justification of results obtained numerically by means of interaction with the computer. The learner was sure of this result, since it agreed with previous empirical results. Thus, the learner was interested in neither verification nor a formal proof, but felt the need to explain in order to gain more insight into the connections in the data.

This kind of need for insight into "why" may vary from person to person and from context to context. In Harel and Sowder's (1998) terms, it fits the causality proof scheme, a part of the deductive proof scheme. It might appear in situations of contradiction, followed by surprise or uncertainty, that lead students to seek for explanations (e.g., Movshovitz-Hadar and Hadass 1990). Students might seek causality after an experimental investigation convinces them and they are encouraged to explain why (deVilliers 1998). It might appear in the interplay of conjectures and checks of certainty and uncertainty, for example, when students feel the need to find the cause of their assertion's untruthfulness as in Hadas et al. study (2000).

2.3   Needs for Computation, Communication and Structure

The needs for computation and communication are interrelated and often concurrent (Harel in press). Harel refers to the need to compute as humans' propensity to quantify, determine or construct an object, or to determine the property of an object or relations amongst objects (e.g., a number, geometric figure, function) by means of symbolic algebra. Again, this necessity was significant in the development of mathematics and proof, in particular. Computing by means of symbolic algebra enabled a shift in focus from the attributes of spatial figures to nineteenth-century investigations into underlying operations, algebraic representations and their structures (Harel and Sowder 2007). The need for communication refers to both formulating and formalising, rooted in the acts of conveying and exchanging ideas. Students with an intuitive explanation for "why" can be pressed to be systematic in expressions of their reasoning and modify their use of notation in order to better express what they have in mind (Harel in press; Thompson 1992). This need also connects to the role of proof in communicating methodologies and techniques for problem solving.

The need for structure refers to the need to (re-)organise information into a logical structure. Harel (in press) distinguishes between an earlier stage, in which one organises one's own knowledge by assimilating it into her or his existing cognitive structure that may not be logically hierarchical, and a later stage where a need may arise to re-organise the structure into a logical structure. Historically, the need to structure Euclidean geometry in *Euclid's Elements* arose from a need to organise and communicate an accumulated body of knowledge. Furthermore, the need to perfect this structure led to the attempt to prove the parallel postulate (Harel in press). The need for structure may lead from disconnected ideas to unifying principles or concepts.

Proof in classroom instruction is sometimes perceived as an exercise in justifying what is obvious rather than proof as a tool to know with (Healy and Hoyles 2000; Herbst and Brach 2006). When proof is a tool to know with and to communicate methodologies, communication is linked with the need to compute and to structure.

3   Facilitating the Need to Prove: A Teacher's Perspective

According to Harel (1998), "[s]tudents feel intellectually aimless in mathematics courses, because we [teachers] usually fail to present them with a clear intellectual purpose" (p. 501). Students are often introduced to mathematical concepts, without however being assisted to see a need for learning what we intend to teach them. As a result, students tend to make little sense of mathematical concepts, in general, (e.g., Harel et al. 2008) and of the construct of proof, in particular.

As discussed above, students often (a) see no real reason for developing a proof in the context of particular activities that call for a proof (from a mathematical standpoint) or (b) have some deeply rooted misconceptions about what it means to validate mathematical claims (e.g., generalisations). The former reflects a lack of appreciation for the need for proof at a local level, whereas the latter reflects a lack of appreciation for the broader need for proof as a mathematical construct. These two problems are obviously interconnected, though the second seems to be more encompassing: A student who does not understand what counts as evidence in mathematics is unlikely to see a reason for developing a proof (as understood by a mathematician) in the context of particular activities.

Instructors therefore face the challenges of how to facilitate the development of an intellectual need for proof amongst students – both in the context of particular activities and more broadly a need for proof as a mathematical construct. Harel (1998),discussing how instruction might help students see an intellectual need for learning what teachers intend to teach them, pointed out:

> "Intellectual need" is an expression of a natural human behavior: When we encounter a situation that is incompatible with, or presents a problem that is unsolvable by our existing knowledge, we are likely to search for a resolution or a solution and construct, as a result, new knowledge. Such knowledge is meaningful to the person who constructs it, because it is a product of a personal need and connects to prior experience. (p. 501)

Here, Harel essentially lays out the grounds for three main guiding strategies for instructional approaches that may lead to necessitating proof for the learner: evoking uncertainty and cognitive conflict; facilitating inquiry-based learning; and conveying the culture of mathematics.

3.1 Uncertainty and Cognitive Conflict as Driving Forces for Creating a Need for Proof

Introducing *uncertainty* and *cognitive conflict* can serve to motivate people to change or expand their existing ways of thinking about a particular concept or to learn about the concept in the first place. The terms "cognitive conflict" and "uncertainty" have overlapping but distinctive meanings (Zaslavsky 2005); here, we use them interchangeably: We also subsume in these two terms a range of other related terms from the mathematics education literature, such as contradiction, perplexity, and surprise. Zaslavsky (2005) provided a detailed discussion about the roots of the notion of cognitive conflict in Dewey's concept of reflective thinking and its relations to psychological theories such as Piaget's equilibration theory, Festinger's theory of cognitive dissonance, and Berlyne's theory of conceptual conflict. Zaslavsky suggests that these theoretical perspectives support the use of tasks that elicit uncertainty and cognitive conflict.

Some recent research studies on the teaching and learning of proof have examined the role of cognitive conflict as a driving force for creating a need for proof amongst students (e.g., Hadas et al. 2000; Stylianides and Stylianides 2009b; Zaslavsky 2005; cf. also Brown 2003). These studies substantiated and exemplified the claim that, through appropriate didactical engineering, cognitive conflict *can* create in students a need for proof. However, different studies developed this claim in different ways. Of the three illustrative studies we present, two (Hadas et al. 2000; Zaslavsky 2005) were concerned primarily with creating a need for proof in the context of particular activities, whereas the third focused on creating a need for proof as a mathematical construct beyond the context of particular activities (Stylianides and Stylianides 2009b).

Hadas et al. (2000) designed two activities, in the context of Dynamic Geometry (DG) environments, intended to lead students to contradictions between conjectures and findings, thereby motivating a need for proof. According to the researchers, "[t]he two activities exemplify a design in which learning in a DG environment opens opportunities for feeling the need to prove, rather than considering proving as superfluous" (p. 148). In the second activity, the students were left uncertain about the correctness of the result they obtained from the DG software. As a result of this uncertainty (which did not occur in the first activity), the students oscillated between two alterative hypotheses and relied on deductive considerations (amongst other things). Consequently, Hadas et al. explained, the second activity provoked a much higher proportion of deductive arguments amongst students (56%) than the first activity (18%).

Zaslavsky (2005) described a classroom situation that spontaneously evoked cognitive conflict through *competing claims* that two students had expressed. One student proved a certain statement and the other came up with a (supposed) counter-example. The proof that the first student presented had no explanatory power; thus, it was not helpful in resolving the contradiction. Zaslavsky told how the instructor encouraged the students to reach an agreed resolution. This process naturally led to establishing the need to prove not only as a means of resolving the uncertainty but also as a means of determining the cause of the phenomenon. In an iterative process, Zaslavsky analysed this learning experience further, and developed a task to facilitate the need for proof. This type of iterative process, performed by a reflective instructor, could lead to many other similar tasks.

Stylianides and Stylianides (2009b) discussed the theoretical foundation and implementation of an instructional intervention that they developed in a 4-year design experiment in an undergraduate university class. The intervention relied heavily on two deliberately engineered cognitive conflicts that motivated stepwise progressions in students' knowledge about proof. The instructor played a critical role in helping students resolve the emerging cognitive conflicts and develop under-standings that better approximated conventional knowledge. In addition, the instructor facilitated social interactions that enabled students to learn from each other. The process culminated in the students' recognising empirical arguments (of any kind) as insecure validations of mathematical generalisations and developing an intellectual need to learn about secure validations. Thus, the intervention provoked an intellec-tual curiosity about issues of validation that surpassed the context of its activities: in other words, a need for proof.

3.2   Inquiry-Based Learning as Means for Creating the Need for Proof

Recent calls for improved mathematics education recommend incorporating investigative approaches and problem solving whilst valuing students' own mathematical reasoning (e.g., Brown and Walter 2005; Hiebert and Stigler 2000; NCTM 2000; NRC 1996). Such an open learning environment involves students in an ongoing process of exploration, conjecturing, explaining, validating and disproving tentative claims (cf. Lakatos 1976). De Villiers (2003, 2010) extensively discussed and demonstrated the value of an experimentation approach to learning mathematics and proof. This approach lends itself to opportunities that necessitate proof for certitude and at the same time addresses the needs for causality, communication and structure.

Buchbinder and Zaslavsky (2011) designed a generic task titled "Is this a coincidence?" that leads students to explore the nature of generality. Each specific task presents a hypothetical student's observation about a single geometrical example. It asks whether the observed outcome is a coincidence or not; that is, whether it holds for every relevant case or just coincidently for some specific cases. The task implicitly encourages either proving that the described geometrical phenomenon is general or constructing a counterexample showing that it is not. The task contains no explicit requirement to prove any claim.

Buchbinder and Zaslavsky (2011) had two groups of participants: six pairs of high school students and six experienced mathematics teachers. For both groups, the task created a need to prove and convince, due either to a sense of uncertainty regarding the mathematical phenomenon in question or to an over-confidence in a false conjecture. In the latter case, the eagerness to prove and convince resulted in flawed arguments and inferences. Hence, facilitating the need to proof should proceed carefully.

3.3   The Teacher as an Agent of the Culture of Mathematics

Hanna and Jahnke (1996) emphasised the limitations that arise when teachers assume completely passive roles in the teaching of proof:

> A passive role for the teacher… means that students are denied access to available methods of proving. It would seem unrealistic to expect students to rediscover sophisticated mathematical methods or even the accepted modes of argumentation. (p. 887)

Similarly, it seems unrealistic to expect students to feel a need for proof without the teachers taking focused actions to provoke that need whether by creating uncertainty and cognitive conflict, or by employing inquiry-based learning activities.

The teacher's actions have the ultimate goal of inducing students to explore and employ conventional mathematical knowledge. In this case, the teacher serves as an agent of the culture of mathematics in the classroom (Lampert 1992; Stylianides 2007; Yackel and Cobb 1996). The teacher assists students to resolve emerging problems by modifying their existing understandings about proof to better approximate mathematical conventions (Stylianides and Stylianides 2009b). This approach to teaching places high demands on the teacher's mathematical knowledge.

4 Concluding Remarks

This chapter provides an overview of the intellectual needs associated with learning to appreciate mathematical proof as a fundamental and purposeful mathematical construct. It examines learners' barriers to understanding and appreciating proof and discusses teachers' possible ways of addressing these barriers. Teachers can lead classrooms of inquiry, orchestrating opportunities for uncertainty and cognitive conflict whilst serving as 'brokers' of mathematical proof practices. Such practices place strong demands on teachers in terms of the required mathematical knowledge and degree of confidence as well as the challenging and time-consuming task of instructional design. Just as it is unrealistic to expect students to see a need for proof without purposeful and focused actions by the teacher, it is unrealistic to expect teachers to be able to attend to this element of teaching without appropriate preparation and support. We hope this chapter contributes to understanding the needs on both sides.

Acknowledgements We wish to thank Larry Sowder and the reviewers for their helpful comments on an earlier version of this paper.

References

Arsac, G. (2007). Origin of mathematical proof: History and epistemology. In P. Boero (Ed.), *Theorems in schools: From history epistemology and cognition to the classroom practice* (pp. 27–42). Rotterdam: Sense Publishers.

Balacheff, N. (1988, April). *A study of students' proving processes at the junior high school level.* Paper presented at the 66th annual meeting of the National Council of Teachers of Mathematics, Chicago.

Balacheff, N. (1991). The benefits and limits of social interaction: The case of mathematical proof. In A. Bishop, F. Melin-Olsen, & J. van Dormolen (Eds.), *Mathematical knowledge: Its growth through teaching* (pp. 175–192). Dordrecht: Kluwer.

Brown, S. (2003). *The evolution of students' understanding of mathematical induction: A teaching experiment.* Unpublished doctoral dissertation, University of California at San Diego and San Diego State University, San Diego.

Brown, S. I., & Walter, M. I. (2005). *The art of problem posing.* Mahwah: Lawrence Erlbaum Associates.

Buchbinder, O., & Zaslavsky, O. (2011). Is this a coincidence? The role of examples in creating a need for proof. *ZDM – Zentralblatt fuer Didaktik der Mathematik, 43,* 269–281.

Coe, R., & Ruthven, K. (1994). Proof practices and constructs of advanced mathematics students. *British Educational Research Journal, 20,* 41–53.

Dawson, J. W. (2006). Why do mathematicians re-prove theorems? *Philosophia Mathematica, 14*, 269–286.

de Villiers, M. D. (1990). The role and function of proof in mathematics. *Pythagoras, 24*, 17–24.

de Villiers, M. (1998). An alternative approach to proof in dynamic geometry. In R. Lehrer & D. Chazan (Eds.), *Designing learning environments for developing understanding of geometry and space*. Mahwah: Erlbaum (pp. 369–394).

de Villiers, M. D. (2003). *Rethinking proof with the Geometer's sketchpad, version 4*. Emeryville: Key Curriculum Press.

de Villiers, M. D. (2010). Experimentation and proof in mathematics. In G. Hanna, H. N. Jahnke, & H. Pulte (Eds.), *Explanation and proof in mathematics: Philosophical and educational perspectives* (pp. 205–221). New York: Springer.

Fischbein, E. (1982). Intuition and proof. *For the Learning of Mathematics, 3*(2), 9–18.

Fischbein, E., & Kedem, I. (1982). Proof and certitude in the development of mathematical thinking. In A. Vermandel (Ed.), *Proceedings of the Sixth International Conference of the Psychology of Mathematics Education* (pp. 128–131). Antwerp: Universitaire Instelling Antwerpen.

Fujita, T., & Jones, K. (2003, July). *Critical review of geometry in current textbooks in lower secondary schools in Japan and UK*. Paper presented at the 27th annual meeting of the International Group for Psychology of Mathematics Education, Honolulu.

Goetting, M. (1995). *The college students' understanding of mathematical proof*. Unpublished doctoral dissertation, University of Maryland, College Park.

Hadas, N., Hershkowitz, R., & Schwarz, B. (2000). The role of contradiction and uncertainty in promoting the need to prove in dynamic geometry environments. *Educational Studies in Mathematics, 44*, 127–150.

Hanna, G. (2000). Proof, explanation, and exploration: An overview. *Educational Studies in Mathematics, 44*, 5–23.

Hanna, G., & Barbeau, E. (2010). Proofs as bearers of mathematical knowledge. In G. Hanna, H. N. Jahnke, & H. Pulte (Eds.), *Explanation and proof in mathematics: Philosophical and educational perspectives* (pp. 85–100). New York: Springer.

Hanna, G., & Jahnke, H. N. (1993). Proof and application. *Educational Studies in Mathematics, 24*, 421–438.

Hanna, G., & Jahnke, H. N. (1996). Proof and proving. In A. J. Bishop, K. Clements, C. Keitel, J. Kilpatrick, & C. Laborde (Eds.), *International handbook of mathematics education* (pp. 877–908). Dordrecht: Kluwer Academic Publishers.

Harel, G. (1998). Two dual assertions: The first on learning and the second on teaching (or vice versa). *The American Mathematical Monthly, 105*, 497–507.

Harel, G. (in press). Intellectual need. In K. R. Leatham (Ed.), *Vital directions for mathematics education research*. New York: Springer.

Harel, G., & Sowder, L. (1998). Students' proof schemes: Results from exploratory studies. In A. H. Schoenfeld, J. Kaput, & E. Dubinsky (Eds.), *Research in collegiate mathematics education III* (pp. 234–283). Providence: AMS, CBMS.

Harel, G., & Sowder, L. (2007). Toward comprehensive perspectives on the learning and teaching of proof. In F. Lester (Ed.), *Second handbook of research on mathematics teaching and learning* (pp. 805–842). Charlotte: Information Age Publishing.

Harel, G., Fuller, E., & Rabin, J. M. (2008). Attention to meaning by algebra teachers. *The Journal of Mathematical Behavior, 27*, 116–127.

Healy, L., & Hoyles, C. (2000). A study of proof conceptions in algebra. *Journal for Research in Mathematics Education, 31*, 396–428.

Herbst, P. (2002). Establishing a custom of proving in American school geometry: Evolution of the two-column proof in the early twentieth century. *Educational Studies in Mathematics, 49*, 283–312.

Herbst, P., & Brach, C. (2006). Proving and doing proofs in high school geometry classes: What is it that is going on for students? *Cognition and Instruction, 24*(1), 73–122.

Hersh, R. (2008). Mathematical practice as a scientific problem. In B. Gold & R. A. Simons (Eds.), *Proof and other dilemmas: Mathematics and philosophy* (pp. 95–108). Washington, DC: Mathematical Association of America.

Hiebert, J., & Stigler, J. W. (2000). A proposal for improving classroom teaching: Lessons from the TIMSS Video Study. *The Elementary School Journal, 101*, 3–20.

Hilbert, T. S., Renkl, A., Kessler, S., & Reiss, K. (2008). Learning to prove in geometry: Learning from heuristic examples and how it can be supported. *Learning and Instruction, 18*, 54–65.

Iannone, P. & Inglis, M. (2010). Self-efficacy and mathematical proof: Are undergraduate students good at assessing their own proof production ability? In *Proceedings of the 13th Annual Conference on Research in Undergraduate Mathematics Education*. Downloaded at: http:// sigmaa.maa.org/rume/crume2010/Abstracts2010.htm

Jahnke, H. N. (2010). The conjoint origin of proof and theoretical physics. In G. Hanna, H. N. Jahnke, & H. Pulte (Eds.), *Explanation and proof in mathematics: Philosophical and educational perspectives* (pp. 17–32). New York: Springer.

Kidron, I., & Dreyfus, T. (2010). Justification enlightenment and combining constructions of knowledge. *Educational Studies in Mathematics, 74*(1), 75–93.

Kleiner, I. (1991). Rigor and proof in mathematics: A historical perspective. *Mathematics Magazine, 64*(5), 291–314.

Knuth, E. J., Choppin, J., Slaughter, M., & Sutherland, J. (2002). Mapping the conceptual terrain of middle school students' competencies in justifying and proving. In D. S. Mewborn, P. Sztajn, D. Y. White, H. G. Weigel, R. L. Bryant, & K. Nooney (Eds.), *Proceedings of the 24th annual meeting of the North American Chapter of the International Group for the Psychology of Mathematics Education* (Vol. 4, pp. 1693–1700). Athens: Clearinghouse for Science, Mathematics, and Environmental Education.

Krantz, S. G. (2007). *The history and concept of mathematical proof.* Downloaded from the Internet. http://www.math.wustl.edu/~sk/eolss.pdf

Kunimune, S., Fujita, T., & Jones, K. (2008). Why do we have to prove this?: Fostering students understanding of proof in geometry in lower secondary school. In F. L. Lin, F. J. Hsieh, G. Hanna, & M. de Villiers (Eds.) (2009). *ICMI study 19: Proof and proving in mathematics education* (pp. 877–980). Taipei: The Department of Mathematics, National Taiwan Normal University.

Lakatos, I. (1976). *Proofs and refutations: The logic of mathematical discovery.* Cambridge: Cambridge University Press.

Lampert, M. (1992). Practices and problems in teaching authentic mathematics. In F. K. Oser, A. Dick, & J. Patry (Eds.), *Effective and responsible teaching: The new synthesis* (pp. 295–314). San Francisco: Jossey-Bass Publishers.

Movshovitz-Hadar, N., & Hadass, R. (1990). Preservice education of math teachers using paradoxes. *Educational Studies in Mathematics, 21*, 265–287.

National Council of Teachers of Mathematics (2000). *Principles and standards for school mathematics.* Reston: Author.

National Research Council (1996). *Mathematics and science education around the world: What can we learn from the survey of mathematics and science opportunities (SMSO) and the third international mathematics and science study (TIMSS)?* Washington, DC: National Academy Press.

Rasmussen, C., Zandieh, M., & Wawro, M. (2009). How do you know which way the arrows go? The emergence and brokering of a classroom mathematics practice. In W.-M. Roth (Ed.), *Mathematical representations at the interface of the body and culture* (pp. 171–218). Charlotte: Information Age Publishing.

Rav, Y. (1999). Why do we prove theorems? *Philosophia Mathematica, 7*, 5–41.

Siu, M.-K. (1993). Proof and pedagogy in ancient China: Examples from Liu Hui's commentary on Jiu Zhang Suan Shu. *Educational Studies in Mathematics, 24*, 345–357.

Siu, M.-K. (2008). Proof as a practice of mathematical pursuit in a cultural, socio-political and intellectual context. *ZDM The International Journal on Mathematics Education, 40*, 355–361.

Sowder, L., & Harel, G. (2003). Case studies of mathematics majors' proof understanding, production, and appreciation. *Canadian Journal of Science, Mathematics, and Technology Education, 3*, 251–267.

Stylianides, A. J. (2007). Proof and proving in school mathematics. *Journal for Research in Mathematics Education, 38*, 289–321.

Stylianides, A. J., & Stylianides, G. J. (2009a). Proof constructions and evaluations. *Educational Studies in Mathematics, 72*, 237–253.

Stylianides, G. J., & Stylianides, A. J. (2009b). Facilitating the transition from empirical arguments to proof. *Journal for Research in Mathematics Education, 40*, 314–352.

Thompson, D. (1991, April). *Reasoning and proof in precalculus and discrete mathematics.* Paper presented at the meeting of the American Educational Research Association, Chicago.

Thompson, P. W. (1992). Notations, conventions, and constraints: Contributions to effective uses of concrete materials in elementary mathematics. *Journal for Research in Mathematics Education, 23*(2), 123–147.

Thurston, W. P. (1994). On proof and progress in mathematics. *Bulletin of the American Mathematical Society, 30*(2), 161–177.

Weber, K. (2008). How do mathematicians determine if an argument is a valid proof? *Journal for Research in Mathematics Education, 39*, 431–459.

Weber, K. (2010). Mathematics majors' perceptions of conviction, validity, and proof. *Mathematical Thinking and Learning, 12*, 306–336.

Williams, E. (1980). An investigation of senior high school students' understanding of the nature of mathematical proof. *Journal for Research in Mathematics Education, 11*(3), 165–166.

Yackel, E., & Cobb, P. (1996). Sociomathematical norms, argumentation, and autonomy in mathematics. *Journal for Research in Mathematics Education, 27*, 458–477.

Zaslavsky, O. (2005). Seizing the opportunity to create uncertainty in learning mathematics. *Educational Studies in Mathematics, 60*, 297–321.

Chapter 10
Contemporary Proofs for Mathematics Education

Frank Quinn

1 Introduction

It is widely known that mathematics education is out of step with contemporary professional practice: Professional practice changed profoundly between about 1890 and 1930, while mathematics education remains modeled on the methodologies of the nineteenth century and before. See Quinn (2011a) for a detailed account.

Professional effectiveness of the new methodology is demonstrated by dramatic growth, in both depth and scope, of mathematical knowledge in the last century. Mathematics education has seen no such improvement. Is this related to continued use of obsolete methodology? Might education see improvements analogous to those in the profession, by appropriate use of contemporary methods?

The problematic word in the last question is "appropriate": Adapting contemporary methods for educational use requires understanding them in a way that relates sensibly to education, and until recently such understanding has been lacking. The thesis here is that the description of contemporary proof in Quinn (2011a) could be useful at any educational level.

According to Quinn (2011a), contemporary proofs are first and foremost an enabling technology. Mathematical analysis can, in principle, give the right answer *every time*, but in practice people make errors. The proof process provides a way to minimise errors and locate and fix remaining ones, and thereby come closer to achieving the abstractly-possible reliability.

This view of proof is much more inclusive than traditional ones. "Show work", for instance, is essentially the same as "give a proof", while the annotations often associated with proofs appear here in "formal proofs" (Sect. 2.2), as aids rather than essential parts of the structure. To emphasise the underlying commonalities,

F. Quinn (✉)
Virginia Tech, Blacksburg, VA 24061, USA
e-mail: quinn@math.vt.edu

© The Author(s) 2021
G. Hanna and M. de Villiers (eds.), *Proof and Proving in Mathematics Education*,
New ICMI Study Series, https://doi.org/10.1007/978-94-007-2129-6_10

the word "proof" is used systematically in this essay, but synonyms such as "show work" are appropriate for use with students.

The first section carefully describes proof and its components, but the essence is: "A transcript of work with enough detail that it can be checked for errors." The second section gives examples of notations and templates designed to let students easily generate effective work transcripts. Good template design depends, however, on deep understanding of student errors. The third section illustrates how carefully designed methods can remain effective for "long problems" well outside the scope of usual classroom work. The final section describes the conflict between contemporary methodology and the way real-world (word) problems are commonly used. Changes and alternatives are suggested.

2 Proofs, Potential Proofs, and Formal Proofs

Too much emphasis on the correctness of proofs tends to obscure the features that help *achieve* correctness. Consequently, I suggest that the key idea is actually "potential proof", which does not require correctness. Variations are described in Sects. 2.1–2.2, and the role of correctness is described in Sect. 2.3. Some educational consequences are discussed in Sect. 2.4; others occur later in the essay.

2.1 Potential Proof

A *potential* proof is a record of reasoning that uses reliable mathematical methods and is presented in enough detail to be checked for errors.

Potential proofs are defined in terms of what they *do* rather than what they *are*, and consequently are context-dependent. At lower educational levels, for instance, more detail is needed. Further, the objective is to enable individual users to get better results, so even in a single class different students may need different versions. Commonalities and functionality are illustrated here, but individual needs must be borne in mind.

2.1.1 Example, Integer Multiplication I

Multiply 24 and 47 using single-digit products.
  Solution:

$$
\begin{array}{r}
24 \\
\underline{47} \\
11 \\
14 \\
16 \\
\underline{8} \\
1111 \\
\end{array}
$$

This is essentially the traditional format, and is designed to efficiently support the algorithm rather than display mathematical structure; see Sect. 2.2.1 for an alternative. It is also not annotated, so it is not a *formal* proof in the sense of Sect. 2.2. Nonetheless, it provides a clear record of the student's work that can be checked for errors, so it is a potential proof that the product is 1111.

2.1.2   About the Example

The example is not a proof because it contains an error. However:
- The error is localised and easily found. Ideally, the student would find and fix it during routine checking.
- The error is not random, and a possible problem can be diagnosed: 11 in the third line is the *sum* of 4 and 7, not the product.
- The diagnosis can be used for targeted intervention. If the error is rare the student can be alerted to watch for it in the future. If it resulted from a conceptual confusion then teachers can work with the student to correct it.

2.1.3   About the Idea

In the last decade I have spent hundreds of hours helping students with computer-based practice tests. In the great majority of cases they more-or-less understand how to approach the problem and have a record of the work they did, but something went wrong and they can't find the error. The goal is to diagnose the error, correct it, and perhaps look for changes in work habits that would avoid similar errors in the future.

Sometimes the student's work is easy to diagnose: Intermediate steps are clearly and accurately recorded; the reasoning used in going from one to the next can be inferred without too much trouble; the methods used are known to be reliable; etc. In other words it is what is described here as a potential proof. In these cases the mistakes are often minor, and the student often catches them when rechecking. Sometimes I can suggest a change in procedure that will reduce the likelihood of similar mistakes in the future (see Sect. 3 on Proof Templates). The occasional conceptual confusions are well-localised and can usually be quickly set right.

In most cases my students' work does not constitute a potential proof. Problems include:

- Intermediate expressions are incomplete or unclear. For instance when simplifying a fragment of a long expression it is not necessary to copy the parts that do not change, but without some indication of what is going on it is hard to follow such steps and there are frequently errors in reassembly.
- Steps are out of order or the order is not indicated, for instance by numbering.
- Too many steps are skipped.
- The student is working "intuitively" by analogy with an example that does not apply.
- Notations used to formulate a problem (especially word problems) are not clear.

All these problems increase the error rate and make finding errors difficult for either the student or a helper. If not corrected they limit what the student can accomplish.

The point here is that "potential proof" is to some extent an abstraction of the work habits of successful students. The same factors apply to the work of professional mathematicians, though their role is obscured by technical difficulty and the fact that checking typically proceeds rapidly and almost automatically once a genuine potential proof is in hand.

2.2   Formal Potential Proof

A *formal* potential proof includes explicit explanation or justification of some of the steps.

The use of justifications is sometimes taken as part of the definition of proof. Here it appears as useful aid rather than a qualitatively different thing: The objective is still to make it possible to find errors, and formality helps with complicated problems and sneaky errors.

The best opportunities for formal proofs in school mathematics are in introducing and solidifying methods that in standard use will not need formality. This process should improve elementary work as well as make the formal-proof method familiar and easily useable when it is really needed. The next example illustrates this.

2.2.1   Example, Integer Multipication II

Multiply 24 and 47 using single–digit products.
   Solution:

Explanation	Result
Write as polynomials in powers of 10	$(2 \times 10^1 + 4 \times 10^0)(4 \times 10^1 + 7 \times 10^0)$
Set up blank form for output	$10^2(\quad) + 10^1(\quad) + 10^0(\quad)$
Enter products in the form, without processing	$10^2(2 \times 4) + 10^1(2 \times 7 + 4 \times 4) + 10^0(4 \times 7)$
Compute coefficients	$10^2(8 \quad) + 10^1(30 \quad) + 10^0(28 \quad)$
Recombine as a single integer	$800 + 300 + 28 = 1128$

2.2.2   Comments

This example uses a "structured" format for proof (see Mannila and Wallin 2009; Peltomaki and Back 2009). I have not had enough experience to judge the benefits of a standardised structure.

The procedure follows the "template" for multiplication of polynomials described in Sect. 3.1. (See Sect. 4.1 for a version used to multiply large numbers.)

Writing in expanded form with explanations clarifies the procedure. Once the procedure is mastered a short-form version can be used:

$$10^2 \underbrace{(2\times4)}_{8} + 10^1 \underbrace{\underbrace{(2 \times 7}_{14} + \underbrace{4 \times 4)}_{16}}_{30} + 10^0 \underbrace{(4\times7)}_{28}$$

$$800 + 300 + 28 = 1128$$

In this form:

- The numbers are not rewritten explicitly as polynomials because the coefficients can be read directly from the decimal form. Some students may have to number the digits to do this reliably.
- The extra space in the outer parentheses after the powers of 10 indicates that the blank template was set up first.
- The products for the coefficients were entered without on-the-fly arithmetic (explained in Sect. 3.1).
- Individual steps in the arithmetic are indicated, as is the final assembly.

Thus, when the method is familiar, a compressed notation provides an effective potential proof that the outcome is correct.

2.2.3 Example, Solutions of Linear Systems

For which values of a is the solution of the system *not* unique?

$$x + ay + 2z = -1$$
$$3y + az = 2 - a$$
$$4x + y = 13$$

Solution:

The solution to a square linear system is not unique exactly when the determinant of the coefficient matrix is zero. The coefficient matrix here is

$$\begin{pmatrix} 1 & a & 2 \\ 0 & 3 & a \\ 4 & 1 & 0 \end{pmatrix}$$

Row operations $R_3 = R_3 - 4R_1$ and $R_3 = R_3 - \dfrac{1-4a}{3}R_2$ do not change the determinant and reduce this to a triangular matrix with $R_3 = (0,0,-8-a\dfrac{1-4a}{3})$.

The determinant of a triangular matrix is the product of the diagonal entries, so the determinant is

$$(1)(3)(-8 - a\frac{1-4a}{3}) = -24 - a(1-4a) = 4a^2 - a - 24$$

This is zero for $a = (-1 \pm \sqrt{385})/8$.

2.2.4 Comments

This example is a bit less detailed than the previous one in that some calculations (effects of the row operations and application of the quadratic formula) are not recorded. Presumably they are on a separate paper, but because the operations themselves are recorded the calculations can be completely reconstructed. At the level of this example, students should be able to reliably handle such hidden steps and explicit display should not be necessary.

An alternative evaluation of the determinant might be: "Cramer's rule applied to the second row gives $(+1)(3)(-4 \times 2) + (-1)(a)(1-4a)...$".

Cramer's rule involves adding up: a sign times the entry times the determinant of the matrix obtained by omitting the row and column containing the entry. The expression reflects this structure, with the 2×2 determinants evaluated. Giving relatively unprocessed expressions like this both reduces errors (by separating organisation from calculation) and allows quick pin-pointing of them when they occur. For example, it would be possible to distinguish a sign error in the second term due to a misunderstanding of Cramer's rule, from a sign error in the evaluation of the sub-determinant.

Students will not give this sort of explanation without examples to copy and quite a bit of guidance. This guidance might include:

- When using a theorem (e.g. nonzero determinant if and only if unique solutions), say enough about it to inspire confidence that you know a precise statement and are using it correctly. Confused statements indicate that conceptual errors are likely in the future, even if this wasn't the problem in this case.
- In particular, mention of the theorem is an essential part of the work and must be included even in short–form versions. (For additional discussion of style in short–form proofs, see *Proof Projects for Teachers* in Quinn (2011b).)
- In lengthy calculations, rather than showing all details, describe the steps and carry out details on a separate sheet. The descriptions should be explicit enough to enable reconstruction of the details. Organising work this way both reduces errors and makes it easier to check.

It can be helpful to have students check each others' work and give explicit feedback on how well the layout supports checking. The eventual goal is for them to diagnose their own work; trying to make sense of others' work can give insight into the process.

2.2.5   Further Examples

For further discussion, and examples of elementary formal proofs concerning fractions and area, see *Proof Projects for Teachers* (Quinn 2011b).

2.3   Proof and Correctness

A *proof* is a potential proof that has been checked for errors and found to be error-free.

Work that does not qualify as a potential proof cannot be a proof even if the conclusion is known to be correct. In education, the goal is not a correct answer but to develop the ability to routinely get correct answers; facility with potential proofs is the most effective way to do this. Too much focus on correctness may undercut development of this facility.

This is usually not an issue with weak students because potential proofs are an enabling technology without which they cannot succeed. Weak students tend to have the opposite problem: the routines are so comforting and the success so rewarding that it can be hard to get them to compress notation (e.g. avoid recopying) or omit minor details even when they have reached the point where it is safe to do so. Similarly, some persist in writing out formal justifications even after they have thoroughly internalised the ideas.

Strong students are more problematic, because the connection between good work habits and correct answers is less direct. I have had many students who were very successful in high-school advanced placement courses, but they got by with sloppy work because the focus was on correctness rather than methodology. Many of these students have trouble with engineering calculus in college:

- The better students figure it out, especially with diagnostic support and good templates (Sect. 3). Most probably never fully catch up to where they might have been, but they are successful.
- Unfortunately a significant number were good enough to wing it in high school and good enough to have succeeded in college with good methodological preparation, but are not good enough to recover from poor preparation.

All students stand to benefit from a potential-proof-oriented curriculum rather than a correctness-oriented one, but for different reasons. Gains by weak and mid-range students are likely to be clearest.

2.4   The Role of Diagnosis

The thesis of this article is that the reliability possible with mathematics can be realised by making mathematical arguments that can be checked for errors, *checking them*, and correcting any errors found. Other sections describe how checkable arguments could become a routine part of mathematics education. However they won't produce benefits unless *checking* also becomes a routine part. To be explicit: Diagnosis and error correction should be key focuses in mathematics education.

- Answers are important mainly as proxies for the work done. Incorrect answers indicate a need for diagnosis and correction. Ideally, *every* problem with a wrong answer should be diagnosed and corrected.
- Mathematics uniquely enables quality, so the emphasis should be on quality not quantity. In other words, doing fewer problems to enable spending more time on getting them right is a good tradeoff.
- An important objective is to teach students to routinely diagnose their own work. The fact that diagnosis is possible and effective is the essence of mathematics, so teaching self-diagnosis is mathematics education in the purest sense.

Ideally, teachers would regularly go through students' work with them so students can see the checking process in action. Students should be required to redo problems when the work is hard to check, not just when the answer is wrong. As explained in the previous section, the goal is to establish work habits that will benefit students; but students respond to feedback from teachers, not to long–term goals.

2.5   Other Views of Proof

There are many other—and quite different—views of the role of proof (cf. Hanna and Barbeau 2008; Rav 1999; Thurston 1994). These generally emphasise proofs as sources of understanding and insight, or as repositories of knowledge.

The basic difference is that I have emphasised proofs as an enabling technology for users. Most other views focus on "spectator proofs": arguments from which readers should benefit, but that are not intended as templates for emulation. Both views are valid in their own way, and this should be kept in mind when considering specific situations.

What counts as user-oriented or spectator-oriented, and the mix in practice, varies enormously with level. In school mathematics—as illustrated here—almost everything is designed for emulation. Spectator proofs play little or no role. Issues that might be addressed with spectator proofs (e.g., how do we know the multiplication algorithm really works?) are simply not addressed at all.

At intermediate levels, college math majors for instance, spectator proofs play a large role. They provide ways for students to learn and develop skills long before they can be emulated. At the research frontier the primary focus is again on user-oriented work. It is a nice bonus if an argument functions as a spectator proof (i.e., is "accessible"), but if the argument cannot be fleshed out to give a fully-precise user-oriented proof it is unsatisfactory.

Misunderstanding these different roles of proof has led to conflict and confusion. For example, Thurston (1994) justified his failure to provide a proof of a major claim by observing that the technology needed for a good spectator proof was not yet available. This point resonated with educators since they have a mainly spectator-oriented conception of proof. However Thurston was responding to criticism (Jaffe and Quinn 1993) that he had failed to provide a user-oriented proof for use in the research community. An inability to provide a spectator proof was not accepted as justifying the failure to provide any proof at all. The problem was later declared unsolved, and complete proofs were eventually provided by others (see Quinn 2011a).

3   Proof Templates

Students learn mainly by abstraction from examples and by imitating procedures. It is important, therefore, to carefully design examples and procedures to guide effective learning.

A "proof template" is a procedure for working a class of problems. Design considerations are:

- Procedures should clearly reflect the mathematical structures they exploit. This makes them more reliable and flexible, and often provides subliminal preparation for more complex work.
- Procedures should minimise problems with limitations of human cognitive abilities. For example, conceptually distinct tasks such as translating word problems, organising a computation, or doing arithmetic, should be separated.
- Efficient short–form versions should be provided.

Examples in this section explain and illustrate these points.

3.1   Polynomial Multiplication

This material is adapted from a polynomial problem developed for a working group of the American Mathematical Society. See *Neuroscience Experiments for Mathematics Education* in Quinn (2011b) for further analysis of cognitive structure.

3.1.1   Problem

Write $(3z^2 - z + 5a)(z^3 + (2-a)z^2 - a)$ as a polynomial in z. Show steps.

3.1.2   Step 1: Organisation

There are three terms in each factor, so there will be nine terms in the product. Some organisational care is needed to be sure to get them all. Further, we would like to have them sorted according to exponent on z rather than producing them at random and then sorting as a separate step. To accomplish this, we set up a blank form in which to enter the terms. A quick check of exponents shows that all exponents from 0 to 5 will occur, so the appropriate blank form is:

$$z^5[\quad]+z^4[\quad]+z^3[\quad]+$$
$$z^2[\quad]+z^1[\quad]+z^0[\quad]$$

Next, scan through all possible combinations of terms, one from each factor. (Use a finger to mark your place in one term while scanning the other.) For each combination, write the product of coefficients in the row with the right total exponent. The result is:

$$z^5[(3)(1)\qquad]+z^4[(3)(2-a)+(-1)(1)]+z^3[(-1)(2-a)+(5a)(1)]+$$
$$z^2[(3)(-a)+(5a)(2-a)]+z^1[(-1)(-a)\qquad]+z^0[(5a)(-a)\qquad]$$

Note the products were recorded with *absolutely no* arithmetic, not even writing $(3)(1)$ as 3. Reasons are:

- Organisation and arithmetic are cognitively different activities. Switching back and fourth increases the error rate in both, with sign errors being particularly common.
- This form can be diagnosed. We can count the terms to see that there are nine of them and the source of each term can be identified. The order of scanning can even be inferred, though it makes no difference.

Note also that every term is enclosed in parentheses. This is partly to avoid confusion, because juxtaposition is being used to indicate multiplication. The main reason, however, is to avoid thinking about whether or not parentheses are necessary in each case. Again, such thinking is cognitively different from the organisational task and may interfere with it.

3.1.3 Step 2: Calculation

Simplify the coefficient expressions to get the answer:

$$3z^5+(5-3a)z^4+(6a-2)z^3+(7a-5a^2)z^2+az+(-5a^2)$$

In this presentation the only written work is the organisational step and the answer. More complicated coefficient expressions, or less experienced students, would require recording some detail about the simplification process. A notation for this is shown in the arithmetic example in Sect. 2.2.1.

3.1.4 Comments

- The separation of organisation and computation makes the procedure reliable and relatively easy to use.

- The close connection to mathematical structure makes the procedure flexible. It is easily modified to handle problems like "Find the coefficient on z^3" or "Write a product involving both x and y as a two-variable polynomial".
- Variations provide methods for by-hand multiplication of integers (Sect. 2.2.1) and multiplication of large integers using a calculator (Sect. 4.1).
- If the baby version in Sect. 2.2.1 is used to multiply integers, then students will find the polynomial version familiar and easy to master.
- Similarly, students who work with polynomials this way will find some later procedures (e.g., products of sums that may not be polynomials, or iterated products like the binomial theorem) essentially familiar and easier to master.

This procedure should be contrasted with the common practice of restricting to multiplication of binomials, using the "FOIL" mnemonic[1]. That method is poorly organised even for binomials, inflexible, and doesn't connect well even with larger products. In particular, students trained with FOIL are often unsuccessful with products like the one in the example.

3.2   Solving Equations

This is illustrated with a very simple problem, so the structuring strategies will be clear.

3.2.1   Problem

Solve $5x - 2a = 3x - 7$ for x.
Annotated Solution:

Explanation	Result
Collect terms: move to other side by adding negatives	$5x - 3x = -7 + 2a$
Calculate	$\underbrace{(5 - 3)}_{2}x = -7 + 2a$
Move coefficient to other side by multiplying by inverse	$x = \dfrac{1}{2}(-7 + 2a)$

3.2.2   Comments

The primary goals in this format are efficiency and separation of different cognitive activities (organisation and calculation).

---

[1] First, Outer, Inner, Last.

The first step is organisational: we decide that we want all x terms on one side and all others on the other. Collecting x terms can be accomplished by adding $-3x$ to each side. However it is inefficient to do this as a separate calculation step because we know ahead of time what will happen on the right side: we have chosen the operation exactly to cancel the $3x$ term. Instead we think of it as a purely organisational step: "move $3x$ to the other side…". To keep it organisational we refrain from doing arithmetic (combining coefficients) and include "by adding negatives" to the mental description.

The second step is pure calculation.

The final step is again organisational, and the description is designed to emphasise the similarity to the first step.

Finally, the steps are guided by pattern–matching: The given expression is manipulated to become more like the pattern $x = ?$. (See the next section for another example.)

3.3 Standardising Quadratics

This is essentially "completing the square" with a clear goal.

3.3.1 Problem

Find a linear change of variables $y = ax + b$ that transforms the quadratic $5x^2 - 6x + 21$ into a standard form $r(y^2 + s)$ with s one of $1, 0 - 1$, and give the standard form.

This is done in two steps, each of which brings the expression closer to the desired form. A short-form version is given after the explanation.

3.3.2 First Step

Eliminate the first-order term with a change of the form $y_0 = x + t$:

Square the general form and multiply by 5 to get $5y_0^2 = 5x^2 + 10tx + 5t^2$, which has the same second-order term as that of the given quadratic. To match the first-order term as well we need $10t = -6$, so $t = -3/5$ and $y_0 = x - 3/5$. Moving the constant term to the other side gives $5y_0^2 - 5t^2 = 5x^2 - 6x$. Use this to replace the first- and second-order terms in the original to transform it to

$$5(y_0)^2 \underbrace{-5(-3/5)^2 + 21}_{-\frac{9}{5} + \frac{105}{5} = \frac{96}{5}} \tag{1}$$

3.3.3   Second Step

Factor out a *positive* number to make the constant term standard.

$$5y_0^2 + \frac{96}{5} = \frac{96}{5}(\underbrace{\frac{5^2}{96}y_0^2}_{(-\frac{5}{\sqrt{96}}y_0)^2} + 1) \tag{2}$$

The number factored out must be positive because we had to take the square root of it.

Comparing with the goal shows the standard form is $\dfrac{96}{5}(y^2 + 1)$ with $y = \dfrac{5}{\sqrt{96}} y_0 = \dfrac{5}{\sqrt{96}}(x - \dfrac{3}{5})$.

3.3.4   Short Form

$$5(\underbrace{x + t}_{y_0})^2 = 5x^2 + \underbrace{10t}_{-6}x + 5t^2$$

So $t = -3/5$.

$$\underbrace{5x^2 - 6x + 21}$$

$$5y_0^2 \underbrace{- 5(3/5)^2 + 21}_{\frac{96}{5}} = \frac{96}{5}(\underbrace{\frac{5}{96}5y_0^2}_{(-\frac{5}{\sqrt{96}}y_0)^2} + 1)$$

So $y = \dfrac{5}{\sqrt{96}} y_0 = \dfrac{5}{\sqrt{96}}(x - \dfrac{3}{5})$ and the form is $\dfrac{5}{96}(y^2 + 1)$.

Methods must be introduced with explanations, but compression is necessary for routine use. It is important for teachers to provide a carefully-designed short format because the compressions which student invent on their own are rarely effective.

For example, it is often necessary to simplify a fragment of an expression. The underbrace notation here indicates precisely which fragment is involved and connects it to the outcome. I have never seen a student do this. Usually, the student either writes fragments without reference or rewrites the whole expression.

Experience often reveals errors that need to be headed off by the notation. In the work above, the notation

$$5y_0^2 \underbrace{- 5(3/5)^2 + 21}_{\frac{96}{5}}$$

clearly indicates that the sign on $-5(3/5)^2$ is part of the fragment being simplified. Many students seem to think of this sign as the connector between the expression fragments, and hence do not include it in the sub-expression. It then gets lost. This is a common source of errors, and may well have resulted in the student making an error in this case. Providing a clear notation and being consistent in examples will avoid such errors.

3.3.5   Pattern Matching

Routine success requires that at any point the student can figure out "What should I do next?" In the problem above there is a direct approach: Plug $y = ax + b$ into the given quadratic, set it equal to $r(y^2 + s)$, and solve for a, b, r, s. This can be simplified by doing it in two steps, as above, but even so it requires roughly twice as much calculation as the method given above. This is a heavy price to pay for not having to think.

By contrast, the suggested procedure uses pattern matching to guide the work. It can be summarised as "What do we have to do to the given quadratic to get it to match the standard pattern?" In the first step we note that the given one has a first-order term and the pattern does not. We get closer to the pattern by eliminating this term, getting something of the form $Ay_0^2 + B$. If B is not 1, 0, or -1 we can get closer to the pattern by factoring something out to get $C(Dy_0^2 + s)$ with standard s. The only thing remaining to exactly match the pattern is to rewrite Dy_0^2 as a square, and whatever result we get is the y we are seeking.

Pattern matching is a powerful technique, a highly-touted feature of computer algebra systems, and humans can be very good at it. Much of the work in a calculus course can be seen as pattern-matching. Students could use it more effectively if teachers presented the idea more explicitly.

3.4   *Summary*

Carefully-designed procedures and templates for students to emulate can greatly improve success and extend the range of problems that can be attempted. Important factors are:

- Procedures should follow the underlying mathematical structure as closely as possible. Doing so reveals connections, provides flexibility, and expands application. It also ensures upward-compatibility with later work, and frequently provides subliminal preparation for this work.
- Ideas that guide the work, pattern matching for example, should be abstracted and made as explicit as possible for the level.
- Procedures should separate different cognitive tasks. In particular, organisational work should be kept separate from computation.
- Short-form formats that show the logical structure (i.e., are checkable) and encourage good work habits should be provided.

Good test design can also encourage good work habits. For example:

- Ask for a single coefficient from a good-sized product like the example in Sect. 3.1. This rewards students who understand the organisational step well enough to pick out only the terms that are needed.
- A computer-based test might ask for an algebraic expression that *evaluates* to give the coefficient[2]. The students could then enter the unevaluated output from the organisational step. This approach rewards careful separation of organisation and calculation, by reducing the time required and reducing the risk of errors in computation.

4 Long Problems

Current pre-college mathematics education is almost entirely concerned with short, routine problems. Advanced-placement courses may include short tricky problems. However, much of the power of mathematics comes from its success with long *routine* problems. Because the conclusions of each step can be made extremely reliable, many steps can be put together and the combination will still be reliable. Further, carefully-designed methods for dealing with short problems will apply to long problems equally well.

Long problems have an important place in elementary mathematics education. They give a glimpse into the larger world and illustrate the power of the methods being learned. They also reveal the need for care and accuracy with short problems. It is not clear how long problems might be incorporated into a curriculum, but group projects are a possibility. The examples here are presented as group problems about multiplication and addition of large integers (with calculators) and logic puzzles.

4.1 Big Multiplications

The goal is to exactly multiply two large (say 14- or 15-digit) integers using ordinary calculators. This cannot be done directly so the plan is to break the calculation into smaller pieces (e.g., 4-digit multiplications) that can each be done on a calculator, and then assemble the answer from these pieces. The method is the same as the by-hand method for getting multi-digit products from single-digit ones, and uses a notation (like that of Sect. 2.2.1) modeled on polynomial multiplication.

The number of digits in each piece depends on the capability of the calculators used. The product of two 4-digit numbers will generally have 8 digits. We will be

---

[2]Tests with this sort of functionality are a goal of the EduTE X project (Quinn 2009).

adding a list of these, but no more than 9, so the outcome will have 9 or fewer digits. Four-digit blocks will therefore work on calculators that can handle nine digits. Eight–digit calculators would require the use of three-digit blocks.

4.1.1 Problem

Multiply 638521988502216 and 483725147602252, using calculators that handle 9 or more digits, by breaking them into 4–digit blocks.

4.1.2 Step 1: Organise the Data

Write the numbers as polynomials:

$$638521988502216 = 2216 + 8850x + 5219x^2 + 638x^3$$
$$483725147602252 = 2252 + 4760x + 7251x^2 + 483x^3$$

where $x = 10^4$.

The power-of-10 notation should be used even with pre–algebra students, because it is a powerful organisational aid. The exponent records the number of blocks of four zeros that follow these digits.

4.1.3 Step 2: Organise the Product

The product of two sums is gotten from all possible products, using one piece from each term. Individual terms follow the rule $(ax^n)(bx^k) = (ab)x^{n+k}$, which we use to organise the work. The product will have terms x^r for $r = 0,\ldots,6$ and seven individuals or teams could work separately on these.

For instance, the x^2 team would collect the pairs of terms whose exponents add to 2: x^0 (x not written) from the first number and x^2 from the second, then x^1 from the first and x^1 from the second, etc. They would record:

$$x^2(2216 \times 7251 + 8850 \times 4760 + 5219 \times 2252)$$

This is an organisational step; no arithmetic should be done. The students can infer how the pieces were obtained, and can double–check each other to see that nothing is out of place and no pieces were left out.

4.1.4 Step 3: Compute the Coefficient

Carry out the arithmetic indicated in the second step, using calculators. If the students can use a memory register to accumulate the sum of the successive products then the output is the answer, $x^2(69947404)$. If the multiplications and addition have to be done separately then the notation of Sect. 2.2.1 can be used:

$$x^2 (\underbrace{2216 \times 7251}_{16068216} + \underbrace{8850 \times 4760}_{42126000} + \underbrace{5219 \times 2252}_{11753188})$$
$$\underbrace{}_{69947404}$$

Again, different students or teams should double-check the outcomes.

4.1.5 Step 4: Assemble the Answer

At this point the group has found the product of polynomials,

$$4990432 + 30478360x + 69947404x^2 + 91520894x^3 +$$
$$45154399x^4 + 7146915x^5 + 308154x^6$$

and the next step is to evaluate at $x = 10^4$, or in elementary terms translate the powers of x back to blocks of zeros, and add the results. The next section gives a way to carry out the addition.

4.2 Big Additions

The goal is to add a list of large integers using ordinary calculators. This cannot be done directly, so the plan is to break the operations into smaller pieces (e.g., 6–digit blocks) that can be done on a calculator and then assemble the answer from these pieces. The procedure is illustrated with the output from the previous section.

4.2.1 Problem

Use calculators to add $4990432 + 30478360 \times 10^4 + 69947404 \times 10^8 + 91520894 \times 10^{12} + 45154399 \times 10^{16} + 7146915 \times 10^{20} + 308154 \times 10^{24}$ using 6–digit blocks.

4.2.2 Step 1: Setup

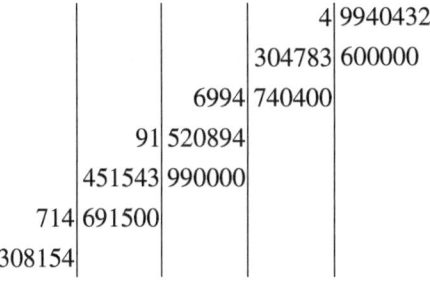

Here we have written the seven numbers to be added in a column with aligned digits. Vertical lines are drawn to separate the 6-digit blocks, and we omit blocks that consist entirely of zeros. We have not, however, omitted zeros at the end of blocks because doing this would mix organisational and arithmetic thinking.

4.2.3 Step 2: Add 6–digit Columns

			4	990432
			304783	600000
		6994	740400	
		91	520894	
	451543	990000		
	714	691500		
308154				
308868	1	517888	1	590432
1	143134	1	045187	

Each column is added separately, for instance by five different students; again, the outcomes should be double–checked.

Most of the sums overflow into the next column. We have written the sums of the even–numbered columns one level lower to avoid overlaps. Since there are fewer than nine entries in each column, the sum can overflow only into the first digit of the next column to the left.

4.2.4 Step 3: Final Assembly

Add the sums of the individual columns:

308868	1	517888	1	590432
1	143134	1	045187	
308869	143135	517889	045188	590432

In this example the final addition is easy, because the overflow from one column only changes one digit in the next. This happens in most cases; if examples are chosen at random, it is very unlikely that students will see more than two digits change due to overflow.

Students should realise, however, that digits in sums are unstable in the sense that, very rarely, an overflow will change *everything* to the left. Teachers should ensure that students encounter such an example, or perhaps challenge them to contrive an example that makes the simple-minded pattern crash. This phenomenon illustrates the difference between extremely unlikely events and mathematically impossible ones, and the "low–probability catastrophic failures[3]" that can occur when the difference is ignored.

4.3   Digits in Big Products

The goal here is to find a specific digit in a product of big numbers, and be sure it is correct. An attractive feature of the formulation is that careful reasoning and understanding of structure are rewarded by a reduction in computational work.

The least-thought/most-work approach is to compute the entire number and then throw away all but one of the digits. I give three variations with increasing sophistication and decreasing computation. In practice, students (or groups) could be allowed to choose the approach that suits their comfort level. More-capable students will enjoy exploiting structure to achieve efficiency. Less-capable ones will be aware of the benefits of elaborate reasoning, but may see additional computation as a safer and more straightforward.

4.3.1   Problem

Find the 18th digit (from the right, i.e. in the 10^{17} place) in the product $52498019913177259058 \times 33208731911634712456$.

4.3.2   Plan A

We approach this as before, by breaking the numbers into 4-digit blocks and writing them as coefficients in a polynomial in powers of $x = 10^4$. These are 20-digit numbers so there are five 4-digit blocks and this gives polynomials of degree 4 (powers of x up to x^4). The product has terms up to degree 8.

The 18th digit is the second digit in the fifth 4-digit block ($18 = 4 \times 4 + 2$). When working with polynomials in $x = 10^4$ this means it will be determined by the terms of degree x^4 and lower (the coefficient on x^5 gets 20 zeros put after it, so cannot effect the 18th digit).

Plan A is to compute the polynomial coefficients up to x^4, combine as before to get a big number, and see what the 18th digit is. This gives a significant savings over computing the whole number because we don't find the $x^5 \ldots x^8$ coefficients.

---

[3] A term from the computational software community, where this is a serious problem.

4.3.3 Plan B

This refinement of Plan A reduces the work done on the x^4 coefficient.

We only need the 18th digit, so only need the second (from the right) digit in the coefficient on x^4. To get this we only need the product of the lowest two digits in each term. To make this explicit, the terms in the coefficient on x^4 are:

$$x^4(9058 \times 3320 + 7725 \times 8731 + 9131 \times 9116 + 8019 \times 3471 + 5249 \times 2456)$$

But we only need the next-to-last digit of this. If we write the first term as $(9000+58) \times (3300+20)$, then the big pieces don't effect the digit we want. It is sufficient just to compute 58×20.

This modification replaces the x^4 coefficient by

$$x^4(58 \times 20 + 25 \times 31 + 31 \times 16 + 19 \times 71 + 49 \times 56)$$

Lower coefficients are computed and the results are combined to give a single number as before. This number will have the same lower 18 digits as the full product, and in particular will have the correct 18th digit.

4.3.4 Plan C, Idea

Plans A and B reduce work by not computing unneeded higher digits. Here, we want to reduce work by not computing unneeded *lower* digits. The overflow problem makes this tricky, and some careful estimation is needed to determine how bad lower-digit overflows can be. This is a nice opportunity for good students to exploit their talents.

1. The coefficients in the product polynomial have at most nine digits (products of 4–digit numbers have at most 8 digits, and we are adding fewer than nine of these in each coefficient). The x^2 term therefore has at most $9+2 \times 4 = 17$ digits. This can effect the 18th digit only through addition overflow.
2. The plan, therefore, is to compute the coefficients on x^4 and x^3, combine these to get a number, and see how large a 17-digit number can be added before overflow changes the 18th. We will then have to estimate the x^2 and lower terms and compare this to the overflow threshold.

   - If the lower-order terms cannot cause overflow into the 18th digit, then the 18th digit is correct.
   - If lower terms might cause overflow, then we will have to compute the x^2 coefficient exactly, combine with the part already calculated, and see what happens. In this case, we will also have to check to see if degree 0 and 1 terms cause overflow that reaches all the way up to the 18th digit. This is extremely unlikely: These terms have at most $9+1 \times 4 = 13$ digits, so overflow to the 18th can only happen if the 14th–17th digits are all 9.
   - In this unlikely worst-case scenario we will have to compute the lower-order terms too.

4.3.5   Plan C, Setup and Compute

The x^3 coefficient and Plan B version of the x^4 coefficient are:

$$x^4(58 \times 20 + 25 \times 31 + 31 \times 16 + 19 \times 71 + 49 \times 56)$$
$$x^3(9058 \times 8731 + 7725 \times 9116 + 9131 \times 3471 + 8019 \times 2456)$$

Computing gives $200894863x^3 + 6524x^4$. Substituting $x = 10^4$ gives

$$(200894863 + 65240000) \times 10^{12} = 266134863 \times 10^{12}.$$

The 18th digit (from the right) is 1. It is not yet certain, however, that this is the same as the digit in the full product.

4.3.6   Plan C, Check for Overflow

The 17th digit in 266134863×10^{12} is 3. If the top (i.e. 9th) digit in the x^2 coefficient is 5 or less then adding will not overflow to the 18th digit. ($3 + 5 = 8$, and overflow from the x^1 and x^0 terms can increase this by at most one).

The next step is to estimate the top digit in this coefficient.

1. The x^2 coefficient has three terms (from x^0x^2, x^1x^1, and x^1x^0).
2. Each term is a product of two 4-digit numbers, so each has at most 8 digits. In other words the contribution of each term is smaller than 10^9. Adding three such terms gives a total coefficient smaller than 3×10^9.
3. When we substitute $x = 10^4$ we get a number less than 3×10^{17}. The top digit is therefore at most 2.
4. Since the top digit of the lower-order term is smaller than the threshold for overflow ($2 \le 5$), we conclude that the 18th digit found above is correct.

We were fortunate: If the 17th digit coming from the higher–order terms had been 7, 8, or 9 then we could not rule out overflow with this estimate. For borderline cases I describe a refined estimate that gives a narrower overflow window.

The actual coefficient on x^2 is 131811939. Knowing this, we see that a 17th digit 7 would not have caused an overflow, while a 9 would have increased the 18th digit by 1, and 8 is uncertain. This conclusion can be sharpened by using more digits: If digits 15–17 are 867 or less, then there is no overflow; if they are 869 or more then there is an overflow of 1; and the small interval between these numbers remains uncertain. As noted above, in rare cases lower–order terms have to be computed completely to determine whether or not overflow occurs.

4.3.7   Grand Challenge

Use this method to find the 25th digit of the product of two 50–digit numbers.

4.4 Puzzles

We will not explore them here but logic puzzles deserve mention as opportunities for mathematical thinking (see Wanko 2009; Lin 2009). These should incorporate an analog of proof: a record of moves that enables reconstruction of the reasoning and location of errors. The notation for recording chess moves (see the Wikipedia entry) may be a useful model.

A minor problem is that the rules of many puzzles are contrived to avoid the need for proof-like activity and should be de-contrived.

For example, the usual goal in Sudoku is to fill entries to satisfy certain conditions. The final state can be checked for correctness and—unless there is an error—would seem to render the record of moves irrelevant. A better goal is to find *all* solutions. If the record shows that every move is forced, then the solution is unique. However, if at some point no forced moves can be found and a guess is made, all branches must be followed. If a branch leads to an error, that branch can be discarded (proof by contradiction). If a branch leads to a solution, then other branches still have to be explored to determine whether they also lead to solutions. This would be made more interesting by a source of Sudoku puzzles with multiple solutions.

Notations and proof also enable collaborative activity. All members of a group would be given a copy of the puzzle, and one appointed "editor". On finding a move, a member would send the notation to the editor as a text message. The editor would check for correctness and then forward the move to the rest of the group. Maintaining group engagement might require a rule like: Whoever submits a move must wait for someone else to send one before submitting another.

5 Word Problems and Applications

This essay concerns the use of contemporary professional methodology in education. Up to this point the ideas have been unconventional and possibly uncomfortable but more-or-less compatible with current educational philosophy.

There are, however, genuine conflicts where both professional methodology and direct experience suggest that educational practices are counterproductive, not just inefficient. Some of the methodological conflicts are discussed in this section. A more systematic comparison is given in *Mathematics Education versus Cognitive Neuroscience* in Quinn (2011b), and conflicts in concept formation are discussed in *Contemporary Definitions for Mathematics Education* in Quinn (2011b). Historical analysis in Quinn (2011a) indicates that many educational practices are modeled on old professional practices that were subsequently found ineffective and were discarded.

5.1 Word Problems and Physical-World Applications

The old view was that mathematics is an abstraction of patterns in the physical world and there is no sharp division between the two. The contemporary view is that there is a profound difference and the articulation between the two worlds is a key issue. The general situation is described in Quinn (2011a); here I focus on education.

5.1.1 Mathematical Models

In the contemporary approach, physical-world phenomena are approached indirectly: a *mathematical model* of a phenomenon is developed and then analyzed mathematically. The relationship between the phenomenon and the model is not mathematical, and is not accessible to mathematical analysis.

5.1.2 Example

A beaker holds 100 ml. of water. If 1 ml of X is added, what is the volume of the result?

Expected solution: $100 + 1 = 101$ ml.

5.1.3 Discussion

The standard expected solution suppresses the modelling step. Including it gives:

Model: volumes add.

Analysis: $100 + 1 = 101$ so the model predicts volume 101 ml.

The analysis of the model is certainly correct, so it correctly predicts the outcome when the model applies: for example, if X is water. If X is sand, salt, or alcohol, then the volume will be more than 100 ml. but significantly less than the predicted value of 101 ml. If X is metallic sodium a violent reaction takes place. When the smoke clears, the beaker will contain considerably less than 100 ml., and may be in pieces.

In the latter instances, the prediction fails because the model is not appropriate. This is not a mathematical difficulty. In particular no amount of checking the written work can reveal an error that accounts for the failure. One might try to avoid the problem in this case by specifying that X should be water, but discrepancies could result from differences in temperature. Even elaborately legalistic descriptions of the physical circumstances cannot completely rule out reality/model disconnects.

The point is that the reality/model part of real-world applications is essentially non-mathematical. Applications have an important place in mathematics courses, but the reality/model aspect should not be represented as mathematics.

Equally important, modelling, and analysis of the model are different cognitive activities. Failing to separate them increases error rates, just as happens with organisation and calculation (see Sect. 3.1). Educational practice is to make success possible by making the mathematical component trivial rather than separating the components. This, however, makes significant applications impossible.

5.2   Applications

Mathematics is brought to life through applications. In this context the word "application" is usually understood to mean "physical-world application". However, such applications alone do a poor job of bringing elementary mathematics to life. After explaining why, I suggest that there are better opportunities using applications from within mathematics.

5.2.1   Difficulties with the Real World

The main difficulty with physical-world applications is a complexity mismatch. In one direction, there are impressive applications of elementary mathematics, but they require significant preparation in other subjects. On the other hand there are easily-modeled real-world problems but these tend to be either mathematically trivial or quite sophisticated.

Examples:

- One can do interesting chemistry with a little linear algebra, but the model-building step requires a solid grasp of atomic numbers, bonding patterns, etc. The preparation required is probably beyond most high-school chemistry courses and certainly beyond what one could do in a mathematics course.
- There are nice applications of trig functions to oscillation and resonance in mechanical systems, electric circuits, and acoustics. Again, subject knowledge requirements makes these a stretch even in college differential equations courses.
- Multiplication of big integers, as in Sect. 4.1, plays an important role in cryptography, but it is not feasible to develop this subject enough to support cryptographic "word problems".

Problems with easily-modeled situations include:

- It is difficult to find simple problems that are not best seen as questions in calculus or differential equations (or worse).
- Special cases may have non-calculus solutions, but these solutions tend to be tricky and rarely give insight into the problem.
- Even as calculus problems, most "simple" models lead to mathematical questions too hard for use in college calculus.
- Our world is at least three-dimensional. Many real problems require vectors in all but the most contrived and physically-boring cases.

In other words, real-world problems should be part of a serious development of a scientific subject in order to be genuinely useful. The next section describes difficulties that result when this constraint is ignored.

5.2.2 Bad Problems

The practical outcome of the complexity mismatch described above is that most word problems—in the US anyway—have trivial or very constrained mathematical components and the main task is formulation of the model (e.g., the example in Sect. 5.1.1).

Some elementary–education programmes exploit this triviality with a "keyword" approach: "When a problem has two numbers, then the possibilities are multiplication, division, addition or subtraction. Addition is indicated by words 'added', 'increased by' ...". The calculator version is even more mindless, because the operations have become keystrokes rather than internalised structures that might connect to the problem: "Press the "+" key if you see 'added', 'increased by', ..."

The higher-level version of this can be thought of as "reverse engineering": Since only a few techniques are being tested, one can use keywords or other commonalities to figure out which method is correct and where to put the numbers.

Other problem types amount to translating jargon: Replace "velocity" with "derivative", "acceleration" with "second derivative", ...

- In other words, there is so little serious contact with any real-world subject that translation and reverse-engineering approaches that *avoid* engagement are routinely successful, and are fast and reliable. Students who master this skill may enjoy word problems, because the trivial math core makes success easy.
- The errors I see make more sense as translation problems than conceptual problems. A common example: When one is modelling the liquid in a container, liquid flowing *out* acquires a negative sign, because it is being *lost* from the system. Translators miss the sign, students who actually envision the situation should not.
- Some of my students despise word problems, regarding them as easily-solved math problems made hard by a smokescreen of terminology and irrelevant material. These students may be weak at this cognitive skill, or they may be thinking too much and trying to engage the subject. In any case, the most effective help I can offer is to show them how to think of it as an intelligence-free translation problem.
- Finally, many problems are so obviously contrived that they cannot be taken seriously. The one that begins "If a train leaves Chicago at 2:00..." has been the butt of jokes in comic strips.

Conventional wisdom holds that word problems engage students and provide an important connection to real-world experience. This notion is abstractly attractive, but the difficulties described above keep it from being effective in practice. Further, a curriculum justified by, or oriented toward, word problems is likely to be weak, because weak development is good enough for immediately-accessible problems.

5.3   Mathematical Applications

A common justification for word problems is that mathematics is important primarily for its applications, and math without applications is a meaningless formal game. I might agree, with the following reservations:

- Goals should include preparation for applications that will not be accessible for years, not just those that are immediately accessible.
- "Application" should be interpreted to include applications *in mathematics* as well as real-world topics.

The application of polynomial multiplication to multiplication of big integers in Sect. 4.1, and the refinements developed in Sect. 4.3 to minimise the computation required to find individual digits, are examples:

- These two topics clearly have genuine substance, and they support extended development.
- Unlike physical-world topics, they are directly accessible, because they concern mathematical structure that has already been extensively developed.
- The multiplication algorithm (Sect. 4.1) does have real-world applications, even if these are not accessible to students. In any case, it is a good example of the kind of mathematical development that has applications.
- The single-digit refinement (Sect. 4.3) is a very good illustration of a major activity in computational science: carefully exploiting structure to minimise the computation required to get a result.
- The Plan C variation (Sect. 4.3.4) provides an introduction to numerical instability and "low-probability catastrophic failure" of algorithms. This is a major issue in approximate (decimal) computation but is completely ignored in education.
- Both projects significantly deepen understanding of the underlying mathematical structure, and develop mathematical intuition.

The usual educational objection to mathematical applications is that, because they lack contact with real-world experience, they do not engage students. I believe this underestimates the willingness of students to engage with almost anything if they can succeed with it. Further, the more obviously nontrivial the material, the more pride and excitement they get from successful engagement.

Student success is the key, and the key to success is methods and templates carefully designed to minimise errors. In other words, methods informed by contemporary approaches to proof.

References

Hanna, G., & Barbeau, E. (2008). Proofs as bearers of mathematical knowledge. *Mathematics Education, 40*, 345–353.
Jaffe, A., & Quinn, F. (1993). Theoretical Mathematics: Towards a synthesis of mathematics and theoretical physics. *Bulletin of the American Mathematical Society, 29*, 1–13.

Lin, C.-C. (2009). *How can the game of hex be used to inspire students in learning mathematical reasoning?* Proceedings of ICMI Study 19 Conference, National Taiwan Normal University, Taipei, Taiwan.

Mannila, L., & Wallin, S. (2009). *Promoting students' justification skills using structured derivations.* Proceedings of ICMI Study 19 Conference, National Taiwan Normal University, Taipei, Taiwan.

Peltomaki, M., & Back, R.-J. (2009). *An empirical evaluation of structured derivations in high school mathematics.* Proceedings of ICMI Study 19 Conference, National Taiwan Normal University, Taipei, Taiwan.

Quinn, F. (2009). *The EduTE X project.* Wiki at http://www.edutex.tug.org.

Quinn, F. (2010). *Education web page.* http://www.math.vt.edu/people/quinn/education/.

Quinn, F. (2011a). *Towards a science of contemporary mathematics.* Draft February 2011, Retrieve from: http://www.math.vt.edu/people/quinn/education/.

Quinn, F. (2011b). *Contributions to a science of mathematics education,* Draft February 2011. Retrieve from: http://www.math.vt.edu/people/quinn/education/.

Rav, Y. (1999). Why do we prove theorems? *Philosophia Mathematica, 7*(3), 5–41.

Thurston, W. (1994). On proof and progress in mathematics. *Bulletin of the American Mathematical Society, 30,* 161–177.

Wanko, J. J. (2009). *talking points: experiencing deductive reasoning through puzzle discussions.* Proceedings of ICMI Study 19 Conference, National Taiwan Normal University, Taipei, Taiwan.

Part IV
Proof in the School Curriculum

Chapter 11
Proof, Proving, and Teacher-Student Interaction: Theories and Contexts

Keith Jones and Patricio Herbst

1   Introduction

This chapter takes up the challenge of theorising about proof, proving, and teacher-student interactions in mathematics classrooms across diverse contexts around the world. We aim to contribute to what Hanna and de Villiers (2008, p. 331) identify as the need to review "what theoretical frameworks ... are helpful in understanding the development of proof" and what Balacheff (2010, p. 133) argues is "the scientific challenge ...to better understand the didactical characteristics" of proof and proving. The theme of the chapter is *the role of the teacher* in teaching proof and proving in mathematics, with a particular focus on theories that illuminate *teacher-student interaction* in the context of mathematics teachers' day-to-day instructional practice.

By using phrases like 'teacher-student interaction in the mathematics classroom' and 'the teaching of proof in the context of the day-to-day instructional practice of teachers', we are deliberately choosing to avoid terms such as *pedagogy* or *didactics*. Both terms come with significant theoretical baggage and neither is unproblematic in English. As Hamilton (1999), for example, shows, some Anglo-American usage of the term *pedagogy* mirrors, in many ways, the use of term *didactics* in mainland

K. Jones (✉)
Mathematics and Science Education Research Centre, School of Education,
University of Southampton, Highfield, Southampton, UK
e-mail: d.k.jones@soton.ac.uk

P. Herbst
School of Education, University of Michigan, Ann Arbor, MI, USA
e-mail: pgherbst@umich.edu

© The Author(s) 2021
G. Hanna and M. de Villiers (eds.), *Proof and Proving in Mathematics Education*,
New ICMI Study Series, https://doi.org/10.1007/978-94-007-2129-6_11

Europe (c.f., Best 1988; Chevallard 1999a; Murphy 2008). The word *instruction*, as used by Cohen et al. (2003) to refer to the interactions amongst teacher-students-content in classroom environments, is probably a better word to designate the locus of the phenomena we target. In focusing on teacher-student interaction, we acknowledge that what learners bring to the classroom (from developmental experiences prior to schooling, to ongoing experiences across varied out-of-school contexts) impacts on such interactions, just as, most certainly, can the diversity of countries, of instructional courses, of student ages, of levels of teacher knowledge, and so on, around the world. Whatever the terminology, our over-arching focus is on *the teacher* – and, in particular, on the teacher's part in the *teacher-student interactions* that occur day-to-day in mathematics classrooms.

In theorising about proof, proving, and teacher-student interaction, we are aware that theories can appear in different guises and operate at different levels and grain sizes. As Silver and Herbst (2007) identify in their analysis, there can be "grand theories", "middle-range theories", and "local theories": where "grand theories" aim at the entirety of phenomena within, say, mathematics education; "middle-range theories" focus on subfields of study; and "local theories" apply to specific phenomena within the field. We also note Kilpatrick's (2010, p. 4) observation: "To call something a theory … is an exceedingly strong claim". It is not our intention to consider whether or not some proposed approach is, or is not, a "theory"; rather, we use the term "theory" as short-hand for 'theoretical framework', 'theoretical perspective', 'theoretical model', or other equivalent terms.

Across all these considerations, we take proof and proving to be "an activity with a social character" (Alibert and Thomas 1991, p. 216). As such, mathematics classroom communities involve students in communicating their reasoning and in building norms and representations that provide the necessary structures for mathematical proof to have a central presence. Hence, our focus on the role of the teacher in teaching proof and proving in mathematics encompasses the teacher managing the work of proving in the classroom even when proof itself is not the main object of teaching. Clearly, in such situations proofs may be requested, and offered, even when proof itself is not the object of study; such possibilities hinge on customary practices (including matters of language) that the teacher has the responsibility to establish and sustain. Balacheff (1999), Herbst and Balacheff (2009) and Sekiguchi (2006), for examples, have studied these forms of classroom practices, and the role of the teacher in establishing and sustaining the practices.

As Balacheff (2010, pp. 116–117) shows, basing classroom practices on "grand theories" such as those of Piaget or Vygotsky has not worked very well. Balacheff argues "The responsibility for all these failures does not belong to the theories which supposedly underlie the educational designs, but to naive or simplifying readers who have assumed that concepts and models from psychology can be freely transferred to education". Balacheff goes on to consider the didactical complexity of learning and teaching mathematical proof by analysing the gap between knowing mathematics and proving in mathematics. In contrast, our approach in selecting relevant theories to review is to choose ones that represent ongoing and current foci for classroom-based research and, importantly, that start from the abstraction of observations in existing school mathematics classrooms. Using these criteria, we review the *theory of socio-mathematical norms*, the *theory of teaching with variation*, and the *theory of instructional exchanges*. We conclude by giving pointers to future research – both empirical and theoretical – that we hope can advance the field.

2   The Teaching of Proof and Proving in Diverse Contexts

The contexts within which proof and proving are taught around the world vary enormously in terms of curriculum specification, student age-level, teacher knowledge, and so on. In this connection, Stigler and Hiebert (1999) have argued that the teaching of mathematics lessons in different countries follows different lesson scripts. Furthermore, Clarke et al. (2006, p. 1) report on "the extent to which students are collaborators with the teacher.... in the development and enactment of patterns of participation that reflect individual, societal and cultural priorities and associated value systems". Such research recognises the impact that diversity worldwide can have on the form of instructional courses in mathematics, on the student age-levels at which educational ideas of proof and proving are introduced, on the scale and nature of teachers' mathematical knowledge, and so on.

Hoyles (1997) uses the term *curricular shaping* to refer to the ways in which school and curriculum factors influence and shape students' views of, and competency in, proof and proving in mathematics. Knipping's (2002, 2004) research comparing classroom proof practices in France and Germany stands out as an attempt to understand the role of culture in shaping classroom proof and proving practices. Other studies include the work of Jones and colleagues on the teaching of proof in geometry at the lower secondary school level in the countries of China, Japan and the UK, some of which is summarised in Jones, Kunimune, Kumakura, Matsumoto, Fujita and Ding (2009) and Jones, Zheng and Ding (2009).

Within this diversity in the teaching of proof and proving, we can nevertheless discern some common elements. Proof in elementary school, for example, is generally viewed in terms of informal reasoning and argumentation. In middle school, students continue exploring proof as argumentation whilst at the same time being exposed to forms of symbolic notation and representation. At the high school level, proofs begin to take on a more formalised character, often (but not always) within topics in geometry – and in some places in a manner commonly called two-column proofs (e.g., Herbst 2002a; Weiss et al. 2009). For an international overview of proof and proving across the stages of education, see, for example, Ball et al. (2002).

Given such diversity, building theory that might help us understand and explain the teacher's role in the classroom teaching of mathematical proof and proving is a complex proposition. In this context, in the next section we consider three carefully selected theories of mathematics teacher-student interaction in more detail, focusing on their relevance to proof and proving.

3   Theories of Teacher-Student Interaction

3.1   Introduction

Mathematics education includes a range of theories that in one way or another concern themselves with proof and proving. As Silver and Herbst (2007) note, mathematics education theories can be classified by their 'grain size'. Some are *grand theories*; theories that attempt to organise the whole field, like Chevallard's (1999b) *théorie anthropologique du didactique* within which it would be possible to give an account of proof and the work of the teacher. Others are *local theories*; they take on specific roles articulating the relationships between problems, research, and practice. An example can be found in Martin and Harel's (1989) study of prospective elementary teachers, where the authors theorise about 'inductive verification types' and 'deductive verification types' to design an instrument they use to study participants' views of proof. Yet a third class of theory is what Merton (see Silver and Herbst 2007) termed a *middle range* theory; this starts from an empirical phenomenon, rather than with broad organising concepts, and builds up abstract concepts from the phenomenon whilst accumulating knowledge about the phenomenon through empirical research. Our three examples below – the theory of *socio-mathematical norms*, the theory of *teaching with variation*, and the theory of *instructional exchanges* – are all middle range theories.

3.2   The Theory of Socio-Mathematical Norms

The notion of *socio-mathematical norm* is a component of what Cobb and Bauersfeld (1995) term an "emergent theory" (in that, in coordinating individual and group cognitions within classroom settings, it seeks systematically to combines various "mini-theories"). The theory of *socio-mathematical norms* aims to describe and explain the construction of knowledge in inquiry-based mathematics classrooms (Cobb et al. 1992) by complementing a constructivist account of how individuals learn with a sociological account of those classrooms where teachers promote learning by inquiry. Taking the notion of *norm* (as conceptualised by Much and Schweder 1978), Cobb and his colleagues made the observation that students engage in acts of challenge and justification during the process of holding each other accountable for their assertions. The authors proposed that the notion that learners should justify their assertions constituted a social norm in the observed inquiry-based classrooms.

Voigt (1995) and Yackel and Cobb (1996) then argued that teachers, in their role as representatives of the discipline of mathematics in the classroom, could promote *socio-mathematical norms* associated with those social norms. In particular, teachers could promote normative understandings of what counts as an appropriate mathematical justification. In proposing this theory of *socio-mathematical norms*, Yackel and Cobb (1996) provided means to understand how the notion of a proof as an explanation accepted by a community at a given time could result from the interaction and negotiation amongst individuals who are both adapting their cognitive schemes in the face of perturbations and responding to the values and practices of the discipline of mathematics. Specific studies, such as that by Sekiguchi (2006) have shown how it is possible to track the development of a socio-mathematical norm for what counts as a proof in an inquiry-based classroom.

Martin et al. (2005), in their study of the interplay of teacher and student actions in the teaching and learning of geometric proof, use the notion of socio-mathematical norms to show how the teacher's instructional choices are key to the type of classroom environment that is established and, hence, to students' opportunities to hone their proof and reasoning skills. More specifically, Martin et al. (2005) argue that in order to create a classroom climate in which participating students make conjectures, provide justifications, and build chains of reasoning, the teacher should "engage in dialogue that places responsibility for reasoning on the students, analyse student arguments, and coach students as they reason" (Martin et al. 2005, p. 95). These instructional choices create a classroom environment in which teacher and students can negotiate socio-mathematical norms such as what counts as an acceptable proof.

This emergent theory with its notion of socio-mathematical norms exemplifies a middle-range theory. In an effort to characterise inquiry-based mathematics classrooms, it uses microanalysis of classroom interactions to track the development of shared norms of classroom mathematics practice. The notion of socio-mathematical norm results from abstracting the directions towards which teachers push classroom norms through social negotiation, not only of what is acceptable mathematical justification but also of other mathematical values.

3.3   The Theory of Teaching with Variation

In the 1990s, the theory of *teaching with variation* emerged from two different, though related, academic fields: the work of Gu (1992, 1994) in mathematics education in mainland China, and the work of Marton (Marton 1981; Marton and Booth 1997) on phenomenology in Sweden. The meeting of these two ideas in the form of the theory of *teaching with variation* is presented by Gu et al. (2004); see also Ko and Marton (2004, especially pp. 56–62). In this section we review the theory of *teaching with variation* and illustrate how it is beginning to be applied to studies of proof and proving in the mathematics classroom.

Teaching with variation has a long tradition in mainland China. For example, Kangshen et al. (1999), in their presentation of *Jiuzhang Suanshu* or *The Nine Chapters on the Mathematical Art* (a Chinese mathematics corpus compiled by several generations of scholars from the tenth century BCE to the first century CE) document the use of methods of varying problems dating back at least 2,000 years. In contemporary school classrooms in China, Gu (1992, 1994) conducted a large-scale study that examined how mathematics teachers varied the tasks that they used with their students. At about the same time, Marton and colleagues (Marton and Booth 1997) were focusing on the variation in ways in which people are capable of experiencing different situations or phenomena.

In the *theory of teaching with variation* (in Chinese, *bianshi*– see Bao et al. 2003a, b, c; Sun and Wong 2005), classroom teaching is seen as aiming to promote learning through the students experiencing two types of variation deemed helpful for meaningful learning of mathematics. Gu et al. (2004) classified the first form of variation as *conceptual variation*, in which the teacher highlights the key features of a mathematical concept by contrasting examples of the concept with counter- or non-examples. The teacher aims thus to provide students with multiple experiences of the selected mathematical concept from different perspectives. The second form of variation, called *procedural variation*, is the process of forming concepts not from different perspectives (as in *conceptual variation*) but through step-by-step changes. An example of *procedural variation* provided by Gu et al. (2004, pp. 320–321) concerns the concept of equation. With *procedural variation*, the teacher might begin with examples of representing an unknown by concrete items, such as when solving a problem involving the purchase of three pencils. The next step might be the use of symbols in place of the concrete items. A third step might be fully symbolic.

However, Park and Leung (2006) argue that the terms *conceptual variation* and *procedural variation* may not be the best way of capturing how contemporary mathematics teachers in China promote student learning through teaching with variation. A key reason, even according to Gu et al.'s own definition, is that *procedural variation* is also related to the formation of mathematical concepts for learners. As such, Park and Leung suggest replacing *conceptual variation* with *multi-dimensional variation* (thus capturing teaching through multiple representations) and *procedural variation* with *developmental variation* (since students learn to construct concepts through step-by-step acquisition). Sun (2011) adds that other Chinese researchers use still other terms (e.g., *explicit variation, implicit variation, form variation, solution variation,* and *content variation*).

Whatever the chosen terminology, Gu et al. (2004) helpfully provide a diagram (Fig. 11.1) to illustrate how a teacher structures a series of classroom problems through the use of variations. The variations serve as means to connect something the learners know how to solve (the known problem) to something that they are to solve (the unknown problem). Through this way of varying problems in class, "students' experience in solving problems is manifested by the richness of varying problems and the variety of transferring strategies" (Gu et al. 2004, p. 322).

To date, a number of researchers have used the *theory of teaching with variation* to analyse mathematics teaching. Some have aimed to provide an account of mathematical problem-solving in Chinese mathematics teaching (e.g., Cai and Nie 2007; Wong 2002), whilst others utilise *teaching with variation* when accounting for the classroom teaching of mathematics (e.g., Huang and Li 2009; Park and Leung 2006). Some research is beginning to use the theory of variation to research the teaching and learning of proof and proving. For example, Sun (2009) and Sun and Chan (2009) provide reports illustrating that the teaching approach of using

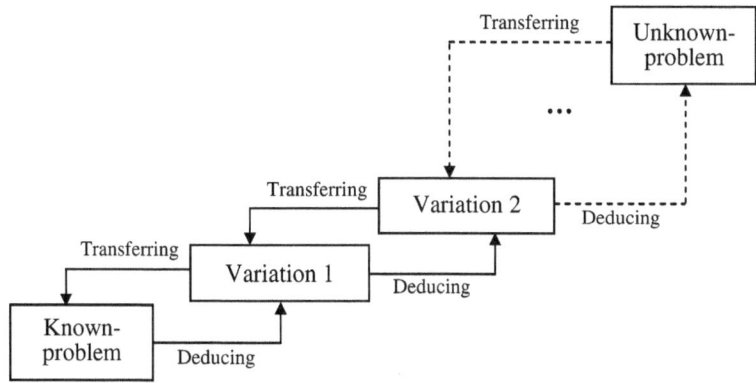

Fig. 11.1 Variation for solving problems (Source: Gu et al. 2004, p. 322)

'problem variations with multiple solutions' (where one problem has a number of solutions) successfully helped older students to reconstruct their own proof solutions by regenerating their past proving experience.

Ding and Jones (2009) and Jones, Zheng and Ding (2009) report on the instructional practices of a sample of expert teachers of geometry at Grade 8 (pupils aged 13–14) in Shanghai, China. From an analysis of the data collected, the research found that two factors characterise the instructional strategies used by the teachers to help their students to understand the discovery function of proof in geometry: the *variation of mathematical problems*, and the *variation of teaching questions*. In the variation of mathematical problems, the teachers started by guiding their students to understand the principles of a '*problem to find*' in order to begin engaging them in seeking the logical connections to the principal parts of a '*problem to prove*'. The data also provided evidence of the variation of teacher questions, in which the teacher used questions both to encourage students to formulate plausible reasons for the properties and relations of a chosen geometric figure and to increase students' awareness of the discovery function of deductive proof.

Whilst such studies provide a start, we need more empirical data on using the theory of teaching with variation. Researchers such as Mok et al. (2008) have raised the issue that whilst students might master the target mathematical ideas being taught, teaching with variation can mean that they miss opportunities for independent exploration. When the difference between one variation and another is rather small, the students have little room to think independently. Thus, teaching with variation does not *necessarily* lead to the full development of mathematical competency (c.f. Huang et al. 2006). Similarly, the type of engagement the teacher creates in the lesson may fail to foster students' higher-order thinking in terms of proof and proving. We need further research on instructional designs that use teaching with variation to develop the appropriate range of mathematical skills and approaches. One such avenue for research is on how teachers can provide for students' mathematical exploration in a way that supports proof and proving whilst not limiting the students' thinking by making the variations of problems too small.

3.4   The Theory of Instructional Exchanges

The *theory of instructional exchanges* proposed by Herbst, and based on prior work by Brousseau (1997) and by Doyle (1988) on the study of classrooms, is a descriptive theory of the role of the teacher in classroom instruction. It is another example of a "middle range theory" (Silver and Herbst 2007) in that it does not attempt to account for all practices related to mathematical thinking, learning, and teaching but rather concentrates on understanding the phenomena associated with the teaching and learning of prescribed knowledge in school classrooms as they exist (i.e. not only of 'inquiry-type' mathematics classrooms). It proposes that mathematics instruction proceeds as a sequence of exchanges or transactions between, on the one hand, the moment-to-moment, possibly interactive, work that students do with their teacher and, on the other hand, the discrete claims a teacher can lay on what has been accomplished.

Central to this theory of instructional exchanges is the notion of *didactical contract* (Brousseau 1997): the hypothesis that a bond exists that makes teacher and students mutually responsible vis-à-vis their relationships with knowledge; in particular, a contract that makes the teacher responsible for attending not only to the students as learners of mathematics but also to mathematics as the discipline that needs to be represented to be learned. Particular classrooms may have specific customary ways of negotiating and enacting that contract and these may vary quite a bit, but in general these various ways will always amount to establishing the teacher's accountability not only to the students but also to the discipline of mathematics.

A second, related hypothesis that is helpful when analysing the teacher's instructional work derives from the observation that classroom activity takes place over multiple timescales. For example, whilst meaningful classroom interactions (e.g., utterances) can be detected at a timescale of the fraction of a second, progress in the syllabus and consequent examinations take place over a larger timescale of weeks and months. Thus, the second key hypothesis is that the work of the teacher includes managing activities and objects within two different timescales: the work done moment-to-moment (at the scale of the utterance) and the mathematical objects of knowledge that exist at the larger scales of the week, month, or year-long curriculum (Lemke 2000, p. 277). In other words, the teacher needs to operate symbolic transactions or exchanges between activities in one timescale and objects of knowledge in the other: moment-to-moment activities serve the teacher to deploy or instantiate large-scale mathematical objects of knowledge; reciprocally, objects of knowledge serve to account for the moment-to-moment activities.

Herbst (2003, 2006) has proposed two basic ways in which that exchange can be facilitated. One, "negotiation of task", describes how a teacher needs to handle 'novel' tasks, ones that are completely new to the students. In these tasks the teacher needs to engage students in identifying, perhaps deciding upon, how the didactical contract applies to the task at hand. In particular, the negotiation includes identifying what aspects of the task embody the target knowledge and what aspects of students' work on that task attest that they are learning the knowledge or know it already.

The second way in which that exchange is facilitated is by 'default to an instructional situation', by framing the exchange according to norms that have framed other exchanges (possibly set up previously through negotiation). In this case, the work done is not one of identifying the mathematics in the task as much as identifying the situation, or cueing into the situation, by acting in compliance with the norms that constitute the situation. Thus, the situation frames that exchange, saving the effort of having to negotiate what needs to be done and what is at stake.

Negotiation of task, and default to a situation, are two 'ideal types' (in the Weberian sense) of teacher-student interaction about content. In practice, there would always be some amount of default and some amount of negotiation of how to handle breaches to the default situation. Nevertheless, this theorisation helps describe how regularities in interaction about content structure much of the workings of the didactical contract. More importantly, the hypothesis explains that novelty is constructed against a background of customary situations; specifically, that novel interaction is constructed by negotiating how to handle a breach in a customary situation.

Some of the tasks in which students might engage, and which (according to the second hypothesis above) the teacher needs to exchange for items of knowledge, involve mathematical moves like those identified by Lakatos (1976) as part of the method of proofs and refutations. Those operations could include deriving a logical consequence from a given statement; proposing a statement whose logical consequence is a given statement; reducing a given problem into smaller problems whose solutions logically entail that problem's solution; bringing new, warranted mathematical objects to a problem in order to translate or reduce the problem; translating strings of symbols into other, equivalent, strings of symbols; operating on one set of objects as if they behaved like other similar set of objects, and so forth. Hence, the mathematical work of proving involves a host of actions that students could perform as transient moves when working on tasks, and for which, the theory anticipates, a teacher might need to find exchange values within the elements of the target knowledge.

In the US high school curriculum, as well as in other countries, proof has traditionally appeared as an element of target knowledge in the context of the study of Euclidean geometry (González and Herbst 2006; Herbst 2002b). Teachers of geometry create work contexts in which students have the chance to experience, learn, and demonstrate knowledge of 'proof'. The notion of an instructional situation as a 'frame' (a set of norms regulating who does what and when) for the exchange between work done and knowledge transacted was initially exemplified in what Herbst and associates called the 'doing proofs' situation (Herbst and Brach 2006; Herbst et al. 2009). That work of modelling classroom interaction as a system of norms produced the observation that many of the operations in the work of proving (e.g., those listed in the previous paragraph) are not accommodated in classroom work contexts where knowledge of proof is exchanged. In other words, 'doing proofs' has become a stable work context where students can learn some of the work of proving but this, at the same time, excludes other important mathematical actions of proving, perhaps by exporting them to other instructional situations where they are disconnected from the functions of proof in the discipline of mathematics.

An important question is whether the practical rationality (Herbst and Chazan 2003) that underpins the teacher's work contains resources that could be used to give value to classroom work that embodies the different functions of proof in mathematics (which contain all the actions that constitute the work of proving). To study that rationality, Herbst and Chazan, and their associates (see http://grip.umich.edu/themat) have created classroom scenarios (complete with animated cartoon characters) depicting mathematical work that create contexts for the work of proving; the latter is sometimes explicitly executed and other times glaringly absent. The researchers have used those animations to engage groups of geometry teachers in conversations about instruction. They have found that, as a group, teachers have resources to justify positive appraisals of certain elements of the work of proving: the use of an unproven conjecture as a premise in proving a target conclusion; the identification of new mathematical concepts and their properties from objects introduced and observations made in justifying a construction; the deductive derivation of a conditional statement connecting two concomitant facts about a diagram; the prediction of an empirical fact by operating algebraically with symbols representing the quantities to be measured; the breaking up of a complicated proof problem into smaller problems (lemmas); the application of a specific proving technique (e.g., reduction to a previously proven case); and the establishment of equivalence relationships amongst a set of concomitantly true statements.

Herbst et al. (2010) have proposed that teachers might use the various functions of mathematical proof documented in the literature (e.g., verification, explanation, discovery, communication, systematisation, development of an empirical theory, and container of techniques) (de Villiers 1990; Hanna and Barbeau 2008; Hanna and Jahnke 1996) to attach contractual value to actions like those listed above. There remain two questions; whether classroom exchanges are possible (manageable) between these actions and the elements of currency; and whether the exchanges can be contained within instances of the 'doing proofs' situation or otherwise whether they require more explicit negotiations of the didactical contract. The theory of instructional exchanges thus illustrates another middle range theory that starts from abstracting from observations in mathematics classrooms where there has been no special instructional intervention (in other words, *intact* mathematics classrooms) and uses those observations to probe into how teachers manage and sustain those work contexts and also how these might be changed.

4   Directions for Future Research

The development of each of the three theories above began with abstraction from observations of mathematics classrooms. Simon (1987, p. 371) wrote that pedagogy (or didactics) is:

> the integration in practice of particular curriculum content and design, classroom strate-
> gies and techniques, and evaluation, purpose and methods. All of these aspects of educa-
> tional practice come together in the realities of what happens in classrooms. Together they
> organize a view of how a teacher's work within an institutional context specifies a particu-
> lar version of what knowledge is of most worth, what it means to know something, and
> how we might construct representations of ourselves, others and our physical and social
> environment.

This passage, famously taken up by McLaren (1998, p. 165), returns us not only to the complexity of developing theory about the role of the teacher in the teaching and learning of proof and proving, but also the diversity of contexts within which proof and proving are taught around the world – for example, in terms of curriculum specification, student age-level, teacher knowledge and so on.

Pollard (2010, p. 5) offers the representation in Fig. 11.2 (slightly amended here) as a way of capturing teacher-student interaction as a science, a craft and an art. This representation might point to a way to take into account the complexity and diversity of classroom teaching strategies when "All of these aspects of educational practice come together in the realities of what happens in classrooms" Simon (1987, p. 371).

Of the three theories reviewed here, the theory of teaching with variation (Gu et al. 2004) appears closer to "craft" (the "craft" vertex of the triangle in Fig. 11.2) than the other two, in that teaching with variation entails teacher mastery of an appropriate repertoire of classroom teaching skills and processes. In Cobb and colleagues' (Cobb and Bauersfeld 1995; Cobb et al. 1992) account of teaching in dif-
ferent mathematical traditions and its use of the idea of socio-mathematical norms to examine the work that a mathematics teacher does in an inquiry-based approach to teaching, this encompasses a responsive and creative capacity, a way in which the teacher responds both to mathematical demands and to students' cognitive demands at the same time. As such, the theory of socio-mathematical norms might fit with the "art" vertex of the triangle (Fig. 11.2). The theory of instructional exchanges, with

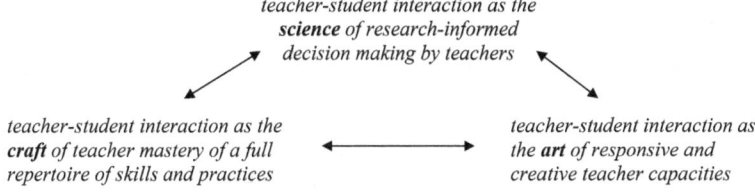

Fig. 11.2 Teaching as a science, a craft and an art (Adapted from Pollard 2010, p. 5)

its pretence of universality (to describe all kinds of teaching and to focus on concepts that are general enough to describe all observations) may come closer to the "science" vertex.

All three theories are *middle range theories*; and it remains an open question whether a *grand theory* of the teacher's role in teaching proof and proving in mathematics is a reasonable longer-term goal, especially in terms of accounting for the nature of teacher-student interaction. Pollard's (2010, p. 5) representation (Fig. 11.2) may provide some ideas towards a way of encompassing all the complexity and diversity of classroom proof teaching.

It is worth reflecting on Sfard's (2002) caution about the over-proliferation of theories. Prolific theorising may signify a "young and healthy scientific discipline" (p. 253), but, in contrast, it may mean that "theories are not being sufficiently examined, tested, refined and expanded" (op cit.). Sfard elaborates that "one of the trademarks of a mature science is that it strives for unity; that it directs its collective thought toward unifying theories and frameworks" (op cit.), at the same time noting that this is "neither a quick nor an easy process" (op cit.). As to directions for future research on proof, proving, and teacher-student interaction in the mathematics classroom, we list Sfard's challenges for the mathematics education community as ones which might inform further research work:

- To carry out research studies within frameworks determined by existing theories with the intention to establish the range of applicability or validity or usefulness of these theories.
- To carry out comparative surveys of several theories, in particular of theories that purport to provide frameworks for dealing with the same or related areas, topics and questions.
- To compare the terminologies used by different theories in order to identify cases where different terms are used for essentially the same idea or where the same term is used to designate ideas that are essentially different.
- To attempt to see the common ideas between different theories and work toward their partial unification; this might be particularly promising in cases where the theories deal with different but closely related issues or areas (op cit.).

Rising to Sfard's challenge, we suggest that one goal for further research in the field of research on the teaching and learning of mathematical proof and proving is to probe the existing theories, perhaps by focussing on what each allows us to accomplish as far as describing, explaining and reconciling novel phenomena in the mathematics classroom. The methodologies for such studies might adopt approaches reviewed by Herbst and Chazan (2009) and might look to using Pollard's (2010, p. 5) representation of teacher-student interaction as a science, a craft, and an art (Fig. 11.2).

5   Concluding Comments

At the start of this chapter, we chose to avoid using terms such as *pedagogy* or *didactics*, instead using phrases such as *teacher-student interaction in the mathematics classroom* and *the teaching of proof in the context of teachers' day-to-day instructional practice*. In the US, at least, the promotion of the term *instruction* (following Cohen et al. 2003) has had the good effect of getting people to see that the interactions that mathematics educators need to examine are ternary (teacher-student-content) rather than binary (teacher-student). In that sense, the term *instruction* has been able to achieve what *didactics* (at least in the Anglo-American world) has not. However, a lingering problem is that 'instruction' can conjure up notions of giving orders. In this sense, rather than ternary (teacher-student-content), or even binary (teacher-student), 'instruction' might evoke the idea of the teacher unilaterally issuing orders.

In another starting point to this chapter, we recognised the diversity of countries worldwide and the impact that this has on forms of instructional courses, on the student age-level at which educational ideas are introduced, on teacher knowledge, and so on. Our prior comments about terms like *pedagogy*, *didactics* and *instruction* reminded us of the influence of language on the ways in which people express themselves. Linguists predominantly think that the fundaments of language are somehow encoded in human genes and are, as such, the same across the human species. From such a perspective, all languages share the same *Universal Grammar*, the same underlying concepts, the same degree of systemic complexity, and so on. The resulting conclusion is that the influence of one's mother tongue on the way one thinks is negligible or trivial. Yet recent work (e.g., Deutscher 2010) is challenging this conclusion, arguing that cultural differences are reflected in language in profound ways, and that emerging evidence indicates that mother-tongue can affect how individuals in different cultural settings think and perceive the world (concurring with the longstanding view of some anthropologists of language).

How such cultural influences might impact on collective work towards a theory of the role of the teacher in teaching proof and proving in mathematics (possible *grander* than the *middle range* theories covered in this chapter), especially in terms of teacher-student interaction, remains to be seen. As Stylianou et al. (2009, pp. 5–6) point out, to date there have not been enough research studies "focused on the teaching of proof in the context of teachers' day-to-day instructional practice". More is currently known about the *learning* of proof (e.g., Harel and Sowder 1998, 2007); the *teaching* of proof warrants equally close attention (e.g., Harel and Rabin 2010a, b). Our review of a carefully selected trio of theoretical frameworks is offered as support for further theorising about teaching proof and proving in mathematics classrooms worldwide.

References*

Alibert, D., & Thomas, M. O. J. (1991). Research on mathematical proof. In D. O. Tall (Ed.), *Advanced mathematical thinking* (pp. 215–230). Dordrecht: Kluwer.

Balacheff, N. (1999). Contract and custom: Two registers of didactical interactions. *The Mathematics Educator, 9*, 23–29.

Balacheff, N. (2010). Bridging knowing and proving in mathematics: An essay from a didactical perspective. In G. Hanna, H. N. Jahnke, & H. B. Pulte (Eds.), *Explanation and proof in mathematics: Philosophical and educational perspectives* (pp. 115–136). Berlin: Springer.

Ball, D. L., Hoyles, C., Jahnke, H. N., & Movshovitz-Hadar, N. (2002). The teaching of proof. In L. I. Tatsien (Ed.), *Proceedings of the International Congress of Mathematicians* (Vol. III: Invited Lectures, pp. 907–920). Beijing: Higher Education Press.

Bao, J., Huang, R., Yi, L., & Gu, L. (2003a). Study in bianshi teaching I. *Mathematics Teaching (Shuxue Jiaoxue), 1*, 11–12 [in Chinese].

Bao, J., Huang, R., Yi, L., & Gu, L. (2003b). Study in bianshi teaching II. *Mathematics Teaching (Shuxue Jiaoxue), 2*, 6–10 [in Chinese].

Bao, J., Huang, R., Yi, L., & Gu, L. (2003c). Study in bianshi teaching III. *Mathematics Teaching (Shuxue Jiaoxue), 3*, 6–12 [in Chinese].

Best, F. (1988). The metamorphoses of the term 'pedagogy'. *Prospects, 18*, 157–166.

Brousseau, G. (1997). *Theory of didactical situations in mathematics: Didactique des Mathématiques 1970–1990* (N. Balacheff, M. Cooper, R. Sutherland, & V. Warfield, Eds. & Trans.). Dordrecht: Kluwer.

Cai, J., & Nie, B. (2007). Problem solving in Chinese mathematics education: Research and practice. *ZDM-The International Journal on Mathematics Education, 30*, 459–473.

Chevallard, Y. (1999a). Didactique? You must be joking! A critical comment on terminology. *Instructional Science, 27*(1–2), 5–7.

Chevallard, Y. (1999b). L'analyse des pratiques enseignantes en théorie anthropologique du didactique. *Recherches en Didactique des Mathématiques, 19*(2), 221–266.

Clarke, D. J., Emanuelsson, J., Jablonka, E., & Mok, I. A. C. (2006). The learner's perspective study and international comparisons of classroom practice. In D. J. Clarke, J. Emanuelsson, E. Jablonka, & I. A. C. Mok (Eds.), *Making connections: Comparing mathematics classrooms around the world* (pp. 1–22). Rotterdam: Sense.

Cobb, P., & Bauersfeld, H. (Eds.). (1995). *The emergence of mathematical meaning: Interaction in classroom cultures*. Hillsdale: Erlbaum.

Cobb, P., Wood, T., Yackel, E., & McNeal, B. (1992). Characteristics of classroom mathematics traditions: An interactional analysis. *American Educational Research Journal, 29*, 573–604.

Cohen, D. K., Raudenbush, S. W., & Ball, D. L. (2003). Resources, instruction, and research. *Educational Evaluation and Policy Analysis, 25*, 119–142.

de Villiers, M. D. (1990). The role and function of proof in mathematics. *Pythagoras, 24*, 17–24.

Deutscher, G. (2010). *Through the language glass: How words colour your world*. London: Heinemann.

Ding, L., & Jones, K. (2009). Instructional strategies in explicating the discovery function of proof for lower secondary school students (Vol. 1, pp. 136–141).*

Doyle, W. (1988). Work in mathematics classes: The context of students' thinking during instruction. *Educational Psychologist, 23*, 167–180.

González, G., & Herbst, P. (2006). Competing arguments for the geometry course: Why were American high school students supposed to study geometry in the twentieth century? *International Journal for the History of Mathematics Education, 1*(1), 7–33.

Gu, L. (1992). *The Qingpu experience.* Paper presented at the 7th International Congress of Mathematical Education, Quebec.

Gu, L. (1994). 青浦实验的方法与教学原理研究 [*Qingpu shiyan de fangfa yu jiaoxue yuanli yanjiu*] [*Theory of teaching experiment – the methodology and teaching principle of Qinpu*]. Beijing: Educational Science Press [in Chinese].

Gu, L., Huang, R., & Marton, F. (2004). Teaching with variation: An effective way of mathematics teaching in China. In L. Fan, N. Y. Wong, J. Cai, & S. Li (Eds.), *How Chinese learn mathematics: Perspectives from insiders* (pp. 309–345). Singapore: World Scientific.

Hamilton, D. (1999). The pedagogic paradox (or why no didactics in England?). *Pedagogy, Culture and Society, 7*(1), 135–152.

Hanna, G., & Barbeau, E. (2008). Proofs as bearers of mathematical knowledge. *ZDM: The International Journal on Mathematics Education, 40*(3), 345–353.

Hanna, G., & de Villiers, M. (2008). ICMI study 19: Proof and proving in mathematics education. *ZDM-The International Journal of Mathematics Education, 40*(2), 329–336.

Hanna, G., & Jahnke, H. N. (1996). Proof and proving. In A. Bishop, K. Clements, C. Keitel, J. Kilpatrick, & C. Laborde (Eds.), *International handbook of mathematics education* (pp. 877–908). Dordrecht: Kluwer.

Harel, G., & Rabin, J. M. (2010a). Teaching practices associated with the authoritative proof scheme. *Journal for Research in Mathematics Education, 41*, 14–19.

Harel, G., & Rabin, J. M. (2010b). Teaching practices that can promote the authoritative proof scheme. *Canadian Journal of Science, Mathematics, and Technology Education, 10*(2), 139–159.

Harel, G., & Sowder, L. (1998). Students' proof schemes: Results from exploratory studies. In A. H. Schoenfeld, J. Kaput, & E. Dubinsky (Eds.), *Research in collegiate mathematics education* (Vol. 3, pp. 234–283). Providence: American Mathematical Society.

Harel, G., & Sowder, L. (2007). Toward comprehensive perspectives on the learning and teaching of proof. In F. Lester (Ed.), *Second handbook of research on mathematics teaching and learning* (pp. 805–842). Greenwich: Information Age Publishing.

Herbst, P. G. (2002a). Establishing a custom of proving in American school geometry: Evolution of the two-column proof in the early twentieth century. *Educational Studies in Mathematics, 49*, 283–312.

Herbst, P. (2002b). Engaging students in proving: A double bind on the teacher. *Journal for Research in Mathematics Education, 33*, 176–203.

Herbst, P. (2003). Using novel tasks to teach mathematics: Three tensions affecting the work of the teacher. *American Educational Research Journal, 40*, 197–238.

Herbst, P. (2006). Teaching geometry with problems: Negotiating instructional situations and mathematical tasks. *Journal for Research in Mathematics Education, 37*, 313–347.

Herbst, P., & Balacheff, N. (2009). Proving and knowing in public: What counts as proof in a classroom. In M. Blanton, D. Stylianou, & E. Knuth (Eds.), *Teaching and learning proof across the grades: K-16 perspective* (pp. 40–63). New York: Routledge.

Herbst, P., & Brach, C. (2006). Proving and 'doing proofs' in high school geometry classes: What is 'it' that is going on for students and how do they make sense of it? *Cognition and Instruction, 24*, 73–122.

Herbst, P., & Chazan, D. (2003). Exploring the practical rationality of mathematics teaching through conversations about videotaped episodes: The case of engaging students in proving. *For the Learning of Mathematics, 23*(1), 2–14.

Herbst, P., & Chazan, D. (2009). Methodologies for the study of instruction in mathematics classrooms. *Recherches en Didactique des Mathématiques, 29*(1), 11–33.

Herbst, P. G., Chen, C., Weiss, M., Gonzales, G., Nachieli, T., Hamlin, M., & Brach, C. (2009). "Doing proofs" in geometry classrooms. In D. A. Stylianou, M. L. Blanton, & E. J. Knuth

(Eds.), *Teaching and learning proof across the grades: K-16 perspective* (pp. 250–268). New York: Routledge.

Herbst, P., Miyakawa, T., & Chazan, D. (2010). *Revisiting the functions of proof in mathematics classrooms: A view from a theory of instructional exchanges.* Deep Blue at the University of Michigan. http://hdl.handle.net/2027.42/78168

Hoyles, C. (1997). The curricular shaping of students' approaches to proof. *For the Learning of Mathematics, 17*(1), 7–16.

Huang, R., & Li, Y. (2009). Pursuing excellence in mathematics classroom instruction through exemplary lesson development in China: A case study. *ZDM-The International Journal on Mathematics Education, 41*(3), 297–309.

Huang, R., Mok, I. A. C., & Leung, F. K. S. (2006). Repetition or variation: Practising in the mathematics classroom in China. In D. J. Clarke, C. Keitel, & Y. Shimizu (Eds.), *Mathematics classrooms in twelve countries: The insider's perspective* (pp. 263–274). Rotterdam: Sense.

Jones, K., Kunimune, S., Kumakura, H., Matsumoto, S., Fujita, T., & Ding, L. (2009). Developing pedagogic approaches for proof: Learning from teaching in the East and West (Vol. 1, pp. 232–237).*

Jones, K., Zheng, Y., & Ding, L. (2009). *Developing pedagogic theory: The case of geometry proof teaching.* Invited plenary paper presented at the 3rd International Symposium on the History and Pedagogy of Mathematics, Beijing, May 2009. Available online at: http://eprints.soton.ac.uk/173033/

Kangshen, S., Crossley, J. N., & Lun, A. W.-C. (1999). *The nine chapters on the mathematical art: Companion and commentary.* Beijing: Science Press.

Kilpatrick, J. (2010). Preface to part 1. In B. Sriraman & L. English (Eds.), *Theories of mathematics education: Seeking new frontiers* (pp. 3–5). New York: Springer.

Knipping, C. (2002). Proof and proving processes: Teaching geometry in France and Germany. In H.-G. Weigand (Ed.), *Developments in mathematics education in German-speaking countries: Selected papers from the annual conference on didactics of mathematics (Bern 1999)* (pp. 44–54). Hildesheim: Franzbecker Verlag.

Knipping, C. (2004). Argumentations in proving discourses in mathematics classrooms. In G. Törner et al. (Eds.), *Developments in mathematics education in German-speaking countries: Selected papers from the annual conference on didactics of mathematics (Ludwigsburg, March 5–9, 2001)* (pp. 73–84). Hildesheim: Franzbecker Verlag.

Ko, P.-Y., & Marton, F. (2004). Variation and the secret of the virtuoso. In F. Marton & A. Tsui (Eds.), *Classroom discourse and the space of learning* (pp. 43–62). Mahwah: Lawrence Erlbaum.

Lakatos, I. (1976). *Proofs and refutations: The logic of mathematical discovery.* (J. Worrall & E. Zahar, Eds.). Cambridge: Cambridge University Press.

Lemke, J. (2000). Across the scales of time: Artifacts, activities, and meanings in eco-social systems. *Mind, Culture, and Activity, 7*, 273–290.

Martin, W. G., & Harel, G. (1989). Proof frames of pre-service elementary teachers. *Journal for Research in Mathematics Education, 20*(1), 41–51.

Martin, T. S., McCrone, S. M. S., Bower, M. L. W., & Dindyal, J. (2005). The interplay of teacher and student actions in the teaching and learning of geometric proof. *Educational Studies in Mathematics, 60*, 95–124.

Marton, F. (1981). Phenomenography: Describing conceptions of the world around us. *Instructional Science, 10*, 177–200.

Marton, F., & Booth, S. (1997). *Learning and awareness.* Mahwah: Lawrence Erlbaum Associates.

McLaren, P. (1998). *Life in Schools: An introduction to critical pedagogy in the foundations of education* (3rd ed.). New York: Longman.

Mok, I., Cai, J., & Fong Fung, A. (2008). Missing learning opportunities in classroom instruction: Evidence from an analysis of a well-structured lesson on comparing fractions. *The Mathematics Educator, 11*(1–2), 111–126.

Much, N., & Schweder, R. (1978). Speaking of rules: The analysis of culture in breach. *New Directions for Child Development, 2,* 19–39.

Murphy, P. (2008). Defining pedagogy. In K. Hall, P. Murphy, & J. Soler (Eds.), *Pedagogy and practice: Culture and identities* (pp. 28–39). London: Sage.

Park, K., & Leung, F. K. S. (2006). Mathematics lessons in Korea: Teaching with systematic variation. In D. J. Clarke, C. Keitel, & Y. Shimizu (Eds.), *Mathematics classrooms in twelve countries: The insiders' perspective* (pp. 247–262). Rotterdam: Sense.

Pollard, A. (2010). *Professionalism and pedagogy: A contemporary opportunity (a commentary by TLRP and GTCe).* London: TLRP.

Sekiguchi, Y. (2006). Mathematical norms in Japanese mathematics classrooms. In D. Clarke, C. Keitel, & Y. Shimizu (Eds.), *Mathematics classrooms in twelve countries: The insiders perspective* (pp. 289–306). Rotterdam: Sense.

Sfard, A. (2002). Reflections on educational studies in mathematics. *Educational Studies in Mathematics, 50*(3), 252–253.

Silver, E. A., & Herbst, P. G. (2007). Theory in mathematics education scholarship. In F. Lester (Ed.), *Second handbook of research on mathematics teaching and learning* (pp. 39–67). Charlotte: Information Age Publishing.

Simon, R. I. (1987). Empowerment as a pedagogy of possibility. *Language Arts, 64*(4), 370–382.

Stigler, J. W., & Hiebert, J. (1999). *The teaching gap: Best ideas from the world's teachers for improving education in the classroom.* New York: Free Press.

Stylianou, D. A., Blanton, M. L., & Knuth, E. J. (2009). Introduction. In D. A. Stylianou, M. L. Blanton, & E. J. Knuth (Eds.), *Teaching and learning proof across the grades: K-16 perspective* (pp. 1–12). New York: Routledge.

Sun, X. (2009). An experiment for the enhancement of a trapezoid area formula proof constructions of student teachers by 'one problem multiple solutions' (Vol. 2, pp. 178–183).*

Sun, X. (2011). "Variation problems" and their roles in the topic of fraction division in Chinese mathematics textbook examples. *Educational Studies in Mathematics, 76*(1), 65–85.

Sun, X. & Chan, K. H. (2009). Regenerate the proving experience: an attempt for improvement original theorem proof construction of student teachers by using spiral variation curriculum (Vol. 2, pp. 172–177).*

Sun, X. H. & Wong, N.-Y. (2005). *The origin of Bianshi problems: A cultural background perspective on the Chinese mathematics teaching practice.* Paper presented at EARCOME-3: ICMI Regional Conference: The Third East Asia Regional Conference on Mathematics Education, Shanghai.

Voigt, J. (1995). Thematic patterns of interaction and sociomathematical norms. In P. Cobb & H. Bauersfeld (Eds.), *The emergence of mathematical meaning: Interaction in classroom cultures* (pp. 163–201). Hillsdale: Lawrence Erlbaum Associates.

Weiss, M., Herbst, P., & Chen, C. (2009). Teachers' perspectives on "authentic mathematics" and the two-column proof form. *Educational Studies in Mathematics, 70*(3), 275–293.

Wong, N.-Y. (2002). Conceptions of doing and learning mathematics among Chinese. *Journal of Intercultural Studies, 23*(2), 211–229.

Yackel, E., & Cobb, P. (1996). Socio-mathematical norms, argumentation, and autonomy in mathematics. *Journal for Research in Mathematics Education, 27*(4), 458–477.

Chapter 12
From Exploration to Proof Production

Feng-Jui Hsieh, Wang-Shian Horng, and Haw-Yaw Shy

The authors of this chapter, collectively, have significant experience as teacher educators, more than 20 years each on average, and have also authored national mathematics textbooks which emphasised integrating exploration in the teaching of proof (Hsieh 1997). Drawing on this experience, and in light of the data presented below, we attempt in this chapter to introduce how exploration, especially hands-on exploration, is integrated into the teaching of proof in Taiwan. In the first section, we briefly discuss the role of exploration with different media in the teaching of proof in Taiwanese schools. We present two extracts from a Taiwanese textbook which demonstrate the integration of hands-on exploration in proving. In the second and third sections below, we describe our position with regard to exploration in the context of proving and why it is valuable in this context. In the fourth section, we propose and discuss a conceptual model for the relationship of exploration, problem solving, proving and proof, which will be illustrated through the use of two exploratory teaching experiments. In the concluding section, we briefly compare the use of dynamic geometry software (DGS) and hands-on exploration, summarise some of the positive and negative issues raised by integrating exploration, and suggest areas worthy of future research.

F.-J. Hsieh (✉) • W.-S. Horng
Department of Mathematics, National Taiwan Normal University, Taipei, Taiwan
e-mail: hsiehfj@math.ntnu.edu.tw; horng@math.ntnu.edu.tw

H.-Y. Shy
Department and Graduate Institute of Mathematics,
National Changhua University of Education, Changhua, Taiwan
e-mail: shy@cc.ncue.edu.tw

© The Author(s) 2021   279
G. Hanna and M. de Villiers (eds.), *Proof and Proving in Mathematics Education*,
New ICMI Study Series, https://doi.org/10.1007/978-94-007-2129-6_12

1 Background

It is impossible to list with certainty all the factors necessary for the successful construction of a mathematics proof; however, they certainly include cognition of the necessary theories to be used, the ability to employ sequencing steps according to accepted logical rules, and the competence to use accepted mathematics registers to express the steps to be conveyed. Developing these abilities poses a range of difficulties for students which mathematics educators have attempted to overcome.

One such attempt has been the use of dynamic geometric software (DGS), such as *Geometer's Sketchpad* (Goldenberg and Cuoco 1998) and *Cabri-géomètrie* (Jones 2000). However, this method is not always convenient and viable, because of its time-intensive nature and the requirement in many school systems that students must be able to construct a proof manually (without the aid of electronic devices) for school examinations, as in Taiwan's demanding mathematics curriculum.

A recent national random-sample survey of more than two thousands each of junior and senior high school students in Taiwan reveals a relative disapproval of the use of computers in teaching mathematics. When asked what teaching media an ideal mathematics teacher should use, both senior and junior high school students regarded the use of computer aids such as DGS or PowerPoint to be less ideal media (approximately a 55% approval rating) than concrete teaching-aids such as paper cards or models (approximately a 90% approval rating; Hsieh 2010). Further, the vast majority of students thought that an ideal mathematics teacher should engage students in exploration (approximately 80%) as well as hands-on activities (approximately 81%) in order for them to appreciate mathematics (Hsieh et al. 2008).

The students' preferences grow out of, and are reflected in, the incorporation of hands-on exploration tasks, especially those related to proving, in many standard Taiwanese mathematics textbooks.

1.1 Taiwanese Textbook Examples

Figure 12.1 is an image from the most commonly used mathematics textbook in Taiwan (Hung et al. 2009). It is a modified replica of an example used previously in the national textbook. The property to prove is "the segment connecting the two midpoints of the two sides of a trapezoid is parallel to the upper and lower bases and is equal to a half of the sum of the bases." The publisher provides paper cards required to perform the actions in the exploration in the back of the textbook. The text poses several questions to guide the exploration: "(1) are D, C, and Q on the same line? Are P, N, and Q on the same line? Why?", "(2) are the quadrilateral DAPQ and DMNQ parallelograms?", and "(3) what are the relationships of DQ, MN, and AP?" This exploration precedes the proof of the property.

Figure 12.2 is another image from the same textbook (Hung et al. 2009). Hsieh (1994) first introduced this example when demonstrating how to use paper-folding

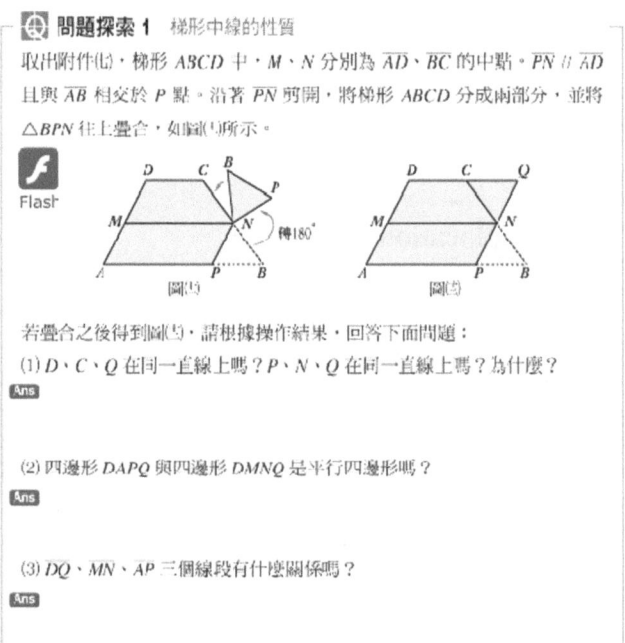

Fig. 12.1 Trapezoid exploration activity in Taiwanese mathematics textbook (Copyright 2009 by Kang Hsuan Educational Publishing. Adopted with permission of the author)

因為 $\overline{AB} > \overline{AC}$，我們把 $\overline{AC}$ 摺疊到 $\overline{AB}$ 上，$\overline{AD}$ 是摺痕，且摺疊後 C 點的位置一定會落在 A、B 之間，假設 C 點落在 E 點上，如圖(五)。然後將紙攤平，並把相關的點和線畫出來，如圖(六)。

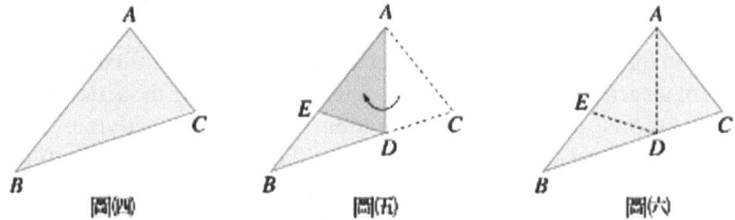

Fig. 12.2 Exploration about the size of angles in a triangle in Taiwanese mathematics textbook (Copyright 2009 by Kang Hsuan Educational Publishing. Adopted with permission of the author)

in helping students to construct auxiliary lines. Hsieh raised an operational principle that "one can compare an object with another object of the assumptions of a proving problem through paper-folding in order to construct auxiliaries". She demonstrated that this principle could also be applied to the objects in the results. For example, in Fig. 12.2, the assumption is "AB > AC" and the result to be claimed is "∠C > ∠B".

Seeing that AB is greater than AC, one can try to compare AB with AC by folding AC to AB; the auxiliary lines then become obvious.

These excerpts exemplify the extent to which hands-on exploration is entrenched in the teaching of mathematics, including proof, in Taiwan.

2 Position on Exploration

The importance of integrating exploration in proving has been the topic of considerable academic discussion. Various scholars consider exploration in general (Arzarello et al. 2009), and hands-on exploration in particular, a valuable component in the learning or construction of proof (MacPherson 1985; Semadeni 1980). Some studies have proposed including exploration via the use of dynamic computer environments (De Villiers 2004; Gonza´lez and Herbst 2009; Larios-Osorio and Acuña-Soto 2009; Mariotti 2000; Yerushalmy and Chazan 1990). However, advocates of "exploration" have often not explicitly defined it. Ponte (2007) attempted to clarify the term, claiming "explorations" have an open nature. From her perspective, even "the problems are not completely formulated beforehand and the student has a role to play in defining the mathematical question to pursue" (p. 421). Notwithstanding, the term "exploration" remains open to multiple interpretations.

In our opinion, two different positions of interpreting "exploration" provide an operational heuristic for considering the relationship between exploration and proving. The first position views exploration as a mental process, the second as an activity that involves manipulation of and interaction with external environments (e.g., hands-on or DGS tools). The former position sees exploration as a necessary component in proving due to its problem-solving nature, which requires the use of observation, connection, inquiring and reasoning. The latter position encourages mental exploration reinforced through the interaction with external objects.

Piaget and Inhelder (1967) claimed that some students have difficulty in imagining the result of an action on objects without actually performing the action themselves. For such students, undertaking exploration via hands-on activities can aid in actually *seeing* the image after its construction. The use of concrete objects increases the possibility of students performing actions due to the invitation provided by these materials' "perceived affordance" (will be discussed in a later section; Norman 1988, 1999). The actions may start either randomly or start in order and progress towards the students' goal; either process helps to make the transformation of objects such as figures more accessible.

In addition, exploration, if used in proving, should be devoted not only to discovering mathematical properties but also to the discovery or mental construction of logical steps required in proof. Because manipulating objects also requires mental exploration, the combination of mental and hands-on exploration encourages both intuitive and analytical thinking which, in turn, promotes the discovery or mental construction of logic steps.

This paper attempts to merge the two by first employing the mental process and then embedding additional hands-on exploration as a means to enhance it. The term "hands-on exploration" refers to organised activity that adopts the use of concrete objects to build an interactive environment, thereby inviting manipulation and encouraging discovery. In what follows, the terms "exploration" and "hands-on exploration" refer to the same concept unless otherwise specified.

3   Exploration's Value in Proving

Here, we focus on analysing the characteristics of exploration and the various notions of proving, as well as the teaching and learning of proving for the classroom. The discussion includes not only the extrinsic notions of exploration but also some intrinsic features, such as discovery and "perceived affordance". We then use the results of our analysis as evidence supporting the value of exploration in the context of proving.

3.1   *Exploration and Discovery*

Discovery is an essential element in exploration. Bruner (1960) argues that mastery of the fundamental ideas of a field, including mathematics, involves not only grasping general principles but also developing an approach to guesses and hunches, as well as realising the possibility of solving problems on one's own. To instil this attitude, Bruner proposes the including an important ingredient in curricula, namely, creating "excitement about discovery–discovery of regularities of previously unrecognized relations and similarities between ideas" (p. 20). According to Bruner, what students learn through discovery will be both useful and meaningful in their thinking, features considered as extremely significant in the learning of mathematics proof. Exploration, thus predicated on the notion of discovery, comprises both meaningful and usable elements in the context of proving.

3.2   *Characteristics of Exploration*

In addition to a sense of discovery, the notion of exploration has to be analysed to clearly demonstrate why exploration is valuable in the process of proving (Chazan 1990; Lakatos 1976; Pólya 1981). Here, we concentrate on hands-on exploration. However, the issue of the characteristics of hands-on exploration remains an open question; various articles (e.g., Boakes 2009; Rosen and Hoffman 2009) have mentioned these, either explicitly or implicitly. From the literature and our experiences and observations in mathematics or mathematics education, the concepts that we especially noticed are interaction, dynamics, tangibility, manipulation, immediate feedback, intuition and divergent thinking. To sum up, hands-on exploration provides an opportunity for an individual to (a) construct mathematics objects, (b) transform figures, (c) probe in multiple directions, (d) perceive divergent visual information, and (e) receive immediate feedback on one's actions. These five factors contribute to the success of constructing a proof.

3.3   Perceived Affordance and Hands-on Exploration

The above aspects of exploration may take effect when a mental exploration is implemented with the medium of only paper-and-pencil. Therefore, one may ask "Why do we need to supply a hands-on environment with concrete objects?" The reason can be clearly explained using Norman's (Norman 1999) idea of "perceived affordance". Gibson (1977, 1979) had originally defined "affordance" as all "action possibilities available" to an individual in the environment. Norman subsequently revisited the concept and divided it into two: actual affordance and perceived affordance. The latter refers to those fundamental properties of an object that determine action possibilities readily perceivable by an individual. That is, the action possibilities are dependent not only on an object's actual affordance, but also on one's visual perception of the object based on one's goals, plans, and learned conventions about the object's logical and cultural constraints. The concept of perceived affordance explains why a hands-on environment enhances exploration. For example, take two sheets of paper, one featuring a graph accompanied by text describing what proof is desired and a space to respond and the other featuring only a figure with no description about what to do. The perceived affordances of these two sheets are different. The first sheet's perceived affordance stimulates students to perform a paper-and-pencil task, whilst the second's invites students to perform various motive actions, such as to move, rotate, flip, fold, cut and draw. If they employ some of these latter actions, the visual feedback they receive encourages them to subsequently react to their artefacts; this process would be less possible if the students were constructing proof by employing only mental exploration.

3.4   Concept of Proving

Without an operational concept of proving or proof, any discussion as to why exploration is valuable in the process of proving will be ambiguous. What constitutes an acceptable mathematical proof has been the subject of considerable academic debate; however, a consensus is yet to be reached (Arzarello 2007), because mathematicians and researchers cannot ignore the various turning points in the evolution of the concept of proof (Kleiner 1991). Two concepts of proof have received considerable attention from mathematics educators: the cognitive aspect, which relates to the conceptual understanding or semantic notion of proof, and, the social aspect, which relates to the syntactic notion (forms and languages) used to express it (Hanna 2000). We regard the conceptual or semantic notion as an essential element of proof and the syntactic aspect as an indispensable element of proof with variant requirements of its forms.

The Chinese language expresses both "proving" and "proof", related but distinct terms in English, by a single term, "zhèng míng" (證明): The first word, "zhèng" (證), literally means "to prove" or "to demonstrate", and the second word, "míng" (明), means "clear" or "to understand". In other words, "proving", as expressed within the Chinese lexicon, is to demonstrate as well as to make things clear and understandable. This idea represents a focus on both the conceptual and semantic notions of proving, which characterises the dominant approach to proving in Taiwan's system of mathematics education.

When asked "What is proof?", a student at one of Taipei's elite high schools referred to a term attributed to Confucius in the *Analects of Confucius*. The term – also pronounced "zhèng míng" (正名) – literally translates as "to justify the name [title];" that is, ensuring that one's social role and rank are correctly labelled, thus lending credibility to one's statements. The student claimed that "zhèng míng" (proving, 證明) is "zhèng míng" (justifying the name, 正名).

Although, by citing Confucius, the student's response diverts focus away from mathematics, it is surprisingly accurate in delineating one aspect of mathematical proof. Justification is one of the most important elements of proof, as justification validates what is being claimed and thus allows it to be used to solve follow-up problems. The student's analogy not only applies to mathematics but also highlights an intrinsic human desire to justify one's claims and statements. This point is important from a pedagogical point of view. When teaching how to construct a proof, justifying "why a property is supposed to be proved in such a way" is as significant as justifying the statement itself.

3.5   Reason for Embedding Exploration in Proving in the Classroom

Exploration is important for learners in the process of proving. One reason may be found in mathematics history: mathematicians have devoted their energies to exploration, making mathematical discoveries and then justifying them. Exploration activates their intuitions and encourages their thinking. The processes of discovery and justification are essential work that one can view as the *real* task in mathematics. Such processes, with necessary learning adaptations, are within the range of what most students can do and should not be exclusive to mathematicians (Ponte 2007); students dealing with mathematics also deserve a chance to experience this real mathematics task. Learners, like expert mathematicians, can apply their intuitions to seek patterns – following hunches, testing ideas and formulating generalisations that may become conjectures (Lakatos 1976; Pólya 1981).

A review of the relevant literature (Fischbein 1987; Hanna 2000; Harel and Sowder 1998; Lakatos 1976; Pólya 1981) and our own experiences and observations working in the field make it clear that constructing mathematics proofs via exploration may additionally benefit learners because the process itself has positive psychological effects, such as enhancing motivation, self-confidence and cognition. Here, we concentrate on notions that relate especially to cognition.

3.5.1 Exploration Reveals Information Necessary to Prove

Educators have introduced certain methods – such as working backwards or forwards in a graphic proof representation – to enhance students' approach to proving (e.g. Croy 2000). However, one of the major problems students encounter in proving is not which direction to proceed, but rather where and how to start. One reason may lie in the notion of learner ability; in this case, the learner lacks the ability for formalised perception of mathematical materials (Krutetskii 1976). Such perception is crucial for understanding information presented by a proving problem and for initiating a solution.

In many cases, information required to solve a proving problem is either hidden or not immediately obvious to students, for example, when starting or continuing a proving exercise requires the construction of auxiliary lines. In this case, exploration, with its interactive and manipulative nature, allows students to probe in multiple directions and transform figures effortlessly. These actions, in turn, reveal additional information that prompts students to think about how to start the process of proving. Furthermore, when students perform actions with objects, such as folding a piece of paper, they receive immediate feedback on those actions and can then react to the new figures accordingly. This crucial process allows students to continue with the remaining steps of proving.

3.5.2 Exploration Facilitates the Understanding of Proving

When teaching proof, educators aim to help students gradually develop an understanding of proof consistent with the approach practised in contemporary mathematics (Harel and Sowder 2007). In addition, Hanna (2000) points out that the best proof of the theorem is one that not only allows students to see "that it is true, but also why it is true" (p. 8). In other words, it is important for students to know not only a particular proof but also why it validates the statement. The process of exploration may facilitate the acquisition of this dual knowledge because it tends to activate the use of intuition. As Bruner (1960) noted, "Intuition implies the act of grasping the meaning, significance, or structure of a problem or situation without explicit reliance on the analytic apparatus of one's craft." When they activate their intuitions in proving, students regard their constructions of certain steps or structures as self-evident and as something of which they have an inherent understanding.

Therefore, by encouraging the use of intuition, exploration contributes to the understanding of the process of proving.

Furthermore, the interactive and manipulative environment used in exploration invites students to actively work with available materials. These actions allow students to have a multitude of incidental experiences through observing such phenomena as the correlations of mathematical objects, the derivative properties of objects, and the relationship between their artefacts and the proving problem in question. Incidental experiences, according to Ausubel (1963), may reshape students' cognitive structures and increase their readiness to learn; that is, "the adequacy of existing capacity in relation to the demands of a given learning task" (p. 31). Both readiness and the cognitive structure are always relevant and crucial variables for meaningful learning (ibid.). The incidental experiences acquired during exploration for the purpose of constructing proof embody Ausubel's theory. With an appropriate design, steps of proving may be closely related to, or even generated from, those incidental experiences, giving rise to meaningful learning.

3.5.3   Exploration Encourages the Generation of Conjectures

Pedemonte (2007), after reviewing some experimental studies, concluded teaching proof primarily by presenting proofs to students without asking them to construct conjectures and argumentations makes proof less "accessible" to students. Exploration is characteristic of encouraging students to probe in multiple directions and perceive large amounts of divergent information prompts students to come up with guesses and provides them with many examples. These effects encourage inductive reasoning which, in turn, gives rise to the generation of conjectures. This benefit of integrating exploration into learning tasks has been addressed repeatedly in the literature (Mariotti 2000).

3.5.4   Exploration Supports Justification for the Process of Proving

Many mathematical educators agree that justification is the very essence of mathematical proof (Pedemonte 2007). Taking this position, Harel's and Sowder's (1998) conducted a study aimed "to help students refine their own conception of what constitutes justification in mathematics" through the developing of their understanding of proof, proof production, and proof appreciation. Harel and Sowder's concept of justification resembles the concept presented here; namely, mathematical justification is based on intuition, internal conviction and necessity.

When presented with a proving task, students try to initiate a solution on their own and then complete the task in a way they believe will be acceptable to their teacher. This process involves using mathematical information, causal relations of mathematical information, and mathematical theorems to support their justifications.

As mentioned above, students profit from exploration by discovering just such mathematical information and causal relations; they then connect these to mathematical theorems in justification.

Furthermore, the process of justification requires making arguments. Exploration, by providing an interactive environment that encourages students to manipulate concrete objects (or computer graphics) and probe in multiple directions, creates a setting for students to adapt their mental structures to proving. Students' actions, artefacts and discoveries provide a point of reference for this adaption of their cognitive structures; this, in turn, strengthens the foundation of students' argumentation, because it is based on their own experiences and constructions. Moreover, reshaping their cognitive structures provides students with chances to construct more structured arguments.

4   The EP-Spectrum

4.1   *Introduction to the EP-Spectrum*

The spectrum we are going to propose centres on the concept of justification. We regard the process of proving in the classroom as a process of justifying a conjecture or a property, and proof as the production of the process of justification, requiring general statements structured with legitimate logical steps.

We propose this spectrum principally for pedagogical purposes, though we have also considered historical, cognitive and social aspects of proof. A proof for the classroom can be viewed as the product of a spectrum of activities starting with exploration, and progressing to the stages of conjecturing, informal explanation, and justification. All these activities closely relate problem-solving in a wide sense and seek to generate a proof at the end (see Fig. 12.3).[1] We hereafter refer to this spectrum as the *EP-spectrum* (*E* for "exploration" and *P* for "proof"). The stages in the spectrum are not mutually exclusive; students can simultaneously experience or work at more than one stage. In addition, not all proving activities require going through all the stages in sequence; for example, a quick solution may only contain the stages of exploration and proof production. However, in designing proving activities, each stage in the EP-spectrum has its own value in terms of aiding students' learning and constructing proof. Therefore, when deciding whether to include or omit any stage of the EP-spectrum in a proving activity, one must take into account the activity's substantive content, the intended participants' mathematics level, and the activity's pedagogical goals.

---

[1] This spectrum was initially developed by the working group WG5. It was then revised by the authors.

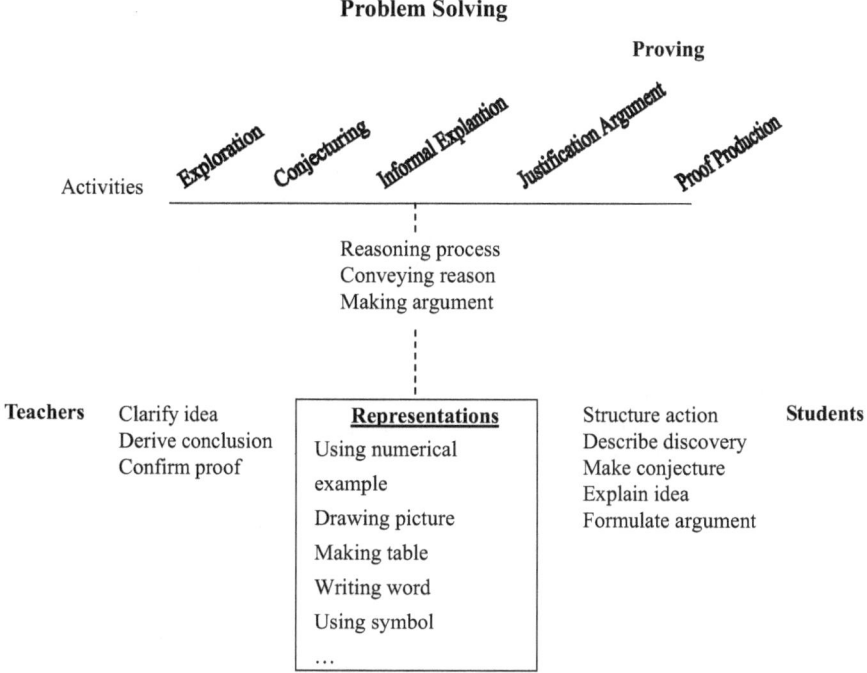

Fig. 12.3 Constructs of EP-spectrum

The process of solving a problem that requires the production of proof is referred to as the process of proving. It necessitates a variety of mental explorations to discover the relationship between mathematical objects, to make connections between objects/relations and concepts/properties, to switch between different reasoning modes, and to organise deductive inferences with general statements. Accordingly, drawing on personal experiences and observations as well as the results of teaching experiments, we propose three factors as crucial to the success of the process of proving in the EP-spectrum: creation, inference and connection. Each factor corresponds to a certain capacity within each individual.

The creation factor relates to an individual's intuition and divergent thinking in adaptation to new environments (Csikszentmihalyi 1996; Poincaré 1956). A creative action may also be initially unconscious and become conscious later on. The connection factor refers to the cognitive domain, where students have to link the content they have internalised – such as concepts, procedures, problem-solving schemes, and languages in some particular registers – in order to cope with problem situations. The inference factor refers to the capacity of reasoning and making judgements in accordance with available information. It is not restricted to inference using valid logic rules that imply certainty; in addition, it includes inferences applied in daily life such as analogical reasoning (Juthe 2005) or inference to the most plausible explanation. These three factors are the *primary* factors of the EP-spectrum. When students engage in tasks at any stage of the EP-spectrum, their capacities in the three primary factors are combined through an *additive model*.[2] Each combination produces a variety of powers for students to deal with the incidences they face.

Figure 12.3 shows how the main constructs of the EP-spectrum relate to each other in action. First, representation plays an important role in all kinds of activities and processes in the EP-spectrum. Representations such as providing numerical examples, drawing pictures, making tables, writing words, and using symbols are potential ways to mentally organise and physically express the actions of reasoning, conveying reasons, and making arguments. Teachers use these representations to clarify ideas, derive conclusions, and confirm proofs, whilst students use them to structure actions, describe discoveries, make conjectures, explain ideas, and formulate arguments.

In order to demonstrate the EP-spectrum in a practical setting, we first introduce two teaching experiments, which demonstrate the integrating of exploration in proof.

4.2   Teaching Experiment Applying the EP-Spectrum

Two teaching experiments applying two processes of the EP-spectrum activities were conducted with 31 ninth-graders in a regular class in Taiwan. On average, the students ranked in the 38th percentile of a national pool of ninth-graders in terms of overall academic competence[3]; approximately 60% of the participants ranked below the 50th percentile. In other words, the class's academic level was below average for Taiwan. At the time the experiments were conducted, the students in question had not yet been formally introduced to the concept of proof, which would be taught two months later. Nevertheless, they had seen their teacher solving problems with rigorous solution steps (in fact, proofs) and had been asked to do the same thing throughout their learning of geometric concepts, such as congruent triangles.

---

[2] The term *additive model* is borrowed from colour theory. In contrast to a *subtractive model*, which employs an assumption of a *minimum whole* requirement to be successful in the process of proving, there is no requirement in an additive model for a fixed minimum whole. We have chosen the additive model since its combinations – produced by the intertwining of varying capacities in each of the three primary factors – have potentially unlimited results, many of which work well in the process of proving. Conversely, according to the subtractive model, lacking a certain capacity in a primary factor will inhibit an individual from being able to successfully complete the proving process, due to the minimum whole requirement.

[3] The students' academic competence was measured by their performance in the simulated national senior high school entrance examinations for ninth-graders. Often, the results of these examinations can accurately predict students' performance in actual national entrance examinations.

Fig. 12.4 Graph of
Tangent-segment-property

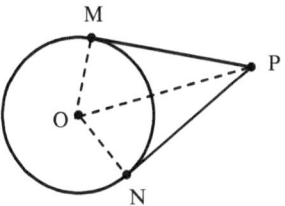

4.2.1   EP-Spectrum Tasks

Two properties were to be introduced to the students. In cooperation with the authors, the students' usual mathematics teacher conducted two sequences of activities with the class, applying the EP-spectrum in each. We designed these activities to provide students with a chance to actively prove the properties instead of just passively listening to the teacher.

4.2.2   Property to be Proved

The first property we asked students to prove was the *Tangent-line-property*, which states:

Let A be a point on circle O. If line L intersects the radius OA at A at a right angle, then L is a tangent line of O.

The second property we asked students to prove was the *Tangent-segment-property*, which states:

Let P be a point outside circle O. If PM and PN are two tangent lines of O which intersect O at M and N, respectively, then PM = PN.

4.2.3   Desired proof

The teacher's desired proofs are as follows:

Tangent-line-property
   $\because$ L $\perp$ OA
   $\therefore$ OA is the distance of point O to L
   <u>Let B be an arbitrary point on L other than A</u>
   <u>Then OB > OA</u> (OA is the distance)
   $\therefore$ B is outside of circle O (OA is the radius)
   $\therefore$ L intersect O at only A
   $\therefore$ L is a tangent line of O

Tangent-segment-property
   Connecting OM, ON, and OP (see Fig. 12.4)
   In $\triangle$POM and $\triangle$PON,
   ...

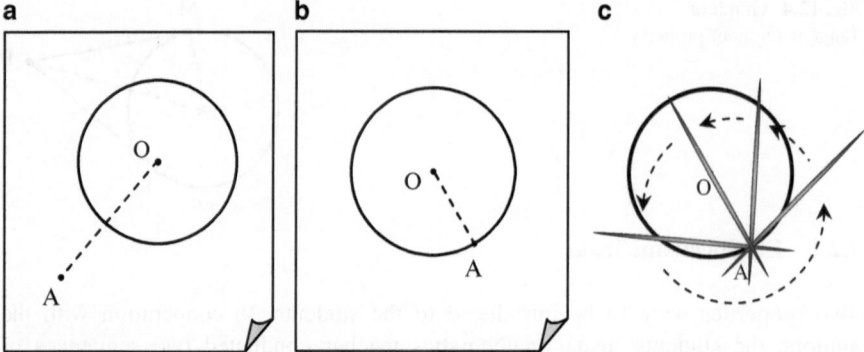

Fig. 12.5 (**a**) Paper for exploration (1). (**b**) Paper for exploration (2). (**c**) Way to use toothpick

Fig. 12.6 Paper for exploration (3)

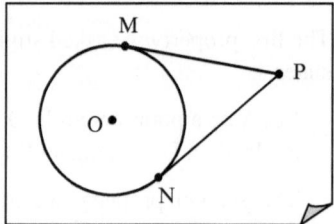

4.2.4 Exploration Activity

The *Tangent-line-property* activity required students to use a toothpick to represent line L to manipulate on paper including Fig. 12.5a, b respectively in sequence. Open questions were posed to students and they were instructed to put the toothpick on A and rotate it to observe the instances (see Fig. 12.5c).

The *Tangent-segment-property* activity asked students to work on Fig. 12.6 by paper-folding. Open questions were again posed to students. The desired creases are the dashed lines in Fig. 12.4 if fold correctly.

4.3 EP-Spectrum Activities

The results of classroom experiments showed the students' progress through the series of "Activities" in the EP-spectrum (see Fig. 12.3).

4.3.1   Exploration

In our *Tangent-line-property* experiment, a toothpick was used to represent a line for students to manipulate on the top of the sheets of paper with Fig. 12.5a, b. Open questions such as "What location of L makes OA the distance of O to L?" were posed to students. The use of this toothpick, of which the perceived affordance encourages concrete actions such as moves and rotations, transforms the original figures to multiple new figures for students to discover and examine. If students were only given the paper (with Fig. 12.5a, b) but without using a toothpick to represent a line, the perceived affordance of this object (i.e., the paper with the figures) does not encourage students to draw multiple lines (such that each line is a rotation of the line drawn in previous step) in reaction to the figures.

In the exploration with respect to Fig. 12.5a, the students were asked, inter alia: "If B is any point on L other than A (when L is the line such that OA is the distance from O to L), then compare the distance between B and O to the length of OA". Experienced teachers have noted that this exploration, which focuses on building the logic scheme that is required in a logic step (the step underlined in the *Tangent-line-property* proof above), is difficult for students to undertake. However, the results of our experiment saw 55% of students employing this logic step in their arguments or proofs, with 76% of this group executing it correctly and in completion. Students with their level of academic competence would not have easily achieved these percentages without using an exploration aid.

In the *Tangent-segment-property* experiment, a sheet of paper with a picture (see Fig. 12.6) was given to each student. The perceived affordance of this sheet with the picture invites students to perform various actions, such as fold, flip, and draw. These actions add new creases or segments to the paper, which transforms the original figure; and the visual feedback students received encourages them to subsequently react to their artefacts to create new creases or segments, and which, in turn, produces a sequence of transformations. This process would not be effortless if students only employ mental exploration.

Our experiment asked students to (a) use paper-folding to confirm if M and N are really tangent points (see Fig. 12.6), and (b) compare PM and PN to check if they have the same length. We found that 58% of students successfully constructed all the auxiliary lines (dashed segments in Fig. 12.4) in their justification arguments or proofs to the *Tangent-segment-property* and, amongst this group, 93% completed them correctly. Both research and experience have shown that many students do not know how to start a proof when auxiliary objects are required in that proof (Ding and Jones 2006; Matsuda and Okamoto 1998). However, this proof had not previously been taught to the students and, further, requires auxiliary lines. The results of the experiment appear to demonstrate that both the manipulative feature of exploration, supported by a hands-on environment, and the procedural feature of exploration, supported by a chain of procedures, enhance students' willingness to solve a proving problem.

4.3.2   Conjecturing

Two completely different approaches may guide the direction of exploration. The first, an abducted situation, provides many opportunities for producing incomplete and erroneous conjectures, as in Douek's (2009) study. The second, an adducted situation, provides an opportunity for producing *one-shot* errorless conjecture. Taiwanese textbooks and classroom instruction usually use the adductive approach because it saves time and works well in terms of heuristics.

In our *Tangent-line-property* case, we utilised the adducted situation. In the exploration of how to determine when L (represented by the toothpick) turned out to be a tangent line, 23 (74%) of the students used the definition of tangents and secants as their method of determination. This high percentage shows that in such cases students can easily imitate what their teachers have often done; that is, apply definitions to make judgements. The expected conjecture, an innovative one, relates to the size of the included angle of OA and L. Only two students noticed the included angle of OA and L. After being cued by the teacher with the question "Is it possible to determine using the included angle", 35% of the students were still unable to write down a conjecture by themselves based on the included angle. These phenomena demonstrate that to make an innovative conjecture is much more difficult than to make an imitative one.

4.3.3   Informal Explanation

The informal explanation stage, rather than that of justification argument, conveys what teachers usually ask students to do and what students are usually aware of and intend to do in the classroom. Its less rigorous character is an important factor; in formal mathematics it has led to the development of irrational numbers and infinity, which have proved invaluable in solving physical problems and modelling physical phenomena (Harel and Sowder 1998; Kleiner 1991).

During the *Tangent-segment-property* experiment, when asked to explain why the two tangent segments were equivalent, some students came up with explanations about their actions using words that had personal relevance and meaning, rather than the mathematical terms or symbols expected by the teacher. We interpret this observation to mean that students informally explaining why they believe a certain conjecture is true are attempting to explain to others what they have discovered in their explorations (with or without certainty); namely, they begin the process of persuading. This process is connected to (a) the retrieval and recollection of what they have done with the external materials; (b) the search for appropriate vocabulary to represent the visual objects and their relations; (c) the quest for a suitable logic rule that chains the causal relations of the objects or operations in their discovery; (d) the elimination from their answers of redundant actions performed in the exploration; and (e) the attempt to make and organise sentences. These five issues demonstrate the significance of this stage in helping students' transition from exploration to argumentation. Students have to switch their attention from manipulative actions to linguistic/visual constructions through presenting explanations. Factors in the social domain, such as teachers' interventions, also play an important role in the success of helping students to gradually transit from the preliminary explanation to argumentation.

At this informal explanation stage, teachers must impose some control on the classroom discourses in order to get effective results in both cognitive and affective domains. These include controls on the scope and depth of, and time allocated to: (a) the substantive content to be proved; (b) the ideas and thoughts relating to the exploration; (c) the usage of ordinary language and symbolic notations relating to the representation; and (d) the adoption and transition of different reasoning types. Though in the cognitive domain, these four issues also relate to the affective domain, as the controls relate to students' motivation to participate in classroom discourses.[4] For example, when a student's short-term goal is to make sure whether a particular procedure or answer (own or peers') works, a lengthy discussion about different strategies will often weaken the student's motivation to actively participate in the discussion. Our observations of classroom discourses in Taiwan have revealed that teachers often either focus heavily on all the above issues and devote considerable time to discussing them with students or, in contrast, fail to address any of these issues at all.

4.3.4   Justification Argument

The use of logical steps distinguishes the stage of informal explanation from the stage of justification argument. The latter stage helps students to structure their explanations (when persuading others) by utilising mathematically accepted statements. Based on our experiments and experiences, we found that the features of justification arguments differ from those of proof in at least the following ways [5]:

1. It is not necessary to be "rigorous" when undertaking justification arguments;
2. Justification arguments may have many redundant components;
3. The inclusion of an inferring step in justification arguments does not need to be directly from the preceding step of inference or other axioms;
4. A mixed use of results obtained from inferring steps and from exploration steps often appears in justification arguments; and
5. Inferences made in the justification argument stage may be context- or content-bound (for example, relating to specific pictures, objects, or actions, rather than a general statement).

During our experiments, it was common for some students to use formally unproved data they found in exploration as evidence to warrant their justification

---

[4] Although the affective domain is not our focus, it is relevant to the effectiveness of this stage in the EP-spectrum.

[5] Some literature or studies do not distinguish argumentation from proof.

arguments; that is, they used abductive reasoning. For example, in the *Tangent-segment-property* case, some students constructed OP by connecting points O and P, although, at this moment, OP was only a segment connecting two points, students used "OP as the internal bisector of ∠MPN", because, to make inferences in their proofs, they needed this information, which they perceived as correct on the basis of their paper-folding. Though this phenomenon can be seen as a defect in the students' approach, it may also be viewed in a positive sense, insofar as a partly completed of a proof is better than no proof at all. Further, by merely changing their source of data to a proved one, students will be able to correctly complete the proof deductively.

At the justification argument stage, students often use verbal statements to express their intuitive ideas or definitions of mathematical concepts. Historically, mathematicians have initially done the same with concepts they invented, such as Cauchy's definition of limit. This verbal form, not the symbolic form, carries and conveys intuitive ideas. But, we emphasise that it may also be appropriate for students to use symbolic representations during the justification argument stage if they are comfortable doing so.

4.3.5 Proof Production

It has been recommended by some researchers that less emphasis be placed on demanding rigour in students' writing of proof (Usiskin 1980). However, the historical development of mathematics has illustrated the effectiveness of symbolic notation as a method of discovery (Kleiner 1991). The use of symbols in proof also appears to be commonplace in mathematics instruction at the high-school level. In our opinion, the requirement for providing deductive reasons marks another difference between justification arguments and proof, suggesting that students are actually providing deductive reasoning instead of putting together major components of proofs (Hsieh et al. 2009). Furthermore, these high standards in the classification of proof are born of a belief that students possess a higher level of mathematics competence if they can justify a property with a clear explanation as to why each step is valid through the use of symbolised notations.

Some research claims that some high school students, even at end of their geometry classes, cannot construct a simple triangle congruence proof (Usiskin 1987). However, our *Tangent-segment-property* experiment turned out differently. With the help of paper-folding exploration for the construction of the auxiliary lines, 54% of students wrote out formal proofs, despite not being advanced students nor having been previously taught the concept of proof formally. This result, coupled with the fact that the students' written proof often closely resembled the symbolic form used by their teacher in previous lessons, provides evidence that:

1. Teachers can instil in students a familiarity with writing mathematical proof by using it in everyday teaching;
2. Sometimes grasping the idea of *how* to prove is much harder for students than the mere writing out of a proof;

3. If students have internalised a proof scheme, (e.g., "to prove the congruence of segments is to prove the congruence of triangles" as used in the *Tangent-segment-property* experiment) they will be able to apply it easily once they discover a part of it (e.g., "to prove the congruence of segments") is required to complete a proof; and

4. Students, like their teacher, prefer to use symbols when writing proofs (if possible), as these are shorter and faster to use than words.

4.4   Summary of EP-Spectrum

4.4.1   Three Primary Factors

Which intertwining of the three primary factors – creation, inference and connection – best contributes to the success of each stage of the EP-spectrum remains an open question. In our opinion, the creation factor is crucial in exploration, as this stage encourages multidirectional inquiries and multiple types of reasoning that strongly relate to divergent thinking and intuition. The reliance on the creation factor decreases, however, in the stages of the EP spectrum closer to proof production.

One might think that, as the stages advance from exploration to proof production, the inference factor becomes progressively more important. However, this is not the case. In practice, the relative success of exploration depends heavily on the diversity of inferences. Inference by analogy, inference to the best explanation, and abductive as well as inductive inferences, all play an extremely important role at this stage. As the stages progress towards proof production, a demand for deductive inference arises.

The connection factor weighs equally on all stages, but may have different connotations. At the exploration stage, connection of mathematical concepts and that of daily life experiences or physical phenomena play equally significant roles. At the stage of informal explanation, connection to mathematical "words" and communication schemes becomes valuable, whilst the stage of justification argument encourages connection to proof schemes. The final production of proof requires a further connection to the register used in proof and to the legitimate way of sequencing and writing a proof.

4.4.2   Justification

Most of the time, instructors introduce proof in a "passive" mode, without students actively participating in its construction. Even if the semantic production is carefully introduced, students may not be convinced of the validity of the meanings and rules introduced, which have been previously invented by others, let alone use them to justify. We employ and name two kinds of justification in the EP-spectrum:

(a) self-justification, namely that one justifies something to oneself with evidence one believes true; and (b) social-justification, namely that one justifies something to others with evidence one believes true and acceptable as true by others. Different EP-spectrum stages highlight the two types of justification differently. At the exploration stage, students direct their actions and construct their own meanings by using self-justification. When they start to explain their ideas informally to others, students reciprocally associate their explanations with their own actions; this requires the operation of both self-justification and social-justification. The evidence used in the self-justification has to be filtered and adapted in order to pass the scrutiny of others. That is, the informal explanation stage provides an opportunity to refine self-justification by merging it with social-justification. This process renders proving accessible to students, because the self-justification constitutes "real" justification for them and the social-justification helps them to construct other people accepted justification. Students learn contemporary, accepted forms of proof through social-justification at both the justification argument and proof production stages.

4.4.3   Representation

Representation plays an important role in mathematics learning and problem solving. We cannot here discuss in detail the functions of representation in the context of integrating exploration in proof, but will examine two important issues using terms and concepts developed by Goldin and Kaput (1996). First, these authors distinguish "inert" versus "interactive" media that embody representation systems. In an inert medium, "the only state-change resulting from a user's input is the display of that input, as when one writes on a piece of paper" (p. 412). In contrast, an interactive medium enables "the addition of something new to the result of a user's actions, something to which the user may then respond" (p. 412). Goldin and Kaput define a representation system as: "an action representation if it contains rules or mechanism for the manipulation of its elements, and a display representation if it does not" (p. 413). Second, they distinguish "imagistic or analogic" systems from the formal systems of representation. They interpret imagistic representation as "'imagined', visualized, represented kinesthetically and/or auditorily" (pp. 414–415). The imagistic representation may include what Bruner (1960) described as "enactive" and "pictorial" representations.

These concepts of representation are applicable to the context of integrating exploration into proving. At the exploration stage in the EP-spectrum, the interactive environment, which allows performing all manner of translations, flips, rotations, cutting and so forth, invites students to construct their internal representations in the form of an action representation. Godlin and Kaput (1996) consider this form of representation leads to powerful internal representations, which students can act on or transform mentally. That is, the manipulative and interactive environment of exploration produces internal representation systems of which mental transformations, often required in problem-solving, are possible. In contrast, an environment using an inert medium, such as paper-and-pencil, to embody representations restricts itself to the development of display representations, which are relatively static and less flexible when used in problem-solving. Moreover, the manipulative feature of exploration generally involves external representations, such as concrete objects and actions with these objects, which include multiple pictorial images and, probably, constructs of the objects and actions. These kinds of external representations, in turn, encourage students to construct imagistic internal representations, which Godlin and Kaput consider "essential to virtually all mathematical insight and understanding" (p. 417).

Equally important, the external representations, the "virtual" ones, appear physically in the classroom. In terms of Kaput's (1991) idea of "representational acts", students performing all activities in the EP-spectrum use notations and languages, including natural languages, "to organize the creation and elaboration of their own mental structures" (p. 56). Prior mental activities conducted in the exploration stage are re-structured both physically and mentally to form physical records (e.g., notations) and actions on these records (e.g., sequencing the notations).

Another point also relates to the issue of representation in the EP-spectrum. Teachers often switch between different forms of representations when they are conducting different activities. For instance, when introducing ideas or clarifying tasks in the exploration stage, they will tend to use a kinesthetically imagistic, action representation; however, as process progresses to the informal explanation stage, they tend to switch to an auditorily imagistic representation; and in the justification argument stage, they often switch to a formal representation to derive conclusions. This switching of forms is often due to the teachers having different instructional goals at different stages. However, neither the teachers nor the students necessarily always notice this transformation of representations.

5   Conclusion

5.1   Computer Assisted Exploration Versus Hands-on Exploration

This chapter focuses on hands-on exploration. Many other studies have instead concentrated on dynamic computer-assisted exploration, due to its unique characteristics not inherent in hands-on exploration: the accuracy of the drawing (Hanna 2000), the effortless construction of a figure, and the possibility of producing unlimited number of figures through dragging (De Villiers 1997).

The consequently emerging view amongst researchers and educators that computer experimental activities are an equally valid form of mathematical confirmation is a milestone in the mathematics community (Hanna 2007). True, dynamic computer software is superior in assisting in the discovery of existing (or sometimes not yet known) properties and making conjectures. However, its ability to deal with the logical sequences required in rigorous proof remains problematic.

Computer exploration also poses problems in practical teaching. In particular, the feature of specification (i.e., easily adapting to specific learning content as well as a learner's individual ability) is not built into the DGS microworld. To resolve this problem, teachers may design a sequence of activities prior to instruction periods by creating a specification program within the computer software system. However, it requires considerable effort (and is sometimes impossible) for teachers to prepare in advance the sequences in a computer environment. In contrast, hands-on activity is superior with respect to specification, but lacks access to resources such as the powerful supporting system embedded in computer software.

Besides, the affordances of "paper figures or physical objects" and "computer screen figures or objects" are quite different. The latter affords pointing, touching, and clicking on computer devices rather than directly on the figures or objects, providing the computer does not feature a touch screen. Even with a touch screen, the motions possible in a three-dimensional space, such as a flipping and folding, can only be simulated rather than physical or actual interactions. However, the exploration using physical materials and actions such as suggested by our research, may significantly increase student learning along the EP-spectrum.

In sum, the issues of when and in what conditions each mode should be used, and whether each mode should be used separately or in tandem with the other, remain open questions requiring further research.

5.2 Further Problems in Exploration

In the exploration stage, some students tend to make justifications based on specific examples or figures resulting from their own actions rather than on general cases. This may reflect what the students have experienced in mathematics instruction; often, instructors introduce a concept or justify a property through a similar approach. This teaching phenomenon, emerging in aspects of Taiwanese mathematics learning, seems to emerge also elsewhere: for example, the United States and Canada, as evidenced by the composition of their mathematics textbooks. The question of how to deal with this problem is a topic worthy of future research.

Another phenomenon also relates to the issue of students' proof and mathematics instruction. For example, in our experiments, many Taiwanese students were able to produce meaningful symbolic proofs. That Taiwanese teachers use symbols regularly in teaching suggests a hypothesis: If teachers use a large amount of symbols, accompanied by thorough explanation of their meanings, in the everyday teaching of mathematics, students may acquire the ability to use these symbols correctly. Furthermore, when asked why so many students were able to use symbols in argumentation and proof, Taiwanese teachers noted that students prefer using symbols if possible, as they believe the method is short and quick. Evidence of these trends currently derives only from observations; further research is necessary to enable conclusive claims.

Our experiment showed that carefully developed exploration can help students in constructing not only conjectures but also proof schemes, as in Pedemonte's (2007) concept of "process pattern generalisation". However, few empirical studies have focused on the factors that should be embedded in an exploration activity in order to advance students' production of proofs. In addition to the factors influencing the spectrum from exploration to proof, many other issues are worthy of investigation: For example, how to overcome some students' habit of providing only partly complete answers to proving problems, using the limited information they have discovered during exploration, rather persisting with completing the proof. A related point also deserves further research: how to deal with students who, encountering an exploration problem, perform a range of actions without thinking they need to justify these actions. Furthermore, the specification feature of hands-on exploration provides the possibility of designing less-error or one-shot tasks, amongst the many alternatives such exploration offers. When or whether an environment should provide only one expedient course or a variety of alternatives also requires a great deal of research.

Acknowledgement   We are grateful to Gila Hanna, John Holt, Roza Leikin, Elena Nardi, and Sarah-Jane Patterson for reviewing and editing our paper. We wish to thank Man Keung Siu, Chia-Jui Hsieh, Ting-Ying Wang, Shia-Jei Tang and Guoheng Chao for providing ideas on the EP-spectrum.

References*

Arzarello, F. (2007). The proof in the 20th century: From Hilbert to automatic theorem proving introduction. In P. Boero (Ed.), *Theorem in schools: From history, epistemology and cognition to classroom practice* (pp. 43–63). Rotterdam: Sense Publishers.

Arzarello, F., Paola, D., & Sabena, C. (2009). Proving in early calculus (Vol. 1, pp. 35–40).*

Ausubel, D. P. (1963). *The psychology of meaningful verbal learning.* New York: Grune & Stratton.

Boakes, N. J. (2009). Origami instruction in the middle school mathematics classroom: Its impact on spatial visualization and geometry knowledge of students. *Research in Middle Level Education Online, 32*(7), 1–12.

Bruner, J. S. (1960). *The process of education.* Cambridge: Harvard University Press.

Chazan, D. (1990). Quasi-empirical views of mathematics and mathematics teaching. *Interchange, 21*(1), 14–23.

Croy, M. J. (2000). Problem solving, working backwards, and graphic proof representation. *Teaching Philosophy, 23*(2), 169–187.

Csikszentmihalyi, M. (1996). *Creativity: Flow and the psychology of discovery and invention.* New York: Harper Collins.

*NB: References marked with * are in F. L. Lin, F. J. Hsieh, G. Hanna, & M. de Villiers (Eds.) (2009). *ICMI Study 19: Proof and proving in mathematics education.* Taipei, Taiwan: The Department of Mathematics, National Taiwan Normal University.

De Villiers, M. (1997). The role of proof in investigative, computer-based geometry: Some personal reflections. In D. Schattschneider & J. King (Eds.), *Geometry turned on!* (pp. 15–24). Washington, DC: MAA.

De Villiers, M. (2004). Using dynamic geometry to expand mathematics teachers' understanding of proof. *International Journal of Mathematical Education in Science and Technology, 35*(5), 703–724.

Ding, L., & Jones, K. (2006). Teaching geometry in lower secondary school in Shanghai, china. *Proceedings of the British Society for Research into Learning Mathematics, 26*(1), 41–46.

Douek, N. (2009). Approaching proof in school: From guided conjecturing and proving to a story of proof construction (Vol. 1, pp. 142–148).*

Fischbein, E. (1987). *Intuition in science and mathematics: An educational approach.* Dordrecht: Reider.

Gibson, J. J. (1977). The theory of affordances. In R. E. Shaw & J. Bransford (Eds.), *Perceiving, acting, and knowing: Toward an ecological psychology.* Hillsdale: Lawrence Erlbaum.

Gibson, J. J. (1979). *The ecological approach to visual perception.* Boston: Houghton Mifflin.

Goldenberg, E. P., & Cuoco, A. (1998). What is dynamic geometry? In R. Lehrer & D. Chazan (Eds.), *Designing learning environments for developing understanding of geometry and space.* Hillsdale: Lawrence Erlbaum.

Goldin, G. A., & Kaput, J. J. (1996). A joint perspective on the idea of representation in learning and doing mathematics. In L. Steffe et al. (Eds.), *Theories of mathematical learning* (pp. 397–430). Mahwah: Lawrence Erlbaum.

Gonza´lez, G., & Herbst, P. G. (2009). Students' conceptions of congruency through the use of dynamic geometry software. *International Journal of Computers for Mathematical Learning, 14*(2), 153–182.

Hanna, G. (2000). Proof, explanation and exploration: An overview. *Educational Studies in Mathematics, 44,* 5–25.

Hanna, G. (2007). The ongoing value of proof. In P. Boero (Ed.), *Theorems in schools: From history, epistemology and cognition to classroom practice* (pp. 3–16). Rotterdam: Sense Publishers.

Harel, G., & Sowder, L. (1998). Students' proof schemes: Results from exploratory studies. In A. H. Schoenfeld, J. Kaput, & E. Dubinsky (Eds.), *Research in collegiate mathematics education III* (pp. 234–283). Providence: American Mathematical Society.

Harel, G., & Sowder, L. (2007). Toward comprehensive perspectives on the learning and teaching of proof. In F. Lester (Ed.), *Second handbook of research on mathematics teaching and learning* (pp. 435–458). Charlotte: Information Age.

Hsieh, F.-J. (1994). 使幾何教學活潑化—摺紙及剪紙篇 [Making the teaching of geometry lively - paper folding and cutting]. *Science Education Monthly, 171,* 29–41.

Hsieh, F.-J. (1997). 國中數學新課程精神與特色 [The spirits and the characteristics of new lower secondary mathematical curriculum]. *Science Education Monthly, 197,* 45–55.

Hsieh, F.-J. (2010). Indicators of teaching ability for ideal mathematics teachers (theoretical framework). In F.-J. Hsieh & J. Chung (Eds.), *Learning from students—sailing for elaborated teaching* (pp. 9–18). Taipei: National Taipei University of Education.

Hsieh, F.-J., Tang, S.-J., Song, Y.-R., & Wang, T.-Y. (2008). 國中埋想數學教師類型探討 [*The types of ideal lower secondary mathematics teachers*]. Paper presented at the meeting of the 24th Science Education Conference, ChungHua, Taiwan.

Hsieh, F.-J., Lee, F.-T., & Wang, T.-Y. (2009). How much proofs are students able to learn in mathematics class from their teachers (Vol. 1, pp. 208–213).*

Hung, Y.-C. et al. (2009) 國民中學第四冊數學 [The fourth volume of lower secondary mathematics textbook]. Taipei: Kang Hsuan Educational Publishing.

Jones, K. (2000). Providing a foundation for deductive reasoning: Students' interpretations when using dynamic geometry software and their evolving mathematical explanations. *Educational Studies in Mathematics, 44,* 55–85.

Juthe, A. (2005). Argument by analogy. *Argumentation, 19*(1), 1–27.

Kaput, J. J. (1991). Notations and representations as mediators of constructive processes. In E. von Glasersfeld (Ed.), *Radical constructivism in mathematics education* (pp. 53–74). Dordrecht: Kluwer.

Kleiner, I. (1991). Rigor and proof in mathematics: A historical perspective. *Mathematics Magazine, 64*(5), 291–314.

Krutetskii, V.A. (1976). *The psychology of mathematical abilities in schoolchildren* (J. Teller, Trans.). Chicago: The University of Chicago Press. (Original work published 1968)

Lakatos, I. (1976). *Proofs and refutations.* Cambridge: Cambridge University Press.

Larios-Osorio, V., & Acuña-Soto, C. (2009). Geometrical proof in the institutional classroom environment (Vol. 2, pp. 59–63).*

MacPherson, E. D. (1985). The themes of geometry: Design of the nonformal geometry curriculum. In C. Hirsch & M. Zweng (Eds.), *The secondary school mathematics curriculum, 1985 yearbook* (pp. 65–80). Reston: National Council of Teachers of Mathematics.

Mariotti, A. (2000). Introduction to proof: The mediation of a dynamic software environment. *Educational Studies in Mathematics, 44*, 25–53.

Matsuda, N., & Okamoto, T. (1998). Diagrammatic reasoning for geometry ITS to teach auxiliary line construction problems. In B. P. Goettl, H. M. Halff, C. L. Redfield, & V. J. Shute (Eds.), *Lecture Notes in Computer Science: Vol. 1452. Intelligent tutoring systems* (pp. 244–253). Berlin, Heidelberg: Springer-Verlag. doi: 10.1007/3-540-68716-5_30

Norman, D. A. (1988). *The psychology of everyday things.* New York: Basic Books.

Norman, D. A. (1999). Affordances, conventions and design. *Interactions, 6*(3), 38–43. May 1999, ACM Press.

Pedemonte, B. (2007). How can the relationship between argumentation and proof be analysed? *Educational Studies in Mathematics, 66*, 23–41.

Piaget, J., & Inhelder, B. (1967). *The child's conception of space.* New York: W. W. Norton.

Poincaré, H. (1956). Mathematical creation. In J. Newman (Ed.), *The world of mathematics* (pp. 2041–2050). New York: Simon & Schuster. Original work published in 1908.

Pólya, G. (1981). *Mathematical discovery: On understanding, learning, and teaching problem solving* (Combined ed.). New York: Wiley.

Ponte, J. P. (2007). Investigations and explorations in the mathematics classroom. *ZDM Mathematics Education, 39*(5–6), 419–430.

Rosen, D., & Hoffman, J. (2009). Integrating concrete and virtual manipulatives in early childhood mathematics. *Young Children, 64*(3), 26–33.

Semadeni, Z. (1980). Action proofs in primary mathematics teaching and in teacher training. *For the Learning of Mathematics, 4*, 32–34.

Usiskin, Z. (1980). What should not be in the algebra and geometry curricula of average college-bound students? *Mathematics Teacher, 73*, 413–424.

Usiskin, Z. (1987). Resolving the continuing dilemmas in school geometry. In M. M. Lindquist & A. P. Shulte (Eds.), *Learning and teaching geometry, K-12* (pp. 17–31). Reston, VA: NCTM.

Yerushalmy, M., & Chazan, D. (1990). Overcoming visual obstacles with the aid of the Supposer. *Educational Studies in Mathematics, 21*(3), 199–219.

Chapter 13
Principles of Task Design for Conjecturing and Proving*

Fou-Lai Lin, Kai-Lin Yang, Kyeong-Hwa Lee, Michal Tabach, and Gabriel Stylianides

This chapter aims to develop principles for designing tasks that teach conjecturing and proving. First, we search for principles by referring to design research and to the literature on designing tasks for learning mathematics. Next, we propose a rationale for conjecturing and proving and formulate a framework for conjecturing and proving. We then develop principles of task design for conjecturing and proving, and for transiting the two in practical tasks. Finally, we present our conclusions and discuss further applications of our principles.

1 Searching for Principles of Designing Tasks

According to www.dictionary.com, the first three definitions of "principle" are:

1. An accepted or professed rule of action or conduct.

---

*With Contributors Jane-Jane Lo, Xuhua Sun, and Kahou Chan

F.-L. Lin (✉) • K.-L. Yang
Department of Mathematics, National Taiwan Normal University, Taipei, Taiwan
e-mail: linfl@math.ntnu.edu.tw; kailinyang3@yahoo.com.tw

K.-H. Lee
Department of Mathematics Education, Seoul National University, Seoul, South Korea
e-mail: khmath@snu.ac.kr

M. Tabach
School of Education, Tel-Aviv University, Tel Aviv, Israel
e-mail: tabach.family@gmail.com

G. Stylianides
Department of Education, University of Oxford, Oxford, UK
e-mail: gabriel.stylianides@education.ox.ac.uk

© The Author(s) 2021
G. Hanna and M. de Villiers (eds.), *Proof and Proving in Mathematics Education*,
New ICMI Study Series, https://doi.org/10.1007/978-94-007-2129-6_13

2. A fundamental, primary, or general law or truth from which others are derived.
3. A fundamental doctrine or tenet; a distinctive ruling opinion.

As can be seen, the concept of principle in task design incorporates elements of all three definitions. As Reigeluth (1999) expressed it, a principle (basic method) is a rule (relationship) that is always true under appropriate conditions, regardless of a specific task (practice) or an approach consisting of a set of tasks (programme). In other words, a principle of task design should be "fundamental," in the sense that such a principle would be theory-based and not too specific to generalise. However, we are also concerned that the principle should not be too theoretical to be practical. Principles for designing tasks are also expected to be used as criteria for task designers, critical friends, or educators to evaluate designed tasks. For us, a principle of task design has both the fundamental function of clearly relating to the learners' roles, learning powers or hypothetical learning trajectories and the practical function of easily evaluating many similar tasks. If these two criteria are satisfied, principles for designing tasks will be useful in identifying diverse goals of conjecturing and proving activities, revealing multiple phases of learning, and serving to mediate theories to practical design.

Thus, Ruthven and his colleagues could position principle in the rank of intermediate theoretical frameworks (Ruthven et al. 2009). In their paper, they demonstrate how intermediate frameworks and design tools serve to mediate the contribution of grand theories to the design process by coordinating and contextualising theoretical insights about the epistemological and cognitive dimensions of a subject for the purposes of designing teaching sequences and studying their operations. In other words, grand theories, intermediate theoretical frameworks and design tools are three essentials of the design approach. Moreover, intermediate theoretical frameworks and the associated design tools are used to make links between grand theories and task design.

Lesh et al. (2000) based their principles for developing thought-revealing activities on the models and modelling perspective. Their six principles for designing productive model-eliciting activities include (1) the model construction principle, (2) the reality principle, (3) the self-assessment principle, (4) the construct documentation principle, (5) the construct shareability and reusability principle, and (6) the effective prototype principle. These suggested principles are crucial in developing model-eliciting tasks and in helping teachers or designers evaluate these activities' appropriateness. Therefore, the six principles serve as an intermediate framework mediating the models and modelling perspective to the design tools; a four-step modelling cycle including description, manipulation, translation and verification could be used as the design tool for sequences of modelling tasks.

For another example, 'what-if-not' (Brown and Walter 1993) could be considered as a tool for designing tasks of conjecturing. A what-if-not task provides a correct statement already known by the students, and asks them to conjecture the consequence of a change in the statement's premise or conclusion. What-if-not could be applied for many statements familiar to students; each statement can be designed as a task of conjecturing. What-if-not tasks could provide students with a cognitive tool to construct their own statements, and dynamic software or other instruments can assist them to reflect on their constructions.

In another example, Wittmann (2009) mentioned a series of tasks about numbers which required transformations between perceptual counters and numerical numbers. These tasks implied that providing manipulative and transformable representations for students is a good strategy for teaching conjecturing. Therefore, thinking tools embedded within tasks could be design tools. However, they still require an intermediate theoretical framework to mediate them to grand theories.

Using a lesson sequence of mathematical activity focused on concept, Bell et al. (1993) discussed perspectives on the nature of the mathematical activity, the conceptual content, and the nature of learning, for the purpose of designing a diagnostic teaching programme. Under the grand perspectives, Bell et al. (1993) suggested principles for the design of diagnostic teaching, including identifying key conceptual points and misconceptions, focusing on these, giving substantial open challenges, provoking cognitive conflicts, and resolving these through intensive discussion. Bell et al. (1993) also studied some teaching experiments according to the principles which connected the grand perspectives and task design through the use of associated design tools (e.g., familiar strategies for resolving misconceptions, questions for challenging students' misconceptions).

2 Conjecturing, Proving and the Transition

The main aim of all science is first to observe phenomena, then to explain them, and finally to predict. The method of explanation in mathematics is proof Gale (1990, 4). Proofs and refutations could be viewed as a dialectical mechanism of mathematical discovery (Lakatos 1976). Lakatos stressed the similarity and dissimilarity between the scientific and mathematical nature of these elements. In speaking of the contrast between science and mathematics, Lakatos states:

> Mathematical heuristic is very like scientific heuristic – not because both are inductive, but because both are characterized by conjectures, proofs, and refutations. The important difference lies in the nature of the respective conjectures, proofs, (or, in science, explanations), and counterexamples (p. 26).

He points out that the similarity between science and mathematics involves heuristic processes of producing a conjecture and looking for plausible arguments to support it; the dissimilarity involves what counts as a conjecture and a proof.

We have identified the characteristics of mathematical conjectures on the basis of two famous mathematical conjectures: Poincaré's Conjecture and Fermat's Conjecture. Poincaré conjectured, "if a space that locally looks like ordinary three-dimensional space but is finite in size and lacks any boundary has the additional property that each loop in the space can be continuously tightened to a point; then it is just a three-dimensional sphere". Fermat's Conjecture says, "if an integer n is greater than 2, then the equation $a^n + b^n = c^n$ has no solutions in non-zero integers". These two examples – "if a space..., then it is..."; and "if an integer n ..., then the equation ...." – show that conjectures are explicitly (sometimes implicitly) delineated as propositions or conditional statements. These conjectures' generality was extended as far as possible. For instance, "no solution in non-zero integers" is more general than "no solution in positive integers". Inevitably, conjectures provide conclusions specified under the tension of simplifying conditions; that is, conjectures denote propositions in mathematics. The main reason conjectures are so named is because they have not been formally proven.

In addition to the function of validating the truth of a conjecture, proof has many other functions in mathematics. For example, Hanna (2000) emphasised proof that provides a satisfactory explanation of why the conjecture is true; de Villiers (1999) argued that proof has several important functions including verification, discovery, explanation, communication, intellectual challenge, and systematisation. Moreover, one value of proof is that it provides a form for critical debate (Davis 1986); hence, proof is a norm for communicating mathematical results, which develops the criteria for acceptable arguments besides the negotiation of meaning.

Mathematical tasks, especially conjecturing and proving tasks, entail intellectual challenge which can be conquered with self-realisation and fulfilment (de Villiers 1999). Without proofs, it would be impossible to organise the results of mathematical research into a deductive system of axioms, definitions and theorems. For example, the primary function of proofs for secondary students to learn to prove some properties of parallelograms is that of systematisation; the purpose is not to check whether the results are true but to logically organise these related results – already known to be true – into a coherent, unified whole (de Villiers 2004).

Conjecturing and proving intertwine in human activities not only for discovering and verifying mathematical knowledge (Lakatos 1976; Pólya 1981) but also for other educational purposes (e.g. Boero 1999; Koedinger 1998; Smith and Hungwe 1998). Conjecturing and proving can also initiate mathematical thinking and develop mathematical methods; therefore, tasks of conjecturing and proving should be designed to be embedded into any level of mathematics classes in order to enhance students' conceptual understanding, procedural fluency, or problem solving (see Kilpatrick et al. 2001). However, for teachers to practise the tasks of conjecturing and proving in their classes is still not simple, in part because of the vagueness of the principles of task design.

For us, a task can be a solitary question or a sequence of questions. In addition to content, tasks can include instructional descriptions, which could be fundamental guidelines or supportive hints, in the task settings. For example, the description could be: "After you join two segments from two vertices in a square to their opposite sides, what new elements can you observe (shapes, segments, angles, etc.)?" The description will lead students to think in certain directions; it is a kind of general intervention.

Design principles of tasks for the learning of conjecturing and proving serve an essential learning goal of promoting ways of thinking, namely, proof schemes. Students are led to internalise diverse proof schemes through developing them and become aware of their plausibility and feasibility. Harel and Sowder (1998) offered a framework for examining students' conceptions of proof (schemes). The proof scheme framework we introduce here is a revision of Harel and Sowder's (2007). The new framework labels three classes of proof schemes: external conviction, empirical and deductive. The external conviction category involves authoritative, ritual and non-referential symbolic proof schemes; the empirical includes inductive and perceptual proof schemes; and the deductive covers transformational and axiomatic proof schemes. The plausibility and feasibility of proof schemes depend on the purposes of the task, either conjecturing or proving. For example, a perceptual proof scheme is plausible and feasible for conjecturing rather than for formally proving. If they can engage learners in perceiving various proof schemes with different levels of validity and utility, tasks will further strengthen learners' proficiency of conjecturing and proving.

In curriculum, students are asked to prove many propositions that have already been shown to be true (Hoyles 1997; Lin et al. 2004); the students may then conceive of proving as a ritual. In order to engage them in understanding multiple functions of proof and internalising proof schemes, teachers need to ask students to come up their own conjectures and to prove conjectures that are not already known by them to be true (Zaslavsky 2005). In the conjecturing task setting, students are asked to formulate conjectures according to some given information which could include either ill-defined or well-defined problems. Proving tasks provide a setting where students need to prove conjectures which could be either ill-formed or well-formed propositions.

Furthermore, proving tasks are classified by the sources of conjectures, either students' own conjectures or others'. The two sources may result in different transitional difficulties. In the first case, students may have unclear boundaries between conjecturing and proving – for example, they may consider empirical arguments as deductive proof (Stylianides and Stylianides 2009); they need to learn about each mode of reasoning and its appropriate level of validity. In the second, students may feel it unnecessary to prove others' conjectures, because they have accepted the truth on the basis of epistemic values (see Duval 1998). Therefore, tasks for helping students smoothly transit either from conjecturing to deductive proofs or from epistemic values to logical values are distinguished. Figure 13.3 shows the classification of conjecturing, proving and transitional tasks.

Below, each section is identified on the basis of Fig. 13.1's classification of tasks. In each, we will formulate principles for designing appropriate tasks, using examples.

3 Developing Principles for Conjecturing Task Design

Harel and Sowder (1998) defined a conjecture as an "observation" made by a person who has no doubt about its truthfulness; that observation ceases to be a conjecture and becomes a fact, from the person's viewpoint, once the person is certain of its truth (p. 241). Pólya (1954) showed how a mathematical conjecture may be generated after observation of one or several examples for which the conjecture is true. In addition, de Villiers (1997, 1999) discussed how students may propose conjectures

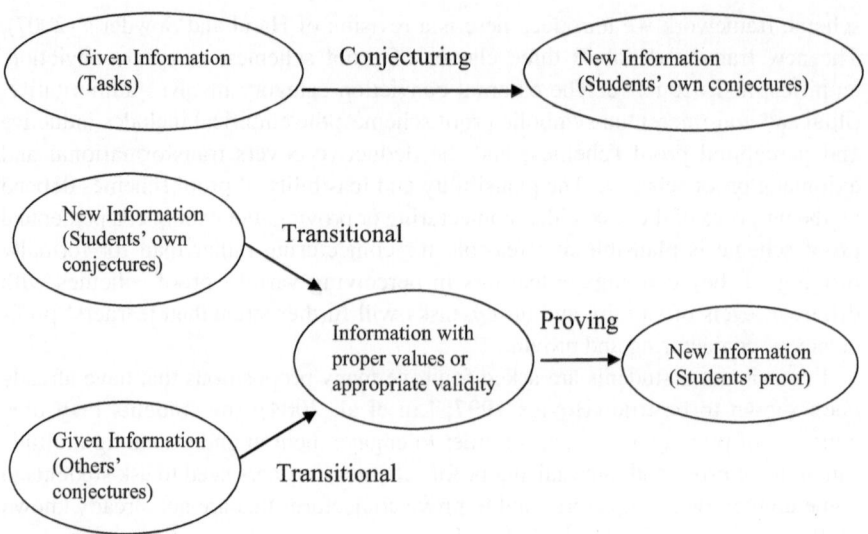

Fig. 13.1 Classification of tasks

whilst engaging in generalisations in technology-enhanced math classrooms. According to de Villiers, students not only generate conjectures but also search for proofs to verify their truthfulness, using tools available in technology-enhanced learning environments. The tasks discussed in these aforesaid studies provide us a foundation for our conjecturing task "principles" – general standards or guidelines for effective or efficient conjecturing task design. Besides the literary sources, the principles are founded on the provision of opportunities to: (1) observe, (2) construct, (3) transform, and (4) reflect.

Principle 1: Promote conjecturing by providing an opportunity to engage in observation.

As Harel and Sowder (1998) illustrated, a conjecture is the result of constant observation through which one sees regularity or a pattern. For this chapter, the term "observation" refers to activities that involve purposeful and/or systematic focus on specific cases in order to understand and/or make a generalisation about the cases. Opportunities for observation may include assessments of finite examples; that is, each student can be asked to systematically observe a particular example, following specific directions, as in the following task:

– Take any two-digit number and reverse the digits
– You now have two numbers
– Subtract the smaller number from the larger number
– What can you conclude from the result?

(Pedemonte 2008, p. 390)

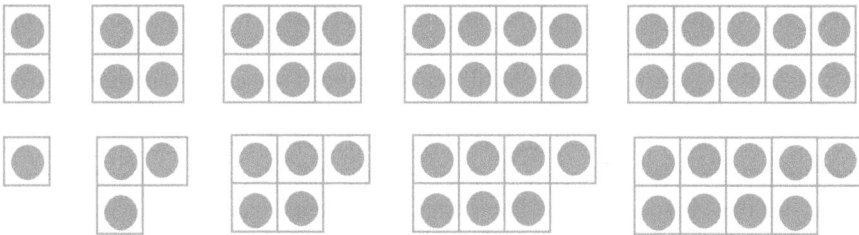

Fig. 13.2 Visual representation (Wittmann 2009, p. 251)

A learning environment that embraces technology may facilitate students' observation opportunities dynamically (Arzarello et al. 2002). For example, the "dragging" function in dynamic geometry software makes conjecturing more accessible. Students are able to explore drawings by moving them and viewing form changes or the absence of form changes, which facilitates discovery of invariant properties. The following task demonstrates how dragging modalities are used to make a conjecture:

> You are given a triangle ABC. Consider any point P on AB and the two resultant triangles APC and PCB. Hypothesize about the properties of ABC essential to guarantee APC and PCB are isosceles. (Arzarello et al. 2002, p. 67)

Observation-based conjecture production may also be stimulated by effective visualisation or mental representation. For instance, odd and even numbers could be introduced through a special pattern of counters (Fig. 13.2). The pattern can be painted on cardboard and cut out so that children can manipulate the pattern pieces and form sums of numbers (Wittmann 2009, p. 251). Using this visual and operational representation, children can make conjectures regarding the sum of two even numbers, the sum of two odd numbers, and the sum of an even number and an odd number. In addition to translating observed arrays into a number language, observation of the resultant arrays for even numbers and odd numbers, as well as of the emergent array produced by combining the two number counters, may increase learning pleasure and ease conjecturing for students.

Principle 2: Promote conjecturing by providing an opportunity to engage in construction.

When learning fundamental rules or structures, encouraging students to construct new mathematical knowledge based on prior knowledge may lead to conjectures. For example, the question "What do you think is the answer to $(a+b)^2$?" asks students to participate in a very simple conjecturing task that gives them the opportunity to construct multiplication principles related to algebraic expressions. This opportunity is lost if the teacher opts to teach the principles only through explanation; that is, choosing not to implement the opportunity as a student activity or task. Giving students the opportunity to complete an open-sentence question via conjecturing could increase their interest in conjecture postulation along with conjecture truth analysis.

Moreover, if students were to incorrectly conjecture that "$(a+b)^2=a^2+b^2$", the teacher could use the conjecture to encourage consideration of the general property as their starting point (Lin 2006). In this sense, incorrect conjectures emerging during construction of new knowledge can also be useful starting points in class.

The task below, suggested by Cañadas et al. (2007), asks students to construct relationships between geometric figures. To solve the task, students need to select particular lines based on learnt knowledge, create tentative relationships, and test creations in order to make a conjecture.

> Given a triangle ABC and a point P inside the triangle construct the three lines from each vertex A, B, C to the point P. What can you say about the relationships between the lines and the sides of the triangle? (Cañadas et al. 2007, p. 59)

Cañadas et al. (2007) proclaim that students improve understanding by construct- ing and testing conjectures; yet they also recognise that students could generate false conjectures. For the above task, students might incorrectly conjecture, "if two lines cut the sides in a 2:1 ratio, the third one will, too" (p. 60); nevertheless, that conjecture can be productive if used as a means to investigate other new properties of a triangle (Cañadas et al. 2007). Besides, conjecturing is typically not an isolated event; it is connected, at least potentially, to the learners' prior or new knowledge. Whilst reflecting on their prior knowledge from different perspectives, learners also maintain an eye for new knowledge. Hence, the "construction" principle also pro- vides opportunities to (a) construct new knowledge through interpolation and extrapolation based on conjecture creation; and (b) to make further conjectures based on the constructed new knowledge.

Principle 3: Promote conjecturing by providing an opportunity to transform prior knowledge.

This third design principle, "transformation", means that the task gives the students opportunities to generate conjectures by transforming given statements, formulae, algorithms, principles, and so on. Partly altering a hypothesis or conclusion in a statement and transforming a dimension or elements in a formula are useful approaches to making conjectures. For example, the task below, developed by Lin (2006), encourages students not only to transform the given formula but also to find meaning in their transformation(s).

> Try to make sense of the following formula. What do you think "A" stands for in the for- mula? Notice the beauty of the formula. What conjectures can you think of for a quadrilat- eral with sides a, b, c, d?

$$A = \sqrt{s(s-a)(s-b)(s-c)}, \ \ where \ \ s = \frac{1}{2}(a+b+c)$$

Probably the students will start by transforming the formula, perhaps proposing quadrilateral conjectures like:

$$B = \sqrt{(s-a)(s-b)(s-c)(s-d)}, \ \ where \ \ s = \frac{1}{2}(a+b+c+d)$$

Then, they may embark on investigating the truthfulness of their conjectures by examining them in terms of a square, rectangle, and so on, which may lead to their learning new knowledge about the formula for the area of an inscribed quadrilateral of circle with side lengths of a, b, c, and d.

Carpenter et al. (2003, p. 54) detail how easily students elicit mathematical conjectures by generalising given number sentences. More interesting, they discuss how students, after establishing a conjecture's truthfulness, make similar statements by transforming it. For example, after formulating the conjecture, "when you add zero to a number, you get the number you started with", students are likely to transform it to (a) "when you subtract zero from a number, you get the number you started with;" (b) "when you multiply a number by 1, you get the number you started with;" (c) "when you divide a number by 1, you get the number you started with;" and so on.

Principle 4: Promote conjecturing by providing an opportunity for reflection.

Learners' conjectures based on observation, of several examples or even one particular example, can be incorrect and meaningless, as Cañadas et al. (2007) discussed. Conjecturing by transformation may also lead students to incorrect and meaningless statements, as may conjecturing through the "what if not" strategy without proper understanding or goals (e.g., Lavy and Bershadsky 2003). Thus, learners' reflecting on their conjectured constructs and on the conjecturing process is essential to the teaching of mathematics via conjecturing. Reflection gives students the chance to further explore mathematical problems and improve their conjectures. Therefore, this last principle for designing a conjecturing task is "reflection", in the sense that opportunities for reflection need to be integrated into the task.

Pedemonte (2008) claims that constructive argumentation contributes to constructing a conjecture, whilst structural argumentation aids in justifying it (p. 390). These two types of argumentation, as they relate to conjecturing, can be combined into a single criterion: reflection on both the conjecturing process and the resultant conjectures. The following task, discussed by Lee (2011), illustrates how reflection on both the soccer ball model conjecturing process and the generated conjectures may lead to discovery of Descartes' formula for semi-regular polyhedrons.

- Observe a virtual soccer ball picture and then construct a soccer ball using tangible material. Make a list of mathematical questions and/or statements about the generated soccer ball model. From the list, choose several items such as "Why do manufacturers make soccer balls from regular pentagons and regular hexagons?" and "What other possible shapes could be used to create a soccer ball?" that you would like to investigate.
- Examine your proposed soccer ball model conjectures and their corresponding justifications by reflecting on actions carried out on the emerged models. Include your creation of definitions, representations, and conjectures when presenting solutions to your chosen questions.

To motivate and facilitate reflection, several exemplary questions could be incorporated in a conjecturing task:

1. Is your conjecture clear? How is it possible to change your conjecture wording so that unnecessary ambiguity is avoided?
2. Explain why you believe your conjecture holds true for the given condition. Does your conjecture still hold true when conditions change?
3. Is there any case for which your conjecture would not hold true? Is it possible to generalise your conjecture?
4. What is the basis for your conjecture? Is there any argumentation for it?

4 Developing Principles for Designing Transiting Tasks Between Conjecturing and Proving

Mathematicians do not construct proofs for the sake of proofs but for specific reasons: for example, to establish new knowledge by verifying or refuting conjectures. Several mathematics education researchers (e.g., Ellis 2007; Hadas et al. 2000; Jones 2000; Mariotti 2000; Marrades and Gutiérrez 2000; Stylianides and Stylianides 2009) have developed tasks and conducted studies, including teaching and design experiments, aimed to engage students, including prospective teachers, in tasks related to proofs and conjectures that authentically reflect the "wider mathematical culture" (Lampert 1992). These investigations raised the issue of how teachers, including teacher educators, can help students make a smooth transition between conjecturing and proving.

In school mathematics, students' generation of conjectures frequently results from examinations of specific cases through patterning tasks (Ellis 2007). Unfortunately many students of all levels of education have the persistent, robust misconception that empirical arguments[1] are proofs (e.g., Coe and Ruthven 1994; Goetting 1995; Harel and Sowder 1998; Healy and Hoyles 2000; Lannin 2005; Martin and Harel 1989). Teachers need to help these students understand that confirming evidence from an examination of cases does not constitute conclusive evidence for the truth of a general mathematical claim; that is, proof. Such evidence supports the development of a reasoned hypothesis about the truth of the claim that requires further investigation; namely, a conjecture.

The increased availability in school mathematics instruction of dynamic mathematics software, such as dynamic geometry software (DGS), raised the concern that such programmes would make the boundaries between conjecturing and proving even less clear for students and would reinforce their misconception that empirical arguments constitute proofs. Different dynamic mathematics softwares allow

---

[1] *Empirical arguments* are arguments that purport to show the truth of a claim by validating the claim in a proper subset of all the possible cases it covers.

students to check easily and quickly a very large number of cases, thus helping students "see" mathematical properties more easily and potentially "killing" any need for students to engage in actual proving.

In a 2000 *Educational Studies in Mathematics* (*ESM*) special issue, four papers (Hadas et al. 2000; Jones 2000; Mariotti 2000; Marrades and Gutiérrez 2000) examined whether and how DGS can be used to help students make the transition from conjecturing and constructing empirical arguments to developing deductive arguments (including proofs). These papers, using various theoretical frameworks, "present results of empirical research which demonstrates that the judicious use of dynamic geometry software in heuristics, exploration and visualisation can foster an understanding of proof" (Hanna 2000, p. 20). In addition to the *ESM* special issue, two more recent studies (Ellis 2007; Stylianides and Stylianides 2009) also directly relate to our focus on conjecturing and making the transition to proof. Both studies reinforce and extend the findings of the *ESM* special issue studies, reporting findings that can be useful for teachers trying to help students make a smooth transition between conjecturing and proving through the development of more sophisticated "justification schemes" (Harel and Sowder 1998). We use these six studies as a source for principles for designing transiting tasks between conjecturing and proving, first summarising them.

Ellis (2007) conducted a teaching experiment with seven lower secondary school students (12 year-olds) working on algebra and using SimCalc Mathworlds (Roschelle and Kaput 1996), a speed simulation computer program that allowed students to generate and test conjectures involving two quantitative variables. The study examined how students' generalisations and justifications in answer to linear patterning tasks were connected. The tasks were based on real-world situations, one of them with real physical devices, and were characterised by a request to use more than one representation. The students created some representations with the aid of the SimCalc software; hence, these were dynamic. Ellis (2007) identified four non-hierarchical connective mechanisms that teachers could use to help their students engage in more sophisticated justification schemes: (a) iterative action/reflection cycles, (b) mathematical focus, (c) generalisations that promote deductive reasoning, and (d) influence of deductive reasoning on generalising. Yet, as Ellis noted, research is needed to examine the applicability of these findings in typical classroom settings; the seven students in the study worked in a technology-based, small-group environment.

Jones (2000) discussed an instructional environment in which lower secondary students were asked to use DGS to classify quadrilaterals in a set of tasks of increasing difficulty. The tasks involved visual explorations of similarities and differences between quadrilaterals and then creating hierarchical classifications. The dynamic property of the software was a crucial aspect of the task. Jones observed a shift in the students' "thinking from imprecise, 'everyday' expressions, through reasoning mediated by the software environment to mathematical explanations of the geometric situation" (p. 80). The students showed improvement in their ability for deductive reasoning; they made progress in formulating accurate mathematical statements and valid deductive arguments.

Mariotti (2000) worked with older students than Jones (2000) and Ellis (2007) – with upper secondary students, 15–16 year-olds. Adopting a Vygotskian perspective, she initiated a long-term teaching experiment that aimed to introduce students to theoretical thinking. Mariotti's experiment drew heavily on the fact that Cabri-géomètre (hereafter, Cabri), the DGS used, allowed for the creation of systems of commands in a way that paralleled the way mathematical axioms and theorems are constructed. The set of software tools initially available to the student participants corresponded to the straightedge and compass of traditional paper-and-pencil geometry. As the students built different geometrical constructions, such as the angle bisector, the Cabri menu was enlarged to include corresponding new commands, such as the "angle bisector" command, which then became available for use in subsequent constructions. This progressive enlargement of the Cabri menu paralleled the enlargement of the theoretical system; the new constructions were subsequently added to the theory. Proof served a twofold function in Mariotti's experiment. First, it ensured the validity of new constructions based on the system of commands available at each chronological point. Second, proof stood in a dialectical relationship to the classroom social contract establishing conjectures or constructions that had to be justified and accepted by the classroom community

Proof served a similar function in the teaching experiment that Marrades and Gutiérrez (2000) conducted, also with upper secondary students. Marrades and Gutiérrez summarised the social role assigned to proof thus:

> The agreed didactical contract between teacher and pupils, in reference to what kinds of answers are accepted, is an important element to success in promoting students' progress. In our experiment, the didactical contract made explicit by the teacher can be summarized as the need to organize justifications by using definitions and results (theorems) previously known and accepted by the class. (p. 120)

However, Marrades and Gutiérrez followed a different approach from Mariotti's (2000). Specifically, they engaged students in working with Cabri to solve a series of geometrical tasks in a unit that aimed to help students improve their understanding about proof and their justification schemes. The software allowed the students to test their solutions and then discuss them during the lesson. The authors showed how the facility that Cabri offered for exploration and visualisation helped the students, over time, jointly understand the limitations of informal approaches and develop more sophisticated justification schemes.

Both Hadas et al.'s (2000) study and Stylianides and Stylianides' (2009) experiment added to the issue of creating in students a social need for proof – prominent in Marrades and Gutiérrez's (2000) and Mariotti's (2000) studies – another kind of need, "intellectual need" (Harel 1998) arising from cognitive conflicts: contradictions, disequilibria, and surprise. Both the former studies showed how carefully designed sequences of tasks can encourage students to feel an internal motivation for proof rather than to consider proof as superfluous or something compulsory to please the teacher.

Hadas et al. (2000) had students make conjectures about certain geometrical properties – namely, the sums of interior and exterior angles of convex polygons – and check them with the DGS. The conjectures turned out to be wrong when examined with the DGS. These contradictions between conjectures and findings created a fruitful context for engaging students in the development of deductive arguments about the geometric properties in question. Specifically, cognitive conflict helped students delve into and develop their understanding of a complex comprehensive topic in geometry in deeper, more mathematically appropriate ways with the use of deductive arguments.

Stylianides and Stylianides' (2009) 4 year study included a design experiment that aimed to help prospective elementary teachers develop their understanding of proof; it did not involve the use of any mathematics computer software and was not specific to any single content area. The authors developed a mathematical task sequence which led prospective teachers to face cognitive conflicts. The study used these conflicts to help the participants investigate the mathematics involved in deeper/more mathematically appropriate ways and most important, to help them reflect on and better understand broader issues of mathematical validation. Specifically, they helped the prospective teachers realise the limitations of empirical arguments as validations for mathematical conjectures and feel an intellectual need to learn about secure validations, namely proofs, that they could generalise to similar contexts beyond the particular task sequence.

Though we did not discuss it specifically, the role of the instructor, teacher or teacher educator was a common theme across all six studies. The instructor (in many cases the researcher) was not only responsible to implementing the tasks and facilitating students' engagement in them, but also played a major role in establishing the classroom social norms that guided the acceptance or rejection of participants' mathematical arguments. Hence our first principle.

Principle 1: The task requires that norms be established in the classroom that allow discussions, under the facilitation of the instructor, about accepting/ rejecting mathematical ideas, including conjectures, based on the logical structure of the mathematical system rather than by appeal to the authority of the instructor.

All six studies purposefully developed the participants' need to engage in proving in the course of several lessons, and in some cases along a sequence of tasks. In most cases, this need related to the social norms established in the classroom: social need, but, in some other cases, it also related to cognitive conflict generated by engagement in carefully designed task sequences and the associated desire of the students to resolve the conflict: intellectual need. Also, the need to move between representations may have played in some cases (cf. Ellis 2007; Jones 2000) a role in creating the need for proving.

Principle 2: The task generates a need for the students to engage in proof.

Although five of the six studies used different mathematics instructional computer software, the sixth (Stylianides and Stylianides 2009) did not. So, we hesitate to propose a principle about a computer/technological environment appearing particularly important for the transition between conjecturing and proving. However, the software used in the five studies helped create a dynamic environment that allowed students to explore their hypotheses and become active participants in the lesson.

Finally, we draw no design principle from the specific characteristics of the tasks used in the six studies, because they seem to have no common methodological theme. This observation might relate to Principles 1 and 2 above: Different kinds of tasks have the potential to facilitate the transition from conjecturing to proving, as long as the classroom norms and the instructor play a supportive role and the students can be led to feel the need to prove.

5   Developing Principles for Designing Proving Tasks

We aim to identify design principles that instructors can use both to generate new proving tasks and to identify existing tasks that have good potential to promote proofs and proving. Therefore, we adopt a working definition for the act of proving that is embedded in the classroom context, be it at the elementary school, secondary school, university, or teachers' professional development level:

Proof is mathematical argument, a connected sequence of assertions for or against a mathematical claim, with the following characteristics:

1. It uses statements accepted by the classroom community (set of accepted statements) that are true and available without further justifications;
2. It employs forms of reasoning (modes of argumentation) that are valid and known to, or within the conceptual reach of, the classroom community; and
3. It is communicated with forms of expression (modes of argument representation) that are appropriate and known to, or within the conceptual reach of, the classroom community (Stylianides 2007, p. 107).

Given this definition, we suggest design principles for proving tasks involving the following five aspects of modes of argumentation and modes of argument representation: classifying mathematical statements, expressing arguments in several modes, changing roles in a task, defining sufficient and necessary proof, and creating and sharing proof. We explain each principle and demonstrate it by example from the research literature.

Principle 1: Promote classifying mathematical statements.

This first principle relates to the learners' need to understand that different modes of argumentation are appropriate for different types of statements. Consider the statement: "The sum of three consecutive natural numbers is divisible by six". One may find triples of consecutive natural numbers that fulfil it, like $1+2+3$; one may find triples of consecutive numbers that do not fulfil it, like $2+3+4$. This statement involves a predicate, which is "sometimes true" and "sometimes false". If we add a quantifier to the predicate, we get a mathematical statement, which can be either true or false: Adding the universal quantifier "for any" will result in a false statement – "the sum of any three consecutive natural numbers is divisible by six", however adding the existential quantifier "there exists" will result in a true statement – "there

Table 13.1 Six cell matrix, showing the truth-value of any statement and (in italics) the minimal necessary and sufficient mode of argumentation needed as proof

Predicate quantifier	Always true	Sometimes true	Never true
Universal	Cell 1. True statement *general proof*	Cell 2. False statement *counter example*	Cell 3. False statement *counter example*
Existential	Cell 4. True statement *supportive example*	Cell 5. True statement *supportive example*	Cell 6. False statement *general proof*

(a) Classify the following statements. You may use the six cells matrix.
(b) Create your own mathematical statement for each category you used.
S1. The sum of any arithmetic sequence with four elements and a difference of 5 is divisible by 2.
S2. The sum of any arithmetic sequence with four elements and a difference of 5 is divisible by 3.
S3. The sum of any arithmetic sequence with four elements and a difference of 5 is divisible by 4.
S4. There exists a sum of arithmetic sequence with four elements and a difference of 5 that is divisible by 2.
S5. There exists a sum of arithmetic sequence with four elements and a difference of 5 that is divisible by 3.
S6. There exists a sum of arithmetic sequence with four elements and a difference of 5 that is divisible by 4.

Fig. 13.3 Classification task Supported by The Israel Science Foundation (grant No. 900/6)

exists a sum of three consecutive natural numbers divisible by six". Each of the two statements needs a different mode of argumentation as proof: to refute the universal statement one counter-example is sufficient; to prove the existential statement one supportive example is sufficient. So, the combination of a quantifier and a predicate both results in a mathematical statement and determines the appropriate mode of argumentation needed to prove the statement. Considering the two quantifiers (universal and existential) and the three possible predicates (always true, sometimes true and never true), results in a six-cell matrix; each cell corresponds to one type of statement, which is either true or false, and determines the mode of argumentation needed (Table 13.1).

One way of explicating the variety of statements and the resulting mode of argumentation may be done by the following task (Fig. 13.3). The design principle of asking the learner to classify is not new. Classifying and sorting involves comparing the classified objects, in this case mathematical statements, in order to pinpoint similarities and differences amongst them. This may explicate for the learner that there are several well-defined types of statements. Part b of this task gives students more practice and helps them reinforce/abstract what they learned in part a. We can elaborate the first design principle as to say a task should involve classifying mathematical statements in order to explicate the various types of statements and their appropriate modes of argumentation.

Jamie solved   38x26 = with the following computation steps. Her justification was also included below.

Discuss with your group member whether this justification is "Clear", "Complete" and "Correct."

If not, please revise to make it a better justification so it will meet the three Cs criteria.

Jamie's solution and justification

8x25=200
30x25=750
1x8=8
20x1=30
200+750+30+8=988

We had the original starter at 8 groups of 25 which is equal to 200. Since we had 30 less in each group, we need to add 30 to all 25 groups so we ended up adding 750 to the 200. We still need one more group of 38 to make 26 groups. So we added 8 and one more group of 30. Then when you add together 200, 750, 8 and 30, your final answer is 988.

Fig. 13.4 Evaluation task (Lo 2009)

Principle 2: Expressing arguments in several modes of argument representation.

The second design principle relates to our working definition about representing the same argument via different representations. Given that using symbolic notation in proving is not the first choice of many students (e.g., Healy and Hoyles 2000), it may be a good practice to encourage students to start by using verbal representation, probably with non-mathematical language, as an entry to the world of proofs and proving. The second design principle in practice means a task requires students to express their arguments in several modes of argument representation.

Principle 3: Promote changing roles whilst engaging with a task.

A well-documented difficulty with respect to students' proving at all levels relates to the formulation of the argument. That is, an argument should be clear to the reader, complete and of course correct. The following task, suggested by Jane-Jane Lo (2009) (personal communication; Fig. 13.4), adopts a somewhat unusual approach to promote students' ability to create such arguments.

The role of the learner in this task is to evaluate the justification and improve it if it does not meet the established standards of the community in terms of clarity, completeness and correctness. In order to do so, the learner must first read the proposed justification and understand it, only then becoming able to determine its clarity, completeness and correctness. In fact, the design principle involved may be seen as "changing roles"; that is, reversing the traditional roles of instructor and learner by positioning the learner as evaluator of the proposed justification.

Bell et al. (1993) identified this third principle of changing roles whilst engaging with a task with respect to task design in general. Tasks which allow students to change roles and explain and teach one another may significantly contribute to learning and to knowledge building.

Do you accept the following as justifications for statement: "There exists a sum of three consecutive natural numbers that is divisible by 6"?
Dana claimed:

1+2+3=6	3+4+5=12	5+6+7=18	7+8+9=24
9+10+11=30	11+12+13=36	15+16+17=4	17+18+19=54
19+20+21=60	21+22+23=66	23+24+25=72	25+26+27=78
39+40+41=120	41+42+43=126	55+56+57=168	87+88+89=264
91+92+93=276	101+102+103=306	111+112+113=336	235+236+237=708

All the results are divisible by 6. Therefore the statement is true.

Fig. 13.5 Evaluation task (Tsamir et al. 2009)

Principle 4: Raise the issue of sufficient and necessary proof.

The use of this kind of task may further elicit the learners' "awareness" (Mason 1998) of the problem of sufficiency of an argument. The example in Fig. 13.5 is mathematically a correct proof for the existential statement. However, it is not a minimal proof, and the issue is whether it is a legitimate proof (for more detail, see Tsamir et al. 2009). The fourth principle leads students to decide whether a proof is sufficient and necessary for the type of statement involved.

Principle 5: Learners create and share their own proofs.

Traditional classroom practice, where the instructor lectures to the learners, may cause the learners to think that proofs are "given" and one can only learn to prove by memorising. Tasks which may be categorised as "create your own proof" may be especially appropriate for changing this misconception. Consider, for example, the task "Prove the Midpoint Theorem for the case of triangle" (Sun and Chan 2009). Each learner first creates a proof, and later presents it to the whole class not only for verification of its validity but also for consideration of qualities such as "simple", "efficient" or "elegant". At the same time, each learner is exposed to the proofs raised by the others, this time evaluating their proofs.

Each of the above five tasks aims to promote the practice of proving in class. We have presented them in order to illustrate each of our five underlying design principles for proving tasks. We selected most of the tasks from mathematical fields other then Geometry, in order to highlight that proof and proving could and should be embedded across the mathematics curriculum.

6 Conclusion

We have developed a total of 11 principles for task design for learning the mathematical skills of conjecturing (four), transiting between conjecturing and proving (two), and proving (five), based on some theory and practical tasks in the literature.

Because of different degrees of richness in the literature with respect to each skill area, the consistency of presentation and the number of each set of principles has varied. The principles are:

6.1 Conjecturing

Principle 1: Promote conjecturing by providing opportunity to engage in observation.
Principle 2: Promote conjecturing by providing an opportunity to engage in construction.
Principle 3: Promote conjecturing by providing an opportunity to transform prior knowledge.
Principle 4: Promote conjecturing by providing an opportunity for reflection.

6.2 Transiting

Principle 1: The task requires that norms be established in the classroom that allow discussions, under the facilitation of the instructor, about accepting/ rejecting mathematical ideas, including conjectures, based on the logical structure of the mathematical system rather than by appeal to the authority of the instructor.
Principle 2: The task generates a need for the students to engage in proof.

6.3 Proving

Principle 1: Promote classifying mathematical statements.
Principle 2: Promote expressing arguments in several modes of argument representation.
Principle 3: Promote changing roles whilst engaged with a task.
Principle 4: Raise the issue of sufficient and necessary proof.
Principle 5: Learners create and share their own proofs.

Here, we further analyse a common feature that links these three sets of principles and provide examples of applying them.

One of the educational principles emphasised in Realistic Mathematics Education is that lines of learning are intertwined (Streefland 1991, p. 21). Intertwining is a specific feature common to our three sets of design principles. With the help of the following two examples, we will discuss this intertwining characteristic and show it operating in relation to practical application of our principles. As a first example, the suggested questions for conjecturing tasks based on the "reflection" principle (conjecturing, 4) could also generate many tasks appropriate for the transitional section. The two questions: "Explain why you believe your conjecture holds for condition A? Does your conjecture still hold when A is changed?" and "Is there any case for which your conjecture would not hold true?" initiate a process that leads learners to invoke norms in the classroom that allow discussions about accepting/rejecting mathematical ideas based on the logical structure of the mathematical system, namely, Principle 1 of Transiting task design.

Second, Principle 1 of Proving task design promoting classifying mathematical statements in order to explicate their various types, can generate tasks whose learning process intertwines with the experiences resulting from involvement in tasks based on Transiting Principle 1 and Conjecturing Principle 4. Thus, amongst the three learning categories, the lines of learning intertwined to create a continuity of learning between conjecturing and proving. All our principles for task design have this intertwining feature.

To further validate the functioning of our design principles we encourage more empirical research, particularly investigating their use in conjunction with educational variables, such as different teachers' backgrounds and different educational systems.

References*

Arzarello, F., Olivero, F., Paola, D., & Robutti, O. (2002). A cognitive analysis of dragging practices in Cabri environments. *Zentralblatt fur Didaktik der Mathematik, 34*(3), 66–72.

Bell, A., Swan, M., Crust, R., & Shannon, A. (1993). *Awareness of learning reflection and transfer in school mathematics* (Report of ESRC Project R000-23-2329). Nottingham: Shell Centre for Mathematical Education, University of Nottingham.

Boero, P. (1999). Argumentation and mathematical proof: A complex, productive, unavoidable relationship in mathematics and mathematics education. Retrieved May 2009, from http://www.lettredelapreuve.it/Newsletter/990708Theme/990708ThemeUK.html

Brown, S., & Walter, M. (1993). *Problem posing reflections and applications*. Hillsdale: Lawrence Erlbaum Associates.

Cañadas, M. C., Deulofeu, J., Figueiras, L., Reid, D., & Yevdokimov, O. (2007). The conjecturing process: Perspectives in theory and implications in practice. *Journal of Teaching and Learning, 5*(1), 55–72.

Carpenter, T. P., Franke, M. L., & Levi, L. (2003). *Thinking mathematically: Integrating arithmetic and algebra in the elementary school*. Portsmouth: Heinemann.

Coe, R., & Ruthven, K. (1994). Proof practices and constructs of advanced mathematics students. *British Educational Research Journal, 20*, 41–53.

Davis, P. (1986). The nature of proof. In M. Carss (Ed.), *Proceedings of the fifth international congress on mathematical education* (pp. 352–358). Adelaide: Unesco.

De Villiers, M. (1997). The role of proof in investigative, computer-based geometry: Some personal reflections. In D. Schattschneider & J. King (Eds.), *Geometry turned on!* Washington, DC: MAA.

De Villiers, M. (1999). *Rethinking proof with sketchpad*. Emeryville: Key Curriculum Press.

---

*NB: References marked with * are in F. L. Lin, F. J. Hsieh, G. Hanna, & M. de Villiers (Eds.) (2009). *ICMI Study 19: Proof and proving in mathematics education*. Taipei, Taiwan: The Department of Mathematics, National Taiwan Normal University.

De Villiers, M. (2004). Using dynamic geometry to expand mathematics teachers' understanding of proof. *International Journal of Mathematical Education in Science and Technology, 35*(5), 703–724.

Duval, R. (1998). Geometry from a cognitive point of view. In C. Mamana & V. Villani (Eds.), *Perspectives on the teaching of geometry for the 21st century* (An international commission on mathematical instruction (ICMI) study, pp. 37–52). Boston: Kluwer Academic.

Ellis, A. B. (2007). Connections between generalizing and justifying: Students' reasoning with linear relationships. *Journal for Research in Mathematics Education, 38*(3), 194–229.

Gale, D. (1990). Proof as Explanation. *Mathematical Intelligencer, 12*(2), 4.

Goetting, M. (1995). *The college students' understanding of mathematical proof.* Unpublished doctoral dissertation, University of Maryland, College Park.

Hadas, N., Hershkowitz, R., & Schwarz, B. B. (2000). The role of contradiction and uncertainty in promoting the need to prove in dynamic geometry environments. *Educational Studies in Mathematics, 44*(1/2), 127–150.

Hanna, G. (2000). Proof, explanation and exploration: An overview. *Educational Studies in Mathematics. Special issue on Proof in Dynamic Geometry Environments, 44*(1–2), 5–23.

Harel, G. (1998). Two dual assertions: The first on learning and the second on teaching (or vice versa). *The American Mathematical Monthly, 105*, 497–507.

Harel, G., & Sowder, L. (1998). Students' proof schemes: Results from exploratory studies. In A. H. Schoenfeld, J. Kaput, & E. Dubinsky (Eds.), *Research in collegiate mathematics education. III* (pp. 234–283). Providence: American Mathematical Society.

Harel, G., & Sowder, L. (2007). Toward comprehensive perspectives on the learning and teaching of proof. In F. K. Lester (Ed.), *Second handbook of research on mathematics teaching and learning* (pp. 805–842). Greenwich: Information Age.

Healy, L., & Hoyles, C. (2000). A study of proof conceptions in algebra. *Journal for Research in Mathematics Education, 31*, 396–428.

Hoyles, C. (1997). The curricular shaping of students' approaches to proof. *For the Learning of Mathematics, 17*, 7–16.

Jones, K. (2000). Providing a foundation for deductive reasoning: Students' interpretations when using dynamic geometry software and their evolving mathematical explanations. *Educational Studies in Mathematics, 44*(1/2), 55–85.

Kilpatrick, J., Swafford, J., & Findell, B. (2001). *Adding it up: Helping children learn mathematics.* Washington, DC: National Academy Press.

Koedinger, K. (1998). Conjecturing and argumentation in high-school geometry students. In R. Lehrer & D. Chazan (Eds.), *Designing learning environments for developing understanding of geometry and space* (pp. 319–347). Mahwah: Lawrence Erlbaum.

Lakatos, I. (1976). *Proofs and refutations: The logic of mathematical discovery.* Cambridge: Cambridge University Press.

Lampert, M. (1992). Practices and problems in teaching authentic mathematics. In F. K. Oser, A. Dick, & J. Patry (Eds.), *Effective and responsible teaching: The new synthesis* (pp. 295–314). San Francisco: Jossey-Bass.

Lannin, J. K. (2005). Generalization and justification: The challenge of introducing algebraic reasoning through patterning activities. *Mathematical Thinking and Learning, 7*, 231–258.

Lavy, I., & Bershadsky, I. (2003). Problem posing via "what if not?" strategy in solid geometry - a case study. *The Journal of Mathematical Behavior, 22*, 369–387.

Lee, K. H. (2011). Soccer ball modeling via conjecturing in an eighth grade mathematics classroom. *International Journal of Science and Mathematics Education, 9*(3), 751–769.

Lesh, R., Hoover, M., Hole, B., Kelly, A., & Post, T. (2000). Principles for developing thought-revealing activities for students and teachers. In *Handbook of research design in mathematics and science education* (pp. 591–645). Mahwah: Lawrence Erlbaum.

Lin, F. L. (2006). *Designing mathematics conjecturing activities to foster thinking and constructing actively.* Retrieved May 2009, from http://www.apecknowledgebank.org/resources/downloads/12_3-4_06_3_Lin.pdf

Lin, F. L., Yang, K. L., & Chen, C. Y. (2004). The features and relationships of explanation, understanding proof and reasoning in number pattern. *International Journal of Science and Mathematics Education, 2*(2), 227–256.

Mariotti, M. A. (2000). Introduction to proof: The mediation of a dynamic software environment. *Educational Studies in Mathematics, 44*, 25–53.

Marrades, R., & Gutiérrez, A. (2000). Proofs produced by secondary school students learning geometry in a dynamic computer environment. *Educational Studies in Mathematics, 44*(1/2), 87–125.

Martin, W. G., & Harel, G. (1989). Proof frames of preservice elementary teachers. *Journal for Research in Mathematics Education, 20*, 41–51.

Mason, J. (1998). Enabling teachers to be real teachers: Necessary levels of awareness and structure of attention. *Journal of Mathematics Teacher Education, 1*, 243–267.

Pedemonte, B. (2008). Argumentation and algebraic proof. *Zentralblatt fur Didaktik der Mathematik, 40*(3), 385–400.

Pólya, G. (1954). *Mathematics and plausible reasoning* (Vol. I). Princeton: Princeton University Press.

Pólya, G. (1981). *Mathematical discovery*. New York: Wiley.

Reigeluth, C. M. (1999). The elaboration theory: Guidance for scope and sequence decisions. In C. M. Reigeluth (Ed.), *Instructional-design theories and models: A New paradigm of instructional theory, volume II*. Hillsdale: Lawrence Erlbaum Association.

Roschelle, J., & Kaput, J. J. (1996). SimCalc Mathworlds for the mathematics of change. *Communications of the ACM, 39*(8), 97–99.

Ruthven, K., Laborde, C., Leach, J., & Tiberghien, A. (2009). Design tools in didactical research: Instrumenting the epistemological and cognitive aspects of the design of teaching sequences. *Educational Researcher, 38*(5), 329–342.

Smith, J. P., & Hungwe, K. (1998). Conjecture and verification in research and teaching: Conversations with young mathematicians. *For the Learning of Mathematics, 18*(3), 40–46.

Streefland, L. (1991). *Fractions in realistic mathematics education: A paradigm of developmental research*. Dordrecht: Kluwer Academic.

Stylianides, A. J. (2007). Proof and proving in school mathematics. *Journal for Research in Mathematics Education, 38*, 289–321.

Stylianides, G. J., & Stylianides, A. J. (2009). Facilitating the transition from empirical arguments to proof. *Journal for Research in Mathematics Education, 40*, 314–352.

Sun, X., & Chan, K. (2009). Regenerate the proving experiences: An attempt for improvement original theorem proof constructions of student teachers by using spiral variation curriculum (Vol. 2, pp. 172–177).*

Tsamir, P., Tirosh, D., Dreyfus, T., Tabach, M., & Barkai, R. (2009). Is this verbal justification a proof? (Vol. 2, pp. 208–213).*

Wittmann, E. (2009). Operative proof in elementary mathematics (Vol. 2, pp. 251–256).*

Zaslavsky, O. (2005). Seizing the opportunity to create uncertainty in learning mathematics. *Educational Studies in Mathematics, 60*, 297–321.

Chapter 14
Teachers' Professional Learning of Teaching Proof and Proving

Fou-Lai Lin, Kai-Lin Yang, Jane-Jane Lo, Pessia Tsamir, Dina Tirosh, and Gabriel Stylianides

1   Introduction

This chapter examines the three essential components of teachers' professional learning on teaching proof and proving: teachers' knowledge of proof, practice of proof and beliefs about proof. The challenges teachers may face in teaching proof and proving, as well as current teacher professional learning activities on proof and proving are also discussed.

For instance, the proof method of mathematical induction introduced in secondary school mathematics is proved to be a problematic content for learning and teaching. Briefly, the method of mathematical induction proceeds in two steps: the starting step establishes that P(n) is true for an initial value $n = n_0$; and the implicative step proves that if P(k) is true, then P(k + 1) is true for an arbitrary $k \in \{n | n \in N, n \geq n_0\}$. Some researchers (e.g., Harel 2001; Stylianides et al. 2007) found that students and pre-service teachers had difficulties in understanding the requirement and sufficiency of the two steps for proving propositions with the form 'P(k), $k \in \{n \mid n \in N, n \geq n_0\}$'.

F.-L. Lin (✉) • K.-L. Yang
Department of Mathematics, National Taiwan Normal University, Taipei, Taiwan
e-mail: linfl@math.ntnu.edu.tw; kailinyang3@yahoo.com.tw

J.-J. Lo
Department of Mathematics, Western Michigan University , Kalamazoo, MI, USA
e-mail: jane-jane.lo@wmich.edu

P. Tsamir • D. Tirosh
School of Education, Tel-Aviv University, Tel-Aviv, Israel
e-mail: pessia@post.tau.ac.il; dina@post.tau.ac.il

G. Stylianides
Department of Education, University of Oxford , Oxford, UK
e-mail: gabriel.stylianides@education.ox.ac.uk

© The Author(s) 2021   327
G. Hanna and M. de Villiers (eds.), *Proof and Proving in Mathematics Education*,
New ICMI Study Series, https://doi.org/10.1007/978-94-007-2129-6_14

To make sense of the three essential components, we contrasted how an experienced teacher and a prospective teacher introduce the concept of mathematical induction in their teaching. The experienced teacher, T1, studied by Chin and Lin (2000), fruitfully used the "Hanoi Tower" activity[1] as a basis for introducing the concept of mathematical induction. T1 had a master's degree in mathematics and had taught mathematics in a public senior high school for 20 years. He used two critical questions to invite the students to think inductively:

1. Can you do it [the puzzle] when N [discs] = 3?
2. Do you believe that if N = 3 is possible then N = 4 will also be possible?

Having had the students solve the puzzle with three discs, T1 introduced a strategy of moving the three-disc solution as a unit in solving the puzzle with four discs, then suggesting the same strategy with the four-disc solution to show how the puzzle could be done with five discs. Meanwhile, T1 used a sequence of relevant questions to lead students to conjecture that the puzzle could be solved with any n discs, thus demonstrating the two steps of the principle of mathematical induction. This procedure also revealed two of T1's pedagogical beliefs:

1. mathematics teaching ought to teach students the nature of mathematical knowledge rather than mathematical forms;
2. mathematics teaching ought to motivate students' interests and willingness to learn.

(Chin and Lin 2000)

On the other hand, the prospective teacher, P1, studied by Hsieh (2005), planned a lesson using the "Hanoi Towers" to teach mathematical induction in a teaching practice with classmates acting as "pseudo-learners". However, P1 initially misstated the rules of the puzzle as:

1. Only one disc may be moved at a time;
2. No disc may be placed on top of a smaller disc;
3. A stack of discs may be taken from one peg to the target peg through the third (spare) peg.

His first and third rules contradicted each other, enabling one of this classmates to replicate T1's method of demonstrating how any number of discs might be used. This led another to question why P1 had chosen to consecutively increase the number of discs, rather than reasoning that if, say, four discs was possible, any n was possible. However, P1 did not take this chance to enrich the discussion about the nature of mathematical induction. In fact, P1 scarcely interacted with his pseudo-learners, although he did assign some tasks to them.

---

[1] The "Hanoi Tower" puzzle requires the user to move a conical stack of n discs from one of three pegs (A, B, C) and reconstruct it on another, moving only one disc at a time and never placing a larger disc on a smaller, and using the third (spare) peg as a way station when necessary.

Both T1 and P1 used the "Hanoi Tower" activity on the premise that students could make sense of the two steps of mathematical induction through solving three discs, four discs and so on. However, the same activity conducted by an experienced teacher led to a development of students' understanding but produced a little learning when conducted by a prospective teacher who had confused the given rules with the principle of demonstrating mathematical induction.

Three essential components of teachers' competency in teaching proof emerge from these two examples: knowledge specific to proof content and proof method, belief specific to the nature and didactics of proof, and practice specific to planning and teaching for motivating and guiding students' argumentation and proof. These three essential components offer understanding and guidance for mathematics teachers' professional development in teaching proof.

1.1   Three Essential Components of Teacher Professional Development

As teaching mathematics involves mathematical concepts, strategies, and reasoning methods, mathematics teachers should have "the mathematical knowledge needed to carry out the work of teaching mathematics" (Ball et al. 2008, p. 395). Similarly, teaching proof requires teachers' understanding of the content. Moreover, teaching proof demands specific knowledge to explain why one proof or some proof method is valid and to validate students' proofs.

The choice and design of tasks provide teachers' main approach to the implementation of their mathematical knowledge. Whilst selecting learning tasks, imaging and evaluating students' various modes of argumentation, and responding to students' misunderstanding in proof, mathematics teachers make many decisions according to their beliefs about proofs and the didactics of proofs (Hanna 1995; Knuth 2002; Mingus and Grassl 1999).

Whilst engaging students into mathematical argumentation and proof, mathematics teachers may negotiate new classroom norms for doing mathematics and discuss what counts as justification with students (see Simon and Blume 1996). Teachers' beliefs about mathematics learning – in particular, whether they have a conceptual or a calculational orientation – have a substantial impact on the way that they teach mathematics in their own classrooms (see Philipp 2007).

Just knowing or believing what to do is not sufficient for practice in planning and teaching lessons. Three related components – object, subject, and community – compose practice in planning and teaching (Engeström 1999). For example, how to motivate one student to frame or focus on an object in the classroom community requires mathematics teachers to turn what they know and believe into action to help students develop mathematical understanding by the investigation of different levels of mathematical validity. The literature has paid much attention to knowledge for teaching proof and belief about the nature and didactics of proof (e.g. Knuth 2002; Stylianides and Ball 2008). How teachers turn knowledge and belief into practical instruction has been less investigated.

Traditionally, school mathematics has focused primarily on formal types of proof, such as methods of mathematical induction and algorithmic proofs (e.g., two-column geometric proof at the high-school level). The releases of two Standards by the National Council of Teachers of Mathematics (NCTM 1989, 2000) in the U.S., and of other curriculum documents around the world, represent a growing recognition that proof should be treated as a tool for learning mathematics at all grade levels. However the set of accepted statements, the modes of argumentations and modes of argument representation differ between primary and secondary mathematics classrooms. Thus, teachers might encounter pedagogical challenges specific to their teaching levels. So, the following three sections will elaborate on teachers' knowledge of proof, practice of proof, and belief about proof separately for primary and secondary levels.

2 Teachers' Knowledge of Proof

2.1 Primary Teachers' Knowledge of Proof

Wittmann (2009) argued that elementary-school students, and thus their teachers, should be provided with opportunities to develop "operative proofs" with these characteristics:

1. They arise from the exploration of a mathematical problem;
2. They are based on operations with "quasi-real" mathematical objects;
3. They are communicable in a problem-oriented language with little symbolism.
   (p. 254)

So, representing odd and even numbers with mathematics counters in two rows, e.g. ⁑⁝ and ⁝⁝⁝, instead of formal symbols such as $A = 2n + 1$, would be appropriate for elementary students to use to prove that "the sum of two odd numbers is an even number":

Past research studies on primary teachers' knowledge of proof have focused mainly on pre-service teachers and on modes of argumentation. Lo and McCrory (2009) propose that future elementary teachers need to learn proof at three different levels: as a mathematical tool; as a mathematical object; and as a development, with the level of assumptions, arguments and representations depending on the students' age and grade level. Researchers have found that many primary teachers rely on external authority, such as textbooks, college instructors or more capable peers, as the basis of their conviction (e.g., Lo et al. 2008; Simon and Blume 1996); they also believe it is possible to affirm the validity of a mathematical generalisation through a few examples (Goetting 1995; Goulding et al. 2002; Martin and Harel 1989; Stylianides and Stylianides 2009b). Frequently, the ritualistic aspects of the proof (e.g., it has the look of a proof) rather than the correctness of the argument, influence such teachers' judgement of the validity of a mathematical argument (Martin and Harel 1989).

Stylianides et al. (2004, 2007), studying 70 primary pre-service teachers at the University of Cyprus, found that primary teachers are weak in their understanding of the contraposition equivalence rule (2004) and mathematical induction (2007). These researchers' in-depth analysis revealed pre-service teachers had two main types of difficulties with proof. First, they lacked understanding of the logic-mathematical underpinnings of different modes of argumentation: What contributes to the validity of a particular mode of proof and what's being proved (and not being proved) by a valid application of a particular proof method? Second, many of the participants were unable to use different modes of representations correctly and appropriately, for example, Grant and Lo (2008) found that U.S. pre-service teachers' erroneous understanding of the number line impeded their ability to use it as a tool for constructing a valid proof for justifying computation with whole numbers and fractions.

2.2   Secondary Teachers' Knowledge of Proof

Secondary teachers' necessary knowledge of proof comprises three components: constructing proofs, evaluating proofs, and knowledge of content and students. Often, prospective secondary mathematics teachers have to complete a first degree with a required focus on mathematical content at the tertiary level. Thus, one might assume, secondary teachers are knowledgeable in constructing proofs, a part of the mathematical knowledge which Shulman (1986) called subject-matter knowledge (SMK). Yet, research reports are conflicting on this point. For example, Barkai et al. (2009) asked 50 practising high school teachers to prove or refute six statements within the context of Elementary Number Theory. All teachers gave valid proofs (or refutations) using either a symbolic or numeric mode of argument. For geometry, Sun (2009) and Sun and Chan (2009) found that prospective teachers were able to construct a variety of original and creative proofs for different theorems. On the other hand, Schwarz and Kaiser's (2009) comparative study of future teachers in Germany, Hong Kong, and Australia found that the majority could not construct formal proofs, even though these proofs required only lower-secondary mathematical content. In Zaslavsky and Peled's (1996) study of 36 high-school teachers' SMK, only a third of the teachers could provide counter-examples to the (false) statement "All commutative actions are also associative."

Being able to evaluate whether a proof is valid also forms an element of SMK. Brown and Stillman (2009) found that only some recent graduates could recognise generality in pre-formal proofs, whilst others could not. Other studies found that the mode of argument may also be related to teachers' value judgements of proofs.

Dreyfus (2000), following Healy and Hoyles' (1998) work with high school students, presented 44 secondary school teachers with nine justifications for the claim "The sum of any two even numbers is even". He found that most of the teachers easily recognised formal proofs, but had little or no appreciation for other types of justifications, such as verbal, visual or generic. In contrast, other studies have shown that students may prefer a verbal mode over symbolic representation (Healey and Hoyles 2000). Nevertheless, in a study of 50 high school teachers evaluating verbal proofs, half of the teachers rejected correct verbal proofs, claiming that these justifications lacked generality and were mere examples (Tsamir et al. 2009b). In addition, only about half of the teachers identified the incorrectness of a symbolic justification that was not general (Tsamir et al. 2008). Similarly, Knuth (2002) reported that, when needing to determine an argument's validity, teachers seem to "focus solely on the correctness of the algebraic manipulations rather than on the mathematical validity of the argument" (p. 392). In other words, when presented with an algebraic justification, the teachers focused on the examination of each step, ignoring the need to evaluate the validity of the argument as a whole.

Yet, even if able to do so themselves, teachers also require pedagogical-content knowledge (PCK) in order to be able to teach their students how to construct and evaluate proofs. Ball and colleagues (Ball et al. 2008) differentiated two components of PCK, knowledge of content and students (KCS) and knowledge of content and teachers (KCT): KCS "combines knowing about students and knowing about mathematics," whereas KCT "combines knowing about teaching and knowing about mathematics" (p. 401).

Regarding the teaching of proofs, KCS includes knowing the types of proofs students may construct in different mathematical contexts. For instance, Tabach et al. (2009) asked their 50 high school teachers to suggest correct and incorrect proofs their students might construct for each of six statements within the context of Elementary Number Theory. As to mode of representation, the teachers (predictably) suggested verbal representations less often than numeric or symbolic ones for both correct and incorrect suggested proofs. The suggestions for incorrect proofs most frequently cited mistakes related to the formal content of proof. This result may be related to teachers' SMK of the formal nature of proof and their PCK of students' difficulties with this formality.

Although it may shed light on teachers' SMK, evaluating proofs may also shed light on teachers' KCS. For example, a teacher may reject a student's proof, even though it is mathematically correct, when the proof is not minimal (e.g., Tsamir et al. 2009a). This may reflect the teacher's PCK. A teacher's judgement of a proof may take into consideration not only mathematical validity and the student's ability to validate or refute a statement, but also the student's knowledge of what is sufficient for proving or refuting a given type of statement. In other words, the evaluation may go beyond the correctness of the justification, reflecting the teacher's SMK, to consider the student's familiarity with the relevant "big ideas", reflecting the teacher's PCK.

3   Teachers' Practice of Proof

3.1   Primary Teachers' Practice of Proof

The studies on primary teachers' knowledge of proof focused primarily on individual teachers and treated their knowledge as a static variable. However, the studies on primary teachers' teaching of proof frequently consider the proving activities in the classroom as interactive. Collectively, the classroom community influences what can be accepted as an acceptable explanation or valid justification; it establishes a sociomathematical norm (Yackel and Cobb 1996). Various theoretical frameworks, such as Toulmin's scheme of argumentation that includes claims, data, warrants, and backing, have been used to analyse explanation and disagreement arising in these class discussions (Yackel 2002). As a member of the classroom community, the teacher plays a central role in establishing an inquiry-based classroom environment that honours both mathematics as a discipline and students as learners (e.g., Ball and Bass 2003; Lampert 2001; Stylianides 2007a, b; Wood 1999).

Stylianides (2007a) analysed three proof-learning episodes from Grade 3 classroom to illustrate various moves teachers made after the base argument (the prevailing student argument at the beginning of the proving activity) was established. The analysis led to the development of a framework of instructional practices for proof and proving in school mathematics that includes four possible courses of action (Stylianides 2007a). Upon recognising a proving instance in the classroom, the teacher first evaluates the base argument with respect to the three components of a proof: set of accepted statements, modes of argumentation and modes of argument representation. If the base argument qualifies as a proof, then the teacher can either (a) bring the proving activity to a close, or (b) help the class develop a more advanced proof by further expanding the tools available in one or more of the components. If it does not qualify as a proof, then the teacher can focus the instruction on the component(s) that disqualify it. Subsequently, the class community develops either a proof or a more advanced argument. The latter occurs when one or more of the proof components are beyond the students' current conceptual reach. This framework further emphasises the importance of teachers' taking an active role in managing the proving activities in their classrooms.

3.2   Secondary Teachers' Practice of Proof

The implementation of proof tasks in the secondary classroom is not simple. In a survey of 27 high school teachers in Brazil, most teachers were not able to describe a single proof-related activity they had developed with their students (Healy et al. 2009). Bieda (2009) observed middle school teachers and their students who participated in the Connected Mathematics Project, a curriculum incorporating proof-related tasks. Even though most teachers implemented the tasks as written, not all the students' generalisations were followed up with justifications. When justifications were not forthcoming, many teachers did not provide necessary feedback, missing the opportunities to engage students in proving activities. In other words, even when teachers are provided with tasks that encourage proving, implementing the proof part of the activity does not necessarily follow.

Herbst (2002) analysed high school geometry lessons to identify teachers' actions that help students in proving geometrical statements. He found that focusing on the form of proving, without stressing the ideas of the proof, may lead students to focus on the form but not on the logic. Herbst (2009) hypothesised that a didactical contract exists between the teacher and students about the knowledge of proofs that students are expected to learn. Herbst (2002) identified three proof-related situations: "installing a theorem", "doing a proof", and "calculating (a measure)". The activities in each situation were guided by norms. For example, when "doing proofs", it was the norm for the teacher to state the "given" and provide the conclusion to be reached – "prove". It was then the norm for students to write a sequence of statements and a reason to follow each statement.

4 Teachers' Beliefs/Values About Proof

Teachers' values and beliefs regarding mathematics and the kind of mathematics they bring to the classroom also relate to classroom practice. Thus, a teacher who values proofs and believes that it is important for students to experience proving will provide students with related activities. Philipp (2007) distinguished between value and beliefs: "Whereas beliefs are associated with a true/false dichotomy, values are associated with a desirable/undesirable dichotomy" (p. 259). Although we acknowledge this distinction between the two terms, here we follow the terminology used in the specific studies discussed.

4.1 Primary Teachers' Values About Proof

Bishop (1999) called for more investigations on the values of teaching and learning in mathematics classrooms almost 20 years ago. However, research studies in this area are still few and have tended to deal in larger generalities, like beliefs about mathematics as a scientific discipline; about teaching and learning mathematics, about teaching and learning in general, about teacher education and professional development (cf. Kaiser et al. 2007), or about pedagogical values systems (Chin and Lin 2000).

Raymond (1997) found that many primary teachers still hold traditional views about the nature of mathematics; for example, "Mathematics is an unrelated collection of facts, rules and skills" and "Mathematics is fixed, predictable, absolute, certain, and applicable" (p. 556). This view of mathematics frequently accompanies traditional teaching practices, including teacher-centred and-controlled lectures and demonstrations, ample individual seatwork and no student input other than providing yes/no or short answers.

4.2   Secondary Teachers' Values About Proof

Cirillo (2009) found that a secondary school teacher was able to deflect his students' exasperation at the difficult and time-consuming construction of proofs because he believed that "real math" had to involve proof; his belief about mathematics affected his classroom practice. Beliefs are also related to the teaching of mathematics. For example, some teachers see proof as a means of convincing, whilst others see it as a means of promoting mathematical understanding (Furinghetti and Morselli 2009). Although the manual Principles and Standards for School Mathematics (NCTM 2000) recognises that reasoning and proof are fundamental in all content areas, some teachers believe that geometry is the ideal domain for teaching proofs (Furinghetti and Morselli 2009).

Teachers may also hold beliefs regarding proof methods, which in turn may influence their evaluation of students' proofs. When investigating secondary school teachers' evaluation of arguments based on visual reasoning, Biza et al. (2009) found that some teachers may accept a visual argument that refutes a statement but not one that is used to prove a statement. These teachers believed that in order to prove the validity of a statement an algebraic argument is necessary; one teacher would not accept any graphic argument because she believed that students should be able to provide an algebraic argument. Biza et al.'s results exemplify the ongoing debate within the mathematics community about whether a visual representation may be accepted as a proof (Hanna 2000).

4.3   How do the Three Components of Teachers' Profession Interact with Each Other?

The above three components necessary for proofs and proving in the elementary and secondary schools are interrelated. First, a teacher's SMK, including being able to construct and evaluate proofs, is necessary. Teachers' PCK, including knowledge of students' proving methods and knowledge of appropriate tasks that encourage proofs, is also necessary. Research studies have shown that primary teachers are generally weak in both areas. Even secondary teachers with strong SMK, provided with appropriate tasks, missed proving opportunities. This situation suggests a missing element, perhaps knowledge regarding how to implement tasks or a belief about the value proofs have in mathematics classrooms.

A closer look at the lessons observed by Herbst (2002, 2009) in geometry classes reveals how the three components are interrelated. For example, the teachers' knowledge of proving tasks in geometry may have been limited to a specific form. This limited knowledge then affected their practice so that the specific task form became the "task norm" in their classes. In addition, the teachers observed by Herbst apparently believed that all geometric proofs must be formal and follow a certain path, leading them to focus on form over logic. Similarly, the proving activities may become an exercise of teachers demonstrating specific steps for students to memorise if that the practice is compatible with their beliefs about doing proof in elementary classrooms. Thus, beliefs as well as knowledge affect practice.

Knowledge and beliefs not only affect practice, they affect each other. A teacher familiar only with algebraic proof may in turn believe that it is the only acceptable form. However, a teacher exposed to other proof methods, such as visual proofs, may learn to value those methods as well. On the other hand, a teacher with a hard and fast belief that all proofs must be formal and rigorous may not be open to learning about other proof methods. Teachers' PCK regarding students' difficulties with formal proofs may lead teachers to believe that formal proofs are usually inaccessible and thus inappropriate for students.

Often, teachers' knowledge and beliefs rest on their past experiences as teachers and students. Thus, it is important to give teachers new experiences to build on. Research has shown that collaboration between researchers and school teachers may have a positive impact not only on teachers' knowledge but on their beliefs and practices as well. In one such study, teachers evaluating students' arguments, who initially dismissed generic examples as purely empirical, eventually accepted this type of argument and became cognisant of the general reasoning inherent in it (Healy et al. 2009). In addition, the teachers went from valuing only formal proofs to allowing natural language as well. In other words, the collaboration led teachers to increase their knowledge (regarding generic examples) which in turn affected their values (regarding formal proofs) which in turn affected their practice (regarding the evaluation of students' proofs).

5   Challenges About Teaching Proof

A teacher may face several challenges relating to the domain of proofs and proving: different mathematical domains, different perceptions of the role of proofs, classroom norms, curriculum materials, a variety of statement types to present, and use of exemplification versus deductive proof. Here we aim to highlight the dilemmas and to point out possible directions for resolutions.

5.1 Mathematical Domains

First, consider the choice of mathematical domains involved in teaching proof and proving. On the one hand, curricula often require implementing proof and proving as part of classroom practice in various mathematical domains: "By exploring phenomena, justifying results, and using mathematical conjectures in all content areas and—with different expectations of sophistication—at all grade levels, students should see and expect that mathematics makes sense," as the NCTM Standards put it (NCTM 2000, p. 4). On the other hand, many teachers teach proofs only in secondary school geometry courses. Furinghetti and Morselli (2009) interviewed 10 practising secondary school teachers in order to understand why. These teachers referred to Euclidean geometry as "the most suitable domain for the teaching of proof" (p. 170). The researchers concluded that teachers' beliefs influence how and when teachers incorporate proofs. Berg (2009) called for attention to the importance of the mathematical context within which the idea of proof is addressed. She found a substantial difference in practising middle-school teachers' attitudes towards geometrical and numerical tasks, both of which she considered as proving tasks. However, the teachers related to only the geometrical task as a proving task; they considered the numerical task as merely a pattern inquiry task.

5.2 Proofs for Mathematicians vs. Proofs in School

The teacher's knowledge about proofs in the domain of mathematics, as compared to proofs in school mathematics, constitutes another issue. Sun (2009) contrasts the role of proofs in the two settings: "mathematicians write own proofs in order to establish the truth of a proposition. In contrast, students … receive the ready-made proofs mainly presented by their teachers according to their textbooks and syllabus and then routinely memorize theorems and proofs" (p. 180). Teachers may value the role proofs play for practising mathematicians, and yet teach proofs only by rote learning because they do not believe their students can create proofs. The teacher's beliefs and values interact with their knowledge to result in the enacted practice. For example, Cirillo (2009) followed one high-school teacher for 3 years, observed him in class and interviewed him in order to better understand his conceptions of proofs in the context of school mathematics. Cirillo argued that both the teacher's past experience with rigorous proofs in college level courses and the available curriculum materials were an obstacle to enacting a practice of proving in his class. The teacher recalled the traumatic shock he felt when first asked to create what he called "real proof" at university level; he felt that until then he had never had to. In order to prevent their experiencing the same trauma, he decided to introduce his high-school students to rigorous proofs.

5.3   Classroom Norms

Proofs are usually not the first thing a mathematics teacher presents in class. Hence, by the time the issue of proofs emerges, some other sociomathematical norms may already be established in the class. Douek (2009), drawing on her experience with middle-school teachers and Pythagoras' theorem, identified two difficulties in establishing the habit of proving in class: "One difficulty consists in the fact that 'to produce a conjecture' is a task that does not fit the most frequent didactical contract ... Moreover the presentation and management of the tasks in a way that guides students' work but does not prevent creativity is not easy" (p. 147). According to Douek, incorporating proofs and proving in classroom practice necessitates a profound change in the already established mathematical habits of mind. In other words, a teacher needs to consider whether and how to productively renegotiate the didactic contract and arrive at more amenable sociomathematical norms.

5.4   Available Curriculum Materials

Amongst other things, a teacher planning a learning sequence considers the materials available for both teacher and students. However, the agenda expressed in written curricular materials may not be compatible with the teacher's agenda regarding the issues of proofs and proving. For example, Bieda (2009) reported on a study done in seven middle-grade classrooms. The teachers used a reform-oriented curriculum, the Connected Mathematics Project (CMP), in which at least 40% of the curriculum tasks encourage students to reason and justify. Bieda's analysis of 49 lessons revealed that 71% of the lesson time was devoted to tasks that involved reasoning, but only about half of the possible proving activities were carried out.

5.5   Types of Statements to Present to Students

Traditionally, teachers present valid universal statements in class. However, during inquiry-based activities, students may also raise conjectures and validate or refute them; these conjectures may not be valid universal statements. Prior to inquiry activities, the teacher could 'prepare the ground' for students by systematically introducing a variety of statements as well as their relevant proving methods. For example, Tsamir et al. (2008) suggested a systematic way for categorising statements according to the quantifier, which may be universal or existential, and the predicate, which may be "always true", "sometimes true" or "never true". The combination of predicate and quantifier determine the statement's degree of validity, and hence the possible methods of proving it. This statement organiser can be presented as a six-cell matrix as follows:

Predicate quantifier	Always true	Sometimes true	Never true
Universal	Cell 1.	Cell 2.	Cell 3.
	True statement	*False statement*	*False statement*
	General proof	*Counter example*	*Counter example*
Existential	Cell 4.	Cell 5.	Cell 6.
	True statement	*True statement*	*False statement*
	Supportive example	*Supportive example*	*General proof*

In *italic* – the truth-value of a statement and the minimal necessary and sufficient mode of argumentation needed as a proof

Such a matrix may help the teacher to decide which type of statements to choose. It may also help the teacher in considering mathematical statements during various phases of practice: whilst planning, whilst making sense of curriculum materials, whilst teaching a lesson, whilst assessing students' knowledge, and so on.

5.6   Shifting Students Proofs Towards Deductive Thinking

Many students consider experimental verifications sufficient to demonstrate the validity of a statement. This tendency may dominate even more in their use of dynamic geometry software, where they can generate numerous examples by 'dragging'. Teachers face a dilemma: how to turn general proofs from a ritual imposed on students to a need motivating them to search for general arguments. Kunimune et al. (2009) studied this problem in the context of teachers' practices in secondary schools in Japan. They found that the tendency towards experimental proof is hard to change, especially in lower-level students. Fischbein (1982) also found a similar tendency towards empirical proof in the context of number theory. However, as Kunimune et al. (2009) reported, a class discussion aimed at shaking students' beliefs about experimental verification and making deductive proof meaningful for them can make a crucial difference.

Taking into consideration the above mentioned dilemmas, one may wonder what teacher educators can do to help teachers in the complex domain of implementing proofs and proving as an integral part of everyday practice. One answer is to raise awareness; that is, teacher educators ought to expose teacher trainees to these tensions and dilemmas and provide a secure environment for both practising and prospective teachers to discuss these and other problems.

6 Professional Learning Activities for Teaching Proof and Proving

One early study that focused on teachers' learning of proof, by Simon and Blume (1996), provided useful insights into how mathematics teacher educators can establish norms in mathematics courses for prospective elementary teachers. According to these norms, members of the classroom community need to justify their ideas and the other members need to be involved in evaluating those ideas. More recent work with prospective elementary teachers (Stylianides and Stylianides 2009b) and prospective and practising secondary teachers (Zaslavsky 2005) capitalised on Simon and Blume's concepts and added the notion of cognitive conflict as a mechanism to help teachers develop their knowledge about proof. All three studies used a range of activities to help teachers develop their understanding of proof: solving proof tasks individually or in small groups, having whole group discussions, sharing and critiquing each others' proofs, and so on. Yet, these studies' most important contribution lies in the method they used to implement such activities, which involves three important dimensions: establishing conviction, the role of the teacher educator, and the notion of cognitive conflict.

6.1 Establishing Conviction

When introducing teachers to the notion of proof, it is important to consider how an argument can establish conviction in the truth or falsity of mathematical claim. Mason (1982) distinguished amongst three levels of conviction (in increasing level of sophistication): (a) convince yourself, (b) convince a friend, and (c) convince a sceptic. The first emphasises the cognitive/individual dimension of conviction in relation to the process of "ascertaining" (Harel and Sowder 2007); that is, removing one's own doubts about a claim's truth or falsity. The other two levels emphasise the social dimension of conviction in relation to the process of "persuading" (Harel and Sowder 2007); that is, removing others' doubts about a claim's truth or falsity.

In nonmathematical communities, to justify an idea (i.e., to establish conviction) an argument;

> must proceed from knowledge that is taken-as-shared in the community, must be seen by the community as logical (i.e., each assertion following reasonably from the previous one), and the idea must fit with knowledge that has previously been accepted by that community (Simon and Blume 1996, p. 6).

This excerpt relates to Mason's (1982) second and third levels of conviction. The criteria described in the excerpt and the criteria adopted in our three focal studies (Simon and Blume 1996; Styliandides and Stylianides 2009b; Zaslavsky 2005) relate to the third level of conviction in mathematics. According to the excerpt, in nonmathematical communities the community is the sole judge of whether or not an argument establishes conviction. In the three focal studies, the teacher educators were trying to help teachers develop arguments consistent with the conventional understandings and criteria of mathematics. In Stylianides and Stylianides's study (2009b), for example, attaining the third level of conviction required additionally that the persuasion satisfied certain criteria developed with the teacher educators' guidance (see Stylianides and Stylianides 2009a), and rooted in criteria shared in the mathematical community. Both Simon and Blume (1996) and Stylianides and Stylianides (2009b) observed that arguments to persuade a sceptic had to be seen as logical not only by the standards of the teacher education classroom community but also according to the conventions about proof of the mathematical community. These conventional understandings served as criteria for considering an argument persuasive enough to convince a sceptic.

6.2   The Role of the Teacher Educator

In all three studies (Simon and Blume 1996; Stylianides and Stylianides 2009b; Zaslavsky 2005), teacher educators played an important role in designing and implementing professional activities in the area of proof. They did not communicate conventional understandings to the teachers merely by means of asserting their authority. Rather, they tried to help the teachers develop better understandings of proof through social interactions. In all three studies, the teacher educators implemented scaffolding strategies, such as focusing the class discussion on why a certain approach to a task was appropriate, asking prompting questions, and encouraging the trainees to develop "compelling arguments" for why a general claim "always" held. However, this aid did not always result in the students' recognition of the limitations of "weak arguments" they had developed, arguments that did not meet the standard of proof. For example, Simon and Blume (1996, pp. 10–17) describe an episode when the prospective teachers could not see a problem in their empirical arguments to justify a mathematical claim and resisted developing more "sophisticated" deductive arguments. In other words, the prospective teachers did not engage spontaneously in a "situation of justification" (ibid.).

Such episodes raised a need for the development of ways to motivate teachers' engagement in sophisticated proofs. Yet, there remained situations in which the prospective teachers saw the need for more sophisticated arguments but could not develop them due to time constraints, conceptual barriers or other problems. In such situations, one teacher educator (Stylianides and Stylianides 2009b) took a more active role than scaffolding, offering the prospective teachers access to the conventional mathematical knowledge which they needed but could not access themselves.

6.3   The Notion of Cognitive Conflict

Simon and Blume (1996) suggested the usefulness of creating classroom norms for engaging prospective teachers in proof. More recently, Zaslavsky (2005) and Stylianides and Stylianides (2009b) introduced the notion of cognitive conflict.

F.-L. Lin et al.

Conflict teaching aims "to help students reflect on their current mathematical understandings [...] confront contradictions that arose in situations in which some of these understandings no longer held, and recognize the importance (need) of modifying these understandings to resolve the contradictions" (Stylianides and Stylianides 2009b, p. 319).

Zaslavsky (2005) and Stylianides and Stylianides (2009b) used cognitive conflict (a) to address robust and persistent student misconceptions about proof, for instance that empirical arguments count as proofs (e.g., Coe and Ruthven 1994; Goetting 1995; Harel and Sowder 1998; Healy and Hoyles 2000; Martin and Harel 1989), and (b) to achieve students' spontaneous engagement in situations of justification (Simon and Blume 1996).

More specifically, Zaslavksy (2005) illustrated how cognitive conflict can be used to motivate students' engagement in specific proving tasks, thereby promoting important mathematical and pedagogical knowledge. Stylianides and Stylianides (2009b) suggested engaging teachers in activities that provoked a "cognitive conflict" in them, helped them realise the limitations of their current "justification schemes" (Harel and Sowder 1998), and thus created in them an "intellectual need" (Harel 1998) to develop more sophisticated arguments. Stylianides and Stylianides (2009b) also implemented theoretical ideas through "instructional sequences" (i.e., a series of tasks and associated instructor actions) developed in their "design experiment" (see, e.g., Cobb et al. 2003; Schoenfeld 2006). In order to address problems faced by the cognitive conflict approach:

> An attempt to create uncertainty by confronting a learner with a mathematical contradiction may not necessarily lead to cognitive conflict as appears to be the case in many examples of people living at peace with mathematical inconsistencies (Tall 1990; Tirosh 1990; Vinner 1990). Even when conflict is evoked, it may not be effective, as in the case of intuitive rules that to a large extent are stable and resistant to change (Tirosh et al. 1998). (Zaslavsky 2005, p. 318)

Finally, both groups of researchers used "critical reflection" (Zaslavsky 2005; cf. Stylianides and Stylianides 2009b) in their practice to refine and improve the use of cognitive conflict in their work.

7 Conclusion

We argued that the three components – knowledge, practice and values/beliefs – of teachers' professional development in teaching proof and proving are inter-related. The survey in Sects. 2–6 above of research regarding knowledge, practice and values revealed that the literature on aspects of teachers' knowledge was noticeably larger than that on the other two components. For the most part, this reflects research studies presented at the ICMI Study 19 Conference (2009). Not to say that knowledge is more important than practice and beliefs: rather, we need to further investigate teachers' beliefs and practice, as well as the interrelationship amongst knowledge, beliefs and practice in the teaching of proofs and proving.

Therefore, it is reasonable to invite studies that focus on the three components simultaneously, as well as the interrelationship between them. Taking the current trend of mathematics education research into consideration, we discuss here possible designs for such research studies.

7.1 Teacher as Designer in a Three-Tiered Design Research

Lesh et al. (2007) have addressed the research trend of multi-tiered design experiments. Taking the teacher as designer, we propose "tasks designing" as a learning strategy for teachers' professional development (Lin 2010). Imagine a three-tiered research design: educator level, teacher level and student level. Educators would design learning activities for teachers; participating teachers would design proving task sequences for their students; these tasks would then be tested with the designers' students, first in small groups, and then in whole classes (Lin 2010).

Teachers designing instructional task sequences of proof are involved in both constructing and testing: constructing and sequencing the tasks, and testing these with students and against the principles of conjecturing and proving.

Therefore, teachers acting as designers are expected to gain knowledge on constructing proof, evaluating proofs and about students; to put this knowledge into and to develop their beliefs about proof and its teaching. Researchers should then be able to view all three components of teachers' professional development in interactions amongst teachers who are designers.

References*

Ball, D. L., & Bass, H. (2003). Making mathematics reasonable in school. In J. Kilpatrick, W. G. Martin, & D. Schifter (Eds.), *A research companion to principle and standards for school mathematics* (pp. 27–44). Reston: National Council of Teachers of Mathematics.

Ball, D. L., Thames, M. H., & Phelps, G. (2008). Content knowledge for teaching: What makes it special? *Journal of Teacher Education, 59*(5), 389–407.

Barkai, R., Tabach, M., Tirosh, D., Tsamir, P., & Dreyfus, T. (2009, February). *Modes of argument representation for proving – The case of general proof.* Paper presented at the Sixth Conference of European Research in Mathematics Education - CERME 6, Lyon, France.

Berg, C. V. (2009). A contextualized approach to proof and proving in mathematics education: Focus on the nature of mathematical tasks (Vol. 1, pp. 100–105).*

Bieda, K. N. (2009). Enacting proof in middle school mathematics (Vol. 1, pp. 59–64).*

---

*NB: References marked with * are in F. L. Lin, F. J. Hsieh, G. Hanna, & M. de Villiers (Eds.) (2009). *ICMI Study 19: Proof and proving in mathematics education.* Taipei, Taiwan: The Department of Mathematics, National Taiwan Normal University.

Bishop, A. J. (1999). Mathematics teaching and values education: An intersection in need of research. *Zentralblattfuer Didaktik der Mathematik, 31*, 1–4.

Biza, I., Nardi, E., & Zachariades, T. (2009). Do images disprove but do not prove? Teachers' beliefs about visualization (Vol. 1, pp. 59–64).*

Brown, J., & Stillman, G. (2009). Preservice secondary teachers' competencies in proof (Vol. 1, pp. 94–99).*

Chin, C., & Lin, F. L. (2000). A case study of a mathematics teacher's pedagogical values: Use of a methodological framework of interpretation and reflection. *Proceedings of the National Science Council, Taiwan, Part D: Mathematics, Science, and Technology Education, 10*(2), 90–101.

Cirillo, M. (2009). Challenges to teaching authentic mathematical proof in school mathematics (Vol. 1, pp. 130–135).*

Cobb, P., Confrey, J., DiSessa, A., Lehrer, R., & Schauble, L. (2003). Design experiments in educational research. *Educational Researcher, 32*(1), 9–13.

Coe, R., & Ruthven, K. (1994). Proof practices and constructs of advanced mathematics students. *British Educational Research Journal, 20*, 41–53.

Douek, N. (2009). Approaching proof in school: From guided conjecturing and proving to a story of proof construction (Vol. 1, pp. 142–147).*

Dreyfus, T. (2000). Some views on proofs by teachers and mathematicians. In A. Gagatsis (Ed.), *Proceedings of the 2nd Mediterranean Conference on Mathematics Education* (Vol. I, pp. 11–25). Nicosia: The University of Cyprus.

Engeström, Y. (1999). Activity theory and individual and social transformation. In Y. Engeström, R. Miettinen, & R. L. Punamaki (Eds.), *Perspectives on activity theory* (pp. 19–38). Cambridge: Cambridge University Press.

Fischbein, E. (1982). Intuition and proof. *For the Learning of Mathematics, 3*(2), 9–18, 24.

Furinghetti, F., & Morselli, F. (2009). Teacher's beliefs and the teaching of proof (Vol. 1, 166–171).*

Goetting, M. (1995). *The college students' understanding of mathematical proof.* Unpublished doctoral dissertation, University of Maryland, College Park.

Goulding, M., Rowland, T., & Barber, P. (2002). Does it matter? Primary teacher trainees' knowledge in mathematics. *British Educational Research Journal, 28*, 689–704.

Grant, T. J., & Lo, J. (2008). Reflecting on the process of task adaptation and extension: The case of computational starters. In B. Clarke, R. Millman, & B. Grevholm (Eds.), *Effective tasks in primary mathematics teacher education* (pp. 25–36). New York: Springer Science+Business Media, LLC.

Hanna, G. (1995). Challenges to the importance of proof. *For the Learning of Mathematics, 15*(3), 42–49.

Hanna, G. (2000). Proof, explanation and exploration: An overview. *Educational Studies in Mathematics, 44*, 5–23.

Harel, G. (1998). Two dual assertions: The first on learning and the second on teaching (or vice versa). *The American Mathematical Monthly, 105*, 497–507.

Harel, G. (2001). The development of mathematical induction as a proof scheme: A model for DNR-based instruction. In S. Campbell & R. Zazkis (Eds.), *Learning and teaching number theory.* Norwood: Ablex Publishing Corporation.

Harel, G., & Sowder, L. (1998). Students' proof schemes: Results from exploratory studies. In A. H. Schoenfeld, J. Kaput, & E. Dubinsky (Eds.), *Research in collegiate mathematics education III* (pp. 234–283). Providence: American Mathematical Society.

Harel, G., & Sowder, L. (2007). Toward comprehensive perspectives on the learning and teaching of proof. In F. K. Lester (Ed.), *Second handbook of research on mathematics teaching and learning* (pp. 805–842). Greenwich: Information Age.

Healy, L., & Hoyles, C. (1998). *Justifying and proving in school mathematics.* London: Institute of Education, University of London.

Healy, L., & Hoyles, C. (2000). A study of proof conception in algebra. *Journal for Research in Mathematics Education, 31*, 396–428.

Healy, L., Jahn, A. P., & Frant, J. B. (2009). Developing cultures of proof practices amongst Brazilian mathematics teachers (Vol. 1, pp. 196–201).*

Herbst, P. (2002). Engaging students in proving: A double bind on the teacher. *Journal for Research in Mathematics Education, 33*(3), 176–203.

Herbst, P. (2009). Testing a model for the situation of "doing proofs" using animations of classroom scenarios (Vol. 1, pp. 190–195).*

Hsieh, C. J. (2005). On misconceptions of mathematical induction. *Letter of the History and Pedagogy of Mathematics in Taiwan (HPM), 8*(2–3), 14–21 (in Chinese).

Kaiser, G., Schwarz, B., & Krackowitz, S. (2007). The role of beliefs on future teachers' professional knowledge. *The Montana Mathematics Enthusiast, Monograph, 3*, 99–116.

Knuth, E. J. (2002). Teachers' conceptions of proof in the context of secondary school mathematics. *Journal of Mathematics Teacher Education, 5*, 61–88.

Kunimune, S., Fujita, T., & Jones, K. (2009). "Why do we have to prove this?" Fostering students' understanding of 'proof' in geometry in lower secondary school (Vol. 1, pp. 256–261).*

Lampert, M. (2001). *Teaching problems and the problems of teaching*. New Haven: Yale University Press.

Lesh, R., Hamilton, E., & Kaput, J. (2007). Directions for future research. In R. Lesh, E. Hamilton, & J. Kaput (Eds.), *Foundations for the future in mathematics education* (pp. 449–453). Mahweh: Lawrence Erlbaum Associates.

Lin, F. L. (2010). Mathematical tasks designing for different learning settings. In M. M. F. Pinto & T. F. Kawasaki (Eds.), *Proceedings of the 34th Conference of the International Group for the Psychology of Mathematics Education* (Vol. 1, pp. 83–96). Belo Horizonte: PME.

Lo, J., & McCrory, R. (2009). Proof and proving in mathematics for prospective elementary teachers (Vol. 2, pp. 41–46).*

Lo, J., Grant, T., & Flowers, J. (2008). Challenges in deepening prospective teachers' understanding of multiplication through justification. *Journal of Mathematics Teacher Education, 11*, 5–22.

Martin, W. G., & Harel, G. (1989). Proof frames of preservice elementary teachers. *Journal for Research in Mathematics Education, 20*, 41–51.

Mason, J., Burton, L., & Stacey, K. (1982). *Thinking mathematically*. London: Addison-Wesley.

Mingus, T., & Grassl, R. (1999). Preservice teacher beliefs about proofs. *School Science and Mathematics, 99*(8), 438–444.

National Council of Teachers of Mathematics [NCTM]. (1989). *Curriculum and Evaluation Standards for School Mathematics*. Reston, Virginia: National Council of Teachers of Mathematics.

National Council of Teachers of Mathematics [NCTM]. (2000). *Principles and standards for school mathematics*. Reston: National Council of Teachers of Mathematics.

Philipp, R. A. (2007). Mathematics teachers' beliefs and affect. In F. K. Lester (Ed.), *Second handbook of research on mathematics teaching and learning* (pp. 257–318). Charlotte: Information Age.

Raymond, A. M. (1997). Inconsistency between a beginning elementary school teacher's mathematics beliefs and teaching practice. *Journal for Research in Mathematics Education, 28*, 550–576.

Schoenfeld, A. H. (2006). Design experiments. In J. L. Green, G. Camilli, P. B. Elmore, A. Skukauskaite, & E. Grace (Eds.), *Handbook of complementary methods in education research* (pp. 193–205). Washington, DC: American Educational Research Association.

Schwarz, B., & Kaiser, G. (2009). Professional competence of future mathematics teachers on argumentation and proof and how to evaluate it (Vol. 2, pp. 190–195).*

Shulman, L. S. (1986). Those who understand: Knowledge growth in teaching. *Educational Researcher, 15*(2), 4–14.

Simon, M. A., & Blume, G. W. (1996). Justification in the mathematics classroom: A study of prospective elementary teachers. *The Journal of Mathematical Behavior, 15*, 3–31.

Stylianides, A. J. (2007a). Proof and proving in school mathematics. *Journal for Research in Mathematics Education, 38*, 289–321.

Stylianides, A. J. (2007b). The notion of proof in the context of elementary school mathematics. *Educational Studies in Mathematics, 65*, 1–20.

Stylianides, A., & Ball, D. (2008). Understanding and describing mathematical knowledge for teaching: Knowledge about proof for engaging students in the activity of proving. *Journal of Mathematics Teacher Education, 11*, 307–332.

Stylianides, A. J., & Stylianides, G. J. (2009a). Proof constructions and evaluations. *Educational Studies in Mathematics, 72*(2), 237–253.

Stylianides, G. J., & Stylianides, A. J. (2009b). Facilitating the transition from empirical arguments to proof. *Journal for Research in Mathematics Education, 40*, 314–352.

Stylianides, A. J., Stylianides, G. J., & Philippou, G. N. (2004). Undergraduate students' understanding of the contraposition equivalence rule in symbolic and verbal contexts. *Educational Studies in Mathematics, 55*, 133–163.

Stylianides, G. J., Stylianides, A. J., & Philippou, G. N. (2007). Preservice teachers' knowledge of proof by mathematical induction. *Journal of Mathematics Teacher Education, 10*, 145–166.

Sun, X. (2009). Renew the proving experiences: An experiment for enhancement trapezoid area formula proof constructions of student teachers by "one Problem Multiple Solutions" (Vol. 2, pp. 178–183).*

Sun, X., & Chan, K. (2009). Regenerate the proving experiences: An attempt for improvement original theorem proof constructions of student teachers by using spiral variation curriculum (Vol. 2, pp. 172–177).*

Tabach, M., Levenson, E., Barkai, R., Tsamir, P., Tirosh, D., & Dreyfus, T. (2009). Teachers' knowledge of students' correct and incorrect proof constructions (Vol. 2, pp. 214–219).*

Tall, D. (1990). Inconsistencies in the learning of calculus and analysis. *Focus on Learning Problems in Mathematics, 12*(3/4), 49–63.

Tirosh, D. (1990). Inconsistencies in students' mathematical constructs. *Focus on Learning Problems in Mathematics, 12*(3/4), 111–129.

Tirosh, D., Stavy, R., & Cohen, S. (1998). Cognitive conflict and intuitive rules. *International Journal of Science Education, 20*(10), 1257–1269.

Tsamir, P., Tirosh, D., Dreyfus, T., Barkai, R., & Tabach, M. (2008). Inservice teachers' judgment of proofs in ENT. In O. Figueras, J. L. Cortina, S. Alatorre, T. Rojano, & A. Sépulveda (Eds.), *Proceedings of the 32nd Conference of the International Group for the Psychology of Mathematics Education* (Vol. 4, pp. 345–352). Morélia: PME.

Tsamir, P., Tirosh, D., Dreyfus, T., Barkai, R., & Tabach, M. (2009). Should proof be minimal? Ms T's evaluation of secondary school students' proofs. *Journal of Mathematics Behavior, 28*, 58–67.

Tsamir, P., Tirosh, D., Dreyfus, T., Tabach, M., & Barkai, R. (2009b). Is this verbal justification a proof? (Vol. 2, pp. 208–213).*

Vinner, S. (1990). Inconsistencies: Their causes and function in learning mathematics. *Focus on Learning Problems in Mathematics, 12*(3/4), 85–98.

Wittmann, E. (2009). Operative proof in elementary mathematics (Vol. 2, pp. 251–256).*

Wood, T. (1999). Creating a context for argument in mathematics class. *Journal for Research in Mathematics Education, 30*, 171–191.

Yackel, E. (2002). What we can learn from analyzing the teacher's role in collective argumentation. *The Journal of Mathematical Behavior, 21*, 423–440.

Yackel, E., & Cobb, P. (1996). Sociomathematical norms, argumentation, and autonomy in mathematics. *Journal for Research in Mathematics Education, 27*, 458–477.

Zaslavsky, O. (2005). Seizing the opportunity to create uncertainty in learning mathematics. *Educational Studies in Mathematics, 60*, 297–321.

Zaslavsky, O., & Peled, I. (1996). Inhibiting factors in generating examples by mathematics teachers and student teachers: The case of binary operation. *Journal for Research in Mathematics Education, 27*(1), 67–78.

Part V
Argumentation and Transition to Tertiary Level

Chapter 15
Argumentation and Proof in the Mathematics Classroom

**Viviane Durand-Guerrier, Paolo Boero, Nadia Douek,
Susanna S. Epp, and Denis Tanguay**

This chapter arose out of discussions of the working group on argumentation, logic, and proof and proving in mathematics. It concerns the relationships between argumentation and proof and begins by addressing the question of what we mean by argumentation and whether it includes mathematical proof. For the purposes of education, we regard argumentation as any written or oral discourse conducted according to shared rules, and aiming at a mutually acceptable conclusion about a statement, the content or the truth of which is under debate. It thus includes proof as a special case.

Study of the relationships between argumentation and proof holds great potential for helping teachers and students deal with the tension between the process leading up to the development of a student's proof and the requirements placed on the final

V. Durand-Guerrier (✉)
Département de mathématiques, I3M, UMR 5149, Université Montpellier 2,
Montpellier, France
e-mail: vdurand@math.univ-montp2.fr

P. Boero
Dipartimento di Matematica, Università di Genova, Genova, Italia
e-mail: boero@dima.unige.it

N. Douek
Institut Universitaire de Formation des Maîtres, Université de Nice, Nice, France
e-mail: ndouek@wanadoo.fr

S.S. Epp
Department of Mathematical Sciences, DePaul University, Chicago, IL, USA
e-mail: sepp@depaul.edu

D. Tanguay
Département de mathématiques, Université du Québec à Montréal (UQAM), Montreal,
QC, Canada
e-mail: tanguay.denis@uqam.ca

© The Author(s) 2021
349
G. Hanna and M. de Villiers (eds.), *Proof and Proving in Mathematics Education*,
New ICMI Study Series, https://doi.org/10.1007/978-94-007-2129-6_15

product. Students need to experience freedom and flexibility during an initial exploratory phase, whilst ultimately producing a proof that conforms to specific cultural constraints involving both logical and communicative norms in the classroom and in the mathematical community.

We also discuss whether the activity of developing proofs under a teacher's guidance can be used to introduce students to meta-mathematical concepts, just as ascertaining the truth of a mathematical statement or the validity of a proof can provide an opportunity for increasing students' mastery of the related mathematical concepts.

Finally, we use examples from the contributions of the members of the working group to illustrate aspects of these issues and suggest some possible activities to gradually increase students' awareness about proving and proof. We also examine the theoretical perspectives and educational issues involved in choosing and designing such activities.

1   Relationships Between Argumentation and Proof

From the time of ancient Greece, philosophers have linked argumentation to three disciplines: rhetoric, dialectic and logic. For the Greeks, rhetoric was the art of effective communication through discourse, especially persuasive public speaking on moral, legal or political issues. Dialectic was the art of conducting a discussion between two or more people who hold differing views but are free to express themselves whilst also seeking to reach agreement. Plato's philosophical dialogues are typical examples of dialectic. Logic was the art of correct thinking, of engaging in well-formed reasoning. Logical argumentation was the verbal manifestation of logical thinking, whether the intended interlocutors were specifically identified or fictitious, or were "ideal" or "universal." Aristotle analysed these concepts in his *Organon*, using *propositions* organised into *syllogisms* as elementary units of discourse, and making a distinction between contingent statements (that could happen to be true, but could also be false) and necessary statements (that could not possibly be false).

1.1   Mathematical Argumentation

Historians (e.g. Boyer 1968) generally agree that the concept of mathematical proof in a form closely related to the modern view was initiated by the Greeks, with Euclid's *Elements* (ca. 300 B. C. E.) as the paramount exemplar. Euclid's treatise starts with a collection of 23 definitions, which are used to state a set of postulates and axioms.[1] Propositions are then derived from the definitions, postulates, and

---

[1] For a discussion about possible meanings of the terms *axiom* and *postulate* according to the Greeks, see Jahnke (2010).

axioms through a process of deductive reasoning. In Aristotle's terms, the results thus established are "necessary truths." For many centuries, Euclid's *Elements* was used as the principal model of argumentation for the purpose of determining the truth of mathematical statements.

In the seventeenth century, Pascal described the excellence of the Euclidean method for "demonstrating truths already found, and of elucidating them in such a manner that the proof of them shall be irresistible" (Pascal 2007, p. 428). However, his contemporary Descartes complained that "the Ancients" were more careful about proving than explaining (Barbin 1988). Emphasising methods and processes, Descartes wrote that argumentation should be conducted through an analytical process, which would reveal "... the true way along which something has been methodically invented; ... so that a reader eager to follow it ... would understand the point thus demonstrated and make it his own as if himself were the inventor." (Descartes 1961, quoted in Barbin, p. 603; our translation).

In the nineteenth century, with the discovery of non-Euclidean geometries, the role and status of axioms in mathematical theory changed, passing progressively from assertions that cannot be reasonably doubted, to "rules of the game" that provide relations amongst the objects of a theory, which are defined only through the web of relationships thus posited. For mathematicians, research was no longer a matter of discovering a supra-reality that describes an objectively determined world, but rather a question of creating theories that are intrinsically coherent, possibly pairwise inconsistent, each of which would provide a *model* in the sense given that word by Tarski (1944).

The 1950s marked a renewed interest in the formal study of argumentation, culminating with the works of Toulmin (2008) and Perelman (Perelman and Olbrechts-Tyteca 2008). These authors viewed argumentation as a reasoned (and reasonable) discourse about an issue under debate, in which appeals to emotion are excluded but in which rationality need not be conveyed solely through deductive reasoning. Their work was deeply grounded in the study of language, which distinguishes it from epistemology and the methodology of science (Plantin 2005, p. 74). These considerations lead to even further questions about the relations amongst exploration, argumentation, proof, explanation, and justification in mathematics.

Thus, a fundamental tension, probably related to the very nature of mathematical activity, has pervaded the history of mathematical argumentation. Mathematical theories are constructed and developed along multiple paths. Some researchers lay special emphasis on the process of mathematical invention: taking care to keep their minds open to new ways of viewing topics; exploring, through careful analysis, multiple approaches to problems; and gathering insight from alternative perspectives. Other researchers "put energy into making mathematical arguments more explicit and more formal" (Thurston 1994, p. 169): searching for a synthesis to tighten and strengthen mathematical arguments; and for a broad framework into which to fit individual mathematical results. The work of Bourbaki in the twentieth century is certainly an example of the latter approach, as might also be considered the work of logicians such as Frege, Russell, Tarski, and Gödel. For these logicians, it was not only a matter of strengthening the foundations of mathematics but also of understanding and clarifying the nature of its formalisation at a "meta" level.

Yet a majority of working mathematicians who have written about their own ways of doing research stress the role of formal verification and the logical checking of proofs less than do logicians. For example, in drawing up the list of "brain and mind facilities" researchers use when searching for new results and their justification, Thurston (1994) sets "logic and deduction" side by side with "intuition, association and metaphor," the importance of which he highlights:

> Personally, I put a lot of effort into "listening" to my intuitions and associations, and building them into metaphors and connections. This involves a kind of simultaneous quieting and focusing of my mind. Words, logic, and detailed pictures rattling around can inhibit intuitions and associations. (op. cit., p. 165)

Concerning the phase of checking an argument, Thurston adds, "Reliability does not primarily come from mathematicians formally checking formal arguments; it comes from mathematicians thinking carefully and critically about mathematical ideas" (op. cit., p. 170).

Hadamard and Poincaré (2007) referred to a similar psychological phenomenon. Each described having solutions to problems pop suddenly from their unconscious to their conscious minds, following a period of intense, conscious, but apparently fruitless work. Pólya (1957) also promoted problem-solving methods in which formal deductive reasoning is not the chief consideration. Thom (1974) went further, linking the psychological aspects of the research process to the epistemological role of *meaning* in mathematics, which he referred to as "the 'ontological justification' for mathematical objects" (op. cit., p. 49).

However, the significance these mathematicians gave to the intuitive aspects of their research activity may proceed from their desire to rectify the rigid, formalistic image often associated with the field. None of them rejected the importance of careful deductive reasoning. For example, Thurston (1994) wrote,

> We have some built-in ways of reasoning and putting things together associated with how we make logical deductions: cause and effect (related to implication), contradiction or negation, etc. (op. cit., pp. 164–165).

The importance of rigorous verification of arguments to mathematicians was made clear in the 1990s following the announcement in June 1993 of a proof for Fermat's Last Theorem, which was reported on the front pages of newspapers around the world. Although Andrew Wiles had worked through many of the details of his "proof" with his colleague Nicholas Katz during the months before the dramatic announcement, it was only during the summer and fall of 1993, when he was writing out the proof for publication, that he found a mistake. After almost a year of painstaking effort, during which he and an associate carefully examined every step of the argument, Wiles finally realised that an approach he had previously thought to be a dead end was, in fact, the way to complete the proof (Singh 1997).

Thus, to varying degrees depending on the mathematician and the particular circumstances of the work, it appears that mathematical activity integrates two approaches. Alcock (2009) describes one as *semantical*: the study of examples and counterexamples; experimentation; instantiation of concepts, properties, and theorems; and development of analogies, associations, mental images, metaphors, and so forth. She describes the other approach as *syntactical*: use of formal definitions and symbolism; manipulation of formulas; and production of inferences that have been checked for logical correctness. Wilkerson-Jerde and Wilensky also commented upon this duality of approaches as they observed mathematicians attempting to solve problems they had not seen before:

> For some experts, specific instantiations of the mathematical object being explored serve a central role in building a densely-connected description of the proof; while for others a formal definition or several small components of the mathematical object serve this purpose. (2009, p. 271)

Barrier et al. (2009), Blossier et al. (2009) and Durand-Guerrier and Arsac (2009) proposed that the mathematical work done when searching and developing proofs is essentially dialectical, with frequent movement back and forth between semantic exploration of the mathematical objects under consideration and more formal syntactical analysis of their definitions, properties, theorems, and formulas. To support their thesis, these authors cite examples from Bolzano, Liouville, Cauchy, Euclid, and the Pythagoreans.

The dividing line between semantics and syntax is seldom clearly fixed for individuals; it evolves through their activities and also depends on their levels of mathematical maturity and mastery of formalisms. Fischbein (1982) suggests that the related *intuitions* also evolve, being a springboard *for* argumentation and proof and simultaneously being enriched, refined and strengthened *by* them. Moreover, individuals might themselves regard their activities as semantic, whereas an observer might regard them as syntactic, and vice-versa.

1.2 Argumentation, Proof, and Proving in Mathematics Education

The tension between the syntactic and semantic aspects of mathematical argumentation for mathematicians has its counterpart for mathematics educators. Didacticians (e.g., Balacheff 1987; Duval 1991; Hanna 1989) have pondered the relationships and potential oppositions involved: between proof as a cultural product subject to constraints of consistency and communication and proving as the process aiming at that product: between mathematical outputs that must fit some rules (i.e., some logical and textual models) and the creative and constructive side of mathematicians' activity; and between mathematical proof and ordinary argumentation. They have considered, *inter alia*, whether rigorous proof and ordinary argumentation create obstacles for each other and which educational conditions would foster fruitful use of either or both of the two approaches.

For example, Duval (1991) argued that (informal) argumentation impedes the learning of (mathematical) proof. He claimed that in argumentation propositions assumed as conclusions of preceding inferences or as shared knowledge are continuously reinterpreted (ibid., p. 241), with primary importance being placed on their semantic content. By contrast, in a deductive step in a mathematical proof, "propositions do not enter directly by reason of their content, but rather through their operational status" (ibid., p. 235; our translation). In other words, because it is a deductive chain of inferences, similar to a chain of calculations, the structure of a proof strongly depends on the operational status of the propositions that compose it. Now, in a given inference (or deductive step), the operational status of a proposition – whether premise, *énoncé-tiers*[2] or inferred proposition – is independent of its content, since a proposition may change its status within the same proof: for instance, when an inferred proposition is "recycled" as a premise for the next deductive step. The possible operational status of the propositions involved in an argument are differentiated according to their *theoretical epistemic value*; namely, as hypotheses or previously inferred propositions for the premises, and as definitions, theorems, or axioms for the *énoncé-tiers*. Duval argues that students have difficulty in learning mathematical proof because they do not easily grasp its specific requirements and hence deal with it as simply argumentation.

Other authors (e.g., Bartolini Bussi 1996, 2009; Grenier and Payan 1998; Lakatos 1976) studied the activity of proving from a *problem-solving* perspective. Some, such as Bartolini Bussi (1996, 2009), Bartolini Bussi et al. (1999) and Douek (2009), have differed with Duval (1991), claiming that argumentation can be used as an effective basis for mathematical classroom discussions concerning not only mathematical reasoning in general but also the rules for mathematical proof and the mathematical practices that play important roles in proof production. These include changing the frame of reference (e.g., from geometric to algebraic), using different representation registers, introducing a new object (e.g., a figure), and reviewing the role played by definitions in the production of a proof (cf. Durand-Guerrier and Arsac 2009). Written proofs, of course, do not include discussion of the practices that led up to their production.

The problem-solving perspective considers a "written proof" as the outcome of a process in which it is helpful to take fully into account the initial and intermediate phases. These phases vary, depending on how the problem is presented to the students; in particular, whether the result is stated or must be discovered; and, if stated, whether its truth-value is given ("Show that…") or is left to the students to determine. Students asked to find a result or to establish its truth-value start the process with an exploratory phase, in which they use various heuristics, such as studying examples, searching for counterexamples, and reflecting on definitions and theorems that may be linked to the problem. This phase normally leads to the statement of a conjecture.

---

[2] An *énoncé-tiers* is a statement already known to be true – namely an axiom, a theorem, or a definition – and generally is (or could be put) in the form "if p, then q". In Toulmin's (2008) model (see below), an *énoncé-tiers* corresponds to a warrant in an inference step.

Barrier et al. (2009) described and analysed the relation between the exploratory phase and the organisation of a proof by extending the concepts of *indoor* and *outdoor* games, introduced by Hintikka (1996). They defined outdoor games as concerned with determining the truth of statements within a given interpretative domain, in which the mathematical objects, their properties, and their relationships are investigated and tested. By contrast, indoor games are concerned with establishing the validity of statements inside a given theory; they consist of strategically using mathematical properties and assertions, such as hypotheses, axioms, and theorems available in the theory, in order to combine them syntactically into a proof, using computation, symbolical manipulations, and deductions. Barrier et al. argued that the construction of a proof most often rests upon a back-and-forth process between outdoor games and indoor games, and that semantic (outdoor) and syntactic (indoor) reasoning are both necessary and dialectically related. In addition, Blossier et al. (2009) proposed that quantification, whether explicit or implicit, is crucial for the semantic control realised through outdoor games. Epp (2009) also noted its importance, pointing out and describing students' difficulties with quantification in various contexts involving proof and disproof. Finally, Inglis et al. (2007) described an experiment where an advanced student eventually succeeded in developing a proof through a dialectical interplay between examination of numerical examples and syntactical analysis of definitions.

To characterise the possibility that heuristic argumentation can result in a conjecture with arguments or 'reasons' suitable for constructing a proof, Garuti et al. (1996) introduced the notion of *cognitive unity*, giving evidence for it in an experiment involving 36 students (two classes) in grade 8, where the arguments students gave in support of their conjecture about the "geometry of sun shadows" are essentially the same as those they arranged into a proof. (See also Boero et al. 2007). Pedemonte (2007) showed experimentally (with two geometry theorems proposed to 102 students in grades 12 and 13) that in some cases where cognitive unity seems possible, some students encounter difficulties in constructing a deductive chain from the otherwise suitable arguments that they produced through exploring and conjecturing, especially when the conjecture is obtained through abduction. She then distinguished between *cognitive unity* (concerning the referential system) and *structural continuity*. Structural continuity between argumentation and proof occurs when inferences in argumentation and proof are connected through the same structure. Pedemonte (2008) also used the notion in examining the same students' algebraic work, showing how in that case cognitive unity more easily results in structural continuity. As concerns cognitive unity and structural continuity, the educational and didactical context in which students are asked to produce conjectures and justifications seems crucial. In particular, tasks designed to lead to a conjecture, which include a request for providing reasons to support it, enhance argumentative activities that may facilitate a link between conjecturing and proving.

In order to study more precisely the articulation between argumentation and proof, Douek (2009) made use of Lolli's analysis of proof production (cf. Arzarello 2007). She considered three modes of reasoning: heuristic exploration, reasoned organisation of relevant propositions, and production of a deductive text obeying mathematicians' norms. Her detailed analysis of these modes showed that producing a proof depends not only on deductive reasoning but also on other activities, almost all dependent primarily on semantic content, including exploration using inductive reasoning, empirical verification of results, and abductive reasoning and argumentation. One difficulty of teaching proof is that the different modes of reasoning obey different rules of validity. For example, exploration that relies solely on examples cannot substitute for the kind of argument needed to establish the truth of a general mathematical proposition. According to Douek, effective use of the different modes of reasoning requires a certain amount of meta-mathematical reflection about the "rules of the game" of proving.

In an endeavour to find a unifying frame to analyse all these aspects of argumentation and proof, authors such as Pedemonte (2007), Inglis et al. (2007), Arzarello et al. (2009a), and Boero et al. (2010)[3] refer to the work of Toulmin (2008). Toulmin proposed a ternary model for the inference step in argumentation, which consists of a Claim, Data, and a Warrant, along with auxiliaries such as a Qualifier, a Rebuttal, and Backing. Use of this model allows one to compare argumentation and mathematical proof. For proof, the warrants (and ultimately the "backings") come from a mathematical theory,[4] whereas for argumentation they may consist of visual evidence or properties established empirically, for example.

Considering Toulmin's (2008) model and Duval's (1991) thesis together raises a fundamental didactical problem: namely, how to make students aware that when the final product is to be a "proof," the warrants must be chosen from amongst propositions with a precise theoretical status and selected according to specific criteria. Balacheff (1987) suggested a necessary shift from the pragmatic to the theoretical. Tanguay (2007) described the necessity to refocus students' outlook from *truth* to *validity*: students must understand that it no longer matters whether the involved propositions are intrinsically true or false but whether a sequence of deductive steps is valid. Barrier et al. (2009) and Durand-Guerrier (2008) go still further, considering how students need to coordinate their ideas of truth in an interpretative domain and the concept of validity in a theory. The didactical and pedagogical questions concerning how to move students towards such a shift are addressed in the next section.

2 Paths to Constructing Argumentation and Proof in the Mathematics Classroom

The emergence of proof reflects a variety of cultural characteristics. Amongst these is a desire for justification, for providing rational support to back up arguments, and for organising processes algorithmically. Attempting to assess the extent to

---

[3] More precisely, the frame proposed by Boero et al. (2010) results from the combination of Toulmin's model with Habermas' (2003) criteria of rationality.

[4] This agrees with Mariotti's definition of "theorem" (Mariotti et al. 1997): a system consisting of a statement, a proof – derived from axioms and other theorems according to shared inference rules – and a reference theory.

which argumentation in a wide range of disciplines constitutes logically valid proof is an important means of examining the grounds on which arguments and inferences are conducted in the "real world." It also promotes understanding for the nature and specific qualities of mathematical proof and their connections to various mathematical fields.

Therefore, not only mathematical theories but also the various modes of reasoning used in mathematics are important for students' intellectual development. Especially at the beginning, students are not exposed to any meta-mathematical reflection concerning the terms used in mathematics or to the differences from their ordinary uses and the systemic connections amongst them. As Vygotsky (1986) put it, these ideas are not treated as "scientific concepts" (i.e., as an explicit area of scientific knowledge), whereas organised exposure to them could lead students to master argumentation that extends beyond mathematics to science or further (e.g., to law; cf. Perelman and Olbrechts-Tyteca 2008). Students could thus benefit from familiarity with the wide variety of mathematical, 'meta-' and 'extra-' mathematical practices linked to the activity of proving.

2.1 What Do We Want Students to Experience?

Exploration, validation, and interpretation engender students' need for understanding when they are confronted with unexpected results, contradictions, or ambiguities. Complex tasks, which lead to doubt because of possible contradictions or obvious ambiguities (through solo activities or with peers), produce a didactic situation that favours proof and stimulates the development of several aspects of its practice. By contrast, the simplified tasks all too frequently proposed to students may obscure the need for proof and proving. Similarly, asking for proofs of statements that appear obvious to students' perception (often the case in middle-school geometry) or of which neither the meaning nor the truth is questioned may have a detrimentally effect on students' understanding of the nature and purpose of proof.

For example, students are rarely, if ever, presented with false mathematical statements and asked to determine whether or not they are true, or under what conditions, they might become so. In fact, it is desirable for students to experience uncertainty about (mathematical) objects, statements, and justifications in order to cultivate an attitude of reasonable scepticism. Problems whose truth is in question or open-ended problem situations require students to engage in exploration and hence argumentation (e.g., Arsac and Mante 2007; Grenier and Payan 1998; Legrand 2001).

An emphasis on exploration naturally generates a need to reflect on the place of conjectures. Whether for mathematical reasons (e.g., lack of cognitive unity between the thought processes leading to the conjecture and those required for a proof) or for curricular reasons (cf. Tanguay and Grenier 2009, 2010), conjecturing does not necessarily prompt proving. Crucially, conjecturing should go beyond mere guessing or unbridled speculation and encompass in some way a search for a structural explanation. Only then does it involve meaningful exploration and interpretation and a genuine need for validation.

Previous experiences of doubt about the truth or falsity of mathematical statements can lead students to see the need for validation as meaningful in terms of their own experiences, conjectures, and mathematical backgrounds, as well as in ways that they have shared with others in their class. Also important is the need for students to realise that various levels of validation are appropriate in various situations, which depend on context, the norms of practice of the community, their own and others' reference knowledge, and so forth. In addition, some important meta-mathematical concepts and their systemic interconnections, especially the relationships between axioms and validity of arguments, can be developed through activities using statements that are valid in one "theory" and not in another, and yet are accessible to a broad range of students (e.g., Parenti et al. 2007). Argumentation within class deliberations about what is acceptable (or not) and under what circumstances may open the way towards a more theoretical way of thinking; appreciation for theory may be enhanced by meta-mathematical discussions. These class discussions can address such questions as "Why do we prove?" and "What does it mean to have proved something?" as well as the relation to axioms, definitions, and more broadly the "rules of the game." For example, what is meant by "checking a proof" or "checking a program" may concern the mode of reasoning (whether deduction, induction, or abduction), the respective roles of syntax and semantics, the relationship between truth and validity, and the logic relevant to the given context.

In addition, in learning about theorems and their proofs, students need to experience not only how to validate statements according to specific reference knowledge and inference rules within a given theory, but also how the "truth" of statements depends on definitions and postulates of a reference theory (e.g., in geometry; cf. Henderson 1995).

Later, when reflection on meta-mathematical concepts needs to become deeper, more exhaustive, and more systematic (e.g., in higher secondary school or in the education of mathematics teachers), tasks could be employed that illustrate the central role definitions play in the formulation of statements, in argumentation, and in proof.

Meta-mathematical issues such as the variability of the means and tools for validation, the conventions for definitions, and the ambiguity of everyday speech compared with the precision of mathematical language, underline the central role of interpretation in argumentation and proof. Interpretation calls for various mathematical skills, such as the ability to change the semiotic register or frame of reference; relate certain mathematical objects to others; use theories, whether mathematical or not; use analogies and metaphors; and transfer procedures from one problem situation to another. At a higher, more specialised, level, we can explain these issues in terms of Tarski's (1983) theory of logical interpretation as the process of validating a mathematical construction by embedding it into a *mathematical model*. More pragmatically, they represent instances of going back and forth from a general statement to examples (whether generic or particular), relating schemas to situations, and referring to a precise context to understand a general statement. Students should experience the need to interpret figures, expressions, statements, and representations in various semiotic registers and develop virtuosity in doing so. Such activity would help make them aware that one cannot take effective communication for granted in mathematics.

Developing competence in *communication* is important, but exercising it relies on the presence of one or more listeners to react, contradict, and make use of what is communicated. Success in communicating in the classroom also depends on consciously shared references and on some form of shared theoretical scaffolding, which enhances the importance of debating mathematical issues in the classroom. However, as Duval (2001) pointed out, the form of control needed for (mathematical) proof development requires that students be able to figure out connections or simultaneities between propositions remote in the discourse, to come back to propositions already stated, to reconsider how propositions are organised, and to allow pauses and reflection. All of these activities only take place to the fullest extent when a *written* product is required.

In sum, students need to experience two main practices: a divergent exploratory one and a convergent validating one. They would then become familiar with the openness of exploration, that is, its "opportunistic" character and its flexible validation rules. They would also learn the rigorous rules needed to write a deductive text and the strict usage of words, symbols, and formulas when constructing or organising a theory. Previous curricula have enhanced the latter, but more recent ones have swung towards the former. However, isolating one to the detriment of the other weakens both, because they are related dialectically. Besides, once we consider exposing them to a wide range of experiences, it is crucial to help students face the contradictory differences amongst the diversity of practices used in argumentation and proof and to make them aware of when and how they can use these practices.

2.2 Designing Learning Environments and Activities

Here, we discuss various classroom activities aimed at developing both mathematical and meta-mathematical concepts. The majority were presented at the 19th *ICMI Study Conference, Proof and Proving in Mathematics Education.*

In one example, Arzarello et al. (2009a, b) described a long-term teaching experiment extending over the 5 years of Italian secondary school (grades 9–13). The experimenters gradually introduced students to local linearity through graphical, symbolic and numerical exploration using extensive technology, group work and class discussion. The problem-solving activities were intended to foster cognitive continuity between the exploration and production phases of developing a proof. A main result of the particular experiment reported in Arzarello et al. (2009a) was the creation of explicit links between the pieces of knowledge that were involved.

Perry et al. (2009a, b) implemented an innovative course for pre-service high school mathematics teachers, which features the collaborative construction of an axiomatic system involving points, lines, planes, angles, and properties of triangles and quadrilaterals.

Tanguay and Grenier (2009, 2010; cf. Dias and Durand-Guerrier 2005) reported an experiment conducted with preservice teachers in France and in Quebec, to whom the following classroom situation was submitted: (a) define and describe regular polyhedra; (b) produce them with given materials; and (c) prove that the list produced in (b) is complete. Very few of the students could conceptualise the relationship between the angles of the faces and the dihedral angles in a polyhedron in a way that would have accessed a proof that only five regular polyhedra exist. Instead, the students uncontrollably used intuition and analogy; some erroneously believed in an infinite sequence of regular polyhedra (therefore, e.g., trying to construct a regular polyhedron with hexagons). This error may have resulted because in a typical classroom proof and verification are closely associated with algebra and manipulation of formulas, with conjectures viewed empirically rather than as necessitating structural explanations.

Boero et al. (2009) suggest that random phenomena can provide teachers with opportunities to deal with definitions as well as more subtle aspects of mathematical argumentation and proof. For example, in spite of its potential logical and theoretical weaknesses, the classical definition of probability can be used in primary school as a basis for approaching the notion of an event's probability and of quantitatively estimating it on the basis of the search for an appropriate set of equally-likely cases. In high school and at the university level, the definition can be used for reflecting on the requirements of a good definition within a theory and related meta-mathematical issues concerning proof.

Brousseau (1997) described a didactical situation in which students aged 9–10 were given the task of enlarging the size of a jigsaw puzzle. Engaging in the task led them to discover for themselves that adding the same amount to the dimensions of all the pieces did not work because the pieces no longer fit together! Of course, at primary school, students cannot provide a mathematical proof that multiplication rather than addition is necessary; the teacher is responsible for asserting the result. Nevertheless, the situation promoted students' understanding that the result did not come from an arbitrary decision by an authority and helped them to take responsibility for their own learning. The activity was originally developed in the 1970s by Brousseau in France as part of a study on the teaching of rational numbers and finite decimals at primary school, and the results were subsequently reproduced many times.

In particular, secondary and tertiary level students have significant difficulty in developing an understanding about the nature of the set of real numbers (cf. Barrier et al. 2009). In an earlier experiment, Pontille et al. (1996) analysed the work of several groups of 3–4 students in grade 11. One group was followed for the entire year. The students were asked whether every increasing function from $\{1, 2, ..., n\}$ to $\{1, 2, ...,n\}$ has a fixed point. Once they had resolved this problem, they were asked to study generalisations to functions defined on the set of all finite decimals (i.e., real numbers with finite decimal expansions) in the interval $[0;1]$, and then on or any subset of the real numbers. The students conducted their work outside of the classroom, under the supervision of a researcher. They wrote results, questions, conjectures and proofs in notebooks, which the researchers analysed along with student interviews.

Students initially managed to solve the problem for the integers, using proofs that relied strongly on the fact that every natural number has a successor. When they then considered the further problems, they encountered significant epistemological questions: (a) Is there a fixed distance between any two "consecutive" finite decimals (i.e., can the notion of successor be used here?) (b) Given any positive finite decimal, can one find a smaller such number?[5] (c) Is there a relation between questions (a) and (b)? The students finally concluded that there is no such thing as a smallest positive finite decimal or a fixed distance between any such number and its 'successor.'

The students then examined a more complex problem involving the graph of an increasing function. Through discussion, the students developed various representations of the numbers involved (e.g., representing two kinds of numbers with different-coloured squares; drawing a straight line graph that contained holes). After a number of exchanges, they found a solution, which also led them to discover a counterexample.[6] Pontille et al.'s description of the students' work shows that students can develop reasoning ability and mathematical knowledge in tandem. It also shows that they may need a long period of intensive work before finding a counterexample. That discovery only occurred following the students' reconsidering, reorganising and adapting their knowledge about real numbers. The research supports the desirability of giving students tasks that lead them back and forth between formal and informal approaches, thus deepening their understanding of mathematical objects whilst simultaneously challenging their reasoning powers. Taking into account both the syntactic and semantic aspects of mathematical proof, such tasks promote the development of both mathematical concepts and meta-mathematical concepts.

In Italy, the Modena group led by M. Bartolini Bussi and the Genoa group led by P. Boero have performed several long-term teaching experiments, in Grades 4–8, to introduce students to validation and proof on both the practical and theoretical levels.[7] In order to implement their experiments, they attempted to construct suitable contexts, easily accessible or already familiar to students, where postulates represented basic "obvious" properties but where statements to be validated were not obvious and required reasoning.

For example, Parenti et al. (2007) reported a teaching experiment, performed with 80 students by three teachers in four classes from Grades 6–8, which used the potential of two contexts: the context of the *representation of visual spatial situations* in Grades 6–7 (cf. Bartolini Bussi 1996) and the context of *sun shadows* in Grade 8 (cf. Garuti et al. 1996). In both cases, students moved from a familiar

---

[5] Notice that if we consider decimal numbers with a given fixed number of digits following the decimal point, then these two questions are answered in the affirmative.

[6] The activity was also given to French students at a more advanced mathematical level, where it gave rise to very similar results.

[7] See Hsu et al. (2009) for another example of an instructional experiment concerning validity.

context (e.g., sun shadows) to the geometry of that context and to some of its obvious properties. (E.g., in the geometry of *sun shadows*, straight lines are projected onto a plane as straight lines or points). In both cases, students learned to infer statements from the properties that seemed obvious to them and to move progressively towards the format of standard proofs. At the end of the experiment, in Grade 8, students came to realise that statements validated in the *sun shadows* geometry may be invalidated in the *representation of visual spatial situations* geometry. They also discovered that proof in *sun shadows* geometry was based on an "obvious property" (parallel straight lines are projected onto a plane as parallel or coincident straight lines or points) which in general is not acceptable in the *representation of visual spatial geometry*.

Thus, the students learned that one can regard spatial situations from more than one point of view, each of which may lead to valid conclusions in its own terms. Regarding systems of postulates as obvious properties of spatial representations according to specific ways of thinking about them not only recalls the discovery of non-Euclidean geometries but also resonates with some recent epistemological positions concerning the close connections between geometric axiomatics and related ways of seeing and thinking about space (cf. Berthoz 1997; Longo 2009).

3   Management, Didactical Organisation, and Teachers' Roles

The preceding examples of activities focused on the development of mathematical and meta-mathematical concepts rather than on how students were asked to carry out their work. However, when aiming to develop competence in argumentation and proof, one needs to carefully organise student-teacher interactions and activities.

A main challenge in teaching argumentation and proof is to motivate students to examine whether and why statements are true or false. Mathematics educators broadly agree that we need alternatives to traditional tasks, which ask students to provide proofs for statements already presented as true. Thus, many mathematics educators now promote the development, at every level of the curriculum, of problems where the truth-goal is at stake. Such problems fall into two categories: problems used to introduce and develop a mathematical concept and to prove theorems about it; and problems used to develop competence in argumentation and proof.

Recently, a number of mathematical educators have turned their attention to developing theoretical frameworks for developing teaching experiments that involve students in solving such problems (cf. Lin et al. 2009).

Given increasing scepticism about the once common view that mathematics teachers should simply deliver mathematical truths, Brousseau (1997) created a set of situations leading to the development of all the required knowledge about finite decimals in elementary school. He also provided a theoretical framework for his work whereby students develop intended knowledge through collective engagement in a succession of phases involving action, formulation, validation, and institutionalisation.

Similarly, in France in the 1980s, the concept of *scientific debate* was introduced into tertiary-level mathematics courses so that students would come to feel personal responsibility for the conjectures they formulate (Alibert 1988). Instructors led students to consider themselves as a community of mathematicians (a specialised instance of a *community of practice*; cf. Wenger 1998) needing to engage in scientific debate on topics proposed by the teacher. They were invited to formulate conjectures and take sides as to the relevance and the truth of these (Legrand 2001).

In another theoretical framework (the *didactic of fields of experience*), students' written solutions and related discussions play a crucial role (Boero et al. 2007, 2008, 2009).

Tanguay and Grenier (2009) pointed out that "implementation in the classroom or in the curricula of the findings of didactical studies are almost always accompanied by alterations and even distortions." Effective implementation of the insights from such studies as those above poses a complex challenge for the designers of teaching experiments.

To take one example, Douek (2009) implemented many of the principles discussed above in experiments in a class of 27 students aged 13–14 in a French middle school. Her aim was to involve students in reorganising the arguments in a proof through discussion about their various roles. Theorems with no cognitive unity had been chosen, in order to allow the teacher to put some features of proof into evidence whilst guiding the process of proving. Argumentation leading to a conjecture was promoted through classroom debate (cf. Bartolini Bussi 1996), and then the experimenter, after leading the students through a proof, encouraged them to see its rationale as meaningful by making a story out of the steps in their reasoning and calculations, and to express themselves both by explaining and by raising questions. Creating a story was intended to help students connect the various steps with the reasons that supported the conclusion, sort the 'blocks' of reasoning into which the proof was divided, and explain how the blocks were joined logically and their possible hierarchical relationship (Knipping 2008). It also provided students with opportunities for personal expression and internalisation of newly-encountered mathematical practices.

For example, in a class session about the Pythagorean theorem, several tasks led students to conjecture the theorem. Then, they had to prove their conjecture by verifying statements prepared by the teacher. Six students were able to prove all the statements. A follow-up task invited students to review the reasons for the steps they had been led through. It triggered class discussion, guided by the teacher, that identified why the statements were needed, and how they were linked together to form a general argument. Some blocks used abductive reasoning (What property makes such-and-such claim possible? How can you view *A* so that such-and-such relation appears?). Others used deduction. Therefore, mathematical proof practices, such as the different types of logical reasoning used and movements from geometric to algebraic analysis, became a significant part of the discussion. Then the students were asked to write, individually, the story of the proof. Eleven wrote reasonably coherent stories, but only two explicitly described why all the steps were needed and useful. The experiment showed that some students can accept the logical flow of a proof but lack consciousness about the reasons that enable it.

4 Conclusion

In the first part of this chapter, we tried to elucidate the complex relationships between argumentation and proof in mathematics from mathematical and educational perspectives, noting the general lack of consensus about the interrelations between the two, both amongst mathematicians and amongst mathematics educators. In the second part, we examined the aims, conditions, and constraints that affect planning and implementation of appropriate situations for facilitating argumentation and proof in mathematical classes. First, we considered what we want students to experience and what aspects of learning proof should be the goal of instruction. We need to develop students' awareness about the core aspects of theorems and proofs and thus support their use when needed. We must engage students in mathematical tasks that also foster the development of meta-mathematical concepts. In the third part, we presented short descriptions of examples of learning environments supportive of the goals for the crucial activities of proving identified in the second part. In sum, engaging students in situations which make them aware of the constructive character of mathematical activities, especially those involving conjecture and proof, poses complex challenges. One needs to consider:

1. The types of tasks: open-ended problems, such as statements whose truth or falsity students are to determine; critical study of students' arguments; discussion about the basis for justification; story-telling; work on proof development; and so forth;
2. Management of activities: for example, student discussions; production of written work; use of software, diagrams, and instruments; class management techniques: use of reasoning diagrams; meta-level discourses; managing different semiotic registers; whole-class debate; working in groups; working individually; and so on;
3. The respective roles of students and teachers.

We hope that we have succeeded in making clear the various perspectives underlying the educational proposals, giving insight into their richness, and providing elements for further research and innovation.

Acknowledgements We wish to thank the members of Working Group 2: Thomas Barrier, Thomas Blossier, Paolo Boero, Nadia Douek, Viviane Durand-Guerrier, Susanna Epp, Hui-yu Hsu, Kosze Lee, Juan Pablo Mejia-Ramos, Shintaro Otsuku, Cristina Sabena, Carmen Samper, Denis Tanguay, Yosuke Tsujiyama, Stefan Ufer, and Michelle Wilkerson-Jerde.

We are grateful for the support of the Institut de Mathématiques et de Modélisation de Montpellier, Université Montpellier 2 (France), IUFM C. Freinet, Université de Nice (France), Università di Genova (Italy), DePaul University (USA), and the Fonds québécois de recherche sur la société et la culture (FQRSC, Grant #2007-NP-116155 and Grant #2007-SE-118696).

We also thank the editors and the reviewers for their helpful feedback on earlier versions of this chapter.

References*

Alcock, L. (2009). Teaching proof to undergraduates: Semantic and syntactic approaches (Vol. 1, pp. 29–34).*

Alibert, D. (1988). Toward new customs in the classrooms. *For the Learning of Mathematics, 8*(2), 31–43.

Arsac, G., & Mante, M. (2007). *Les pratiques du problème ouvert*. Lyon: Scéren CRDP de Lyon.

Arzarello, F. (2007). The proof in the 20th century. In P. Boero (Ed.), *Theorems in school* (pp. 43–63). Rotterdam: Sense Publishers.

Arzarello, F., Paola, D., & Sabena, C. (2009a). Proving in early calculus (Vol. 1, pp. 35–40).*

Arzarello, F., Paola, D. & Sabena, C. (2009b). Logical and semiotic levels in argumentation (Vol. 1, pp. 41–46).*

Balacheff, N. (1987). Processus de preuve et situations de validation. *Educational Studies in Mathematics, 18*(2), 147–176.

Barbin, E. (1988). La démonstration mathématique: significations épistémologiques et questions didactiques. *Bulletin de l'APMEP, 36*, 591–620.

Barrier, T., Durand-Guerrier, V., & Blossier, T. (2009). Semantic and game-theoretical insight into argumentation and proof (Vol. 1, pp. 77–82).*

Bartolini Bussi, M. G. (1996). Mathematical discussion and perspective drawing in primary school. *Educational Studies in Mathematics, 31*, 11–41.

Bartolini Bussi, M. G. (2009). Proof and proving in primary school: An experimental approach (Vol. 1, pp. 53–58).*

Bartolini Bussi, M. G., Boni, M., Ferri, F., & Garuti, R. (1999). Early approach to theoretical thinking: Gears in primary school. *Educational Studies in Mathematics, 39*, 67–87.

Berthoz, A. (1997). *Le sens du mouvement*. Paris: Editions Odile Jacob [English translation. The Brain's sense of movement. Cambridge: Harvard University Press, 2000].

Blossier, T., Barrier, T., & Durand-Guerrier, V. (2009). Proof and quantification (Vol. 1, pp. 83–88).*

Boero, P., Garuti, R., & Lemut, E. (2007). Approaching theorems in grade VIII. In P. Boero (Ed.), *Theorems in school* (pp. 261–277). Rotterdam: Sense Publishers.

Boero, P., Douek, N., & Ferrari, P. L. (2008). Developing mastery of natural language: Approach to theoretical aspects of mathematics. In L. English (Ed.), *Handbook of international research in mathematics education* (pp. 262–295). New York/London: Routledge.

Boero, P., Consogno, V., Guala, E., & Gazzolo, T. (2009). Research for innovation: A teaching sequence on the argumentative approach to probabilistic thinking in grades I-V and some related basic research results. *Recherches en Didactique des Mathématiques, 29*, 59–96.

Boero, P., Douek, N., Morselli, F., & Pedemonte, B. (2010). Argumentation and proof: A contribution to theoretical perspectives and their classroom implementation. In M. F. F. Pinto & T. F. Kawasaki (Eds.), *Proceedings of the 34th Conference of the International Group for the Psychology of Mathematics Education* (Vol. 1, pp. 179–205). Belo Horizonte: PME

Boyer, C. B. (1968). *A history of mathematics*. New York: Wiley.

Brousseau, G. (1997). *Theory of didactical situations in mathematics*. New York: Springer.

Descartes, R. (1961). *Les méditations métaphysiques*. Paris: Presses Universitaires de France.

Dias, T., & Durand-Guerrier, V. (2005). Expérimenter pour apprendre en mathématiques. *Repères IREM, 60*, 61–78.

Douek, N. (2009). Approaching proof in school: From guided conjecturing and proving to a story of proof construction (Vol. 1, pp. 148–153).*

---

*NB: References marked with * are in F. L. Lin, F. J. Hsieh, G. Hanna, & M. de Villiers (Eds.) (2009). *ICMI Study 19: Proof and proving in mathematics education*. Taipei, Taiwan: The Department of Mathematics, National Taiwan Normal University.

Durand-Guerrier, V. (2008). Truth versus validity in mathematical proof. *ZDM The International Journal on Mathematics Education, 40*(3), 373–384.

Durand-Guerrier, V., & Arsac, G. (2009). Analysis of mathematical proofs: Some questions and first answers (Vol. 1, pp. 148–153).*

Duval, R. (1991). Structure du raisonnement déductif et apprentissage de la démonstration. *Educational Studies in Mathematics, 22*(3), 233–262.

Duval, R. (2001). Écriture et compréhension: Pourquoi faire écrire des textes de démonstration par les élèves ? In *Produire et lire des textes de démonstration. Collectif coord. par É Barbin, R. Duval, I. Giorgiutti, J. Houdebine, C. Laborde.* Paris: Ellipses.

Epp, S. (2009). Proof issues with existential quantification (Vol. 1, pp. 154–159).*

Fischbein, E. (1982). Intuition and proof. *For the learning of mathematics, 3*(2), 8–24.

Garuti, R., Boero, P., Lemut, E., & Mariotti, M. A. (1996). Challenging the traditional school approach to theorems: A hypothesis about the cognitive unity of theorems. In *Proceedings of PME-XX* (Vol. 2, pp. 113–120). Valencia: PME.

Grenier, D., & Payan, C. (1998). Spécificités de la preuve et de la modélisation en mathématiques discrètes. *Recherches en didactiques des mathématiques, 18*(1), 59–99.

Habermas, J. (2003). *Truth and justification.* Cambridge: MIT Press.

Hadamard, J., & Poincaré, H. (2007). *Essai sur la psychologie de l'invention dans le domaine mathématique – L'invention mathématique.* Paris: Éditions Jacques Gabay.

Hanna, G. (1989). More than formal proof. *For the Learning of Mathematics, 9*(1), 20–25.

Henderson, D. W. (1995). *Experiencing geometry on plane and sphere.* Upper Saddle River: Prentice Hall.

Hintikka, J. (1996). *The principles of mathematics revisited.* Cambridge: Cambridge University Press.

Hsu, H. Y, Wu Yu, J. Y., Chen, Y. C. (2009). Fostering 7th grade students understanding of validity. An instructional experiment (Vol. 1, pp. 214–219).*

Inglis, M., Mejia-Ramos, J.-P., & Simpson, A. (2007). Modelling mathematical argumentation: The importance of qualification. *Educational Studies in Mathematics, 66*, 3–21.

Jahnke, H. N. (2010). The conjoint origin of proof and theoretical physics. In G. Hanna, H. N. Jahnke, & H. Pulte (Eds.), *Explanation and proof in mathematics* (pp. 17–32). New York: Springer.

Knipping, C. (2008). A method for revealing structures of argumentation in classroom proving processes. *ZDM, 40*, 427–441.

Lakatos, I. (1976). *Proofs and refutations.* Cambridge: Cambridge University Press.

Legrand, M. (2001). Scientific debate in mathematics courses. In D. Holton (Ed.), *The teaching and learning of mathematics at university level: An ICMI study* (pp. 127–135). Dordrecht: Kluwer Academic Publishers.

Lin, F. L., Hsieh, F. J., Hanna, G. & de Villiers M. (Eds.) (2009). *ICMI Study 19: Proof and proving in mathematics education.* Taipei, Taiwan: The Department of Mathematics, National Taiwan Normal University.

Longo, G. (2009). Theorems as constructive visions (Vol. 1, pp. 13–25).*

Mariotti M. A., Bartolini Bussi, M. G., Boero, P., Ferri, F., & Garuti, R. (1997). Approaching geometry theorems in contexts: from history and epistemology to cognition. In *Proceedings of the 21th PME Conference*, Lathi.

Parenti, L., Barberis, M. T., Pastorino, M., & Viglienzone, P. (2007). From dynamic exploration to "theory" and "theorems" (from 6th to 8th grades). In P. Boero (Ed.), *Theorems in school: From history, epistemology and cognition to classroom practice* (pp. 265–284). Rotterdam: Sense Publishers.

Pascal, B. (2007). *Thoughts, letters, and minor works. The five foot shelf of classics* (Vol. XLVIII) (C. W. Eliot, Ed. & W. F. Trotter, Trans.). New York: Cosimo Classics.

Pedemonte, B. (2007). How can the relationship between argumentation and proof be analysed? *Educational Studies in Mathematics, 66*(1), 23–41.

Pedemonte, B. (2008). Argumentation and algebraic proof. *ZDM, 40*(3), 385–400.

Perelman, C., & Olbrechts-Tyteca, L. (2008). *Traité de l'argumentation* (6th ed., 1st ed. in 1958). Bruxelles: Éditions de l'Université de Bruxelles.

Perry, P., Samper, C., Camargo, L., Molina, Ó., & Echeverry, A. (2009a). Learning to prove: Enculturation or …? (Vol. 2, pp. 124–129).*

Perry, P., Samper, C., Camargo, L., Molina, Ó., & Echeverry, A. (2009b). Assigning mathematics tasks versus providing pre-fabricated mathematics in order to support learning to prove (Vol. 2, pp. 130–135).*

Plantin, C. (2005). *L'argumentation. Coll. Que sais-je?* (1st ed., in 1996). Paris: Presses Universitaires de France.

Pólya, G. (1957). *How to solve it*. Princeton: Princeton University Press.

Pontille, M. C., Feurly-Reynaud, J., & Tisseron, C. (1996). Et pourtant, ils trouvent. *Repères IREM, 24*, 11–34.

Singh, S. (1997). *Fermat's enigma*. New York: Walker & Company.

Tanguay, D. (2007). Learning proof: From truth towards validity. In *Proceedings of the Xth Conference on Research in Undergraduate Mathematics Education* (RUME), San Diego State University, San Diego, CA. On the Web. http://www.rume.org/crume2007/eproc.html

Tanguay, D., & Grenier, D. (2009). A classroom situation confronting experimentation and proof in solid geometry (Vol. 2, pp. 232–238).*

Tanguay, D., & Grenier, D. (2010). Experimentation and proof in a solid geometry teaching situation. *For the Learning of Mathematics, 30*(3), 36–42.

Tarski, A. (1944). The semantic conception of truth. *Philosophy and Phenomenological Research, 4*, 13–47.

Tarski, A. (1983). *Logic, semantics and metamathematics, papers from 1923 to 1938*. Indianapolis: John Corcoran.

Thom, R. (1974). Mathématiques modernes et mathématiques de toujours, suivi de Les mathématiques « modernes », une erreur pédagogique et philosophique ? In R. Jaulin (Ed.), *Pourquoi la mathématique* ?, (pp. 39–88). Paris: Éditions 10–18.

Thurston, W. P. (1994). On proof and progress in mathematics. *Bulletin of the American Mathematical Society, 30*(2), 161–177.

Toulmin, S. E. (2008). *The uses of argument*. (8th ed., 1st ed. in 1958) Cambridge: Cambridge University Press.

Vygotsky, L. S. (1986). *Thought and language* (2nd ed., 1st English ed. in 1962). Cambridge: MIT Press.

Wenger, E. (1998). *Communities of practices: Learning, meaning and identity*. Cambridge: Cambridge University Press.

Wilkerson-Jerde, M. H., & Wilensky, U. (2009). Understanding proof: Tracking experts' developing understanding of an unfamiliar proof (Vol. 2, pp. 268–273).*

Chapter 16
Examining the Role of Logic in Teaching Proof

Viviane Durand-Guerrier, Paolo Boero, Nadia Douek,
Susanna S. Epp, and Denis Tanguay

In this chapter, we examine the relevance of and interest in including some instruction in logic in order to foster competence with proof in the mathematics classroom. In several countries, educators have questioned of whether to include explicit instruction in the principles of logical reasoning as part of mathematics courses since about the 1980s. Some of that discussion was motivated by psychological studies that seemed to show that "formal logic ... is not a model for how people make inferences" (Johnson-Laird 1975). At the same time, university and college faculty commonly complain that many tertiary students lack the logical competence to learn advanced mathematics, especially proof and other mathematical activities that require deductive reasoning. This complaint contradicts the view that simply doing mathematics at the secondary level in itself suffices to develop logical abilities.

V. Durand-Guerrier (✉)
Département de mathématiques, I3M, UMR 5149, Université Montpellier 2,
Montpellier, France
e-mail: vdurand@math.univ-montp2.fr

P. Boero
Dipartimento di Matematica, Università di Genova, Genova, Italia
e-mail: boero@dima.unige.it

N. Douek
Institut Universitaire de Formation des Maîtres, Université de Nice, Nice, France
e-mail: ndouek@wanadoo.fr

S.S. Epp
Department of Mathematical Sciences, DePaul University, Chicago, IL, USA
e-mail: sepp@depaul.edu

D. Tanguay
Département de mathématiques, Université du Québec à Montréal (UQAM), Montreal, Canada
e-mail: tanguay.denis@uqam.ca

© The Author(s) 2021
G. Hanna and M. de Villiers (eds.), *Proof and Proving in Mathematics Education*,
New ICMI Study Series, https://doi.org/10.1007/978-94-007-2129-6_16

In addition, in many countries prospective elementary and secondary teachers arrive at university with poor logical reasoning abilities. One questions whether, if they do not learn basic logical principles, these candidates will graduate as teachers able to guide their own students' reasoning.

Enhancing the teaching of mathematics thus calls for increasing teachers' awareness of crucial logical aspects of proof. The question is how best to achieve this aim. We discuss this question and offer some examples intended to provide ideas about how to develop teachers' and students' logical competence and how to modify curricula to foster development of these abilities. First, we review various positions on the role of logic in argumentation and proof; we then discuss them from an educational perspective; and finally, we offer suggestions that emerged from the contributions of the members of our working group.

1 Positions on the Role of Logic in Argumentation and Proof

By logic, one may mean common sense, or the principles of predicate calculus, or a mathematical subject or a branch of philosophy. The literature in mathematics education and other related fields (psychology, epistemology, linguistics) allows for several such broad definitions. In order to avoid potential misunderstanding, we define logic as the discipline that deals with both the semantic and syntactic aspects of the organisation of mathematical discourse with the aim of deducing results that follow necessarily from a set of premises. When we refer to logic as a subject, we mainly restrict ourselves to the mathematical uses of the words *and*, *or*, *not*, and *if-then* (the basis for "propositional logic"), especially in statements that involve variables, as well as *for-all*, and *there-exists* (the extension to "predicate logic").

1.1 Positions from Psychology

Much psychological research on human reasoning has primarily focused on the content-free logical structures of the propositional calculus. It has especially emphasised implication, which plays a crucial role in reasoning whatever the context, and certainly in mathematics. For example, a significant portion of the literature concerns the famous Wason selection task (Wason 1966), which tests the ability of adults to recognise the cases that falsify an implication. Its original formulation is as follows:

> Subjects are shown a set of four cards placed on a table, each of which has a number on one side and a letter on the other. The visible faces of the cards show 4, 7, A, and D. Subjects are asked to "decide which cards [you] would *need* to turn over in order to determine whether the experimenter was lying in the following statement: If a card has a vowel on one side, then it has an even number on the other side" (Wason 1966, p. 146).

If the statement is interpreted as a material conditional, then the correct answer is that it is necessary and sufficient to turn over the cards showing the letter A and the number 7. Fewer than 10% of the university student subjects answered correctly, which led Wason (1977) to argue that propositional logic fails to explain aspects of human reasoning. More recently, Johnson-Laird (1986) claimed that humans reason using non-logic-based mental models in which a semantic (content-dependent) point of view predominates.[1] These developments in cognitive psychology resulted in devaluing the role of abstract logic in understanding human reasoning.

Additional studies gave participants tasks logically identical to the Wason task but situated in a familiar setting that involved interpreting the implication as a permission or a requirement. Students did much better on these tasks than on the Wason task. For instance, in one of the first such experiments, Johnson-Laird et al. (1972) gave participants both the original Wason task and a concrete variation in which they imagined they were postal workers examining envelopes on a conveyor belt. They were shown four envelopes: the back of a sealed envelope, the back of an unsealed envelope, the front of an envelope with a 5d stamp, and the front of an envelope with a 4d stamp. Their job was to indicate which envelopes they would need to turn over given the rule that a sealed envelope should have a 5d stamp. Despite the similar logic required to determine the answers in both situations, only seven of the 24 participants determined the correct answer in the abstract Wason situation, whereas 22 out of the 24 answered correctly in the concrete situation. Results similar to this have been replicated in many subsequent experiments. A possible reason for the increase in success may be that subjects' previous experience has familiarised them with the fact that, in situations involving permissions and requirements, $p \rightarrow q$ and not-q implies not-p and $p \rightarrow q$ and not-q does not necessarily imply p.

In a 1986 paper, Cheng, Holyoak, Nisbett and Oliver described one experiment that used pre- and post-tests in two separate introductory logic classes at the University of Michigan and seemed to show that having taken a logic course had little effect on students' ability to succeed with variations of the Wason selection task, but another experiment described in the same paper "indicated that training in standard logic, when coupled with training on examples of selection problems, leads to improved performance on subsequent selection problems." In addition, the reports of Stenning, Cox, and Oberlander (1995) and van der Pal and and Eysink (1999) actually show significant improvement in logical reasoning skill amongst students who used the Barwise & Etchemendy Tarski's World materials.

According to Stenning and Van Lambalgen (2008), "results such as those of the selection task, purportedly showing the irrelevance of formal logic to actual human reasoning, have been widely misinterpreted, mainly because the picture of logic current in psychology and cognitive science is completely mistaken." (p. xiii) They aimed to more accurately depict mathematical logic and show that logic is still a helpful concept in cognitive science. In particular, they considered semantics interpretation as underlying two key processes in deductive reasoning: "interpretation

---

[1] Ufer et al. (2009) propose a cognitive model based on mental models.

372 V. Durand-Guerrier et al.

processes of reasoning to an interpretation *and* derivational processes of reasoning from the interpretation imposed" (ibid, p. 197). Lee and Smith (2009) also considered this distinction relevant for mathematics education. Thus, a strictly syntactic view of logic only inadequately explains human reasoning, which essentially involves an ongoing interaction between syntax and the interpretive role played by semantics.

1.2 Positions from Mathematics Educators

Several mathematics educators' positions on the role of logic in mathematics education seem to have been influenced by the psychological research cited above and by mathematicians' personal accounts of their own work. For example, Hanna (2000) described how Simpson (1995) "differentiates between 'proof through logic,' which emphasises the formal, and 'proof through reasoning,' which involves investigations. The former is 'alien' to students, in his view, since it has no connection with their existing mental structure, and so can be mastered only by a minority" (p. 5). Simpson's view replicates the researchers' statements about the difference in results between the classical Wason test and the versions dependent on familiarity with certain kinds of contexts (e.g. permission or requirements).

The literature contains relatively few references to the work of logicians, despite the fact that logic seems relevant to an epistemology for mathematical proof (e.g., Sinaceur 2001). Mariotti (2006), writing about the research and debate on proof and proving in mathematics education, stated, "From both an epistemological and a cognitive point of view, it seems impossible to make a clear separation between the semantic and the theoretical level, as required by a purely formal perspective." (p. 5). This statement by Mariotti was an answer to Duval's (1991) assertion that proof consists of a logical sequence of implications from which one derives the theoretical validity of a statement and the organisation of which relies only on the theoretical status of the propositions involved, independently of their content. Notably, Duval's position is based on a purely syntactic view of propositional calculus. Thus, unless one considers logic as restricted to the syntax of propositional calculus, arguing against Duval is not the same as arguing against the role of logic. Indeed, Tarski (1944) assumed that the use of predicate logic involves a dialectic between semantics and syntax, through the two crucial notions of truth in an interpretation and validity in a theory (Durand-Guerrier 2008). Barrier et al. (2009), referring to Hintikka (1996), claimed that a model that would articulate syntactic and semantic analysis with game-theoretical tools could contribute to the understanding of the process of proof and shed light on the dialectical nature of proof construction. This last point, also supported by Mamona-Downs and Downs (2009), agrees with Mariotti's (2006) statement, "Proof clearly has the purpose of validation—confirming the truth of an assertion by checking the logical correctness of mathematical arguments—however, at the same time, proof has to contribute more widely to knowledge construction" (p. 24).

Beyond the debate about logic's role in argumentation and proof, questions arise about the possibility of teaching logic in a way that would foster competence in activities of proving. In France, for example, the only rule of inference explicitly taught now is *modus ponens* (*A* and "If *A*, then *B*"; hence *B*) in grade 8 (ages 13–14); in high school, students have few opportunities to encounter challenging activities that would call their modes of reasoning into question and offer occasions to develop their logical skills. Similar curricula could, at least partly, explain why students arriving at university both in France and elsewhere have severe problems with the logic needed for learning more advanced mathematics, especially the logic that involves quantification (Blossier et al. 2009; Epp 2009; Roh 2009; Selden and Selden 1995).

2 Logical Perspectives on Argumentation and Proof

2.1 The Role of Logic in Learning and Teaching Mathematical Proof

As teachers ourselves, we can attest to the considerable difficulties students face in dealing with the logical issues that arise when they attempt to learn about the process of proof in mathematics (e.g., Epp 2003). It is not easy to determine what kind of work with students will help them overcome these difficulties. In this section, we aim to clarify the role that logic may play in the learning and teaching of proof, taking into consideration different aspects of various proving activities and their dependence on the mathematical topics in question.

The exploratory and constructive phases of reasoning involve frequent back-and-forth movement between examining individual objects, properties and relationships and referring to relevant conjectures, definitions, theorems and axioms. Such movement necessitates an intensive use of inference rules. This use may occur subconsciously, but it must be done correctly in order to be developed into an actual proof. In the mathematics curriculum, geometry and number theory provide opportunities for using inference rules from the logic of propositions, such as *modus ponens*,[2] *modus tollens*,[3] hypothetical syllogism,[4] disjunctive syllogism,[5] the contradiction rule,[6] and so forth. They also give students occasion to explore problems by using particular cases; formulating a general conjecture; coming back to specific objects in order to support, reject or refine it; and when they believe it to be true, looking for reasons that enable confidence in that belief.

---

[2] *A*; and "If *A*, then *B*"; hence *B*.

[3] Not *B*; and "If *A*, then *B*"; hence not *A*.

[4] If *A* then *B*; If *B* then *C*; hence If *A* then *C*.

[5] *A* or *B*; not *B*; hence *A*.

[6] If not *A*, then both *B* and not *B*; hence *A*.

In discussing the exploratory stages of problem solving, Pólya (1954) pointed out the crucial importance of both inductive ("if regularity appears, then it could be because of a general result") and abductive ("given that this phenomenon has appeared, then it could be a consequence of this other phenomenon") heuristic reasoning, despite the fact that neither produce necessary deductions from the point of view of logic. When one reasons with conditional statements that are not equivalences, only two valid inference rules are available (*modus ponens* and *modus tollens*); one can deduce no necessary conclusions from a conditional statement when one only knows that the antecedent is false or that the consequent is true. This principle forms the very core of the notion of implication and its distinction from equivalence (Durand-Guerrier 2003).

Four additional inference rules from predicate logic also have crucial importance in mathematical reasoning.

1. Given that a result is known to be true for all objects of a certain kind, *universal instantiation* allows us to conclude that the result is true for any particular object of that kind. This rule is used constantly in the manipulation of algebraic expressions and in providing reasons for most steps in a proof.
2. When an object of a certain type is known to exist, *existential instantiation* allows one to give it a name. This rule is used every time one introduces a letter into a mathematical explanation. Learning how to use such symbols correctly is a significant challenge for most students.
3. When one knows or hypothesises that a particular object of a certain type exists, *existential generalisation* allows one to say that there exists an object of that type. Along with the ability to formulate negations of statements, this rule is used every time one produces a counterexample to a general statement. It is also part of the basis for the logic of solving equations.
4. If one can show that a certain property holds true for a particular but arbitrarily chosen object of a certain type, *universal generalisation* allows us to conclude that the property holds true for all objects of that type. This rule is the basis for almost all mathematical proofs.

Having some level of understanding of the fundamental rules of predicate logic helps students check mathematical statements that are in doubt, avoid invalid deductions, and comprehend the basic structures of both mathematical proof (direct and indirect) and disproof by counterexample. Here an educational question arises: at which level, in which contexts, and to what extent the rules of predicate logic should be taught explicitly. Durand-Guerrier and Arsac (2005) reported a case in which teachers[7] examining a student's proof immediately recognised that the proof was invalid because they knew that the statement was false.[8] The teachers' knowledge of the subject made it unnecessary for them to use reasoning to determine that the

---

[7] In the academic year 1998–1999, 22 mathematics teachers from various French scientific universities were asked to comment on an invalid proof written by an undergraduate (see Durand-Guerrier and Arsac 2005, pp.159–163).

[8] Mathematicians with broad knowledge of mathematical subjects can similarly recall mathematical facts without having to re-derive them, which could be one of the reasons they may underrate the extent to which logic plays a role in their work.

student had made a mistake. However, students learning a new mathematical subject lack their teachers' rich knowledge base. So, to provide additional help for evaluating mathematical statements, at the same time that they give students experience with examples and counterexample, teachers also need to help students develop their deductive reasoning abilities. Indeed, in order to solve mathematical problems that involve deduction, students must be able both to derive statements from other statements through logical rules (the syntactic aspects of proof), to consider counterexamples, to explore supportive examples, to work with a generic example, and so on (the semantic aspects of proof; cf. Barrier et al. 2009).

Knowledge of the principles of logical reasoning becomes most important when familiarity with the mathematical subject matter does not suffice by itself to ascertain truth or falsity in a given situation. Drawing on their research and teaching experiences, Durand-Guerrier and Arsac (2009) and Epp (2009) support the claim that teaching students the basic principles of predicate logic (e.g., Copi 1954; Gentzen 1934) provides them with a relevant tool for successful mathematical activity. Knowledge of these principles makes it possible for students to answer the questions: "What would it mean for such-and-such to be true? What would I have to do to show that it is true? What would I have to do to show that it is false?" The answers to these questions depend entirely on inference rules from predicate logic and the syntactic structure of the definitions involved. Knowing how to answer them gives students a means for focusing on the mathematical subject matter in question. Thus, including instruction in logical principles as one part of mathematics education provides a balance between two extreme positions: that checking the validity of a proof requires that it be completely formalised; and that success in proving requires no explicit knowledge of logical principles.

2.2   *Logic and Language: The Role of Context*

R. Thom (1974) criticised instruction in logical principles where students are primarily asked to work on sentences such as "A New Yorker is bald or tall" or "If a New Yorker has blue eyes, then he is a taxi-driver". However, Epp (2003) has suggested that using examples in which the ordinary interpretation of a statement in natural language is identical to its interpretation in logic can improve students' reasoning skills. The latter approach conforms to that of logicians and philosophers who, since Aristotle, have developed formal systems that avoid ambiguities inherent in natural language and enable determination of the soundness of deductive conclusions, "remaining close both to natural reasoning and to mathematical reasoning and proof " (Durand-Guerrier 2008, p. 382).

The problem of the relationship between natural and mathematical language is crucial in mathematics education, in particular because it is a matter of fact that natural language is commonly used in mathematics classes, both orally, in writing, as well as in textbooks. Nevertheless, the language of everyday conversation differs significantly from that of mathematical discourse. Some of these differences concern the meanings of the words used to define mathematical objects. For example, a "ring," refers to a circular object in the real world but to a certain kind of abstract mathematical structure in the field of abstract algebra. Other differences concern the meanings of terms that convey the logical structure of statements (e.g., Boero et al. 2008). For instance, in everyday language "It is necessary to do A to get B" is most commonly interpreted to mean "If I don't do A then I won't get B" and "If I do A then I will get B," whereas in mathematics it means only the former and not the latter. On the other hand, there are situations in ordinary life where the phrase has exactly the same meaning as in mathematics, for example "It is necessary for France to win the semi-finals in order to win the World Cup."

With just a few exceptions, one can find examples in everyday language where words important for determining a sentence's logic have the same meaning as in mathematics. Recognising this allows one to emphasise those cases where the mathematical interpretation of a statement genuinely differs from its interpretation in ordinary language. For instance, "There is a lid for every pot" is typically understood to mean the same as "For every pot, there is a lid;" that is, different pots typically have different lids. In everyday language, statements where the words "there is" precede the words "for every" are almost always interpreted as if the words "for every" preceded the words "there is." In mathematics, however, the phrases "there is" and "for every" are interpreted from left to right. Thus, "For every positive real number x, there is a positive real number y such that $y<x$" is true, whereas "There is a positive real number y such that for every positive real number x, $y<x$" is false. Because of their experience with ordinary language, many students do not see a difference between these two statements.[9]

Several researchers have studied the difficulties experienced by university students in dealing with the relationship between natural language and mathematical language (e.g., Blossier et al. 2009; Boero et al. 2008; Chellougui 2009; Dubinsky and Yiparaki 2000; Epp 2009; Selden and Selden 1995). In the United States, many colleges and universities offer "transition-to-higher-mathematics courses" to help students communicate effectively in the language of mathematics in order to move from calculation-based courses to proof-based ones. According to Moore (1994), linguistic problems are one of the main difficulties students face in such courses.

In ordinary classroom settings, except when they are themselves writing careful mathematical proofs, teachers often use natural language that typically contains implicit assumptions, hidden quantification, and unstated reasoning rules rather than explicit references to logical principles (Durand-Guerrier and Arsac 2005). This lack of rigour in teachers' discourse may impede students' ability to write

---

[9] In addition, the logical analysis of statements may vary from one natural language to another. For example, in French statements of the form "tous les A ne sont pas B" ("All A are not B") are ambiguous. Depending on the context, they may mean either "There is an A that is not-B" (existential statement) or "For all A, A is non-B" (universal statement). In written Arabic, however, this ambiguity does not exist (Durand-Guerrier and Ben Kilani 2004). Such issues were discussed in a paper presented at *ICMI Study 21: Mathematics Education and Language Diversity*, São Paolo, Brazil, September 16–21, 2011.

precise proofs, and in some cases, there is also a lack of rigour in textbooks (e.g. Durand-Guerrier and Arsac 2005). For example, in one instance, in a French textbook for first year university students, the author used an invalid inference rule (Houzel 1996, p. 27). In the given mathematical context, one could deduce the conclusion from the premises, but only by referring to an extraneous mathematical property (Durand-Guerrier 2008). This situation replicates the unequal positions of teachers (who can bring in additional, unstated knowledge when needed) and students (whose knowledge base is limited).

2.3 Implicit Versus Explicit Assumptions About Logic

Proof and argumentation take place within social and organisational contexts that contain specific implicit assumptions about the appropriateness of various forms of reasoning. However, in making these assumptions, students may depend on naïve thinking processes and/or previous everyday and school experiences,[10] whereas teachers can draw on extensive mathematical experience and knowledge. For instance, students entering secondary school geometry may be disconcerted by teachers' requirements for deductive reasoning because teachers in previous grades considered activities involving visualisation and/or measurement sufficient to establish the truth of a statement. More generally, ordinary language and informal mathematical usage often hide the logical structures of sentences, such as the scope of quantification, the interrelations between connectors and quantification, the hierarchy of connectors, etc. At the university level, recognising the logical structure of mathematical statements is necessary for success in proving, but students have difficulty "unpacking" their logic (Durand-Guerrier 2003; Epp 2003; Selden and Selden 1995).

Teaching should aim to make explicit the assumptions about the uses of logic, in terms of both reasoning and language, that students have not yet learned. Douek (2003) proposed framing, clarifying and implementing this position by combining the work of Vergnaud (1990) and Vygotsky (1986). This synthesis examines how teachers can manage and develop the dialectic between everyday concepts and scientific concepts in order to promote students' conceptualisations of logical principles by placing logical "theorems-in-action"[11] (with their respective domains of validity) in the foreground and by clarifying linguistic representations.

In this connection, deciding on priorities becomes a delicate issue of which theorems-in-action to make explicit and at which grade level. The answer depends on

---

[10] This will be discussed in Sect. 3.4.

[11] A theorem-in-action is a mathematical property that a person may not be consciously aware of but may use in certain situations, such as to find an answer to a mathematical question. However, because the property may not apply to all the situations in which the person might try to use it, it could lead to an invalid deduction.

the needs of the specific mathematical topics addressed in each grade and the extent to which they involve making inferences. As Quine wrote, "The most conspicuous purpose of logic, in its applications to science and everyday discourse, is the justification and criticism of inference" (Quine 1982, p. 45). However, at any grade level, teachers cannot effectively guide their students' reasoning activities if they themselves are not explicitly aware of the basic principles of logical reasoning.

2.4   Syntactic and Semantic Forms of Reasoning

At first glance, one might consider logic as merely relevant to syntactic aspects of proof: for example, how to correctly use inference rules, and how to correctly symbolise quantification or negation and their relations. However, difficulties with logic, whether from a didactic or a psychological perspective, are also linked with semantics and with the interplay between everyday and mathematical language.

The semantic perspective in logic appeared with Aristotle and was significantly further developed in the late nineteenth and early twentieth centuries, by Frege (1882), Wittgenstein (1921), Tarski (1983, 1944), Quine (1982), and others. In particular, Tarski (1944) provided a semantic definition of truth that is formally correct and materially adequate, through the crucial notion of satisfying an open sentence with an object. He thus developed a model theoretic point of view, with semantics at its core, which elucidates the relation between truth in an interpretation and validity in a theory (cf. Durand-Guerrier 2008). According to Sinaceur (2001, p. 52), Tarski showed that "semantic analysis is *furthered*, and not superseded, by syntactic analysis."

Many logicians nowadays share a belief in the importance of semantics. For example, Da Costa (1997) has claimed that, in order to correctly understand a mathematical field, one needs to take into account semantics (the relation between signs and objects), syntax (the rules of integration of signs in a given system), and pragmatics (the relationship between subjects and signs). And Hintikka (1996) introduced a distinction between "indoor" and "outdoor" games that can be used to model the back-and-forth process in proof development.

Barrier et al. (2009) provided some evidence that the two kinds of games play an important role in proof and proving, because the emergence of a proof often rests upon a back-and-forth process between actions on objects (semantics) and exploration in a theory (syntax). They illustrate this with an example from Barallobres (2007, pp. 42–43). During a classroom session, 13–14 year-old students, who were beginning the study of algebra, were placed by their teacher in teams which competed to find the sum of 10 consecutive numbers in the quickest possible way, given an initial number. Then they were asked to describe their strategy in terms of the initial number. During the classroom debate different algebraic expressions and methods emerged. Asked to explain why two different formulations gave the same solution, the students combined both "indoor" syntactic and "outdoor" semantic points of view. For example, one response was, "When you multiply by 10, you've got a zero at the end, and so, adding 45 is the same as adding 4 to the digit in the tens column and then putting a 5 in the ones column," which led students to propose the equivalence "$10n + 45 = 10n + 40 + 5$". Kieran and Drijvers (2006) described a similar interplay between syntactic work on algebraic expressions and semantic work in the numerical domain through evaluation, and its role in fostering students' understanding of the notion of equivalence in algebra.

3   Contexts in Mathematics Education That Foster Understanding of Logic

3.1   Addressing Known and Common Misconceptions

The mathematics education literature provides several examples of how students' mathematical reasoning leads them to false conclusions because of misunderstandings about the ways words such as *if-then*, *and*, *or*, *not*, *all*, and *some* are used in mathematics and about what it means for statements involving these words to be false (e.g., Epp 2003, 2009). Including such examples in mathematics education programmes helps teachers both to better understand their students' work and to improve their own understanding of important aspects of mathematical reasoning.

Take for example the relationship between implication and equivalence. For every *real* number x, $x > 1$ implies $x^2 > x$; whilst there exists x such that, $x^2 > x$ and $x \ngtr 1$ (e.g., $[-2]^2 > -2$, and $-2 < 1$). If, however, the inequalities are viewed over the set of *natural* numbers, then "for every x, $x^2 > x$ implies $x > 1$". Interviews with first-year university students in Genoa about these inequalities have revealed a tendency to generalise incorrectly from a few examples, to confuse real with natural numbers, and to conflate implication with conjunction.

Although logic is only part of students' difficulties with such questions, asking them to check whether such implications are equivalences over different sets of numbers can serve several purposes. It can help them understand what it means for a quantified if-then statement to be false and thereby introduce them to the idea of counterexample; it can sensitise them to the domain over which variables are defined; it can reveal a misunderstanding on their part that implication and equivalence are the same; and it can increase their store of concrete examples for future reference.

As another example, in Euclidean geometry, Varignon's theorems state (1) In any quadrilateral the midpoints of the sides are the vertices of a parallelogram; and (2) If the quadrilateral is convex, then the area of the parallelogram is one half the area of the quadrilateral. Theorem (1) is often "proved" by considering only the case of a convex quadrilateral and examining the pairs of opposite triangles formed by each of its diagonals. However, that "proof" does not work if the quadrilateral is not convex or is crossed, although theorem (1) *is* true in both these cases. On the other hand, convexity is necessary for theorem (2).

Varignon's theorems illustrate the importance of precise definition in mathematics: in this case, quadrilaterals can take three different forms; a statement that is true for one form may not be true for the others. In addition, this example addresses the common incorrect belief that to prove a theorem in Euclidean geometry one only needs to draw a single "suitable" representative of a class of figures that fits the conditions of the definitions and hypotheses in the theorem, whilst excluding any figure that satisfies additional, specific conditions. A related mistaken belief is that a single counterexample suffices to show the necessity of a particular condition in a theorem. For Varignon's theorems, no single *suitable* generic example can be found to prove theorem (1) because the proofs are different for each type of quadrilateral. Moreover, two different counterexamples must be found to show that convexity is needed for theorem (2). Thus the theorems provide a situation, relevant to secondary school and teacher education, in which a given definition must be explored in order to identify the entire range of objects that satisfy it (for other examples, see Deloustal-Jorrand 2004).

Mathematics students at the tertiary level have problems working with definitions, such as the definition of limit, which involve chains of quantifiers. Even a definition involving only two quantifiers (e.g., the definition of a surjective function, which is important at the secondary level in many countries) creates difficulties for many students. Working with such definitions offers opportunities for addressing misconceptions about connectors, quantifiers, and their mutual relationships, whilst dealing with important mathematical subject matter, particularly in linear algebra, advanced calculus, and real analysis.

Instructors can use definitions as a basis for class discussions that explore students' misunderstandings about important logical principles. For example, all of the following misunderstandings (amongst others) are typically revealed in students' work: that the negation of an *and* statement is an *and* statement; that the negation of an *if-then* statement is an *if-then* statement; that negating only the first quantifier suffices to negate a multiply-quantified statement; and that a statement of the form "there exists A for all B" means the same as "for all A there exists B." Many students, even those in advanced courses at universities, continue to hold such misconceptions, even though these may have been hidden when the students were working on standard tasks in less advanced mathematical areas (e.g. Chellougui 2009; Dubinsky and Yparaki 2000; Epp 2003; Selden and Selden 1995).

3.2 Familiar Contexts That can Foster Understanding of Logic

This section includes a few illustrations of how logical concepts can be developed through activities in contexts familiar enough that students' reasoning depends only on "mature knowledge"; that is, their previous experiences combined with knowledge of relevant language and symbolism.

Bartolini Bussi et al. (1999) claimed that, given a suitable sequence of tasks and proper guidance from a teacher, Fourth-Graders can produce experience-based generalisations, which they can express as conditional statements and for which they can construct proof-like justifications. For example, in a long teaching experiment in Italy, one class of 17 9 and 10 year-old students had been studying about gears since Grade 2. In Grade 4, they began to view the operation of the gears in terms of mathematical objects; drawings of toothed wheels became circles, a pointing finger became a symbolic arrow, and so on. The students' reasoning basis shifted from observation and experimentation to exploration of the mathematical objects' organisation and references to shared knowledge.

After observing and describing the functioning of two wheels in gear through observation, the students were asked to conjecture the movement of three hypothetical wheels in gear and to justify their conjecture. They did so verbally, through drawings, and through gestures. Then, the teacher asked them to express a general law governing the movement of any number of gears. To do so, they had to formulate conditional statements. Most students solved the problem on a theoretical level; offering general conclusions based on generally valid arguments belonging to "sure knowledge" that they had established through previous problem solving. For instance, Elisabeta expressed her reasoning as follows, using the shared knowledge that two wheels in gear move in opposite directions: *"The wheels, two by two, if [they are] odd [in number] are in gear but they block [each other's movement] and if they are even [in number], they are in gear and do not block [each other] .... I have done a drawing to make sure ...."*. This task was followed by a collective discussion using a document supplied by the teacher, which compared excerpts of Heron's kinematic on toothed wheels with excerpts of the students' texts. Students compared their proofs to Heron's, gave the status of key statements from Heron's model, showed how these could lead to a proof, and so forth. They could identify key aspects of their reasoning using Heron's words, such as "we postulate" or "we prove". The process offered a genuine opportunity to develop a sense of the meaning and value of proof, to distinguish experimental from theoretical proof, and to express relationships between logical structure and the contents of their own proofs.

Another example that uses a familiar context is the "maze task" discussed by Durand-Guerrier (2003, pp. 7–9). A drawing shows a maze consisting of 20 rooms labelled from A to T, some of which connect with others through open doors (Fig. 16.1). A person named X is said to have passed through the maze, never going through any door twice.

Durand-Guerrier and Epp have presented this task hundreds of times in France and the United States, and there has always been disagreement about the statement, "If X passed through L, then X passed through K." Some have responded that they "can't tell" whether it is true or false, whereas others say it is false; the drawing actually shows that someone who passed through L could have also passed through K, but could also have passed through L without passing through K. Because the context of the question is easily understood, discussion can focus on the interpretation of the conditional, whether as a generalised (universally quantified) statement or as a material conditional statement whose truth value depends only on the truth-value of the two components, and on the logical status of the letter X as referring to

Fig. 16.1 The drawing
of the maze

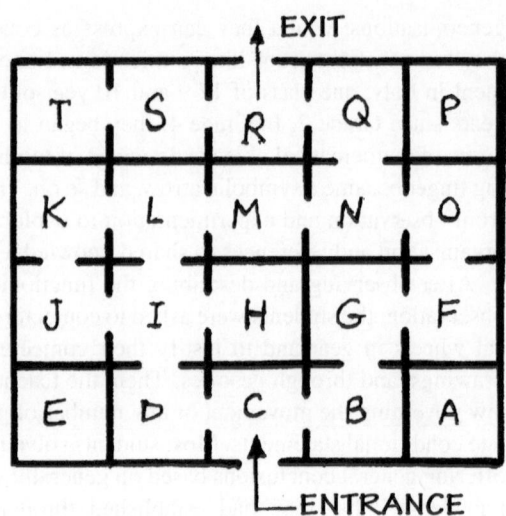

a particular individual or a "generic person." Thus, teachers can use this task to explore the difference between universal and material conditional statements and to alert students to the fact that universal quantification is often implicit.

Relatively familiar mathematical contexts also allow examples that can help students develop their logical skills. For instance, the question of how the sum of p consecutive integers depends on the value of p offers an opportunity for lower-secondary students to compare generalisation through particular examples with generalisation by a "generic" example. It can also give them experience with thinking about counterexamples and determining the set of objects for which a general statement is true. At the upper-secondary-school and university levels, number theory and discrete mathematics provide opportunities to make explicit and formalise various modes of reasoning such as *reductio ad absurdum*, proof by division into cases, proof by contraposition, and recursive reasoning (cf. Battie 2009; Epp 2003; Grenier and Payan 1998).

3.3 Contexts That May Require Understanding Logic on a Syntactic Level

If work on logical concepts done within familiar contexts appears to be relevant for young students, one may ask what logical skills are needed in more advanced mathematical studies. Expert mathematicians, such as Thurston (1994) and Thom (1974), claim that their natural and intuitive approach to logic, whilst necessary, is sufficient for them to succeed in their work. In fact, Thom (1974) wrote that over-emphasising the formal aspects of mathematics may interfere with a mathematician's work, in the same way that individuals may hesitate in "speaking a [foreign] language because they know the grammar too well and are afraid of making mistakes" (p. 45, our translation).

On the other hand, examination of mathematicians' published work reveals a very considerable command of the principles of logical reasoning. It is possible that, through innate ability or unconscious absorption during their school years, many mathematicians reached this level of competence without formal training, but, as indicated previously, large numbers of mathematics students have not attained it by the time they reach the university. Moreover, even prominent mathematicians have been misled because of weak syntactic and formal control over the concepts with which they were dealing. For instance, Hitt (2006) described the efforts of Leibniz, Wolf, d'Alembert and others who tried to solve problems of infinite sums, limits, convergence and divergence at a time when the appropriate mathematical definitions had not yet been formulated. Formal definitions of infinite sum, limit, convergence, divergence and continuity, which relied heavily on quantification, were developed in the early to middle years of the nineteenth century. Cauchy and Abel were amongst the mathematicians who contributed to clarifying and formalising these concepts, and yet their work contains several faulty statements and incorrect proofs. These errors mainly resulted from incorrect management of stacked quantifiers.[12] Durand-Guerrier and Arsac (2005) analysed, from this point of view, Cauchy's proof of the incorrect statement that a limit of a (pointwise) convergent sequence of continuous functions is continuous (e.g., Lakatos 1976), and pointed out that Abel recognised the available counterexamples to Cauchy's proof, but did not identify the mistake. Moreover, he made a similar error in two of his own proofs. (Durand-Guerrier and Arsac 2005, pp. 155–156) The identification of that error ultimately led to the notion of uniform convergence. Analyses like these authors' support the hypothesis that the difficulties faced by these mathematicians involved closely interrelated mathematical and logical matters.

From an educational perspective, these examples suggest that syntactic control is not only particularly important for novices but is also necessary for more advanced students facing concepts about which their semantic control is still shaky. This necessity for syntactic control calls for attention from educators, teachers and textbook writers (cf. Dreyfus 1999). In such contexts, a lessening of rigour, considered by experts as an innocuous shortcut, can result in serious misunderstandings by students.

3.4   Logic and Cultural Contexts: An Insight

We hypothesise the relevance of predicate logic throughout the curriculum on the basis of both epistemological and didactic considerations, as well as on research

---

[12] These stacked quantifiers may also be described as instances of the *dependence rule for variables*: in a statement of the form "For all x, there exists y such that $F(x, y)$," y is dependent on x.

with secondary and undergraduate students. However, the teaching and practice of logic as defined at the beginning of Sect. 1 takes place in cultural contexts where students and teachers may be exposed in other parts of their lives to different ways of organising and communicating reasoning. In their recent work, Boero and colleagues (Boero et al. 2010; Morselli and Boero 2009) have proposed adapting Habermas's (2003) theory of rational behaviour into a framework for situating and comparing different ways of interpreting and communicating logical requirements. According to Habermas, rational behaviour in discourse consists of three interrelated components: epistemic rationality (conscious checking of statements' and inferences' validity within a given culture); teleological rationality (conscious choice and use of tools to achieve the discursive aim); and communicational rationality (conscious choice of means to communicate reasoning according to the given community's rules). All three components of rational behaviour may be related to a cultural context.

For an example concerning teleological and epistemic rationality consider the study by Luria (1976) about Uzbek peasants living in Central Asia country villages of the Soviet Union in the 1930s, who would accept a syllogism only if they knew its major premise was true. (For a summary of that study and of more recent ones in several different cultures and social environments, see Norenzayan et al. 2007). In such a context, argument by contradiction is impossible. On the other hand, there have been psychological studies showing that some children are able to apply modus tollens (e.g. Noveck et al. 1991), and modus tollens was one of the five inference rules explicitly used by the Stoics (Blanché 1970). These results could be relevant to mathematics education because of the cultural diversity amongst students in many countries.

A second example illustrates the fact that, even in a given cultural context, communicational rules, and hence the required degree of explicitness in reasoning, depends on the domain. As an example, the need to stress a statement's content is experienced in several familiar domains such as songs, poetry, and religious texts as well as in ordinary interpersonal communication, and this may affect some students' mathematical writing. For instance, during a teaching experiment reported in Boero et al. (1995), students in a grade six Italian classroom had to justify the fact that the only common divisor of two consecutive numbers is 1. The students had been given no model of mathematical proof before this task, except for a discussion about how general reasoning can avoid the need to check the truth of a statement by examining every case. Lucia, a brilliant student with a powerful mathematical intuition, wrote a correct justification, but as a series of statements without logical connectives.

> The difference between two consecutive numbers is 1.
> The difference between two consecutive multiples of a number is that number.
> Two consecutive numbers are multiple of 1.
> No other number can divide them.

From the interview it was clear that her choice of statements was based on a conscious logical thread, but she did not use connectives because, as she put it, she wanted to "communicate the facts, the true things."

These two examples are intended to suggest the desirability of reflecting with teachers and students on the fact that modes of reasoning, including what is regarded as acceptable mathematical argumentation, are a culturally and historically situated component of our scientific rationality.

4   Conclusion

We have tried in this chapter to provide evidence for the value of integrating instruction in the principles of logic into instruction in mathematical argumentation and proof. Not all mathematics educators share this view, but there is little disagreement that most students, even many in relatively advanced university courses, have serious difficulties with the logical reasoning required to determine the truth or falsity of mathematical statements. In addition, teaching logic as an isolated subject generally appears to be inefficient in developing reasoning abilities.

Initially, we discussed the positions of cognitive psychologists and mathematics educators about the role of logic in argumentation and proof. This brief review reinforced our position that it is important to view logic as dealing with both the syntactic and semantic aspects of the organisation of mathematical discourse. In the second section, after reviewing the role of logic in mathematical argumentation and proof, we concluded that familiarity with logical principles and their application is most useful when students' mathematical knowledge is not sufficient, by itself, to evaluate the truth of a mathematical statement. In order to enable students to check the validity of a proof or disproof, or to deal with potential ambiguities in the language used in the mathematical classroom, instructors should render explicit the logical aspects of mathematical activity in the classroom, taking into account issues both of context and of language. In the third section, we gave some examples of potentially rich contexts for fostering students' understanding of logical concepts and for addressing and discussing common misconceptions. Familiar contexts can be used to focus on the logical aspects of mathematical activity; unfamiliar contexts could reveal the necessity of an interplay between semantic and syntactic perspectives. We also explored examples showing the sensitivity of logic to cultural contexts. A challenge for the future is to develop and implement these suggestions in research programmes.

It would be worthwhile to develop a framework for the comprehensive study of the complexity of the proving process as a whole and to provide the educational groundwork for fostering student competence in argumentation and proof, including issues that concern logic and language – key aspects of proof and proving in mathematics education.

Such a framework should take into account the systemic aspects of proof and proving as related to a given theory, proof and proving within a given theory, and proof and proving in the construction/expansion of a theory. From this perspective, it is desirable to investigate the relationships and tensions between the problem-solving aspect of the proving process and its logical and communicative aspects.

In order to be able to provide teachers with concrete suggestions and solid proposals, it is necessary to research further the variety of logical activities related to proof to help students better produce, organise, and understand arguments and assess their validity. Such research will enable educators to enhance the various means on which teachers may rely for developing students' competence in argumentation and proof: the design of activities, responses to student work, cultural mediation, and so forth.

Here, we have considered detailed questions of when and how only lightly and non-systematically, but the most important step is in "educating awareness," as Mason (2010) has put it, about the issues linked to logic in the teaching and learning of proof. This is the outcome to which we hope this chapter has contributed.

Acknowledgements We wish to thank the members of Working Group 2: Thomas Barrier, Thomas Blossier, Paolo Boero, Nadia Douek, Viviane Durand-Guerrier, Susanna Epp, Hui-yu Hsu, Kosze Lee, Juan Pablo Mejia-Ramos, Shintaro Otsuku, Cristina Sabena, Carmen Samper, Denis Tanguay, Yosuke Tsujiyama, Stefan Ufer, and Michelle Wilkerson-Jerde.

We are grateful for the support of the Institut de Mathématiques et de Modélisation de Montpellier, Université Montpellier 2 (France), IUFM C. Freinet, Université de Nice (France), Università di Genova (Italy), DePaul University (USA), and the Fonds québécois de recherche sur la société et la culture (FQRSC, Grant #2007-NP-116155 and Grant #2007-SE-118696).

We also thank the editors and the reviewers for their helpful feedback on earlier versions of these chapters.

References*

Barallobres, G. (2007). Introduction à l'algèbre par la généralisation: Problèmes didactiques soulevés. *For the learning in mathematics, 27*(1), 39–44.

Barrier, T., Durand-Guerrier, V. & Blossier, T. (2009). Semantic and game-theoretical insight into argumentation and proof (Vol. 1, pp. 77–82).*

Bartolini Bussi, M. G., Boni, M., Ferri, F., & Garuti, R. (1999). Early approach to theoretical thinking: Gears in primary school. *Educational Studies in Mathematics, 39*, 67–87.

Battie, V. (2009) Proving in number theory at the transition from secondary to tertiary level: Between organizing and operative dimensions (Vol. 1, pp. 71–76).*

Blanché, R. (1970). *La logique et son histoire*. Paris: Armand Colin.

Blossier, T., Barrier, T. & Durand-Guerrier, V. (2009). Proof and quantification (Vol. 1, pp. 83–88).*

Boero, P., Chiappini, G., Garuti, R. & Sibilla, A. (1995). Towards statements and proofs in elementary arithmetic. In *Proceedings of PME-XIX* (Vol. 3, pp. 129–136). Recife: Universitade Federal de Pernambuco.

Boero, P., Douek, N., & Ferrari, P. L. (2008). Developing mastery of natural language: Approach to theoretical aspects of mathematics. In L. English (Ed.), *Handbook of international research in mathematics education* (pp. 262–295). New York/London: Routledge.

Boero, P., Douek, N., Morselli, F., & Pedemonte, B. (2010). Argumentation and proof: A contribution to theoretical perspectives and their classroom implementation. In M. F. F. Pinto &

---

*NB: References marked with * are in Lin, F. L., Hsieh, F. J., Hanna, G. & de Villiers M. (Eds.) (2009). *ICMI Study 19: Proof and proving in mathematics education*. Taipei, Taiwan: The Department of Mathematics, National Taiwan Normal University.

T. F. Kawasaki (Eds.), *Proceedings of the 34th conference of the international group for the psychology of mathematics education* (Vol. 1, pp. 179–205). Belo Horizonte: PME.

Chellougui, F. (2009). L'utilisation des quantificateurs universels et existentiels en première année d'université, entre l'implicite et l'explicite. *Recherches en didactique des Mathématiques, 29*(2), 123–154.

Cheng, P., Holyoak, K., Nisbett, R. E., & Oliver, L. (1986). Pragmatic versus syntactic approaches to training deductive reasoning. *Cognitive Psychology, 18*, 293–328.

Copi, I. (1954). *Symbolic logic.* New York: The Macmillan Company.

Da Costa, N. C. A. (1997). *Logiques classiques et non classiques: essai sur les fondements de la logique.* Paris: Masson.

Deloustal-Jorrand, V. (2004). *Studying the mathematical concept of implication through a problem on written proofs, proceedings of the 28th conference of the international group for the psychology of mathematics education* (Vol. 2, pp. 263–270). Bergen: Bergen University College.

Douek, N. (2003). *Les rapports entre l'argumentation et la conceptualisation dans les domaines d'expérience.* Thèse, Université ParisV, Paris.

Dreyfus, T. (1999). Why Johnny can't prove. *Educational Studies in Mathematics, 38*, 85–109.

Dubinsky, E., & Yiparaki, O. (2000). On students understanding of AE and EA quantification. *Research in Collegiate Mathematics Education IV, CBMS Issues in Mathematics Education, 8*, 239–289.

Durand-Guerrier, V. (2003). Which notion of implication is the right one ? From logical considerations to a didactic perspective. *Educational Studies in Mathematics, 53*, 5–34.

Durand-Guerrier, V. (2008). Truth versus validity in mathematical proof. *ZDM The International Journal on Mathematics Education, 40*(3), 373–384.

Durand-Guerrier, V., & Arsac, G. (2005). An epistemological and didactic study of a specific calculus reasoning rule. *Educational Studies in Mathematics, 60*(2), 149–172.

Durand-Guerrier, V., & Arsac, G. (2009). Analysis of mathematical proofs: Some questions and first answers (Vol. 1, pp. 148–153).*

Durand-Guerrier, V., & Ben Kilani, I. (2004). Négation grammaticale versus négation logique dans l'apprentissage des mathématiques. *Exemple dans l'enseignement secondaire Tunisien, Les Cahiers du Français Contemporain, 9*, 29–55.

Duval, R. (1991). Structure du raisonnement déductif et apprentissage de la démonstration. *Educational Studies in Mathematics, 22*(3), 233–262.

Epp, S. (2003). The role of logic in teaching proof. *The American Mathematical Monthly, 110*(10), 886–899.

Epp, S. (2009). Proof issues with existential quantification (Vol. 1, pp. 154–159).*

Frege, G. (1882). Uber die Wissenschaftliche Berechtigung einer Begriffschrift. *Zeitschrift für Philosophie und Philosophische kritik Nf, 81*, 48–56.

Gentzen, G. (1934). Untersuchungen über das logische Schließen. *Mathematische Zeitschrift, 39*(2), 176–210.

Grenier, D., & Payan, C. (1998). Spécificités de la preuve et de la modélisation en mathématiques discrètes. *Recherches en didactiques des mathématiques, 18*(1), 59–99.

Habermas, J. (2003). *Truth and justification.* Cambridge: MIT Press.

Hanna, G. (2000). Proof, explanation and exploration: An overview. *Educational Studies in Mathematics, 44*, 5–23.

Hintikka, J. (1996). *The principles of mathematics revisited.* Cambridge: Cambridge University Press.

Hitt, F. (2006). L'argumentation, la preuve et la démonstration dans la construction des mathématiques: des entités conflictuelles ? Une lettre de Godefroy Guillaume Leibnitz à Chrétien Wolf (1713). In D. Tanguay (Ed.), *Actes du colloque du Groupe des didacticiens des mathématiques du Québec* (pp. 135–146). Montréal: Université du Québec à Montréal.

Houzel, C. (1996). *Analyse mathématique. Cours et exercices.* Paris: Belin.

Johnson-Laird, P. N. (1975). Models of deduction'. In R. J. Falmagne (Ed.), *Reasoning: Representation and process in children and adults* (pp. 7–54). Hillsdale: Lawrence Erlbaum Associates.

Johnson-Laird, P. N. (1986). Reasoning without logic. In T. Meyers, K. Brown, & B. McGonigle (Eds.), *Reasoning and discourse processes* (pp. 14–49). London: Academic.

Johnson-Laird, P. N., Legrenzi, P., & Legrenzi, M. (1972). Reasoning and a sense of reality. *British Journal of Psychology, 63*(3), 395–400.

Kieran, C., Drijvers, P., Boileau, A., Hitt, F., Tanguay, D., Saldanha, L., & Guzmán, J. (2006). Learning about equivalence, equality, and equation in a CAS environment: The interaction of machine techniques, paper-and-pencil techniques, and theorizing. In C. Hoyles, J. Lagrange, LH Son, & N. Sinclair (Eds.), *Proceedings for the Seventeenth ICMI Study Conference: Technology Revisited*, Hanoi : Hanoi University of Technology.

Lakatos, I. (1976). *Proofs and refutations*. Cambridge: Cambridge University Press.

Lee, K. & Smith III, J.P. (2009). Cognitive and linguistic challenges in understanding proving (Vol. 2, pp. 21–26).*

Luria, A. R. (1976). *The Cognitive Development: Its Cultural and Social Foundations*. London: Harvard University Press.

Mamona-Downs, J. & Downs, M. (2009). Proof status from a perspective of articulation (Vol. 2, pp. 94–99).*

Mariotti, A. (2006). Proof and proving in mathematics education. In A. Gutiérrez & P. Boero (Eds.), *Handbook of research on the psychology of mathematics education: Past, present and future* (pp. 173–204). Rotterdam/Taipei: Sense.

Mason, J. (2010). Mathematics education: Theory, practice & memories over 50 years. *For the Learning of Mathematics, 30*(3), 3–9.

Moore, R. C. (1994). Making the transition to formal proof. *Educational Studies in Mathematics, 27*, 249–266.

Morselli, F. & Boero, P. (2009). Habermas' construct of rational behaviour as a comprehensive frame for research on the teaching and learning of proof (Vol. 2, pp. 100–105).*

Norenzayan, C., & Peng. (2007). Perception and cognition. In S. Kitayama & D. Cohen (Eds.), *Handbook of cultural psychology* (pp. 569–594). New York: The Guilford Press.

Noveck, I. A., Lea, R. B., Davidson, G. M., & O'Brien, D. P. (1991). Human reasoning is both logical and pragmatic. *Intellectica, 11*, 81–109.

Pólya, G. (1954). *Mathematics and plausible reasoning: Volume 1: Induction and analogy in mathematics. Vol. 2: Patterns of plausible inference*. Princeton: Princeton University Press.

Quine, W. V. (1982). *Methods of logic* (4th ed., 1st ed. 1950). New York: Holt, Rinehart & Winston.

Roh. (2009). Students' understanding and use of logic in evaluation of proofs about convergence (Vol. 2, pp. 148–153).*

Selden, A., & Selden, J. (1995). Unpacking the logic of mathematical statements. *Educational Studies in Mathematics, 29*, 123–151.

Simpson, A. (1995). *Developing a proving attitude. Conference Proceedings: Justifying and Proving in School Mathematics* (pp. 39–46). London: Institute of Education, University of London.

Sinaceur, H. (2001). Alfred Tarski, semantic shift, heuristic shift in metamathematics. *Synthese, 126*, 49–65.

Stenning, K., & Lambalgen, M. V. (2008). *Human reasoning and cognitive science*. Cambridge: Bradfors Books.

Stenning, K., Cox, R., & Oberlander, J. (1995). Contrasting the cognitive effects of graphical and sentential logic teaching: Reasoning, representation and individual differences. *Language and Cognitive Processes, 10*(3–4), 333–354.

Tarski, A. (1944). The semantic conception of truth. *Philosophy and Phenomenological Research, 4*, 13–47.

Tarski, A. (1983). *Logic, semantics and metamathematics, papers from 1923 to 1938*. Indianapolis: John Corcoran.

Thom, R. (1974). Mathématiques modernes et mathématiques de toujours, suivi de Les mathématiques « modernes », une erreur pédagogique et philosophique ? In R. Jaulin (Ed.), *Pourquoi la mathématique ?* (pp. 39–88). Paris: Éditions 10–18.

Thurston, W. P. (1994). On proof and progress in mathematics. *Bulletin of the American Mathematical Society, 30*(2), 161–177.

Ufer, S., Heinze, A. & Reiss, K.(2009). What happens in students minds when constructing a geometric proof ? A cognitive model based on mental models (Vol. 2, pp. 239–244).*

van der Pal, J., & Eysink, T. (1999). Balancing situativity and formality: The importance of relating a formal language to interactive graphics in logic instruction. *Learning and Instruction, 9*(4), 327–341.

Vergnaud, G. (1990). La théorie des champs conceptuels. *Recherches en didactique des mathématiques, 10*(2), 133–170.

Vygotsky, L. S. (1986). *Thought and language* (2nd ed.). Cambridge: MIT Press.

Wason, P. C. (1966). Reasoning. In I. B. M. Foss (Ed.), *New horizons in psychology*. Harmondsworth: Penguin.

Wason, P. C. (1977). Self-contradictions. In P. N. Johnson-Laird & P. C. Wason (Eds.), *Thinking: Readings in cognitive science* (pp. 114–128). New York: Cambridge University Press.

Wittgenstein, L. (1921). Annalen der naturphilosophie [Tractatus logico-philosophicus], Leipzig (Ed. & Trans.). London: Routledge and Kegan Paul Ltd.

Chapter 17
Transitions and Proof and Proving at Tertiary Level*

Annie Selden

There are many changes in the didactic contract (Brousseau 1997) when students move from secondary school to undergraduate study in mathematics or from undergraduate to graduate study in mathematics. Beginning university students often face "a difficult transition, from a position where concepts have an intuitive basis founded on experience, to one where they [the concepts] are specified by formal definitions and their properties reconstructed through logical deductions" (Tall 1992, p. 495).

The problem of the passage from secondary to tertiary mathematics is not new. In the first volume of the UNESCO series, *New Trends in Mathematics Teaching* (ICMI 1966), there is a conference report devoted to this issue. In addition, a specific section of the ICMI Study volume, *The Teaching and Learning of Mathematics at University Level* (Holton 2001) discusses the secondary-tertiary interface (Wood 2001). Also, a special issue of the *International Journal of Mathematical Education in Science and Technology* (Zazkis and Holton 2005) was devoted to this issue. Further, a substantial amount has been written about the various mathematical, social, and cultural difficulties involved in this transition (Clark and Lovric 2009; Gueudet 2008; Guzman et al. 1998; Hourigan and O'Donoghue 2007; Kajander and Lovric 2005). In particular, it has been noted that in many countries, upon entry to the university, the emphasis changes from a more computational, problem-solving approach to a more proof-based approach to mathematics.

By "tertiary level", we refer to undergraduate students majoring in mathematics or mathematics education, as well as undergraduate students majoring in other subjects, such as engineering or computer science, who take courses containing

---

* *With input from Lara Alcock, Véronique Battie, Geoffrey Birky, Ana Breda, Connie Campbell, Paola Iannone, Nitsa Movshovitz-Hadar, Teresa Neto, Kyeong Hah Roh, Carmen Samper, and John Selden*

A. Selden (✉)
Department of Mathematical Sciences, New Mexico State University, Las Cruces, NM, USA
e-mail: js9484@usit.net

© The Author(s) 2021
G. Hanna and M. de Villiers (eds.), *Proof and Proving in Mathematics Education*,
New ICMI Study Series, https://doi.org/10.1007/978-94-007-2129-6_17

proofs. We will also consider inservice secondary teachers, and other professionals such as engineers, who take additional mathematics courses that include proof and proving. Finally, we include Masters and Ph.D. students in mathematics and mathematics education.

In Sect. 1, after discussing how proof at the tertiary level is different from proof at the school level, we discuss transitions from secondary to undergraduate level and from undergraduate level to graduate. In Sect. 2, we present research on how tertiary students deal with various aspects of proof and proving including logical reasoning, understanding and using definitions and theorems, selecting examples and helpful representations, and knowing how to read and check proofs. In Sect. 3, we provide information, research, and resources on the teaching of proof and proving at tertiary level. In Sect. 4, we summarise the preceding sections and pose questions for future research.

1   The Character of Proof and Proving at Tertiary Level

1.1   Proofs at the Tertiary Level

The nature of proofs and proving at tertiary level, with its increased demand for rigour, constitutes a major hurdle for many beginning university students. At this level, constructing proofs involves understanding and using both formal definitions and previously established theorems, as well as considerable creativity and insight. Understanding and constructing such proofs entails a major transition for students but one that is often supported by relatively little explicit instruction. These changes can be seen as a shift from elementary to advanced mathematical thinking (Selden and Selden, 2005; Tall 1991), and as going "from describing to defining, from convincing to proving in a logical manner based on definitions" (Tall 1991, p. 20). Here Tall is apparently referring to the systematisation function of proof. In addition, the verification, explanation, discovery, and communication functions (de Villiers 1990) of proof are all important at tertiary level.

Proofs that tertiary students are expected to study, and to construct, are more formal than those expected of students at primary or secondary level. In general, mathematicians and aspiring mathematicians, such as university mathematics majors, prove theorems, whereas children or novices tend to justify through less formal forms of argumentation. Tertiary level students may still use informal arguments, examples and intuitive reasoning to think things through initially, but eventually such reasoning must be made more formal for communication purposes (Hanna et al. 2009), whether for assessments, theses, dissertations, or publications.

In addition, proofs at tertiary level generally tend to be longer and more complex than those expected of students at primary or secondary level. If one compares typical secondary-school geometry proofs with proofs in real analysis, linear algebra, abstract algebra, or topology, one sees that the objects in geometry are idealisations of real things (points, lines, planes), whereas the objects in real analysis, linear algebra, abstract algebra, or topology (functions, vector spaces, groups, topological spaces) are abstract reifications.

Furthermore, proofs at tertiary level require a larger knowledge base. For example, real analysis proofs require a much deeper knowledge of the real number system than do secondary-school algebra or geometry proofs. In addition, tertiary professors increasingly use students' original proof constructions, not just the reproduction of textbook or lecture proofs as a means of assessing their students' content understanding. They also place more emphasis on fostering and assessing students' creativity within mathematics, especially at the Masters and Ph.D. levels where professors expect students to conjecture interesting new results and to prove and publish them.

1.2 The Secondary-Tertiary Transition

1.2.1 Looking at the Transition from an Anthropological Point of View

Clark and Lovric (2008, 2009) regard the secondary-tertiary transition as a "modern-day rite of passage". They have considered the role of the university mathematical community in this transition, the change in the didactic contract with its new expectations of student autonomy, the introduction of more abstract concepts, the students' strong emotional reactions to these changes and expectations, the long time such transitions take, and the new responsibilities assumed of students and their teachers. This transition represents a significant stage in the students' enculturation into the practices of mathematicians (Nickerson and Rasmussen 2009; Perry et al. 2009a).

In regard to proof and proving, Clark and Lovric (2008, pp. 763–764) noted that there is a change in the nature of the mathematical objects considered at university, a change in the kinds of reasoning done about those objects, and a shift from informal to formal language. Students must learn to reason from formal definitions, understand what theorems say, apply theorems correctly, and make connections between concepts – all of which cause them difficulties. However, professors unfamiliar with secondary school curricula often mistakenly assume that secondary-school students are already familiar with elements of logic and can work with implications and quantifiers. To help ease the transition, Lovric (2005) has suggested using slightly older peer tutors who can empathise with entering university students' experiences and can act as undergraduate teaching assistants.

1.2.2 Specific Examples of the Secondary-Tertiary Transition

Gueudet (2008) in her survey article discussed many aspects of the secondary-tertiary transition. She noted that studies from many different countries "have shown that only a minority of students are able to build consistent proofs at the end of high school" (Gueudet 2008, p. 243). At university, in addition to expecting the production of rigorous proofs, professors expect students to use symbols, especially quantifiers, properly and to exhibit autonomy and flexibility of thinking. There are also new expectations about what requires justification and what does not. For example, in French secondary school, when students are asked to show a family of vectors forms an orthonormal basis, they need only compute the scalar product for each pair of non-identical vectors. However, in a university linear algebra course, students are expected to justify this by referring to the theorem establishing that orthonormal vectors are independent (Gueudet 2008, p. 246). Such expectations, often not explicitly addressed by professors, constitute a vast change in the didactic contract.

In 1998, elementary number theory was introduced into the French secondary-school curriculum to promote students' mathematical reasoning in an area in addition to geometry. Topics included divisibility, Euclid's algorithm, relatively prime numbers, prime factorisation, least common multiples, Bézout's identity, and Gauss' theorem. Whilst investigating the secondary-tertiary transition within number theory in France, Battie (2009) proposed two complementary, and closely intertwined, epistemological dimensions. The *organising dimension* includes selecting the global, logical structure of a proof and associated techniques, such as contradiction, induction, or reduction to a finite number of cases. The *operative dimension* includes working effectively with specific techniques in the implementation of those global choices. This means performing actions, such as using key theorems and properties or selecting appropriate symbolic representations and algebraic transformations. Battie (2010) found that, due to traditional teaching, French secondary students are given autonomy mainly in the operative dimension, whereas first-year university students are also expected to be responsible for the organising dimension of proofs and to be fluent in both dimensions, and that this was one source of their difficulties with proving.

According to Praslon (1999), who also researched the secondary-tertiary transition in France, proofs at university play a new role. Some results establish methods or provide useful tools, whilst others serve as intermediate steps on a path towards an important theorem. Studying tasks related to the derivative in secondary and university textbooks, Praslon observed that secondary-school tasks were split into simpler subtasks; the derivative was a tool for studying function behaviour rather than itself being an object of study; and there was no need for students to develop any real mathematical autonomy. In contrast, at university, student autonomy and flexibility between mathematical registers (e.g., algebraic, graphical, and natural language) were required.

Also in France, according to Castela (2009), when transitioning from Grades 6–9 in Collège (lower secondary school) to Grades 10 and 11 in Lycée (upper secondary school), students are required to show an increasing autonomy as problem solvers and mathematics learners. Castela analysed this not in terms of transitioning to advanced mathematical thinking, but rather in terms of Chevallard's (1999) anthropological theory of didactics, which considers the constraints of the institutional setting in which the mathematical activity takes place.

Similarly in Spain, Bosch, Fonseca, and Gascon (cited in Gueudet 2008, pp. 246–247) found that novice university students knew only a single technique for a given task and, after applying it, could not interpret the result. The authors found this corresponded to Spanish secondary-school textbooks where a single technique is proposed, interpreting the result is (mostly) not required, and working with mathematical models is rare. In the examined secondary-school textbooks the focus was on narrow tasks with rigid solution methods; topics were also poorly connected.

Knuth (2002) interviewed 16 U.S. secondary-school mathematics teachers, with from 3–20 years' teaching experience, some with master's degrees, participating in a professional development programme. Knuth asked about their conceptions of proof and its place in secondary-school mathematics. All professed the view that a proof establishes the truth of a conclusion, but several also thought it might be possible to find some contradictory evidence to refute a proof. In the context of secondary school, the teachers distinguished formal proofs, less formal proofs, and informal proofs. For some, two-column geometry proofs were the epitome of formal proofs. All considered proof as appropriate only for those students in advanced mathematics classes and those intending to pursue mathematics-related majors at university.

A case study of two typical Irish senior secondary-school mathematics classes (Hourigan and O'Donoghue 2007) found that the (implicit) didactic contract included the following: The terminal exam should be the central aim of the class; the teacher should not depart from the set lesson routine of going over homework quickly, introducing a new topic through a few worked examples, and having students practise similar exercises; the teacher should not ask pupils complicated questions; the teacher should provide pupils a step-by-step breakdown of problem-solving techniques; pupils should not interrupt the lesson unnecessarily; the teacher should not ask questions that require thought and reflection (pp. 471–472). Whilst the authors concentrated on problem solving, and did not specifically mention proof and proving, they concluded that "mathematics-intensive courses at tertiary level need independent learners possessing [the] conceptual and transferable skills required to solve unfamiliar problems [but] the development of these essential skills is not fostered within the [secondary] classrooms studied" (p. 473).

Lithner (2004) stated that in Sweden the beginning undergraduate learning environment is dominated largely by textbooks, teaching and exams. He investigated the reasoning required by the exercises in one calculus textbook and personally worked through, and classified, the solution strategies required by 598 single-variable calculus exercises. He found that 85% were solvable by identification of similarities (essentially mimicking), 8% by local plausible reasoning, and 7% by global plausible reasoning. He concluded that "it is possible in about 70% of the exercises to base the solution not only on searching for similar situations, but on searching only the solved examples" (p. 422) and suggested that "a larger proportion of not so very difficult GPR [global plausible reasoning] exercises should be included in textbooks" (p. 426).

1.2.3 Epistemological and Cognitive Difficulties of the Secondary-Tertiary Transition

Guzman et al. (1998) pointed out, that secondary-school students can (almost) always take it for granted that a given problem has a solution. In considering some epistemological and cognitive difficulties of the secondary-tertiary transition, the authors noted that "existence proofs are notoriously difficult for students" and "it is not easy for them [the students] to recognize their need, as this type of situation is rarely raised in secondary mathematics" (p. 753). Proofs of sufficiency are especially difficult. "Sometimes, a proof requires not only to apply directly a theorem in a particular case, but also to adapt or even to transform a theorem before recognizing and/or using it" (p. 753). Guzman et al. suggested that tertiary students need to learn to distinguish between mathematical knowledge and meta-mathematical knowledge of the correctness, relevance and elegance of proof and take responsibility for their own mathematical learning.

In addition, some concepts learned at secondary school need to be reconstructed at tertiary level; these, too, constitute part of the transition's epistemological and cognitive difficulties. For example, in secondary-school geometry, a tangent is defined globally as the unique line that touches a circle at exactly one point and is perpendicular to the radius. However, in calculus, the tangent to a function at a point is defined locally as the limit of approximating secant lines and, somewhat later, as the line whose slope is given by the value of the derivative at that point (Biza et al. 2008; Biza and Zachariades 2010). This change from a global to a local view of tangents can be difficult for students. For example, even after having studied analysis (i.e., calculus) from the beginning of senior secondary school, French final-year secondary students still had great difficulty determining from a graph whether a given line was tangent to a particular function at an inflection point or a cusp (Castela, cited in Artigue 1992, pp. 209–210).

The treatment of equality in analysis is another instance of a concept that needs to be reconstructed. In secondary-school algebra and trigonometry, students grow used to proving that two expressions are equal by transforming one into the other using known equivalences. However, in analysis one can prove two numbers a and b are equal by showing that for every $\varepsilon > 0$, one has $|a - b| < \varepsilon$. That this idea is not easy to grasp is indicated by a French study in which more than 40% of entering university students thought that if two numbers a and b are less than $1/N$ for every positive integer N, then they are not equal, only infinitely close (Artigue 1999, p. 1379).

1.3 The Transition from Undergraduate to Graduate Study in Mathematics

There are also transitions from undergraduate to Masters and Ph.D. level study, where one eventually has to find and prove some previously unknown theorems of interest to the mathematical community (Duffin and Simpson 2002, 2006; Geraniou 2010; Herzig 2002).

1.3.1 The Transition to a Research Degree in the U.K

Geraniou (2010) noted that, in the U.K., "the transition from a taught degree to a research degree involves significant changes in the way students deal with the subject" (p. 282). Persistence, interest, confidence, and problem-solving skills, as well as the guidance of the thesis/dissertation supervisor, are often cited as crucial for the successful transition to graduate studies. Geraniou (2010) described Ph.D. students as going through three motivational transition stages. The first, in the first year, is the *adjustment* stage, during which a student comes to terms with what a pure research Ph.D. degree is and develops survival strategies. The second, beginning in the first half of the second year and continuing until a student's research is complete, is the *expertise* stage. The student must research an unsolved problem, but first may have to develop the appropriate research and problem-solving skills (which may be lacking due to dependence on memorisation for undergraduate success). The final stage is the *articulation* stage, during which time the student actually writes the Ph.D. dissertation in final form.

Duffin and Simpson (2006) noted that U.K. undergraduates usually experience lecture classes of 100 or more where syllabuses are clearly defined, the pace of new material is set, and the majority of the given problems have predetermined solutions. However, when students reach graduate school:

> there is no formal requirement to attend or pass taught courses during the period of doctoral study and almost all of a student's three years of study is spent in independent research towards the production of a single, substantial dissertation ... [which is] required to be the result of original research, to show an awareness of the relationship of the research to a wider field of knowledge and, potentially, to be publishable. (p. 236)

In a small study of the entire population of 13 Ph.D. students at one medium size U.K. university, Duffin and Simpson (2002, 2006) uncovered three distinct learning styles. There were *natural learners* whose tendency for independent thinking, rather than memorisation, no longer proved a hindrance as it had when they were undergraduates and who found the movement to graduate study smooth. There were *alien learners* who had succeeded in the past by reproducing proofs, learning procedural techniques and accepting new ideas without being overly concerned about their meanings; they found the transition to graduate study far from smooth. Lastly, one student had a flexible learning style, adjusting it according to what he wished to gain from the material.

1.3.2 The Transition to Graduate Study in the U.S

In the U.S., doctoral students spend the first 3 years taking content courses and passing a set of comprehensive examinations. Only after that, do they begin their research. Even so, many students still feel unprepared. Herzig (2002) found in one large U.S. Ph.D-granting university that students were discouraged from filling in gaps by taking Masters level courses, felt they could not ask questions of their professors, and wanted more feedback in their courses.

> Rather than the coursework building on students' enthusiasm for mathematics and involving them in authentic mathematical work, coursework distanced students from mathematics making it more difficult for them to learn what they needed to learn to participate effectively in mathematical practice. (Herzig 2002, p. 192)

Even after students had "proved themselves" by passing their comprehensive examinations, they felt they wanted more interaction with professors.

Doctoral content courses require students to construct original proofs; later for the dissertation, they need to conjecture and prove interesting, new results. However, many entering U.S. mathematics graduate students have a limited ability to construct proofs -- a major stumbling block to success in their doctoral content courses. To alleviate this problem, Selden and Selden (2009b) designed a one-semester beginning graduate course in which students construct a variety of different kinds of proofs, present them at the blackboard, and receive extensive criticism and advice. The Seldens divide proofs into their *formal-rhetorical* parts and their *problem-centred* parts, which they see as somewhat like Battie's (2009) organising and operative dimensions, respectively (cf. Sect. 1.2.2). The Seldens aim to have students automate the formal-rhetorical parts of proofs so they can devote scarce working memory to the problem-centred parts (Selden et al. 2010).

2 How Tertiary Students Deal with Various Aspects of Proof and Proving

Both tertiary students' learning to understand and construct proofs and professors' teaching tertiary students about proof and proving are daunting tasks, not easily approached by lecture alone. The following subsections provide information on tertiary students' documented proving difficulties, along with some suggestions from the literature on helping students avoid or overcome them.

2.1 Language and Reasoning

It is important that students know the difference between the pragmatic everyday use of logical terms and their mathematical use. Some university students find it challenging to keep the everyday and the mathematics registers separate (Lee and Smith 2009). For example, students sometimes confuse "limit" with its everyday meaning of a bound as in "speed limit" (Cornu 1991). Even students for whom such confusions are not a problem, can have other difficulties with mathematical language.

Rather than presenting students with truth tables or having them formally construct valid arguments, Epp (2003) has suggested stressing the difference between everyday and mathematical language and emphasising "exercises where students apply logical principles to a mix of carefully chosen natural language and mathematical statements" because "simply memorizing abstract, logical formulas and learning to apply them mechanically has little impact on students' broader reasoning powers" (p. 894).

2.1.1 Quantifiers

Students have difficulties understanding the importance of, and implications of, the order of existential and universal quantifiers, as well as knowing their often implied scope (Dubinsky and Yiparaki 2000; Epp 2009). For example, in a study 61 U.S. undergraduate mathematics students, many in their third or fourth year, it was found that they could unpack the logical structure of informally worded statements correctly just 8.5% of the time (Selden and Selden 1995). In addition, undergraduate students often consider the effect of an interchange of existential and universal quantifiers as a mere rewording (David and Movshovitz-Hadar 1998; Dubinsky and Yiparaki 2000).

Roh (2009) gave U.S. undergraduate introductory real analysis students two arguments, one to show that the sequence $\{1/n\}$ converges to 0, the other to show that it did not converge to 0. No student noticed that both arguments were fallacious, because ε was chosen to depend on the positive integer N. In general, students have difficulty with proving universally quantified statements, in which one customarily considers an arbitrary, but fixed element, as in $\varepsilon - \delta$ real analysis proofs. They often wonder why the result has been proved for all elements, rather than only for the specific arbitrary, but fixed element selected (Selden et al. 2010).

2.1.2 Proof by Contradiction and Its Link with Negation

Students sometimes find proofs by contradiction difficult to understand and construct (Epp 2003; Reid and Dobbin 1998). One reason may be the difficulties students have with the formulation of negations in mathematics. Antonini (2001), studying Italian first-year calculus students, found they had three ways, or schemes, for dealing with negation, often taken over from natural language. The first scheme was that of the opposite. For example, the negation of "f is increasing" was seen as "f is decreasing".

The second more prevalent scheme was that of considering possibilities. For example, the students saw the negation of "g is strictly decreasing" as having several possibilities, namely, g can be increasing, constant, or decreasing but not strictly decreasing. Antonini (2001) conjectured that one reason for the strength of students' scheme of possibilities is that in natural language one can often only express the negation of p as "not p". Consequently Antonini (2003) later suggested that the tendency to express negation in terms of possibilities may require specific pedagogical attention. The third scheme, that of properties, is the one used in mathematical reasoning. For example, to deny that "f is increasing" one says "there exist x and y such that $x < y$ and $f(x) \geq f(y)$" which expresses the negation in a positive way using a property common to all non-increasing functions. Doing so allows one to proceed with a proof by contradiction.

2.2 Students' Initial Nonstandard Views of Proof

Harel and Sowder (1998) investigated U.S. undergraduate students' views of proof. They defined a person's *proof scheme* to be that which "constitutes ascertaining and persuading for that person" (p. 244). They categorised the proof schemes they found as: *external conviction*, *empirical*, and *analytical*. The first two are somewhat idiosyncratic and nonstandard; the latter are like proofs that mathematicians would accept. In an earlier study, Martin and Harel (1989) asked 101 preservice elementary teachers to judge verifications of a familiar result, *if the sum of the digits of a whole number is divisible by 3, then the number is divisible by 3*, and an unfamiliar result, *if a divides b and b divides c, then a divides c*. Both inductive and deductive arguments were acceptable to these students with fewer than 10% consistently rejecting inductive arguments.

Tertiary students sometimes mistakenly think proofs are constructed in a linear fashion from the "top down" because this is the way they have seen them presented in lecture. Selden and Selden (2009a) report a mathematics education graduate student who thought this based on her undergraduate experience in a real analysis course and was very surprised to learn this was not so. Such ideas may also be a result, at least in the U.S., of being required to construct two-column geometry proofs in a linear way from the "top down" in secondary school. The normative way to construct proofs, as seen by U.S. secondary-school geometry teachers, is to always give a reason for each statement in a two-column proof before continuing, that is, immediately after the statement has been made (Weiss et al. 2009). Further, they view it as unnatural to let a student make an assumption in the middle of constructing a proof in order to investigate whether the proof might be completed that way, and then come back later to reconsider the assumption (Nachlieli and Herbst 2009). Perhaps employing a less rigid style of proving at secondary school and university would help alleviate such linear "top down" views of constructing proofs.

Another suggestion is to have students first write the *formal-rhetorical* part of a proof, that is, the hypotheses and the conclusion and then unpack and examine the conclusion in order to structure the proof. Doing so "exposes the real problem" to be solved and allows one to concentrate on the *problem-centred* parts of a proof (Selden and Selden 2009a). For further information, see Sect. 1.3.2 above.

2.3 Knowing, Using, and Understanding Other Constituents of Proofs and Proving

Although proofs at tertiary level are rigorous deductive arguments which require reasoning with correct logic, students need to learn more than logic in order to become successful provers.

2.3.1 Understanding Formal Definitions and Using Them in Proving

Too often tertiary mathematics students can recite definitions yet fail to use them when asked to solve problems or prove theorems. For example, Alcock and Simpson (2004) reported on 18 British first-year real analysis students, a number of whom argued using prototypical examples of convergent sequences rather than the formal definition of convergence. Definitions need to become *operable* for an individual. Bills and Tall (1998) consider a definition to be *formally operable* for a student if that student "is able to use it in creating or (meaningfully) reproducing a formal argument [proof]" (p. 104).

In addition, as Edwards and Ward (2004) pointed out, instructors need to make clear the distinction between mathematical definitions (*stipulated* or *analytic* definitions) and many everyday or dictionary definitions (*descriptive*, *extracted*, or *synthetic* definitions) of which students may not be aware. One could help them become aware of this distinction by discussing it with them and by engaging them in the act of defining. For example, in a university geometry course, Edwards and Ward suggest beginning with the usual definition of triangle, which is initially useful in all three of the Euclidean plane, the sphere, and the hyperbolic plane; but eventually students will notice that that certain theorems are not true for all triangles on the sphere, a problem which requires a refined definition which they can participate in developing.

Furthermore, Lakatos (1961) suggested that there can be a dialectical interplay between concept formation, definition construction, and proof. He also developed the ideas of *zero-definitions* (initial tentative or "working definitions") and *proof-generated definitions* (analytic definitions related to proof). To bring this interplay to students' awareness, one needs to design problems whose resolution requires definition construction.

Following Lakatos, Ouvrier-Buffet (2002, 2004, 2006) sought to engage entering French university students in zero-definition construction. She selected the mathematical concept of *tree* (giving it a neutral name); in France formal definitions of tree are not presented in secondary school. She first gave the students four examples and two non-examples of trees, asking them to infer a zero-definition. This was followed by a task requiring them to prove that a *given* connected graph contained *a tree*. Ouvrier-Buffet reported that students attempting this proof sometimes, but not always, revised their initial definitions.

2.3.2 Understanding the Statement of a Theorem to Be Proved

At tertiary level, the pace of lecture courses like abstract algebra or real analysis is often very fast. Formal definitions are introduced one after another, followed by theorems proved by the professor or the textbook. Subsequent assignments often consist of requests to prove (moderately) original theorems using the newly introduced definitions and theorems. Upon being given a statement to prove, a student's first job is to understand both the statement's structure and its content.

If a theorem is stated informally, then the students must first ascertain which are the hypotheses and which is the conclusion. For example, they need to know that the word "whenever" is usually a synonym for "if" and hence introduces one of the hypotheses. Whilst one can attempt to teach such linguistic variations, it would probably be easier if professors, at least initially, stated theorems using the familiar *if-then* and *if-and-only-if* forms.

The student's next job is to unpack the conclusion in order to understand it. This can entail looking up and unpacking the definitions of unfamiliar terms. Carefully unpacking these definitions can inform students where to begin a proof. However, instead of unpacking the conclusion, students will often initially focus on the hypotheses, deduce just anything and never arrive at the conclusion (Selden et al. 2010).

2.3.3 Interpreting and Using Previous Theorems and Definitions in Proving

Undergraduate students sometimes fail to use or interpret relevant theorems correctly or fail to verify that the conditions of the hypotheses of a theorem are satisfied.

For example, the Fundamental Theorem of Arithmetic, guaranteeing a unique prime decomposition of integers, forms part of the core mathematics curriculum for preservice elementary teachers. However, in one Canadian study, 8 of 21 preservice teachers appeared to deny the uniqueness. For example, instead of applying this Theorem when asked whether 17^3 was a square number, these students took out their calculators to extract the square root. In addition, many of these students believed that prime decomposition means decomposition into *small* primes (Zazkis and Campbell 1996; Zazkis and Liljedahl 2004).

Undergraduate students often ignore relevant hypotheses or apply the converse of a theorem when it does not hold. In addition they sometimes use theorems, especially named theorems as vague "slogans" that can be easily remembered as pat "solutions" to answer questions to which the theorems apparently apply, whether they do or not (e.g., Hazzan and Leron 1996).

Finally, students sometimes have difficulty with substitution into a definition or into the statement of a theorem to be used or proved. This problem occurs more often when the substitution involves a compound variable (e.g., Selden et al. 2010, p. 212).

2.3.4 Having a Repertoire of Examples and Using Them Appropriately

Constructing all but the most straightforward of proofs involves a good deal of persistence and problem solving to put together relevant concepts. In order to use a concept flexibly, it is important to have a rich *concept image*.[1]

---

[1] The idea of concept image was introduced by Vinner and Hershkowitz (1980) and elaborated by Tall and Vinner (1981). One's concept image is the collection of examples, non-examples, facts, properties, relationships, diagrams, images, and visualisations, that one associates with that concept.

In many upper-level undergraduate mathematics courses, students are given abstract definitions together with a few examples, after which they are expected to use these definitions reasonably flexibly. To do so, they may need to find additional examples or non-examples and to prove or disprove related conjectures, more or less without guidance. Such activities can help build students' concept images, especially for newly introduced concepts.

Dahlberg and Housman (1997) presented 11 third- and fourth-year U.S. undergraduate mathematics students with an unfamiliar formal definition. They asked the students to study the definition, to generate examples and nonexamples, to determine which of several functions satisfied it, and to determine the truth of four conjectures involving it. They found that students used four basic learning strategies: example generation, reformulation, decomposition and synthesis, and memorisation. Students who used example generation (with reflection), made the most progress in understanding the new concept. In contrast, those who employed memorisation or decomposition and synthesis often misinterpreted the definition.

Thus, generating examples and counterexamples can help students enrich their concept images and enable them to judge the probable truth of conjectures. To disprove a conjecture, it takes just one counterexample, so having a repertoire of examples is useful. In addition, examples are helpful in "making sense" of the statements in a proof.

However, students are often reluctant to generate examples. When Watson and Mason (2002) asked U.K. graduate students for an example of a continuous function which is not everywhere differentiable, they all responded $|x|$. When asked for a second example, most suggested a translate of $|x|$. When asked for a third example, they commented that one could generate a vast number of examples from $|x|$, but had trouble coming up with other different examples. Watson and Mason (2005) have since examined using examples as a teaching and learning strategy.

Weber et al. (2005) examined 11 U.S. transition-to-proof course students' use of examples in proving. Some never used examples; those who did, tried to use them correctly to understand a statement, to evaluate whether an assertion was correct, or to construct a counterexample. However, their attempts were largely ineffective; only one in nine attempts produced a valid or mostly valid proof. Frequently, the students selected inappropriate examples and did not check to see whether these satisfied the relevant definitions or the conditions of the statement to be proved.

In sum, students can find it useful to construct both examples and non-examples to clarify a new concept. In addition, "the ideal examples to use in teaching are those that are *only just* examples, and the ideal non-examples are those that are *very nearly* examples" (Askew and Wiliam 1995, p. iii).

2.3.5 Selecting Helpful Representations When Proving

Another aspect of understanding and using a concept is knowing which symbolic representations are likely to be appropriate in certain situations. Concepts can have several (easily manipulated) symbolic representations or none at all. For example, prime numbers have no such representation; they are sometimes defined as those positive integers having exactly two factors or being divisible only by 1 and themselves. Zazkis and Liljedahl (2004) have argued that the lack of an (easily manipulated) symbolic representation makes understanding prime numbers especially difficult, in particular, for preservice teachers.

Some symbolic representations can make certain features *transparent* and others *opaque*. For example, representing 784 as 28^2 makes the property of being a perfect square transparent and the property of being divisible by 98 opaque. Likewise, certain results in linear algebra are more transparent and easier to prove if expressed as linear transformations than as matrices. Students often lack the experience to know when to use a given symbolic representation.

Moving flexibly between representations (e.g., symbolic or graphic for functions) is an indication of the richness of one's understanding of a concept (Even 1998). However, conversion between representations can be both cognitively complex and asymmetric, that is, it can be easier to go from an algebraic to a graphic representation than vice versa; in linear algebra, such conversions are particularly difficult (Duval 2006).

2.3.6 Knowing to Use Factual Knowledge

Two U. S. companion studies asked 19 first-year university calculus students and 28 second-year university differential equations students to solve five moderately non-routine calculus problems (problems slightly different from what they had been taught). Immediately afterwards, the students took a short routine test which showed they had the resources needed to solve the five problems. However, they remained unaware of the resources' relevance and unable to bring them to mind (Selden et al. 2000; Selden et al. 1994). The researchers conjectured that, in studying and doing homework, the students had mainly mimicked worked examples from their text-books; thus, they never needed to consider different ways to solve problems and had no experience at bringing their various resources to mind.

To date, mathematics education research has had little to say about how one brings an idea, formula, definition or theorem to mind when it would be particularly helpful; and probably, there are several ways. Carlson and Bloom (2005) found that mathematicians solving problems frequently did not access the most useful information at the right time, suggesting how difficult it is to draw appropriately from even a vast reservoir of facts, concepts, and heuristics. Instead, the authors found that the mathematicians' progress depended on their approaches; that is, on such things as their ability to persist in making and testing various conjectures.

2.3.7 Knowing Which Theorems Are Important and Useful

Seeing the relevance and usefulness of one's knowledge and bringing it to bear on a problem or a proof is not easy, but seems to become easier with experience and training. For example, Weber and Alcock (2004) observed four U.S. undergraduates and four U.S. doctoral students attempting to prove or disprove whether specific pairs of groups are isomorphic. The undergraduates first looked to see if the groups had the same cardinality; then attempted unsuccessfully to construct an isomorphism between them. In contrast, the doctoral students examined properties preserved under isomorphism. In a somewhat similar study, in trying to prove propositions about groups, four U.S. doctoral students immediately recalled the First Isomorphism Theorem, whereas the four U.S. undergraduates did not. When asked why, the doctoral students typically said "Because this is such a fundamental and crucial fact that it's one of the first things you turn to" (Weber 2001, p. 113).

In yet another U.S. study, four undergraduates who had completed an abstract algebra course and four professional algebraists were interviewed about the ways they think about and represent groups. The algebraists thought about groups in terms of multiplication tables, generators and relations, and could call on specific examples. In contrast, none of the undergraduates could provide a single intuitive description of a group; for them, it was a structure that satisfies a list of axioms (Weber and Alcock 2004).

Perhaps, undergraduates mainly study completed proofs and focus on their details, rather than noticing the importance of certain results and how they fit together. Thus, they may not come to see other theorems as important or useful. The mathematics education literature contains few specific suggestions on how to teach students to know which theorems are likely to be important in various situations.

Knowing the most effective way to proceed in proving a theorem is important. Weber and Alcock (2004) discussed two reasoning strategies: first, *semantic reasoning,* in which the prover uses examples to gain insight and translates that insight into an argument based on appropriate definitions and theorems; and second, *syntactic reasoning,* in which the prover uses the statement of the theorem to structure a proof and draws logical inferences from associated definitions and theorems. Alcock (2009c) described the contrasting perspectives of two professors who teach transition-to-proof courses in the U.S.; one used a semantic approach and emphasised meaning and the generation of examples; the other used a syntactic approach and emphasised precision in notation and structural thinking. Yet both were preparing students for similar proof-based courses.

Iannone (2009) explored U.K. mathematicians' views on this; her participants said that syntactic knowledge is effective or suitable for proofs about concepts not having an initial pictorial representation, such as showing a sequence does or does not converge. However, syntactic knowledge can be used only ineffectively or not at all if the proof requires very complex syntactic representations. The mathematicians considered knowing how to tackle a proof, whether syntactically or semantically, a meta-mathematical ability that students need to acquire.

2.3.8 Knowing How to Read and Check Proofs

An integral part of the proving process is the ability to tell whether one's argument is correct and proves the theorem it was intended to prove. Selden and Selden (2003) conducted a study of how eight U.S. undergraduates from the beginning of a transition-to-proof course validated proofs. When asked how they read proofs, the students said they checked carefully line-by-line to see whether each assertion followed from previous statements, checked to make sure the steps were logical, and looked to see whether any computations were left out. The students evaluated and judged the correctness of four student-generated "proofs" of a very elementary number theory theorem. They made judgements regarding the correctness of each purported proof four times. The first time, these judgements were just 46% correct, whereas the last time they were 81% correct, a difference due to the students' having reconsidered, and reflected on, the purported proofs several times.

Several transition-to-proof textbooks include purported proofs to validate, but Selden and Selden's (2003) results suggest it would probably be helpful to have students validate actual student-generated proofs. When mathematicians at one U.S. university implemented proof-validation group activities once a week in abstract algebra, they found those activities improved their students' proof writing (Powers et al. 2010).

Most secondary-school students have no experience with proof evaluation, because their teachers not only do not teach it but may not be expert at it either. For example, Knuth (2002) gave 16 U.S. secondary-school teachers five sets of statements with three–five arguments purporting to justify them; in all, 13 arguments were proofs and eight were not. In general, the teachers were successful in recognising proofs, with 93% of the proofs rated as such. However, a third of the nonproofs were also rated as proofs.

Having described a variety of difficulties tertiary students have with constructing and comprehending proofs, we next consider various ways of helping students avoid or overcome these.

3 Teaching Proof and Proving at the Tertiary Level

Tertiary students learn about proof and proving in at least two different classroom contexts. First, in small classes, where in addition to or instead of hearing lectures, students may present their own proofs at the blackboard and receive critiques. They may work in small groups with the teacher acting as a resource/coach, or teachers and students may jointly work as a community of practice to develop the mathematics. Second, in large lecture classes, of 40 to 100 (or more), such individual attention is not possible.

3.1 Courses That Teach Proving

3.1.1 Transition-to-Proof Courses

In the U.S., undergraduate mathematics majors typically spend the first 2 years in computationally taught courses like calculus, beginning differential equations, and elementary linear algebra before going on to proof-based courses, such as real analysis and abstract algebra. Consequently, many U.S. universities have instituted *transition-to-proof* or *bridge* courses (Moore 1994); others use linear algebra, number theory (Smith 2006) or a discrete mathematics course (Epp 2004) for this purpose. Mathematicians have written more than 25 transition-to-proof course textbooks: for example, Velleman (1994) and Fendel and Resek (1990). Whilst their contents vary, such courses and books often begin with a decontextualised treatment of logic emphasising truth tables and valid arguments, followed by a discussion of direct and indirect proofs, sets, equivalence relations, functions, and mathematical induction. After that, the books diverge, selecting specific mathematical areas, such as graph theory or number theory, in which students are to practise proving.

The courses are often relatively small, having perhaps 15–40 students; the teaching methods can vary widely: small group learning in class, student presentations at the blackboard, lectures, or a mix of all these. However, not many studies of the effectiveness of such courses have been conducted. Marty (1991) examined the later success of all 120 students in his introductory proof courses (i.e., bridge courses) at one U.S. university over a 10 year period. He compared the students who received his instruction with 190 students taught in a traditional lecture format. He found his students two to three times more likely to pass their subsequent courses in real analysis and four times as likely to continue their studies of advanced mathematics.

A number of studies, other than for effectiveness, have been conducted on transition-to-proof courses. For example, Baker and Campbell (2004) identified problems transition-to-proof course students experience in proof writing. Weber et al. (2005) examined how transition-to-proof course students use examples (see Sect. 2.3.4).

3.1.2 Proving in a Community of Practice

In a second university course in plane geometry for Colombian preservice secondary teachers, Perry et al. (2008) developed a methodology based in part on a reconstructive approach (Human and Nel 1984) that harks back to Freudenthal's (1973) ideas. The content is not from a textbook nor is it presented by the teacher. Instead, the teacher and students as a community construct and develop a reduced Euclidean axiom system involving points, lines, planes, angles, triangles, and quadrilaterals. They jointly define geometric objects, empirically explore problems, formulate and verify conjectures, and write deductive arguments. Consequently, the students participate in axiomatisation and learn which basic elements they can use to justify statements. The teacher as expert orchestrates the process through a carefully developed set of questions and tasks (Perry et al. 2009b), the use of a dynamic geometry programme (*Cabri*) for developing conjectures, and careful management of the class. The approach provides the preservice teachers with experiences that they can use with their future pupils.

3.1.3   Moore Method Courses

This distinctive method of teaching at tertiary level was developed by the accomplished U.S. mathematician R. L. Moore (Parker 2005); it has been continued by his students and their mathematical descendants (Coppin et al. 2009; Mahavier 1999). It has been remarkably successful in Ph.D. level courses, but has also been used at the undergraduate level in small classes of 15–30. Typically, the teacher provides students with a set of notes containing definitions and statements of theorems or conjectures and asks the students to prove or provide counterexamples without help from anyone (except perhaps the teacher). The teacher structures the material and critiques the students' efforts. Students just begin. However, once having proved the first small theorem, a student can often progress very rapidly. Though interesting, Moore Method courses have been only a little researched (Smith 2006). However, the *Journal of Inquiry-Based Learning in Mathematics* publishes university-level courses notes and practical advice for such courses.

3.1.4   Co-construction of Proofs

McKee et al. (2010) implemented a modified Moore Method in a small (at most 10 students) supplementary, voluntary, proving class, for undergraduate real analysis students who felt unsure how to construct proofs. The supplement instructor wrote a theorem entirely new to the students, but similar to one they were to prove for homework, on the board. Then the students, or the supplement instructor if need be, suggested what to do next. For each suggested action, such as writing an appropriate definition, drawing a sketch, or introducing cases, one student was asked to carry out the action at the blackboard. This process aimed to get students to reflect on what had occurred and later perform the same or similar actions autonomously on the assigned homework. All students were encouraged to participate in co-constructing the proof.

The entire co-construction process, with accompanying discussions, was slow – so slow that only one theorem was proved and discussed in detail in each 75-minute supplement period. However, the students reported that they enjoyed this method of learning and the real analysis teacher reported improvements in the students' proof writing.

3.1.5   The Method of Scientific Debate

Daniel Alibert and colleagues at the University of Grenoble in France designed the *method of scientific debate* in which first-year university students are encouraged to become part of a classroom mathematical community. In one implementation (Alibert and Thomas 1991), the class consisted of about 100 students. First, the teacher got the students to make conjectures which were written on the blackboard without an immediate evaluation. Then the students discussed these, supporting their views by various arguments – a proof, a refutation or a counterexample. The conjectures that were proved became theorems. There rest became "false-statements" with a corresponding counterexample. The method has been systematically implemented for many years.

For the method of scientific debate to be successful, it is necessary to renegotiate the didactic contract (Brousseau 1997) so that students come to understand and accept their responsibilities. In addition, the teacher needs to refrain from revealing opinions, allow time for students to develop their arguments, and encourage maximum student participation (for details see Legrand 2001).

3.2 Alternative Ways of Presenting Proofs

Whilst a traditional definition-theorem-proof style of lecture presentation may convey the content in the most efficient way, there are other ways of presenting proofs that may enable students to gain more insight.

3.2.1 Generic Proofs

For certain theorems, a teacher can go through a (suitable) proof using a generic example (particular case) that is neither too trivial nor too complicated. Gauss' proof that the sum of the first n integers is $n(n+1)/2$, done for $n=100$ is one such generic proof. Done with care, going over a generic proof interactively with students can enable them to "see" for themselves the general argument embedded in the particular case. There is one caveat; there is some danger that students will not understand the generic character of the proof. In an attempt to avoid this, one can subsequently have them write out the general proof (Rowland 2002).

In an intensive study with 10 first-year Israeli linear algebra students, Malek and Movshovitz-Hadar (2011) investigated the use of generic proofs, which they have called "transparent pseudo-proofs"(Movshovitz-Hadar 1988). Their students benefited by acquiring "transferable cognitive structures related to proof and proving" (Malek and Movshovitz-Hadar 2009, p. 71). Further, Leron and Zaslavsky (2009) stated that the advantage of generic proofs is that "they enable students to engage with the main ideas of the complete proof in an intuitive and familiar context, temporarily suspending the formidable issues of full generality, formalism, and symbolism", making them more accessible to students (p. 56).

3.2.2 Structuring Mathematical Proofs

Instead of presenting a generic proof and letting students write out the general proof, one might try presenting a proof differently. Proofs are normally presented in lectures or advanced textbooks in a step-by-step linear fashion, which is well suited for checking the proof's validity but is perhaps not as good for communicating its main ideas.

Leron (1983) has suggested that one might arrange and present a proof in levels, proceeding from the top down, where the top level gives the proof's main ideas in general, but precise, terms. The second level elaborates on the top level ideas, supplying both proofs for as yet unsubstantiated statements and more details, including the construction of objects whose existence had merely been asserted before. If the second level is complicated, one may give only a brief description there, and push the details down to lower levels. The top level is normally short and free of technical details, whilst the bottom level is quite detailed, resembling a standard linear proof. Leron (1983) gave three sample structured proofs: one from number theory, one from calculus, and one from linear algebra.

An instructor might use structured proofs in several ways: (a) to present the higher levels of the proof and let the students complete the lower levels; (b) to have students take a standard textbook proof and find its structure; (c) to give students two similar theorems and have them determine to what level the similarity extends, counting from the top down. However, no empirical studies have yet substantiated that structured proof presentations are easier for students to comprehend (Mejia-Ramos et al. 2011).

3.3 Making Expectations Clear When Students Are Asked to Construct Their own Proofs

Sometimes students do not know what they should produce when asked to "explain", "demonstrate", "show", "justify" or "prove". Teachers and textbooks may not make this clear and may use different modes of argumentation (visual, intuitive, etc.) at different times, leaving students confused about which behaviour to imitate. However, sometimes the students just don't have facility with mathematical language. Dreyfus (1999, p. 88) gives the example of a first-year linear algebra student who was asked to determine whether the following statement was true or false and explain. *If $\{v_1, v_2, v_3, v_4\}$ is a linearly independent set, then $\{v_1, v_2, v_3\}$ is also a linearly independent set.* The student's answer was "True because taking down a vector does not help linear dependence". Possibly, the student understood, but the use of "taking down" and "help" points to a lack of linguistic capability and left the teacher to speculate on the extent of that student's understanding. On the other hand, sometimes students give a step-by-step account, somewhat like a travelogue, of how a problem was solved or a proof generated, which is not what most teachers want. Students also give redundant explanations.

Mejia-Ramos and Inglis (2011), in a study of 220 U.S. undergraduate mathematics students who were given arguments to evaluate, found that students responded differently to the noun "proof" than to the verb "prove". The noun form elicited evaluations related to an argument's validity, whereas the verb form elicited evaluations related to how convincing the argument was.

Duval (summarised in Dreyfus 1999) distinguished explanation, argumentation, and proof. The function of explanation is descriptive – to answer why something is so. Both arguments and proofs provide reasons that must be free of contradictions. However, in argumentation, the semantic content of the reasons is important. For proofs, the structure and the content determine the truth of the claim. Douek (1999), commenting on Duval's distinction, added that argumentation allows for a wider range of reasoning than proof; not only deduction but also metaphor and analogy can be used.

Professors also need to be exemplary in their use of quantifiers. Chellougui (2004), in a study involving 97 first-year Tunisian mathematics and informatics university students, observed that many different formulations of quantification were used in the same textbook or by the same teacher. Sometimes quantifiers remained implicit; sometimes they were used very strictly; sometimes they were used incorrectly as shorthand, even by teachers who did not accept such usage from their students.

3.4 Alternative Methods of Assessing Students' Comprehension of Proofs

3.4.1 Writing to Learn About Proofs and Proving

Writing to learn mathematics is not a new idea; Ganguli and Henry (1994) produced an annotated bibliography on the topic. Kasman (2006) uses the following technique in both modern algebra and introductory proofs courses. Students are given a fictitious scenario of two students discussing how they are going to construct a proof, along with their final written proof. Students then write a short expository paper stating whether the fictional proof is valid, identifying any errors in it, commenting on whether the fictitious students had the right idea but were not expressing it well, deciding which fictional student had a better understanding, and justifying their own reactions. Kasman provided sample scenarios, advice for writing more scenarios, suggestions for specific directions to give to students in order to get thoughtful reactions, and suggestions for grading.

3.4.2 Testing Proof Comprehension via a Variety of Questions

Conradie and Frith (2000) have pointed out that mid-term tests and final examinations given by the Mathematics Department of the University of Cape Town in South Africa have successfully used proof comprehension tests for at least 9 years. In such tasks, students are given a proof of a theorem that has already been proven in class. This is followed by questions like: What method of proof is used here? How are the terms and functions defined? What assumptions are justifiable? Variations include: having students fill in gaps in a proof or asking how the proof would be affected if condition *X* were replaced by condition *Y*. The authors argued that such tests motivate students to understand the proofs presented in the course rather than just to memorise them; and give a clear evaluation of a student's understanding. Furthermore, students can be prepared for such comprehension tests by giving them similar exercises as homework. Finally, the students preferred this type of test to the usual requests for the reproduction of proofs.

However, as reported by Mejia-Ramos and Inglis (2009), to date there has been little mathematics education research on the comprehension of proofs as compared with that on the construction and validation of proofs.

3.5   Videos About Proofs and Proving to Use with Students

Raman et al. (2009) have produced videos of undergraduate students engaged in proving theorems that many students find difficult, along with materials to help tertiary teachers use those videos with their own students. The intent is for students who view the videos to reflect on their own thinking by discussing and reflecting on the thinking of the students in the videos. These materials have been tested at four U.S. universities in transition-to-proof courses (see Sect. 3.1.1). In the process of creating these videos, Raman et al. (2009) identified three significant "moments" in constructing a proof: getting a *key idea* that gives one a sense of why the theorem might be true; discovering a *technical handle* needed to translate that key idea into a written proof; and putting the proof in final *standard form* with an appropriate level of rigour. However, successfully negotiating one or two of these moments does not guarantee completing the other(s).

Such videos could be used in relatively small classes where discussion is possible, but not realistically in large classes. Alcock (2009a), who routinely teaches real analysis to classes of over 100, has developed an alternate video technique, called *e-Proofs*. Eight analysis theorems (also proved in lectures) were elucidated by videos with line-by-line explanations of the reasoning used and a breakdown into large-scale structural sections. Students could replay particular sections as often as they wished. The aim was to improve students' comprehension of theorems proved in class. Alcock and others are now working on an e-Proofs authoring tool called ExPOUND, a project of the Joint Information Systems Committee (JISC) part of the U.K. Higher Education Funding Council.

3.6   Resources About Proof and Proving for Mathematicians

3.6.1   A DVD of Students Constructing Proofs

Alcock (2009b) has a DVD, designed to provide mathematicians with an opportunity to watch undergraduate mathematics students attempting to construct two proofs, one on upper bounds of sets of real numbers, the other on increasing functions and maxima. The students' proof attempts are broken up into excerpts, with prompts at the end of each excerpt designed to facilitate reflection or group discussion. The DVD also contains additional material for those wishing to follow up on these ideas in the mathematics education research literature.

3.6.2 Two Booklets Incorporating Ideas from Mathematics Education Research

Nardi and Iannone (2006) have a concise booklet designed for teachers of first-year U.K. undergraduates. There are sections on conceptualising formal reasoning, "proof" by example, proof by counterexample, proof by mathematical induction, and proof by contradiction. Each section contains examples of actual students' written responses, some relevant mathematics education research findings, some suggested pedagogical practices, and references to the research literature.

Alcock and Simpson (2009) produced a booklet containing chapters on the distinction between a student's concept image and the formal concept definition; on the difference between thinking of a mathematical notion, as a process or as an object; and on the distinction between semantic and syntactic reasoning strategies in proving. Whilst both strategies are useful in proving, some students seem to prefer one over the other (see Sect. 2.3.7).

3.6.3 Other Resources

The Mathematical Association of America's Special Interest Group on Research in Undergraduate Mathematics Education has produced a "research into practice" volume, in which researchers describe their research in an expository way for mathematicians (Carlson and Rasmussen 2008). The volume includes five chapters on proving theorems (pp. 93–164). Also discussed in the volume are findings regarding students' learning of concepts, information on teacher knowledge, strategies for promoting student learning, and classroom and institutional norms and values. Other attempts to bring such information to mathematicians include sessions at professional meetings, such as those of the American Mathematical Society and the Mathematical Association of America.

In addition, the International Commission on Mathematical Instruction has produced a study volume (Holton 2001) with sections on the secondary/tertiary interface, teaching practices, uses of technology, project work, assessment, mathematics education research and proof.

The extent to which these resources have been consulted by, and are found to be useful by, those teaching proof and proving to tertiary students needs more investigation.

4   Summary and Questions for Future Research

For a variety of reasons, students often find the transition to tertiary level mathematics study difficult. These reasons can include an expectation of greater student auton- omy, a change in the nature of the objects studied, and the seeming unapproachability of professors. However, one of the major reasons for students' difficulties seems to be the requirement that they understand and construct proofs. A variety of difficult aspects confront tertiary students struggling with proof and proving: the proper use of logic; the necessity to employ formal definitions; the need for a repertoire of examples, counterexample, and nonexamples; the requirement for a deep under- standing of concepts and theorems; the need for strategic knowledge of important theorems, and the important ability to validate one's own and others' proofs.

In addition, there are ways to teach proof other than by presenting proofs in a finished, linear top-down fashion. One can use generic proofs or employ structured proofs. There are a variety of courses and strategies that may help: transition-to- proof courses, communities of practice, the Moore Method, the co-construction of proofs, and the method of scientific debate. There are also resources in the form of videos, DVDs, and books to help university teachers.

Despite this, many questions remain about how students at tertiary level come to understand and construct proofs and how university teachers might help them do so. Here are some issues that could be, but have not yet been, the object of much research.

1. How instructors' expectations about tertiary students' performance in proof- based mathematics courses, such as real analysis or abstract algebra, differ from those in courses students have previously experienced.
2. To what extent tertiary instructors use the ability to construct proofs as a measure of a student's understanding.
3. How effective tertiary instructors' proof presentations are.
4. Whether there are effective ways to use technology, such as clickers and comput- ers, to help tertiary students learn to construct proofs.
5. What amount and types of conceptual knowledge are directly useful in making proofs in subjects such as abstract algebra and real analysis.
6. How tertiary students conceive of theorems, proofs, axioms, and definitions, and the relationships amongst them.
7. Whether and how teaching proof to future mathematics teachers should differ from teaching proof to future mathematicians or students of other mathematics- oriented disciplines.
8. How secondary preservice teachers can acquire the abilities necessary to effec- tively teach argumentation, proof, and proving to their future pupils.
9. Which previous experiences have students had with argumentation and proof that tertiary teachers can and should take into consideration.

Acknowledgements Thanks to Michèle Artigue, Michael de Villiers, and Tommy Dreyfus for their helpful suggestions on an earlier draft and to Kerry McKee for her input on two-column geometry proofs.

References*

Alcock, L. (2009a). E-Proofs: Student experience of online resources to aid understanding of mathematical proofs. In *Proceedings of the Twelfth Special Interest Group of the Mathematical Association of America Conference on Research on Undergraduate Mathematics Education*, Raleigh, NC. Retrieved May 24, 2010, from http://mathed.asu.edu/CRUME2009/Alcock1_LONG.pdf.

Alcock, L. (2009b). *Students and proof: Bounds and functions* (DVD). Essex: The Higher Education Academy (Maths, Stats & OR Network).

Alcock, L. (2009c). Teaching proof to undergraduates: Semantic and approaches (Vol. 1, pp. 29–34).*

Alcock, L., & Simpson, A. (2004). Convergence of sequences and series: Interactions between visual reasoning and the learner's beliefs about their own role. *Educational Studies in Mathematics, 57*, 1–32.

Alcock, L., & Simpson, A. (2009). *Ideas from mathematics education: An introduction for mathematicians*. Birmingham: The Higher Education Academy: Maths, Stats & OR Nework. Also available free at http://www.mathstore.ac.uk/index.php?pid=257.

Alibert, D., & Thomas, M. (1991). Research on mathematical proof. In D. Tall (Ed.), *Advanced mathematical thinking* (pp. 215–230). Dordrecht: Kluwer.

Antonini, S. (2001). Negation in mathematics: Obstacles emerging from an exploratory study. In M. van den Heuvel-Panhuizen (Ed.), *Proceedings of the 25th Conference of the International Group for the Psychology of Mathematics Education* (Vol. 2, pp. 49–56). Utrecht: Freudenthal Institute.

Antonini, S. (2003). Non-examples and proof by contradiction. In N. A. Pateman, B. J. Dougherty, & J. Zilliox (Eds.), *Proceedings of the 2003 Joint Meetings of PME and PMENA* (Vol. 2, pp. 49–55). Honolulu: CRDG, College of Education, University of Hawai'i.

Artigue, M. (1992). The importance and limits of epistemological work in didactics. In W. Geeslin & K. Graham (Eds.), *Proceedings of the Sixteenth Conference of the International Group for the Psychology of Mathematics Education* (Vol. 3, pp. 195–216). Durham: University of New Hampshire.

Artigue, M. (1999). What can we learn from educational research at the university level? Crucial questions for contemporary research in education. *Notices of the American Mathematical Society, 46*, 1377–1385.

Askew, M., & Wiliam, D. (1995). *Recent research in mathematics education, 5–16 (OFSTED reviews of research)*. London: HSMO.

Baker, D., & Campbell, C. (2004). Fostering the development of mathematical thinking: Observations from a proofs course. *Primus, 14*, 345–353.

Battie, V. (2009). Proving in number theory at the transition from secondary to tertiary level: Between the organizing and operative dimensions (Vol. 1, pp. 71–76).*

Battie, V. (2010). Number theory in the national compulsory examination at the end of the French secondary level: Between organizing and operative dimensions. In V. Durand-Guerrier, S. Soury-Lavergne, & F. Arzarello (Eds.), *Proceedings of CERME 6, January 28th-February 1st, 2009* (pp. 2316–2325). Lyon: INRP. Available at: www.inrp.fr/editions/cerme6.

Bills, L., & Tall, D. (1998). Operable definitions in advanced mathematics: The case of the least upper bound. In A. Olivier & K. Newstead (Eds.), *Proceedings of the 22nd Conference of the International Group for the Psychology of Mathematics Education* (Vol. 2, pp. 104–111). Stellenbosch: University of Stellenbosch.

Biza, I., & Zachariades, T. (2010). First year mathematics undergraduates' settled images of tangent line. *The Journal of Mathematical Behavior, 29*(4), 218–229.

---

*NB: References marked with * are in F. L. Lin, F. J. Hsieh, G. Hanna, & M. de Villiers (Eds.) (2009). *ICMI Study 19: Proof and proving in mathematics education*. Taipei, Taiwan: The Department of Mathematics, National Taiwan Normal University.

Biza, I., Christou, C., & Zachariades, T. (2008). Student perspectives on the relationship between a curve and its tangent in the transition from Euclidean geometry to analysis. *Research in Mathematics Education, 10*(1), 53–70.

Brousseau, G. (1997). *Theory of didactical situations in mathematics: Didactique de mathématiques 1970–1990* (N. Balacheff, M. Cooper, R. Sutherland, & V. Warfield, Eds. & Trans.). Dordrecht: Kluwer.

Carlson, M. P., & Bloom, I. (2005). The cyclic nature of problem solving: An emergent multidimensional problem solving framework. *Educational Studies in Mathematics, 58*, 45–76.

Carlson, M. P., & Rasmussen, C. (Eds.). (2008). *Making the connection: Research and teaching in undergraduate mathematics education* (MAA notes, Vol. 73). Washington, DC: Mathematical Association of America.

Castela, C. (2009). An anthropological approach to a transitional issue: Analysis of the autonomy required from mathematics students in the French lycée. *NOMAD (Nordisk Mathematikkdidaktikk), 14*(2), 5–27.

Chellougui, F. (2004) *Articulation between logic, mathematics, and language in mathematical practice.* Paper presented at the 10th International Congress on Mathematical Education (ICME 10), Copenhagen, Denmark.

Chevallard, Y. (1999). L'analyse des pratiques enseignantes en théorie anthropologique du didactique. *Recherches in Didactique des Mathématiques, 19*(2), 221–266.

Clark, M., & Lovric, M. (2008). Suggestion for a theoretical model for secondary-tertiary transition in mathematics. *Mathematics Education Research Journal, 20*(2), 25–37.

Clark, M., & Lovric, M. (2009). Understanding secondary-tertiary transition in mathematics. *International Journal of Mathematical Education in Science and Technology, 40*(6), 755–776.

Conradie, J., & Frith, J. (2000). Comprehension tests in mathematics. *Educational Studies in Mathematics, 43*, 225–235.

Coppin, C. A., Mahavier, W. T., May, E. L., & Parker, G. E. (2009). *The Moore Method: A pathway to learner-centered instruction* (MAA notes, Vol. 75). Washington, DC: Mathematical Association of America.

Cornu, B. (1991). Limits. In D. Tall (Ed.), *Advanced mathematical thinking* (pp. 153–166). Dordrecht: Kluwer.

Dahlberg, R. P., & Housman, D. L. (1997). Facilitating learning events through example generation. *Educational Studies in Mathematics, 33*, 283–299.

David, H., & Movshovitz-Hadar, N. (1998). The effect of quantifiers on reading comprehension of mathematical texts. In *Proceedings of the International Conference on the Teaching of Mathematics* held at the University of the Aegean. Samos: Wiley.

de Guzman, M., Hodgson, B. R., Robert, A., & Villani, V. (1998). Difficulties in the passage from secondary to tertiary education. *Documenta Mathematica, Extra Volume: Proceedings of the International Congress of Mathematicians, Berlin*, 747–762.

de Villiers, M. (1990). The role and function of proof in mathematics. *Pythagoras, 24*, 17–24.

Douek, N. (1999). Some remarks about argumentation and mathematical proof and their educational implications. In I. Schwank (Ed.), *European Research in Mathematics Education, I, Proceedings of the First Conference of the European Society for Research in Mathematics Education* (Vol. 1, pp. 128–142). Osnabrück: Forschungsinstitut für Mathematikdidaktik.

Dreyfus, T. (1999). Why Johnny can't prove. *Educational Studies in Mathematics, 38*, 85–109.

Dubinsky, E., & Yiparaki, O. (2000). On student understanding of AE and EA quantification. In E. Dubinsky, A. H. Schoenfeld, & J. Kaput (Eds.), *Research in collegiate mathematics education, IV* (CBMS Series Issues in Mathematics Education, pp. 239–289). Providence, RI: American Mathematical Society.

Duffin, J., & Simpson, A. (2002). Encounters with independent graduate study: Changes in learning style. In A. D. Cockburn & E. Nardi (Eds.), *Proceedings of the 26th Conference for the Psychology of Mathematics Education* (Vol. 2, pp. 305–312). Norwich: University of East Anglia School of Education and Professional Development.

Duffin, J., & Simpson, A. (2006). The transition to independent graduate studies in mathematics. In F. Hitt, G. Harel, & A. Selden (Eds.), *Research in collegiate mathematics education, VI*

(CBMS Series Issues in Mathematics Education, pp. 233–246). Providence, RI: American Mathematical Society.

Duval, R. (2006). A cognitive analysis of problems of comprehension in a learning of mathematics. *Educational Studies in Mathematics, 61*, 103–131.

Edwards, B. S., & Ward, M. B. (2004). Surprises from mathematics education research: Student (mis)use of mathematical definitions. *The American Mathematical Monthly, 111*, 411–424.

Epp, S. S. (2003). The role of logic in teaching proof. *The American Mathematical Monthly, 110*, 886–899.

Epp, S. S. (2004). *Discrete mathematics with applications* (3rd ed.). Florence: Brooks/Cole (Cengage Learning).

Epp, S. S. (2009). Proof issues with existential quantification (Vol. 1, pp. 154–159).*

Even, R. (1998). Factors involved in linking representations. *The Journal of Mathematical Behavior, 17*(1), 105–121.

Fendel, D., & Resek, D. (1990). *Foundations of HIGHER mathematics: Exploration and proof.* Reading: Addison-Wesley.

Freudenthal, H. (1973). *Mathematics as an educational task.* Dordrecht: Reidel.

Ganguli, A., & Henry, R. (1994). *Writing to learn mathematics: An annotated bibliography* (Technical report, Vol. 5). Minneapolis: University of Minnesota, The Center for Interdisciplinary Studies of Writing. Retrieved May 29, 2010 from http://writing.umn.edu/docs/publications/Ganguli_Henry.pdf.

Geraniou, E. (2010). The transitional stages in the Ph.D. Degree in terms of students' motivation. *Educational Studies in Mathematics, 73*, 281–296.

Gueudet, G. (2008). Investigating the secondary-tertiary transition. *Educational Studies in Mathematics, 67*, 237–254.

Hanna, G., Jahnke, H. N., & Pulte, H. (Eds.). (2009). *Explanation and proof in mathematics: Philosophical and educational perspectives.* New York: Springer.

Harel, G., & Sowder, L. (1998). Students' proof schemes: Results from exploratory studies. In A. H. Schoenfeld, J. Kaput, & E. Dubinsky (Eds.), *Research in collegiate mathematics education, III* (CBMS Series Issues in Mathematics Education, pp. 234–283). Providence, RI: American Mathematical Society.

Hazzan, O., & Leron, U. (1996). Students' use and misuse of mathematical theorems: The case of Lagrange's theorem. *For the Learning of Mathematics, 16*(1), 23–26.

Herzig, A. H. (2002). Where have all the students gone? Participation of doctoral students in authentic mathematical activity as a necessary condition for persistence towards the Ph.D. *Educational Studies in Mathematics, 50*, 177–212.

Holton, D. (Ed.). (2001). *The teaching and learning of mathematics at university level: An ICMI study.* Dordrecht: Kluwer.

Hourigan, M., & O'Donoghue, J. (2007). Mathematical under-preparedness: The influence of the pre-tertiary mathematics experience on students' ability to make a successful transition to tertiary level mathematics courses in Ireland. *International Journal of Mathematical Behavior in Science and Technology, 38*(4), 461–476.

Human, P. G., & Nel, J. H. (1984). *Alternative instructional strategies for geometry education: A theoretical and empirical study.* (In cooperation with M. de Villiers, T. P. Dreyer, & S. F. G. Wessels with M. de Villiers, Trans., USEME Project Report). Stellenbosch: University of Stellenbosch.

Iannone, P. (2009). Concept usage in proof production: Mathematicians' perspectives (Vol. 1, pp. 220–225).*

ICMI (International Commission on Mathematical Instruction). (1966). Report of the conference, "Mathematics at the coming to university: Real situation and desirable situation". In *New Trends in Mathematics Teaching* (Vol. I). Paris: UNESCO.

Kajander, A., & Lovric, M. (2005). Transition from secondary to tertiary mathematics: McMaster University experience. In R. Zazkis & D. Holton (Eds.), New trends and developments in tertiary mathematics education: ICME-10 perspectives [special issue] (pp. 149–160). *International Journal of Mathematical Education in Science and Technology, 36*(2–3), 149–160.

Kasman, R. (2006). Critique that! Analytical writing assignments in advanced mathematics courses. *Primus, 16*(1), 1–15.

Knuth, E. J. (2002). Secondary school mathematics teachers' conceptions of proof. *Journal for Research in Mathematics Education, 33*(5), 379–405.

Lakatos, I. (1961). *Essays in the logic of mathematical discovery.* Unpublished Ph.D. thesis, Cambridge University, Cambridge.

Lee, K., & Smith III, J. (2009). Cognitive and linguistic challenges in understanding proving (Vol. 2, pp. 21–26).*

Legrand, M. (2001). Scientific debate in mathematics classrooms. In D. Holton (Ed.), *The teaching and learning of mathematics at university level: An ICMI study* (pp. 127–135). Dordrecht: Kluwer.

Leron, U. (1983). Structuring mathematical proofs. *The American Mathematical Monthly, 90*(3), 174–184.

Leron, U., & Zaslavsky, O. (2009). Generic proving: Reflections on scope and method (Vol. 2, pp. 53–58).*

Lithner, J. (2004). Mathematical reasoning in calculus textbook exercises. *The Journal of Mathematical Behavior, 23*(4), 405–427.

Lovric, M. (2005). Learning how to teach and learn mathematics – "Teaching mathematics" course at McMaster University. In *Proceedings of the 1st Africa Regional Congress of the International Commission on Mathematical Instruction (ICMI), 22–25 June 2005.* Johannesburg: University of the Witwatersrand.

Mahavier, W. S. (1999). What is the Moore method? *Primus, 9,* 339–354.

Malek, A., & Movshovitz-Hadar, N. (2009). The art of constructing a transparent proof (Vol. 2, pp. 70–75).*

Malek, A., & Movshovitz-Hadar, N. (2011). The effect of using transparent pseudo-proofs in linear algebra. *Research in Mathematics Education, 13*(1), 33–57.

Martin, G. W., & Harel, G. (1989). Proof frames of preservice elementary teachers. *Journal for Research in Mathematics Education, 29*(1), 41–51.

Marty, R. H. (1991). Getting to eureka! Higher order reasoning in math. *College Teaching, 39*(1), 3–6.

McKee, K., Savic, M., Selden, J., & Selden, A. (2010). Making actions in the proving process explicit, visible, and "reflectable". In *Proceedings of the 13th Annual Conference on Research in Undergraduate Mathematics Education,* Raleigh, NC. Retrieved June 2, 2010, from http://sigmaa.maa.org/rume/crume2010/Archive/McKee.pdf.

Mejia-Ramos, J. P., & Inglis, M. (2009). Argumentative and proving activities in mathematics education research (Vol. 2, pp. 88–93).*

Mejia-Ramos, J. P., & Inglis, M. (2011). Semantic contamination and mathematical proof: Can a non-proof prove? *The Journal of Mathematical Behavior, 30,* 19–29.

Mejia-Ramos, J. P., Fuller, E., Weber, K., Samkoff, A., Rhoads, A., Doongaji, D., & Lew, K. (2011). Do Leron's structured proofs improve proof comprehension? Retrieved January 27, 2011 from http://sigmaa.maa.org/rume/crume2011/RUME2011/FinalSchedule.htm

Moore, R. C. (1994). Making the transition to formal proof. *Educational Studies in Mathematics, 27,* 249–266.

Movshovitz-Hadar, N. (1988). Stimulating presentation of theorems followed by responsive proofs. *For the Learning of Mathematics,* 8(2), 12–19, 30.

Nachlieli, T., & Herbst, P. (2009). Seeing a colleague encourage a student to make an assumption while proving: What teachers put in play when casting an episode of instruction. *Journal for Research in Mathematics Education, 40,* 427–459.

Nardi, E., & Iannone, P. (2006). *How to prove it: A brief guide to teaching proof to year 1 mathematics undergraduates.* Norwich: School of Education and Lifelong Learning, University of East Anglia. Also available free at http://mathstore.ac.uk/publications/Proof%20Guide.pdf.

Nickerson, S., & Rasmussen, C. (2009). Enculturation to proof: A theoretical and practical investigation (Vol. 2, pp. 118–123).*

Ouvrier-Buffet, C. (2002). An activity for constructing a definition. In A. D. Cockburn & E. Nardi (Eds.), *Proceedings of the 26th Conference of the International Group for the Psychology of Mathematics Education* (Vol. 4, pp. 25–32). Norwich: School of Education and Professional Development, University of East Anglia.

Ouvrier-Buffet, C. (2004). Construction of mathematical definitions: An epistemological and didactical study. In M. J. Høines & A. B. Fuglestad (Eds.), *Proceedings of the 28th Conference of the International Group for the Psychology of Mathematics Education* (Vol. 3, pp. 473–480). Bergen: Bergen University College.

Ouvrier-Buffet, C. (2006). Exploring mathematical definition construction processes. *Educational Studies in Mathematics, 63*, 259–282.

Parker, J. (2005). *R. L. Moore: Mathematician and teacher*. Washington, DC: Mathematical Association of America.

Perry, P., Samper, C., Camargo, L., Echeverry, A., & Molina, Ó. (2008). Innovación en la enseñanza de la demostración en un curso de geometría para formación inicial de profesores. In Electronic book of the XVII Simposio Iberoamericano de Enseñanza de las Matemáticas: "Innovando la enseñanza de las matemáticas". Toluca: Universidad Autónoma del Estado de México.

Perry, P., Camargo, L., Samper, C., Molina, Ó. & Echeverry, A. (2009a). Assigning mathematics tasks versus providing pre-fabricated mathematics in order to support learning to prove (Vol. 2, pp. 130–135).*

Perry, P., Samper, C., Camargo, L., Molina, O., & Echeverry, A. (2009b). Learning to prove: Enculturation or … ? (Vol. 2, pp. 124–129).*

Powers, A., Craviotto, C., & Grassl, R. M. (2010). Impact of proof validation on proof writing in abstract algebra. *International Journal of Mathematical Education in Science and Technology, 41*(4), 501–514.

Praslon, F. (1999). Discontinuities regarding the secondary/university transition: The notion of derivative as a specific case. In O. Zaslavsky (Ed.), *Proceedings of the 23rd Conference of the International Group for the Psychology of Mathematics Education* (Vol. 4, pp. 73–80). Haifa: Israel Institute of Technology.

Raman, M., Sandefur, J., Birky, G., Campbell, C., & Somers, K. (2009). "Is that a proof?": Using video to teach and learn how to prove at university level (Vol. 2, pp. 154–159).*

Reid, D., & Dobbin, J. (1998). Why is proof by contradiction so difficult? In A. Olivier & K. Newstead (Eds.), *Proceedings of the 22nd Conference of the International Group for the Psychology of Mathematics Education* (Vol. 4, pp. 41–48). Stellenbosch: University of Stellenbosch.

Roh, K. H. (2009). Students' understanding and use of logic in evaluation of proofs about convergence (Vol. 2, pp. 148–153).*

Rowland, T. (2002). Generic proofs in number theory. In S. R. Campbell & R. Zazkis (Eds.), *Learning and teaching number theory: Research in cognition and instruction* (pp. 157–183). Westport: Ablex Publishing.

Selden, J., & Selden, A. (1995). Unpacking the logic of mathematical statements. *Educational Studies in Mathematics, 29*(2), 123–151.

Selden, A., & Selden, J. (2003). Validations of proofs written as texts: Can undergraduates tell whether an argument proves a theorem? *Journal for Research in Mathematics Education, 34*(1), 4–36.

Selden, A., & Selden, J. (2005). Advanced mathematical thinking [special issue.]. *Mathematical Thinking and Learning, 7*(1), 1–73.

Selden, J., & Selden, A. (2009a). Teaching proving by coordinating aspects of proofs with students' abilities. In D. A. Stylianou, M. L. Blanton, & E. J. Knuth (Eds.), *Teaching and learning proof across the grades: A K-16 perspective* (pp. 339–354). New York: D.A. Routledge/Taylor & Francis.

Selden, J., & Selden, A. (2009b). Understanding the proof construction process (Vol. 2, pp. 196–201).*

Selden, J., Selden, A., & Mason, A. (1994). Even good calculus students can't solve nonroutine problems. In J. Kaput & E. Dubinsky (Eds.), *Research issues in undergraduate mathematics learning: Preliminary analyses and results* (MAA notes, Vol. 33, pp. 19–26). Washington, DC: Mathematical Association of America.

Selden, A., Selden, J., Hauk, S., & Mason, A. (2000). Why can't calculus students access their knowledge to solve non-routine problems? In A. H. Schoenfeld, J. Kaput, & E. Dubinsky (Eds.), *Research in collegiate mathematics education, IV* (CBMS Series Issues in Mathematics Education, pp. 128–153). Providence, RI: American Mathematical Society.

Selden, A., McKee, K., & Selden, J. (2010). Affect, behavioral schemas, and the proving process. *International Journal of Mathematical Education in Science and Technology, 41*(2), 199–215.

Smith, J. C. (2006). A sense-making approach to proof: Strategies of students in traditional and problem-based number theory courses. *The Journal of Mathematical Behavior, 25*, 73–90.

Tall, D. (Ed.). (1991). *Advanced mathematical thinking.* Dordrecht: Kluwer.

Tall, D. (1992). The transition to advanced mathematical thinking: Functions, limits, infinity, and proof. In D. A. Grouws (Ed.), *Handbook of research on mathematics teaching and learning* (pp. 495–511). New York: Macmillan.

Tall, D., & Vinner, S. (1981). Concept image and concept definition with particular reference to limits and continuity. *Educational Studies in Mathematics, 12*, 151–169.

Velleman, D. J. (1994). *How to prove it: A structured approach.* New York: Cambridge University Press.

Vinner, S., & Hershkowitz, R. (1980). Concept images and common cognitive paths in the development of some simple geometrical concepts. In R. Karplus (Ed.), *Proceedings of the 4th International Conference for the Psychology of Mathematics Education* (pp. 177–184). Berkeley: International Group for the Psychology of Mathematics Education.

Watson, A., & Mason, J. (2002). Extending example spaces as a learning/teaching strategy in mathematics. In. A. D. Cockburn & E. Nardi (Eds.), *Proceedings of the 26th Conference for the Psychology of Mathematics Education* (Vol. 4, pp. 378–385). Norwich: University of East Anglia School of Education and Professional Development.

Watson, A., & Mason, J. (2005). *Mathematics as a constructive activity: Learners generating examples.* Mahwah: Lawrence Erlbaum Associates.

Weber, K. (2001). Student difficulty in constructing proofs: The need for strategic knowledge. *Educational Studies in Mathematics, 48*(1), 101–119.

Weber, K., & Alcock, L. (2004). Semantic and syntactic proof productions. *Educational Studies in Mathematics, 56*(3), 209–234.

Weber, K., Alcock, L., & Radu, I. (2005). Undergraduates' use of examples in a transition-to-proof course. In G. M. Lloyd, M. Wilson, J. L. M. Wilkins, & S. L. Behm (Eds.), *Proceedings of the 27th Annual Meeting of the North American Chapter of the International Group for the Psychology of Mathematics Education.* Roanoke: Virginia Tech.

Weiss, M., Herbst, P., & Chen, C. (2009). Teachers' perspectives on "authentic mathematics" and the two-column proof form. *Educational Studies in Mathematics, 70*, 275–293.

Wood, L. (2001). The secondary-tertiary interface. In D. Holton (Ed.), *The teaching and learning of mathematics at university level: An ICMI study* (pp. 87–98). Dordrecht: Kluwer.

Zazkis, R., & Campbell, S. (1996). Prime decomposition: Understanding uniqueness. *The Journal of Mathematical Behavior, 15*(2), 207–218.

Zazkis, R., & Holton, D. (2005). New trends and developments in tertiary mathematics education: ICME-10 perspectives [special issue]. *International Journal of Mathematical Education in Science and Technology, 36*(2–3), 129–316.

Zazkis, R., & Liljedahl, P. (2004). Understanding primes: The role of representation. *Journal for Research in Mathematics Education, 35*(3), 164–186.

Part VI
Lessons from the Eastern
Cultural Traditions

Chapter 18
Using Documents from Ancient China to Teach Mathematical Proof

Karine Chemla

Discussions amongst mathematics educators on how to teach proof touch on many issues of importance to those working on the history of mathematical proof in the ancient world: How does generality matter when carrying out a proof? What is transparency for a proof? What is the connection between solving a problem and proving? All these questions are relevant to discussions of ancient source materials and call for further exchanges between Mathematics Education and History of Mathematics. That the two disciplines share such common questions clearly indicates why we may expect those exchanges to be fruitful.

As a historian of mathematics in ancient China, specifically interested in the subject of proof, I offer here some source material for further discussion between the two disciplines with at least two agenda in mind. One is to show that proofs existed in ancient China. It is quite important, in the world in which we now live, to convey to students that mathematical proof was not merely a Western product but has roots in many different parts of the planet. Second, I shall also suggest that mathematical source material from ancient China provides interesting ideas for teaching aspects of proof in the classroom, specifically algebraic proof. I shall focus on documents that, as far as I know, show types of proof specific to China, at least in early times, and represent an early history of algebraic proof, whose relation to later history awaits further research. These proofs are specific but have a universal dimension.

Let me outline some elements of context in order to introduce my sources. The earliest known Chinese mathematical book that was handed down through the

K. Chemla (✉)
CNRS, Université Paris Diderot, Sorbonne Paris Cité, Research Unit SPHERE,
team REHSEIS, UMR 7219, CNRS, F-75205 Paris, France
e-mail: chemla@univ-paris-diderot.fr

© The Author(s) 2021   423
G. Hanna and M. de Villiers (eds.), *Proof and Proving in Mathematics Education*,
New ICMI Study Series, https://doi.org/10.1007/978-94-007-2129-6_18

written tradition is *The Nine Chapters on Mathematical Procedures* (henceforward, *The Nine Chapters*). Specialists still dispute the date of this book's completion, but the book probably took the form in which we can read it at the latest in the first century C.E. *The Nine Chapters* is in the main composed of problems and general algorithms solving them. From early on, Chinese scholars considered the book a 'Classic.' Some wrote commentaries on it. Liu Hui completed the earliest extant commentary in 263 C.E., the document in which we are interested here. No ancient edition of *The Nine Chapters* has survived without Liu Hui's commentary, indicating that the Classic was always read in relation to commentaries selected by the compilers to be handed down with it.[1]

In contrast to *The Nine Chapters*, Liu Hui's commentary not only contains problems and algorithms but also includes reflections on mathematics, philosophical developments and, more important, proofs that systematically establish that the algorithms given in the Classic are correct. This commentary is the earliest extant source material handed down from ancient China in which one finds explicit proofs. In contrast to, say, those in Euclid's *Elements*, these proofs are specific, in that they all prove the correctness of algorithms. This feature makes them interesting for the classroom, since today algorithms have become one of the subject matters with which mathematical teaching deals.

Proving the correctness of an algorithm means establishing that the algorithm yields the desired magnitude and an exact value for it (or an approximate value, in a sense that the proof must make clear — see below how Liu Hui handles the value for π used in *The Nine Chapters*). We know that Liu Hui perceives part of his commentary as proofs, because he regularly concludes them by recurring formulas such as "this is why one obtains the result [sought-for]". Before I concentrate on one type of proof for which the commentary yields evidence, let me illustrate by example the kind of reflections on mathematical objects introduced in *The Nine Chapters* that the commentators develop.

Chapter 4 of *The Nine Chapters* contains an algorithm to compute square roots. This algorithm relies on a decimal place-value system and yields the digits of the root one by one, starting from the one corresponding to the highest power of 10. At the end of this algorithm, however, we find an interesting assertion: "If, by extraction, the [number] is not exhausted, that means that one cannot extract the [its] root, hence, one must name it [i.e., the number] with 'side'." Several historians independently established that the meaning of the latter clause amounts to suggesting that, in such cases, the result be given as "square root of" the number whose root is sought for. The argument relies on Liu Hui's commentary on the sentence quoted

---

[1] Chemla and Guo Shuchun (2004) present a critical edition and a French translation of *The Nine Chapters* and the two ancient commentaries with which it was handed down. The book also contains a complete bibliography and discussion of all statements made here about *The Nine Chapters* and the commentaries, except those incorporating results published after the book's publication.

above. More important, Liu Hui comments on why *The Nine Chapters* ought to yield the result in such cases in the form of quadratic irrationals.

To introduce this reflection, Liu Hui first considers a way of giving the result of root extraction as a quantity of the type of an integer increased by a fraction, but then discards this possibility. He asserts: "One cannot determine its value [i.e., the value of the root]. *Therefore, it is only when "one names it* [i.e., the number *N*] *with 'side'" that one does not make any mistake* [or, *that there is no error*]" (my emphases). This assertion leads him to make explicit the constraints that, he thinks, the result of the root extraction should satisfy and thus the reason why *The Nine Chapters* gives the result as the book does. He writes: "*Every time one extracts the root of a number-product* [a number that has been produced by a multiplication] *to make the side of a square,* the multiplication of this side by itself must in return *restore* [this number-product]" (my emphasis).

Thus Liu Hui sees a relationship between the way in which the result should be given and a property that operations must have. The result to be used is the one that guarantees a property for a sequence of reverse operations. In fact, he establishes here a link between the kinds of numbers to be used as results and the possibility of transforming a sequence of two reverse operations. For him, the exactness of the result of the square root extraction ensures the fact that the sequence of two reverse operations annihilates their effects and restores the original data; this sequence can thereby be deleted.

Having emphasised the importance of this connection, Liu Hui goes on to stress that the same link between one's achieving exact results and reverse operations cancelling each other holds true in other cases. He writes: "*This is* analogous to *the fact, when one* divides *10 by 3, to take its rest as being 1/3*, one is hence again able to restore [*fu*] its value" (my emphases). This statement also shows the perspective from which Liu Hui finds a similarity between quadratic irrationals and integers with fractions, despite his understanding, made explicit in other parts of his commentary, that, as kinds of numbers, they differ. He compares the fact of yielding the result of a root extraction as a quadratic irrational and that of giving the result of division with fractions. It is, he stresses, also when the result of divisions are given as exact, in the form of integers increased by a fraction, that the reverse operation of multiplication restores the number originally divided. Such are the kind of reflections the commentators made on mathematics and on *The Nine Chapters*. Note how different Liu Hui's perspective on quadratic irrationals is from the one expressed in Greek texts, whether Plato's *Theaetetus* or Book 10 of Euclid's *Elements*.

The property of reverse operations that Liu Hui stressed in his commentary brings us back to the issue of proof, and precisely to the type of proof on which I concentrate here. This appears clearly, when one looks for places in the commentaries where that property is put into play. One finds that the property – the operation inverse to a square root extraction or a division *restores* (*fu*) the original number and the meaning of the magnitude to which the extraction or the division was applied – is mentioned *only* in relation to these proofs which I have designated "algebraic proof in an algorithmic context."

Let me clarify what I mean by that expression.[2] Problem 5.28 in *The Nine Chapters* gives the values of the volume (*V*) of a circular cylinder and its height (*h*), asking one to determine the value of the circumference of its basis (*C*). The algorithm formulated after the problem asserts that the following list of operations yields the result[3]:

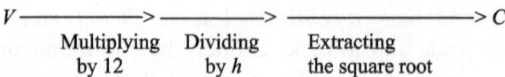

In his commentary, Liu Hui establishes that this algorithm correctly yields the result, that is to say, he makes clear why this list of operations yields the correct result. He does so, within the framework of the hypothesis that the authors of *The Nine Chapters* used a value for π equal to 3.[4] The reasoning he unfolds is essential to argue the above thesis. He starts from an algorithm that he had proved correct earlier in his commentary. This algorithm computes the volume of a cylinder, when one has the values of the circumference of its base and its height. It reads as follows:

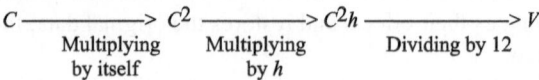

In order to prove that the initial algorithm given after problem 5.28 is correct, Liu Hui must establish that it yields the circumference *C*. In other terms, the question is to determine the meaning of the following sequence of operations applied to *V*:

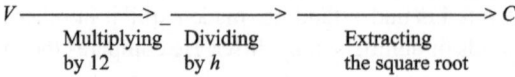

Liu Hui approaches the problem by making clear which magnitude and which value each of the operations of the algorithm to be proved correct yields; in his own terms, he determines in turn the "intention," the "meaning" (in Chinese, *yi*) of each operation. He does so by relying on the analysis of *V* provided by the algorithm whose correctness he has already established.

---

[2] For more extensive argumentation and a discussion of other cases, see Chemla (in press).

[3] I represent the algorithm as follows: On the left of an arrow, appear the data entered in an operation and, under the arrow, the operation applied. Two successive arrows indicate that the results yielded by the left operation are entered as terms in the right operation.

[4] It is only when one uses this value for π that one can explain why the factor appearing in the algorithm is 12. The key point is that the area of a circle, whose circumference is *C*, is equal to $C^2/4\pi$. *The Nine Chapters* uses this value of π throughout the book.

The meaning of the result of the first two steps can be determined as follows:

$$\text{C} \xrightarrow[\substack{\text{Multiplying} \\ \text{by itself}}]{} \text{C}^2 \xrightarrow[\substack{\text{Multiplying} \\ \text{by } h}]{} \text{C}^2\text{h} \xrightarrow[\substack{\text{Dividing} \\ \text{by 12}}]{} \text{V} \xrightarrow[\substack{\text{Multiplying} \\ \text{by 12}}]{} \text{C}^2\text{h}$$

then

$$C \xrightarrow[\substack{\text{Multiplying} \\ \text{by itself}}]{} C^2 \xrightarrow[\substack{\text{Multiplying} \\ \text{by } h}]{} C^2h \xrightarrow[\substack{\text{Dividing} \\ \text{by } h}]{} C^2$$

In this reasoning, Liu Hui successively applies the operation inverse to the last operation which produced the result for each in a sequence of algorithms. Doing so, he states, restores the meaning and value of the last intermediary step. This statement is correct, because, in Liu Hui's terms, in the examples above multiplying by 12 *restores* that to which the division by 12 had been applied. Thereafter, dividing by *h restores* that to which multiplying by *h* had been applied. The property that the operations cancel each other is put into play in the reasoning. This property allows Liu Hui to formulate the meaning and value of each of the successive operations.

Now, again, because of the property of square root discussed, we have:

$$C \xrightarrow[\substack{\text{multiplying} \\ \text{by itself}}]{} C^2 \xrightarrow[\substack{\text{extracting the} \\ \text{square root}}]{} C$$

At this point, the meaning of the result of the following algorithm is established:

$$V \xrightarrow[\substack{\text{Multiplying} \\ \text{by 12}}]{} \xrightarrow[\substack{\text{Dividing} \\ \text{by } h}]{} C^2 \xrightarrow[\substack{\text{Extracting} \\ \text{the square root}}]{} \sqrt{12V/h} = C$$

This is how the correctness of the algorithm is established. In the case of problem 5.28, the inverse operations successively applied are a multiplication, a division and a squaring. At each step, Lui Hui stresses that "restoring" was achieved. The "restoring" occurs correctly, because of the fact that exact results were secured for each operation.

This example illustrates the relationship between the property of numbers which permits restoration and the conduct of a kind of argumentation that relies on lists of operations and transforms them into other lists of operations in order to establish the correctness of an algorithm. The argument goes as follows:

We know that:

$$V = \frac{1}{12}C^2 \cdot h$$

Therefore:

$$12V = C^2 . h$$

Thus:

$$C^2 = 12V/h$$

$$C = \sqrt{\frac{12V}{h}}$$

and

These transformations are correct for exactly the same reasons as those Lui Hui made explicit. This fact explains why there is a correlation between this property shared by various kinds of numbers and the conduct of such types of proof. The kind of argument developed by Liu Hui in such cases, within an algorithmic context, corresponds to an algebraic proof. Further, Liu Hui accounts for the correctness of the steps of this proof – the transformations of lists of operations – through his commentary on the necessary exactness of the results of the operations involved. This is how he can establish the meaning (yi) of the result of a list of operations.

One can derive several conclusions from the outline above. First, what was presented suggests questions about the history of mathematical proof. The argument above, amongst others (see Chemla and Guo Shuchun, 2004; Chemla in press) raises the issue of the part played in the history of mathematical proof by the activity of proving the correctness of algorithms. Research on this question has just begun; one hopes to have clearer answers to the question in the near future. In the example here, we have seen how proof and algorithm interacted. The proof is constituted by an algorithm and by transformations of this algorithm as well as by the interpretation of the successive results that the algorithms allow. The algorithms entering into the proof and the algorithm to be proved have interesting relationships that provoke further thought.

Second, to come back to the introduction to this note, these algorithmic versions of algebraic proofs appear as potentially fruitful tools to help students work with, or even think about, algebraic proof. The practice of keeping track of the meaning of operations along an algorithm, of pondering transformations of algorithms *qua* algorithms of the type encountered above, of bringing to light the relationship between the set of numbers with which one operates and the correctness of transformations applied to lists of operations: all these procedures for which the Chinese sources provide evidence offer ideas for teaching algebraic proofs.

References*

Chemla, K. (In press). 'Reading proofs in Chinese commentaries: Algebraic proofs in an algorithmic context'. In K. Chemla (Ed.), *The history of mathematical proof in ancient traditions.*Cambridge: Cambridge University Press.

Chemla, K., & Guo S. (2004). *Les neuf chapitres. Le Classique mathématique de la Chine ancienne et ses commentaires*. Paris: Dunod.

---

*My sincere thanks to Gila Hanna, John Holt and Sarah-Jane Patterson for the generosity with which they helped me improve the text. Naturally, I am responsible for all remaining shortcomings.

Chapter 19
Proof in the Western and Eastern Traditions: Implications for Mathematics Education

Man Keung Siu

There is something about mathematics that is universal, irrespective of race, culture or social context. For instance, no mathematician will accept the following "proof", offered as a "joke-proof" by Oscar Perron (1880–1975) but not without pedagogical purpose:

"Theorem": *1 is the largest natural number.*
"Proof": Suppose N is the largest natural number, then N^2 cannot exceed N, so $N(N-1) = N^2 - N$ is not positive. This means that $N-1$ is not positive, or that N cannot exceed 1. But *N* is at least 1. Hence, $N = 1$. Q.E.D.

Likewise, well-known paradoxes on argumentation exist in both the Western and the Eastern world. The famous Liar Paradox, embodied in the terse but intriguing remark "I am a liar", is ascribed to the fourth century B.C.E. Greek philosopher Eubulides of Miletus. A similar flavour is conveyed in the famous shield-and-halberd story told by the Chinese philosopher Hon Fei Zi (Book 15, Section XXXVI, *Hon Fei Zi*, c. third Century B.C.E.):

"My shields are so solid that nothing can penetrate them. My halberds are so sharp that they can penetrate anything."

"How about using your halberds to pierce through your shields?"

In the Chinese language the term "*mao dun*", literally "halberd and shield", is used to mean "contradiction". Indeed, Hon Fei Zi used this story as an analogy to prove that the Confucianist School was inadequate while the Legalist School was effective and hence superior.[1] His proof is by *reductio ad absurdum*.

---

[1] The Confucianist School and the Legalist School were two streams of thought in ancient China, which would be too vast a subject to be explained, even in brief, here. If suffices to point out that the Legalist School maintained that good government was based on law and authority instead of on special ability and high virtue of the ruler who set an exemplar to influence the people. In particular, the story of shields and halberds was employed to stress that the two legendary leaders, Yao and Shun, whom the Confucianist School extolled as sage-kings, could not be both held in high regard.

M.K. Siu (✉)
Department of Mathematics, University of Hong Kong, Hong Kong SAR, China
e-mail: mathsiu@hkucc.hku.hk

© The Author(s) 2021

G. Hanna and M. de Villiers (eds.), *Proof and Proving in Mathematics Education*,
New ICMI Study Series, https://doi.org/10.1007/978-94-007-2129-6_19

In his book *A Mathematician's Apology*, English mathematician Godfrey Harold Hardy (1877–1947) said that "*reductio ad absurdum*, which Euclid loved so much, is one of a mathematician's finest weapons" (Hardy 1940/1967, p. 94). Many people are led by this remark to see the technique of proof by contradiction as a Western practice, even to the extent that they wonder whether the technique is closely related to Greek, and hence Western, culture. I was once asked whether Chinese students would have inherent difficulty in learning proof by contradiction, because such argumentation was absent from traditional Chinese mathematics. My immediate response was that this learning difficulty shows up in a majority of students, Chinese or non-Chinese, and does not seem to be related to a student's cultural background. Nonetheless, this query urged me to look for examples of proof by contradiction in traditional Chinese thinking. Since then, I have gathered some examples, many of which are in a non-mathematical context. One mathematical presentation that approaches a proof by contradiction is Liu Hui's (c. third century C.E.) argument in his commentary on Chapter 1 of *Jiu Zhang Suan Shu* (*Nine Chapters on the Mathematical Art*) explaining why the ancients were wrong in taking 3 to be the ratio of the perimeter of a circle to its diameter (Siu 1993, p. 348). Still, I have not yet found a written proof in an ancient Chinese text that recognisably follows prominently and distinctly the Greek fashion of *reductio ad absurdum*.

However, the notion of a proof is not so clear-cut when it comes to different cultures as well as different historical epochs. Mathematics practised in different cultures and in different historical epochs may have its respective different styles and emphases. For the sake of learning and teaching it will be helpful to study such differences.

Unfortunately, many Western mathematicians have come to regard Eastern mathematical traditions as not 'real' mathematics. For example, take Hardy's assessment:

> The Greeks were the first mathematicians who are still 'real' to us to-day. Oriental mathematics may be an interesting curiosity, but Greek mathematics is the real thing. The Greeks first spoke a language which modern mathematicians can understand; as Littlewood said to me once, they are not clever schoolboys or 'scholarship candidates', but 'Fellows of another college'. (Hardy 1940/1967, pp. 80–81)

However, proper study of the different traditions leads one to disagree with Hardy's assessment.

A typical example of the cross-cultural difference in style and emphasis is the age-old result known in the Western world as Pythagoras' Theorem. Compare the proof given in Proposition 47, Book I of Euclid's *Elements* (c. third century B.C.E.) (Fig. 19.1) and that given by the Indian mathematician Bhaskara in the twelfth century C.E. (Fig. 19.2). The former is a deductive argument with justification provided at every step. The latter is a visually clear dissect-and-reassemble procedure, so clear that Bhaskara found it adequate to simply qualify the argument by a single word, "Behold!"

The notion of proof permeates other human endeavour in the Western world. Indeed, one finds the following passage in Book1.10 in *Institutio Oratoria* by Marcus Fabius Quintilianus (First century C.E.):

> Geometry [Mathematics] is divided into two parts, one dealing with Number, the other with Form. Knowledge of numbers is essential not only to the orator, but to anyone who has had

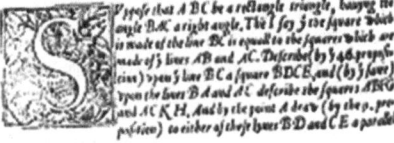

The elements of geometrie of the most auncient philosopher Euclide of Megara. Faithfully (now first) translated into the Englishe toung, by H. Billingsley, citizen of London. (1570)

Proposition 47,
Book I of Euclid's
Elements
(c.3rd century B.C.)

Fig. 19.1 Euclid's proof of Pythagoras' theorem

Fig. 19.2 Bhaskara's proof of Pythagoras' theorem

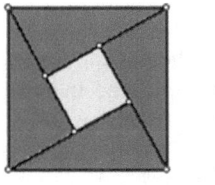 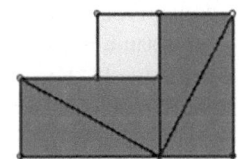

Behold!

Bhāskara (12th century)

even a basic education. (...) In the first place, order is a necessary element in geometry; is it not also in eloquence? Geometry proves subsequent propositions from preceding ones, the uncertain from the certain: do we not do the same in speaking? Again: does not the solution of the problems rest almost wholly on Syllogisms? (...) Finally, the most powerful proofs are commonly called "linear demonstrations". And what is the aim of oratory if not proof? Geometry also uses reasoning to detect falsehoods which appear like truths. (...) So, if (as the next book will prove) an orator has to speak on all subjects, he cannot be an orator without geometry [mathematics]. (Quintilian 2001, pp. 231, 233, 237)

Stephen Toulmin, in examining "how far logic *can* hope to be a formal science, and yet retain the possibility of being applied in the critical assessment of actual arguments" (Toulmin 1958, p.3), opines that one source from which the notion of proof arose is argument on legal matters. He propounds a need for a rapprochement between logic and epistemology, for a re-introduction of historical, empirical and even anthropological considerations into the subject which philosophers have prided themselves on purifying:

> The patterns of argument in geometrical optics, for instance (…) are distinct from the patterns to be found in other fields: e.g. in a piece of historical speculation, a proof in the infinitesimal calculus, or the case for the plaintiff in a civil suit alleging negligence. Broad similarities there may be between arguments in different fields, (…) it is our business, however, not to insist on finding such resemblances at all costs but to keep an eye open quite as much for possible differences. (Toulmin 1958, p. 256)

This year (2009) is the 200th anniversary of the birth of the great English naturalist Charles Darwin (1809–1882) and the 150th anniversary of the publication of *On the Origin of Species* (1859). Not many may have noted what Darwin once said in his autobiography about mathematics:

> I attempted mathematics, and even went during the summer of 1828 with a private tutor (a very dull man) to Barmouth, but I got on very slowly. This work is repugnant to me, chiefly from my not being able to see any meaning in the early steps in algebra. This impatience was very foolish, and in after years I have deeply regretted that I did not proceed far enough at least to understand something of the great leading principles of mathematics, for men thus endowed seem to have an extra sense. (Darwin 1887, Chapter II, Volume I, p. 46)

This kind of *extra sense* shows up in another important historical figure, the American polymath Benjamin Franklin (1706–1790). He thought in a precise, rational way even about seemingly non-mathematical issues and used mathematical argument for a social debate (Pasles 2008, Chapter 1, Chapter 4).

The same use of mathematical argument in other contexts also happens in the Eastern world. For example, the Indian-British scholar and recipient of the 1998 Nobel Prize in Economics, Amartya Sen, presents an interesting discussion of the case in India in his book *The Argumentative Indian: Writings on Indian Culture, History and Identity* (2005).

Next, I draw your attention to two styles in doing mathematics, using terms borrowed from Peter Henrici (Henrici 1974), who labels the two styles as "dialectic" and "algorithmic". Broadly speaking, dialectic mathematics is a rigorously logical science, in which "statements are either true or false and objects with specified properties either do or do not exist." (Henrici 1974, p.80) On the other hand, algorithmic mathematics is a tool for solving problems, in which "we are concerned not only with the existence of a mathematical object but also with the credentials of its existence" (Henrici 1974, p. 80). In a lecture (July, 2002), I attempted to synthesise the two aspects from a pedagogical viewpoint with examples from historical mathematical developments in Western and Eastern cultures. In this 19th ICMI Study Conference, I reiterated this theme, focusing on proof, and discussed how the two aspects complement and supplement each other in proof activity (Siu 2009b). A procedural (algorithmic) approach helps to prepare more solid ground on which to build up conceptual understanding; conversely, better conceptual (dialectical) understanding enables one to handle algorithms with more facility, or even to devise improved or new algorithms. Like *yin* and *yang* in Chinese philosophy, these two aspects complement and supplement each other, each containing some part of the other.

Several main issues in mathematics education are rooted in understanding these two complementary aspects, "dialectic mathematics" and "algorithmic mathematics". Those issues include: (1) procedural versus conceptual knowledge; (2) process versus object in learning theory; (3) computer versus computerless learning environments; (4) "symbolic" versus "geometric" emphasis in learning and teaching; and (5) "Eastern" versus "Western" learners/teachers. In a seminal paper, Anna Sfard explicates this duality and develops it into a deeper model of concept formation through interplay of the "operational" and "structural" phases (Sfard 1991).

Tradition holds that Western mathematics, developed from that of the ancient Greeks, is dialectic, while Eastern mathematics, developed from that of the ancient Egyptians, Babylonians, Chinese and Indians, is algorithmic. Even if it holds an element of truth as a broad statement, under more refined examination this thesis is an over-simplification. Karine Chemla has explained this point in detail (Chemla 1996). I will discuss the issue with examples from Euclid's *Elements*.

Saul Stahl has summarised the ancient Greek's contribution to mathematics:

> Geometry in the sense of mensuration of figures was spontaneously developed by many cultures and dates to several millennia B.C. The science of geometry as we know it, namely, a collection of abstract statements regarding ideal figures, the verification of whose validity requires only pure reason, was created by the Greeks. (Stahl 1993, p. 1)

A systematic and organised presentation of this body of knowledge is found in Euclid's *Elements*.

Throughout history, many famed Western scholars have recounted the benefit they received from learning geometry through reading Euclid's *Elements* or some variation thereof. For example, Bertrand Russell (1872–1970) wrote in his autobiography:

> At the age of eleven, I began Euclid, with my brother as tutor. This was one of the great events of my life, as dazzling as first love. (…) I had been told that Euclid proved things, and was much disappointed that he started with axioms. At first, I refused to accept them unless my brother could offer me some reason for doing so, but he said, 'If you don't accept them, we cannot go on', and as I wished to go on, I reluctantly admitted them *pro temp.* (Russell 1967, p. 36)

Another example, Albert Einstein (1879–1955), wrote in his autobiography:

> At the age of twelve I experienced a second wonder of a totally different nature: in a little book dealing with Euclidean plane geometry, which came into my hands at the beginning of a school year. (…) The lucidity and certainty made an indescribable impression upon me. (…) it is marvelous enough that man is capable at all to reach such a degree of certainty and purity in pure thinking as the Greeks showed us for the first time to be possible in geometry. (Schilepp 1949, pp. 9, 11)

That axiomatic and logical aspect of Euclid's *Elements* has long been stressed. However, reasoning put forth by S.D. Agashe (1989) leads one to look at an

Fig. 19.3 Proposition
14 of *Elements Book II*

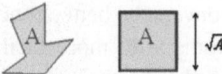

> Proposition 14 of Euclid's *Elements Book II*:
> **To construct a square equal to *a* given rectilineal figure**.

| Motivation: | Comparison of two line segments |

a

b

Proposition 3 of *Book I*: Given two
unequal straight lines, to cut off
from the greater a straight line
equal to the less. (relying on
Postulates 1,2,3)

Comparison of two rectilineal
figures

a

b

(reduce each to a square and
compare, relying on **Postulates 4**)

alternative feature of the *Elements*; namely, right from the start metric geometry plays a key role, not just in the exposition but even in the motivation of the book's design. In addition, there is a procedural flavour to the reasoning.

For example, Proposition 14 of *Elements, Book II* proposes, "To construct a square equal to a given rectilineal figure." The problem of interest is to compare two polygons. To achieve the one-dimensional analogue, comparing two straight line segments, is easy; one simply overlays one segment on the other and checks whether one segment lies completely inside the other or whether the two are equal. This is in fact what Proposition 3 of *Book I* attempts: "Given two unequal straight lines, to cut off from the greater a straight line equal to the less." To justify the result, one relies on Postulates 1, 2 and 3. The two-dimensional problem is not so straightforward, except for the special case when both polygons are squares; in this case, one can compare their areas through a comparison of their sides, by placing the smaller square at the lower left corner of the larger square. Incidentally, here one needs to invoke Postulate 4. What Proposition 14 of *Book II* sets out to do is to reduce the comparison of two polygons to that of two squares (Fig. 19.3).

The proof of Proposition 14 of *Book II* can be divided into two steps: (1) construct a rectangle equal (in area) to a given polygon (Fig. 19.4); (2) construct a square equal (in area) to a given rectangle (Fig. 19.5). Note that (1) is already explained through Propositions 42, 44 and 45 of *Book I*, by triangulating the given polygon then converting each triangle into a rectangle of equal area. Incidentally, one has to rely on the famous (notorious?) Postulate 5 on (non-)parallelism to prove those results. To achieve the solution in (2), one makes the preliminary step of converting the given rectangle into an L-shaped gnomon of equal area. This is

Fig. 19.4 Construction
of a rectangle equal (in area)
to a given polygon

How to construct a rectangle
equal (in area) to a given
rectilineal figure?

Decompose into triangles and
construct a rectangle (more
generally a parallelogram with
one given angle) equal in area
to each triangle. (Proposition 42,
44, 45 of *Book I*, relying on
Postulate 5)

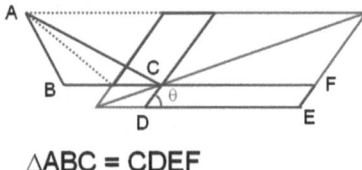

$\triangle ABC = CDEF$

Fig. 19.5 Squaring a
rectangle

How to square a rectangle?
Reduce a rectangle to a gnomon
(L-shaped figure).

This is the content of Proposition
5 of *Book II*.

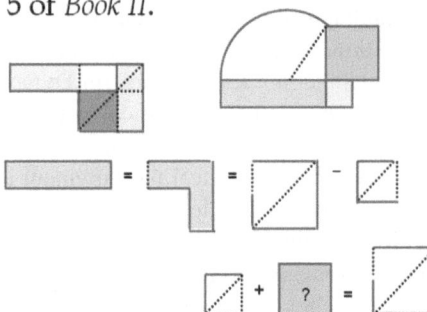

gnomon = difference of two squares
sum of two squares = square?
(Proposition 47 of *Book I*)

illustrated in Proposition 5 of *Book II*, "If a straight line be cut into equal and unequal segments, the rectangle contained by the unequal segment of the whole together with the square on the straight line between the points of section is equal to the square on the half."

Proposition 5 of *Book II* asserts that a certain rectangle is equal (in area) to a certain gnomon which is a square (c^2) minus another square (b^2). To finalise step (2), one must construct a square (a^2) equal to the difference between two squares ($c^2 - b^2$); or equivalently, the square (c^2) is a sum of the two squares ($a^2 + b^2$). This leads naturally to Pythagoras' Theorem, Proposition 47 of *Book I*, which epitomises the interdependence between shape and number, between geometry and algebra. (For an enlightening exposition of Pythagoras' Theorem in Clairaut's *Eléments de géométrie* [1741, 1753], see (Siu 2009a, pp. 106–107)). In studying this problem to compare two polygons we see how algorithmic mathematics blends in with dialectic mathematics in Book I and Book II of *Elements*.

However, despite such evidence of parallels between the Western and Eastern mathematical traditions, some teachers hesitate to integrate history of mathematics with the learning and teaching of mathematics in the classroom. They cite their concern that students lack enough knowledge on culture in general to appreciate history of mathematics in particular. This is probably true, but one can look at the problem from the reverse, seeing the integration of history of mathematics into the day-to-day mathematics classes as an opportunity to let students know more about other cultures in general and other mathematical traditions in particular. They can thus come into contact with other variations in the development of proof and proving. Proof is such an important ingredient in a proper education in mathematics that we can ill afford to miss such an opportunity.

Earlier, I suggested (Siu 2008) four examples that might be used in such teaching. The first examines how the exploratory, venturesome spirit of the 'era of exploration' in the fifteenth and sixteenth centuries C.E. centuries influenced the development of mathematical practice in Europe. It resulted in a broad change of mentality in mathematical pursuit, not just affecting its presentation but, more important, bringing in an exploratory spirit. The second example deals with a similar happening in the Orient, though with more emphasis on the aspect of argumentation. It describes the influence of the intellectual milieu in the period of the Three Kingdoms and the Wei-Jin Dynasties from the third to the sixth centuries C.E. in China on mathematical practice as exemplified in the work of Liu Hui. The third example, the influence of Daoism on mathematics in ancient China, particularly astronomical measurement and surveying from a distance, examines the role religious, philosophical (or even mystical) teachings may play in mathematical pursuit. The fourth example, the influence of Euclid's *Elements* in Western culture compared to that in China after the first Chinese translation by the Ming Dynasty scholar-minister Xu Guang Qi (1562–1633) and the Italian Jesuit Matteo Ricci (1552–1610) in 1607 points out a kind of 'reverse' influence; namely, how the mathematical thinking may stimulate thinking in other areas of human endeavour. As a 'bonus', these examples sometimes suggest ways to enhance understanding of specific topics in the classroom.

Finally, one benefit of learning proof and proving is important but seldom emphasised in Western education, namely, its value in character building. This point had been emphasised in the Eastern world rather early, perhaps as a result of the influence of the Confucian philosophical heritage.

In an essay on the Chinese translation of the *Elements*, the co-translator Xu wrote:

> The benefit derived from studying this book [the *Elements*] is many. It can dispel shallow-ness of those who learn the theory and make them think deep. It can supply facility for those who learn the method and make them think elegantly. Hence everyone in this world should study the book. (…) Five categories of personality will not learn from this book: those who are impetuous, those who are thoughtless, those who are complacent, those who are envious, those who are arrogant. Thus to learn from this book one not only strengthens one's intellectual capacity but also builds a moral base. (cited in Siu (2009a, p. 110))

Such emphasis on proof for a moral reason still sometimes echoes in modern times. As the late Russian mathematics educator Igor Fedorovich Sharygin (1937–2004) once put it, "Learning mathematics builds up our virtues, sharpens our sense of justice and our dignity, and strengthens our innate honesty and our principles. The life of mathematical society is based on the idea of proof, one of the most highly moral ideas in the world." (cited in (Siu 2009a, p. 110)).

Acknowledgement I wish to thank Gila Hanna, John Holt and Sarah-Jane Patterson for the careful editing of my contribution.

References

Agashe, S. D. (1989). The axiomatic method: Its origin and purpose. *Journal of the Indian Council of Philosophical Research, 6*(3), 109–118.

Chemla, K. (1996). Relations between procedure and demonstration. In H. N. Jahnke, N. Knoche, & M. Otte (Eds.), *History of mathematics and education: Ideas and experiences* (pp. 69–112). Göttingen: Vandenhoeck & Ruprecht.

Darwin, F. (1887). *The life and letters of Charles Darwin including an autobiographical chapter* (3rd ed.). London: John Murray.

Hardy, G. H. (1940/1967). *A mathematician's apology.* Cambridge: Cambridge University Press.

Henrici, P. (1974). Computational complex analysis. *Proceedings of Symposia in Applied Mathematiics, 20,* 79–86.

Pasles, P. C. (2008). *Benjamin Franklin's numbers: An unsung mathematical Odyssey.* Princeton: Princeton University Press.

Quintilian. (2001). The Orator's education [Institutio Oratoria], Books 1–2 (Donald A. Russell). Cambridge/London: Harvard University Press.

Russell, B. (1967). *The autobiography of Betrand Russell* (Vol. 1). London: Allen & Unwin.

Schilepp, P. A. (1949). *Albert Einstein: Philosopher-scientist* (2nd ed.). New York: Tudor.

Sfard, A. (1991). On the dual nature of mathematical conceptions: Reflections on process and objects as different sides of the same coin. *Educational Studies in Mathematics, 22,* 1–36.

Siu, M. K. (1993). Proof and pedagogy in ancient China: Examples from Liu Hui's Commentary on Jiu Zhang Suan Shu. *Educational Studies in Mathematics, 24,* 345–357.

Siu, M. K. (2008). Proof as a practice of mathematical pursuit in a cultural, socio-political and intellectual context. *ZDM-The International Journal of Mathematics Education, 40*(3), 355–361.

Siu, M. K. (2009a). The world of geometry in the classroom: Virtual or real? In M. Kourkoulos, C. Tzanakis (Eds.), *Proceedings 5th International Colloquium on the Didactics of Mathematics* (Vol. II, pp. 93–112). Rethymnon: University of Crete.

Siu, M. K. (2009b). The algorithmic and dialectic aspects in proof and proving. In F. L. Lin, F. J. Hsieh, G. Hanna, M. de Villiers. (Eds.), *Proceedings of the ICMI Study 19 Conference: Proof and Proving in Mathematics Education* (Vol. 2, pp. 160–165). Taipei: National Taiwan Normal University.

Stahl, S. (1993). *The Poincaré half-plane: A gateway to modern geometry*. Boston/London: Jones and Bartlett Publishers.

Toulmin, S. (1958). *The uses of argument*. London/New York: Cambridge University Press.

Correction to: Proof and Proving in Mathematics Education

Correction to:
G. Hanna and M. de Villiers (eds.), *Proof and Proving in Mathematics*
Education, **New ICMI Study Series,**
https://doi.org/10.1007/978-94-007-2129-6

The updated version of this book can be found at https://doi.org/10.1007/978-94-007-2129-6

Acknowledgements

The contribution of the following is acknowledged and greatly appreciated.

Plenary speakers: Jonathan Borwein, Judith Grabiner, Giuseppe Longo, and Frank Quinn

Panelists: Karine Chemla, Wann-Sheng Horng, and Man Keung Siu

Book reviewers: Michèle Artigue, Hyman Bass, Ed Barbeau, Tommy Dreyfus, Fulvia Furinghetti, Gila Hanna, Matthew Inglis, Hans Niels Jahnke, Keith Jones, Boris Koichu, Margo Kondratieva, Roza Leikin, Maria Alessandra Mariotti, John Monaghan, Nitsa Movshovitz-Hadar, Elena Nardi, John Olive, Annie Selden, David Tall, Dina Tirosh, Michael de Villiers, Walter Whiteley, and Oleksiy Yevdokimov.

Manuscript editor: John Holt

Graduate assistants: Sarah-Jane Patterson and Gunawardena Egodawatte.

Presenters at the ICMI 19 conference in Taipei.

International Program Committee: Gila Hanna and Michael de Villiers (Co-Chairs); Ferdinando Arzarello, Tommy Dreyfus, Viviane Durand-Guerrier, Wann-Sheng Horng, Hans Niels Jahnke, Fou-Lai Lin, Annie Selden, David Tall, Oleksiy Yevdokimov and Bernard Hodgson (ex-officio).

ICMI advisors: Michèle Artigue, Hyman Bass and Mariolina Bartolini Bussi.

Local Organizing Committee (LOC): Wann-Sheng Horng, and to the members of the LOC: Feng-Jui Hsieh, Fang-Chih Cheng, Yu-Ching Hung, Yu-Hsien Chang, Chuang-Yih Chen, Tai-Yih Tso, Shao-Tung Chang, Po-Son Tsao (Dennis), Chien Chin, Rung-Chin Tsai, Kai-Lin Yang and Yu-Ping Chang.

The generous support from the following institutional sponsors is also acknowledged.

The National Science Council, Taiwan

The National Taiwan Normal University

The Ontario Institute for Studies in Education, University of Toronto, Canada

The Social Sciences and Humanities Research Council (SSHRC) of Canada

Universität Duisburg-Essen, Germany

Laboratoire de Recherche en Didactique et en Histoire des Sciences et des Techniques, Lyon 1, France

IREM de Paris 7, France

The International Commission on Mathematical Instruction (ICMI)

G. Hanna and M. de Villiers (eds.), *Proof and Proving in Mathematics Education*, New ICMI Study Series, https://doi.org/10.1007/978-94-007-2129-6

Appendix 1
ICMI Study 19: Proof and Proving in Mathematics Education: Discussion Document

Gila Hanna, Michael de Villiers, Ferdinando Arzarello, Tommy Dreyfus, Viviane Durand-Guerrier, Hans Niels Jahnke, Fou-Lai Lin, Annie Selden, David Tall, and Oleksiy Yevdokimov

Rationale

Mathematics educators face a significant task in getting students to understand the roles of reasoning and proving in mathematics. This challenge has now gained even greater importance as proof has been assigned a more prominent place in the mathematics curriculum at all levels. The recent National Council of Teachers of Mathematics (NCTM) Principles and Standards document and several other mathematics curricular documents have elevated the status of proof in school mathematics in several educational jurisdictions around the world.

This renewed curricular emphasis on proof has provoked an upsurge in research papers on the teaching and learning of proof at all grade levels. This re-examination of the role of proof in the curriculum and of its relation to other forms of explanation, illustration and justification (including dynamic graphic software) has already produced several theoretical frameworks, giving rise to many discussions and even heated debates. An ICMI Study on this topic would thus be both useful and timely.

An ICMI Study on proof and proving in mathematics education would necessarily discuss the different meanings of the term proof and bring together a variety of viewpoints. Proof has played a major role in the development of mathematics, from the Euclidean geometry of the Greeks, through various forms of proofs in different cultures, to twentieth-century formal mathematics based on set-theory and logical deduction. In professional mathematics today, proof has a range of subtly different meanings: for example, giving an axiomatic formal presentation; using physical conceptions, as in a proof that there are only five Platonic solids; deducing conclusions from a model by using symbolic calculations; or using computers in experimental mathematics. For mathematicians, proof varies according to the discipline involved, although one essential principle underlies all its varieties:

> To specify clearly the assumptions made and to provide an appropriate argument supported by valid reasoning so as to draw necessary conclusions.

© The Author(s) 2021
G. Hanna and M. de Villiers (eds.), *Proof and Proving in Mathematics Education*, New ICMI Study Series, https://doi.org/10.1007/978-94-007-2129-6

This major principle at the heart of proof extends to a wide range of situations outside mathematics and provides a foundation for human reasoning. Its simplicity, however, is disguised in the subtlety of the deep and complex phrases "to specify the assumptions clearly", "an appropriate argument" and "valid reasoning".

The study will consider the role of proof and proving in mathematics education, in part as a precursor for disciplinary proof (in its various forms) as used by mathematicians but mainly in terms of developmental proof, which grows in sophistication as the learner matures towards coherent conceptions. Sometimes the development involves building on the learners' perceptions and actions in order to increase their sophistication. Sometimes it builds on the learners' use of arithmetic or algebraic symbols to calculate and manipulate symbolism in order to deduce consequences. To formulate and communicate these ideas require a simultaneous development of sophistication in action, perception and language.

The study's conception of "developmental proof" has three major features:

1. Proof and proving in school curricula have the potential to provide a long-term link with the discipline of proof shared by mathematicians.
2. Proof and proving can provide a way of thinking that deepens mathematical understanding and the broader nature of human reasoning.
3. Proof and proving are at once foundational and complex, and should be gradually developed starting in the early grades.

A major classroom role for proof is essential to maintaining the connection between school mathematics and mathematics as a discipline. Although proof has not enjoyed the same degree of prominence in mathematical practice in all periods and contexts, and although standards of rigour have changed over time, proof undoubtedly lies at the heart of mathematics.

Similarly proof and proving are most properly used in the classroom to promote understanding, which in no way contradicts their role in mathematics. Mathematical proof consists, of course, of explicit chains of inference following agreed rules of deduction, and is often characterised by the use of formal notation, syntax and rules of manipulation. Yet clearly, for mathematicians proof is much more than a sequence of correct steps; it is also and, perhaps most importantly, a sequence of ideas and insights with the goal of mathematical understanding – specifically, understanding why a claim is true. Thus, the challenge for educators is to foster the use of mathematical proof as a method to certify not only that something is true but also why it is true.

Finally, the learning of proof and proving in school mathematics should be developmental and should start in the early grades. The success of this process would clearly depend on teachers' views about the essence and forms of proofs, on what teachers do with their students in classrooms, on how teachers interpret and implement curricular tasks that have the potential to offer students opportunities to engage in proving, and on how they diagnose students' difficulties in proving and design instructional interventions to help overcome these difficulties.

Themes of the Study

The ICMI Study will be organised around themes that provide a broad range of points of view on the teaching and learning of proof in various contexts, whether symbolic, verbal, visual, technological or social. Within each of the themes, the following issues are of utmost importance:

1. Teachers' views and beliefs
2. Teachers' preparation and professional development
3. Curriculum materials and their role in supporting instruction

Below, we describe some of the themes and suggest a number of related research questions. Contributions on each theme should address these specific questions but need not be limited to them, so long as any additional questions raised are relevant to that theme.

Cognitive Aspects

Cognitive aspects of proof cover the entire development of proof and proving, from the young child to the research mathematician. They range from the manner in which the growing person develops a proving attitude to convince the self and others, through the initial use of specific examples, through prototypical numerical and visual examples representing broader classes of instances, to formal axiomatic proofs widely acceptable to the mathematical community. While proofs are considered either valid or invalid, the development of proof, both in the growing child and in the research of mathematicians, involves arguments that carry various levels of conviction that are not absolute. For example, Tall's framework of worlds – of conceptual embodiment, proceptual symbolism and axiomatic formalism – suggests a dynamic development of proof through embodiment and symbolism to formalism. For instance, the formula for the sum of the first n whole numbers can be proved from a specific or generic picture, from a specific, generic or algebraic sum, from a practical potentially infinite form of induction, from a finite axiomatic form of induction from the Peano postulates, or even from a highly plausible visual demonstration. This part of the study will consider various theories of cognitive aspects of proof.

Possible questions about cognitive aspects:

1. Is it possible/preferable to classify forms of proof in terms of *cognitive development*, rather than just in terms of *type of proof* (e.g., by exhaustion, contradiction, induction)?
2. When we classify proof cognitively, can we look from the learners' viewpoint as they grow from the elementary grades to university, rather than just from the expert's viewpoint, and appropriately value their current ways of proving?

3. How do we encompass empirical classifications of proof processes within a coherent cognitive development (which may differ for different individuals)?
4. How can teachers and mathematics educators use our knowledge about learners' cognitive development to develop ways of teaching proof that take account of each learner's growing ways of proving?
5. What are learners' and teachers' beliefs about proof, and how do they affect the teaching and learning of proof?
6. What theoretical frameworks and methodologies are helpful in understanding the development of proof from primary to tertiary education, and how are these frameworks useful in teaching?

Argumentation and Proof

Understanding the relationship between argumentation (a reasoned discourse that is not necessarily deductive but uses arguments of plausibility) and mathematical proof (a chain of well-organized deductive inferences that uses arguments of necessity) may be essential for designing learning tasks and curricula that aim at teaching proof and proving. Some researchers see mathematical proof as distinct from argumentation, whereas others see argumentation and proof as parts of a continuum rather than as a dichotomy. Their different viewpoints have important didactical implications. The first group would focus mainly on the logical organization of statements in a proof and would aim to teach a conceptual framework that builds proof independent of problem solving. On the other hand, the second group would focus primarily on the production of arguments in the context of problem solving, experimentation and exploration, but would expect these arguments to later be organized logically so as to form a valid mathematical proof.

From a very young age, students show high degrees of ability in reasoning and in justifying their arguments in social situations; however, they do not naturally grasp the concept of mathematical proof and deductive reasoning. Therefore, educators must help students to reason deductively and to recognize the value of the concept of mathematical proof. Some educators hold the traditional assumption that teaching students elements of formal logic, such as first-order logic with quantifiers, would easily translate into helping them to understand the deductive structure of mathematics and to write proofs. However, research has shown that this transfer doesn't happen automatically. It remains unclear what benefit comes from teaching formal logic to students or to prospective teachers, particularly because mathematicians have readily admitted that they seldom use formal logic in their research. Hence, we need more research to support or disconfirm the notion that teaching students formal logic increases their ability to prove or to understand proofs.

Possible questions about argumentation and proof:

1. How can we describe the argumentative discourses developed in mathematics teaching? What is the role of argumentation and proof in the conceptualization process in mathematics and in mathematics education?

2. Within the context of argumentation and proof, how should mathematics education treat the distinction that logicians and philosophers make between truth and validity?

3. To what extent could focussing on the mathematical concept of implication in both argumentation and proof contribute to students' better grasp of various kinds of reasoning?

4. How can educators make explicit the different kinds of reasoning used in mathematical proof and in argumentative discourse (e.g., *Modus Ponens*, exhaustion, disjunction of cases, *Modus Tollens*, indirect reasoning etc.)?

5. Quantification, important in reasoning as well as in mathematics, often remains implicit. To what extent does this lead to misconceptions and to lack of understanding?

6. How can teachers deal with the back-and-forth between conjectures and objects or between properties and relations involved in the exploration of mathematical objects? To what extent does this exploration help students understand the necessity of mathematical proof rather than just argumentation?

7. Are we justified in concluding that logic is useless in teaching and learning proof just because many mathematicians claim that they do not use logic in their research? What kind of research program could be developed to answer this question?

8. What are the relationships between studies on argumentation and proof by researchers from other disciplines, e.g., logicians, philosophers, epistemologists, linguists, psychologists and historians, and research in mathematics education?

9. What conditions and constraints affect the development of appropriate situations for the construction of argumentation and proof in the mathematics classroom?

10. Which learning environments and activities help to improve students' ability in argumentation and proof?

Types of Proof

Some aspects of the study might deal with types of proof characterized by their mathematical or logical properties, such as specific proof techniques, (e.g., proof by exhaustion, proof by mathematical induction, proof by contradiction) or proofs of specific types of claims (e.g., existence proofs, both constructive and nonconstructive).

These different types of proof (or techniques of proving) may have many diverse pedagogical properties and didactic functions in mathematics education. A case in point is inductive proof (proof by example), which is frequently the only type of proof comprehensible to beginners; it may be mathematically valid (e.g., for establishing existence or for refutation by counterexample) or invalid (e.g., supportive examples for a universal statement). Another type, generic (or transparent) proof, is infrequently used but may have high didactic potential.

The various ways of proving, such as verbal, visual or formal, may be a factor in understanding proofs and in learning about proving in general. Specific proofs may lend themselves particularly well to specific ways of proving.

Possible questions about types of proof:

1. To what extent, and at which levels of schooling, is it appropriate to introduce specific proof techniques? What are the particular cognitive difficulties associated with each type of proof?
2. Is it important to introduce proof in a diversity of mathematical domains and which proofs are more appropriate in which domains?
3. At which level and in which curricula is it relevant to introduce the notion of refutation? In particular, when should one raise the question of what is needed to prove or refute an existential claim as opposed to a universal one?
4. How and at which stage should teachers facilitate the transition from inductive proof (proof by example) to more elaborate forms of proof?
5. What status should be given to generic proof? How can the properties of generic proofs be used to support students' transition from inductive to deductive proof?
6. At which level, and in which situations, should the issue of the mathematical validity or lack of validity of inductive proofs be discussed, and how?
7. To what extent and how is the presentation of a proof (verbal, visual, formal etc.) relevant in understanding it and in learning about the notion of proof generally?
8. To what extent is the presentation of a proof (in)dependent of the nature of the proof? Do some proofs lend themselves particularly well to specific presentations? For example, can visual theorems have non-visual proofs?
9. Do students perceive different types of proofs as more or less explanatory or convincing?

Dynamic Geometry Software and Transition to Proof

Both philosophers and psychologists have investigated the connection between deductive reasoning and argumentation. However, there is still no consensus on the exact nature of this connection. Meanwhile some researchers have looked for possible mediators between plausible argumentation and mathematical proof. The main didactical problem is that at first glance there seems to be no natural mediator between argumentation and proof. Hence, the problem of continuity or of discontinuity between argumentation and proof is relevant for research and for teaching of proof.

Dynamic Geometry Software (DGS) fundamentally changes the idea of what a geometric object is. DGS can serve as a context for making conjectures about geometric objects and thus lead to proof-generating situations. Specifically, it can play the role of mediator in the transition between argumentation and proof through its 'dragging function', thanks to its instant feedback and to the figures created on the screen as a result of the dragging movements. The dragging function opens up new routes to theoretical knowledge within a concrete environment that is meaningful to students. For example, it can introduce seemingly infinite examples to support a conjecture or it can help in showing students degenerate examples or singular counterexamples to a statement (e.g., when a given construction that works for building a figure degenerates into singular cases, producing a different figure). Moreover, while dragging, pupils often switch back and forth from figures to concepts and from abductive to deductive modalities, which helps them progress from the empirical to the theoretical level. The different modalities of dragging can be seen as a perceptual counterpart to logical and algebraic relationships. In fact, dragging makes the relationships between geometric objects accessible at several levels: perceptual, logical and algebraic.

Possible questions about DGS environments:

1. To what extent can explorations within DGS foster a transition to the formal aspects of proof? What kinds of didactical engineering can trigger and enhance such support? What specific actions by students could support this transition?
2. How could the issues of continuity/discontinuity among the different phases and aspects of the proving processes (exploring, conjecturing, arguing, proving etc.) be addressed in DGS environments?
3. To what extent can activities within DGS environments inhibit or even counter the transition to formal aspects of proof?
4. What are the major differences between proving within DGS environments and proving with paper and pencil?
5. How can the teacher handle the different modalities of proving (induction, abduction, deduction etc.) that explorations in DGS environments may generate?
6. How can DGS help in dealing with proofs by contradiction or proofs by example, given that through dragging one could get 'infinite examples', degenerate examples or the singular counterexamples to a statement?
7. How can DGS environments be used for approaching proofs not only in geometry but also in other subjects, such as algebra and elementary calculus?
8. What are the significant differences among different DGSs used in teaching proof?
9. What are the main differences between DGS environments and other technological environments (software other than DGS, concrete materials, mathematical machines, symbolic computation systems etc.) in tackling the issue of proof in the classroom? Can a multiple approach, which suitably integrates different environments, be useful for approaching proof?

The Role of Proof and Experimentation

The traditional view of proof has ignored the role of experimentation in mathematics and has perceived the verification of mathematical statements as the only function of proof. However, in recent years several authors have emphasized the intimate relationship between proof and experimentation, as well as the many other important functions of proof within mathematics besides verification: explanation, discovery, intellectual challenge, systematization etc. Moreover, research in dynamic geometry has shown that, despite obtaining a very high level of conviction by dragging, students in some contexts still display a strong cognitive need for an explanation of a result; that is, why it is true. Such a need gives a good reason for the introduction of proof as a means of explaining why a result is true.

However, not all new results in mathematics are discovered through experimentation. Deductive reasoning from certain givens can often directly lead to new conclusions and to new discoveries through generalization or specialization. In this context, proof takes on a systematizing role, linking definitions, axioms and theorems in a deductive chain. Likewise, experimentation in mathematics includes some important functions relevant to proof: conjecturing, verification, refutation, understanding, graphing to expose mathematical facts, gaining insights etc. For example, mathematicians can formulate and evaluate concept definitions on the basis of experimentation and/or formal proof, as well as comparing and selecting suitable definitions on the basis of criteria such as economy, elegance, convenience, clarity etc. Suitable definitions and axioms are necessary for deductive proof in order to avoid circular arguments and infinite regression. Thus, the establishment of a mathematical theorem often involves some dynamic interplay between experimentation and proof.

The relationship between proof and experimentation poses a general didactical and educational research question: How can we design learning activities in which students can experience and develop appreciation for these multi-faceted, inter-related roles of proof and experimentation? This in turn comprises several additional questions.

Possible questions about proof and experimentation:

1. How can teachers effectively use the explanatory function of proof to make proof a meaningful activity, particularly in situations where students have no need for further conviction?
2. How can students' abilities to make their own conjectures, critically evaluate their validity through proof and experimentation, and produce counter-examples if necessary be stimulated and developed over time?
3. How can teachers and mathematics educators develop effective strategies to help students see and appreciate the discovery function of proof – for example, deriving results deductively rather than experimentally or from deriving further unanticipated results and subsequent reflections on those proofs?
4. What are students' natural cognitive needs for conviction and verification in different mathematical contexts, with different results and at different levels? How can these needs be utilized, changed and developed through directed instructional activities so that students appreciate the verification function of proof in different contexts?
5. What arguments can teachers use in school and university to foster students' appreciation of the meaning of proof and to motivate students to prove theorems?
6. What type of 'guidance' is needed to help students eventually produce their own independent proofs in different contexts?

7. Rather than just providing them with pre-fabricated mathematics, how do we involve students in the deductive systematization of some parts of mathematics, both in defining specific concepts and in axiomatizing a piece of mathematics? How able are students to identify circular arguments or invalid assumptions in proofs and how do we develop these critical skills?

Proof and the Empirical Sciences

Frequently, students do not see a connection between argumentation in empirical situations and mathematical proof. They consider proof a mathematical ritual that does not have any relevance to giving reasons and arguments in other circumstances or disciplines. However, mathematical proof is not only important in mathematics itself but also plays a considerable role in the empirical sciences that make use of mathematics.

Empirical scientists put up hypotheses about certain phenomena, say falling bodies, draw consequences from these hypotheses via mathematical proof and investigate whether the hypotheses fit the data. If they do, we accept the hypotheses; otherwise we reject them. Thus, in the establishment of a new empirical theory the flow of truth provided by a mathematical proof goes from the consequences to the assumptions; the function of a proof is to test the hypotheses. Only at a later stage, after a theory has been accepted, does the flow of truth go from the assumptions to the consequences as it usually does in mathematics. These considerations suggest a series of questions for investigation.

Possible questions about proof and the empirical sciences:

1. To what extent should mathematical proofs in the empirical sciences, such as physics, figure as a theme in mathematics teaching so as to provide students with an adequate and authentic picture of the role of mathematics in the world?
2. Would insights about the role of proof in the empirical sciences be helpful in the teaching of geometry, given that geometry deals with empirical statements about the surrounding space as well as with a theoretical system about space?
3. Could insights about the complex role of proof in the empirical sciences be helpful in bridging students' perceptual gap between proof and proving in mathematics and argumentation in everyday life?
4. To what extent and how should philosophers of mathematics, mathematics educators and teachers develop a unified picture of proving and modelling, which are usually considered completely separate topics in mathematics?
5. Could a stronger emphasis on the process of establishing hypotheses (in the empirical sciences) help students better understand the structure of a proof that proceeds from assumptions to consequences and thus the meaning of axiomatics in general?
6. To what extent does a broader conception of proof require the collaboration of mathematics and science teachers?

Proof at the Tertiary Level

At the tertiary level, proofs involve considerable creativity and insight as well as both understanding and using formal definitions and previously established theorems. Proofs tend to be longer, more complex and more rigorous than those at earlier educational levels. To understand and construct such proofs involves a major transition for students but one that is sometimes supported by relatively little explicit instruction. Teachers increasingly use students' original proof constructions as a means of assessing their understanding. However, many questions remain about how students at the tertiary level come to understand and construct proofs. Here we lay some of the questions out clearly, proposing to examine them in the light of both successful teaching practices and current research.

Possible questions about proof at the tertiary level:

1. How are instructors' expectations about students' performance in proof-based mathematics courses different from those in courses students experienced previously?
2. Is learning to prove partly or even mainly a matter of enculturation into the practices of mathematicians?
3. How do the students conceive theorems, proofs, axioms, definitions and the relationships among them? What are the students' views of proof and how are their views influenced by their experiences with proving?
4. What are the roles of problem solving, heuristics, intuition, visualization, procedural and conceptual knowledge, logic and validation?
5. What previous experiences have students had with proof that teachers can take into consideration?
6. How can we design opportunities for student teachers to acquire the knowledge (skills, understandings and dispositions) necessary to provide effective instruction about proof and proving?

Appendix 2
Conference Proceedings: Table of Contents

Table of Contents

---

http://140.122.140.1/~icmi19/files/Volume_1.pdf

http://140.122.140.1/~icmi19/files/Volume_2.pdf

© The Author(s) 2021 453
G. Hanna and M. de Villiers (eds.), *Proof and Proving in Mathematics Education*,
New ICMI Study Series, https://doi.org/10.1007/978-94-007-2129-6

VOLUME 2

Author Index

© The Author(s) 2021
G. Hanna and M. de Villiers (eds.), *Proof and Proving in Mathematics Education*,
New ICMI Study Series, https://doi.org/10.1007/978-94-007-2129-6

Subject Index

© The Author(s) 2021
G. Hanna and M. de Villiers (eds.), *Proof and Proving in Mathematics Education*,
New ICMI Study Series, https://doi.org/10.1007/978-94-007-2129-6